PROGRESS IN COLLOID & POLYMER SCIENCE

Editors: H.-G. Kilian (Ulm) and G. Lagaly (Kiel)

Volume 79 (1989)

Trends in Colloid and Interface Science III

Guest Editors: P. Bothorel and E. J. Dufourc (Talence)

Springer-Verlag Berlin Heidelberg GmbH

ISBN 978-3-662-16129-6   ISBN 978-3-7985-1690-8 (eBook)
DOI 10.1007/978-3-7985-1690-8
ISSN 0340-255 X

© 1989 by Springer-Verlag Berlin Heidelberg

Originally published by Dr. Dietrich Steinkopff Verlag GmbH & Co. KG, Darmstadt in 1989

Softcover reprint of the hardcover 1st edition 1989

Chemistry editor: Dr. Maria Magdalena Nabbe; Copy editor: James Willis; Production: Holger Frey.

Preface

In 1988, the yearly meeting of the European Colloid and Interface Society (ECIS) was held in Arcachon, France, from the 19th to the 22nd of September. The first such meeting took place in Como, Italy, in 1987, and was attented by 130 participants coming from 17 European countries. The Arcachon meeting welcomed 220 scientists from almost all European countries and from North America. This increase in participants shows the good healths of our still young society.

About 40 oral communications and 130 poster presentations were given in Arcachon. The present volume collects full papers taken from communications and posters. This volume is arbitrarily divided into 5 sections:

A. Colloids of Biological Interest, B. Colloids of Industrial Interest, C. Wetting, Adsorption and Interfaces, D. Structure and Stability of Colloids, and, E. Theoretical Studies of Colloids.

On behalf of ECIS, we would like to thank all the participants for their contributions and for their very stimulating discussions, all along the meeting; the members of the Scientific Committee: L. Burlamacchi, B. Jonsson, H. Lekkerkerker, E. Sackmann and Th. Tadros who helped us for the preparation of the scientific program; our colleagues from the Centre de Recherche Paul Pascal (CNRS) for their collaboration in the organization of the 1988 meeting, and all the generous meeting sponsors.

Pierre Bothorel
Erick J. Dufourc

Contents

Progress in Colloid & Polymer Science   Progr Colloid Polym Sci 79:1-5 (1989)

A. Colloids of biological interest

Dielectrophoresis, instability, and electrofusion in membrane systems

D. S. Dimitrov and P. Doinov

Central Laboratory of Biophysics, Bulgarian Academy of Sciences, Sofia, Bulgaria

Abstract: An approach which is based on knowledge from the colloid chemistry of surfaces and thin films is presented. It is useful for understanding the mechanisms of membrane approach, instability, and fusion, as well as for optimization of cell electrofusion. It focuses on the use of external electric fields to induce membrane approach (by dielectrophoresis), instability (by reversible electroporation), and fusion. Experimental data for these processes and possible theoretical explanations are discussed. The basic conclusions are: 1) methods and results from thin-film dynamics combined with laws of membrane motion and deformation induced by external electric fields can be useful for understanding mechanisms of membrane electrofusion and fusion in general; 2) electrofusion can be optimized by measuring cell polarizability by dielectrophoresis and the critical voltage of reversible electroporation of membranes.

Key words: _D_ielectrophoresis, _i_nstability, _e_lectrofusion, _p_rotoplasts, _l_iposomes.

Introduction

In colloid chemistry, the phenomena of aggregation and coalescence of solid particles, bubbles, and drops are mainly investigated on the basis of principles of physical chemistry of surfaces and thin films. It has been shown that the specific thermodynamic and dynamic properties of the thin liquid films, formed between the approaching surfaces, largely determine the aggregation and coalescence, in particular, the kinetics of approach of deformable bubbles and drops and stability of thin-liquid films [1-3].

The theoretical and experimental results for colloid systems in most cases cannot be used directly to describe the complex biological membrane systems. The basic physicochemical approaches and results from that work may prove useful if combined with knowledge of the specific physicochemical properties of membranes. It has been proposed that the kinetics of membrane adhesion and fusion can follow the basic stages of aggregation and coalescence of bubbles and drops [2-7]. This may be true at least because of the very similar, in some cases identical, equations which describes the approach of tangentially immobile liquid/liquid interfaces and membranes. However, several essential differences between single interfaces and membranes should be taken into account:

1) The intermembrane interactions, which are the driving forces of membranes adhesion and fusion, may be different from those for single interfaces. While the electrostatic and hydration forces depend mainly on the type of the surface and the liquid film between them, the van der Waals interaction forces depend also on the membrane material and thickness and can be quite different from those for liquid surfaces, in particular with respect to the functional dependence on the separation distance; 2) the membranes can be permeable for the solution between them, while single interface are not. The effect of membrane hydraulic permeability can strongly increase the rate of approach; 3) the membrane are tangentially immobile, while liquid surfaces are mobile. Adding surfactants can decrease and almost stop the tangential mobility of liquid interfaces; 4) the membrane tension depends on membrane area changes while liquid surface tension does not. However, when surfactants are adsorbed onto the interface the surface tension depends on area change;

5) equilibrium deformations of membranes are described by membrane tension, shear elasticity, bending elasticity, and osmotic effects due to volume changes, while those of liquid interfaces are characterized only by surface tension. In many cases only the effect of membrane tension is important and then, mathematically, this case is identical to the case of a single interface; the surface tension should be replaced by the membrane tension; 6) cell membranes have complex structure, characterized by lateral and transmembrane inhomogeneity, and geometry where energy-consuming processes (active membrane motions, villi formation, etc.) are dominant, unlike the case of "non-living" interface.

Membrane fusion requires close approach and destabilization [8]. In electrofusion [9], the membrane contact is achieved by dielectrophoresis and the destabilization is induced by DC pulses. The basic advantage of the physical methods, in particular, of the electric fields, over conventional ones, e.g., by using chemicals, is the possibility for fast control of the entire process. This is due mainly to the much higher speed of propagation of the electric field than that of the diffusion of chemical substances. Dielectrophoresis is motion of particles in non-homogeneous AC fields [10]. It arises because of the interaction of the external electric field with the dipoles due to particle, polarization induced by the field. At close approach the induced dipoles interact with each other. Actually, this mutual interaction attaches the particles to each other and overcomes the repulsion forces. The electric-field-induced membrane destabilization is due to the transmembrane potential and surface tangential stresses which tend to decrease the thickness of the membrane and increase its area. When the electric forces overcome the elastic and viscous resistance of the membrane material, the membrane can break, which results, in many cases, in pore formation. The destabilized membranes fuse at close approach. The membrane shape changes. The cell protoplasm intermingles. The cells spherize to form the final product — the cell hybrid (in case of cell fusion). Similar processes may occur with other membrane systems.

Consequently, cell electrofusion, dielectrophoresis, membrane instability, in particular electroporation, are strongly interrelated, at the least because of the similar physiochemical mechanisms. The basic physical phenomena involved in these processes are known in principal. However, due to the complexity of the biological systems, in most cases the optimum parameters of the external constraints and the systems have been found empirically.

This paper discusses some of our recent work, and tries to make use of knowledge from colloid chemistry to explain and optimize some processes of motion and fusion in membrane systems.

The model

Motion of membranes leading to fusion can be split into four main stages: 1) continuous approach of two membranes; 2) destabilization of the membrane system, which can result in rupture of the intervening liquid film, bending of the membranes, rupture of the membranes or rupture of the film and the membranes; 3) membrane fusion itself, and 4) post-fusion phenomena such as expansion of the liquid film between the membranes and membrane shape relaxation. It is realized that some of the stages may not exist or may occur in a different way. In addition, this consideration seems to ignore molecular mechanisms. The phenomenological approach has the basic advantage that the theoretical formulae are functions of macroscopic quantities (membrane tension, viscosity, etc.), which can be measured experimentally and it is valid for any particular molecular mechanism.

The total time of the fusion process is the sum of the times for the separate stages. The duration of the membrane fusion itself (stage 3), which is due to rather fast molecular rearrangements, is commonly very short. The durations of the other three stages can be approximately described by the solution of the following non-linear, partial differential equation [2]:

$$\partial H/\partial t + 2Lp = \nabla\left[H^3(\partial p/\partial r)/12\mu\right] \tag{1}$$

$$p + p_e + \Pi(H) = B\Delta^2 H/2 - T\Delta H/2. \tag{2}$$

In the case of axisymmetric approach $\nabla = r^{-1} + \partial/\partial r$ and $\Delta = \nabla(\partial/\partial r)$, r being the radial coordinate. Here $H(r, t)$ is the local separation of the membranes, t — time, L — hydraulic permeability of the membranes, μ — liquid viscosity, p — dynamic pressure in the liquid between the membranes, T — membrane tension, B — membrane bending elasticity modulus, Π — disjoining pressure, p_e — equilibrium pressure in the liquid. These equations should be combined with appropriate boundary conditions. The mathematical difficulties arise from the highly non-linear equation (1). We will use several simple solutions for the case of spherical membranes, modeling, e.g., cells and liposomes. We will focus on cases where the driving force for membrane approach is due to external electric fields as is in dielectrophoresis.

Dielectrophoresis

By solving Eq. (1) for the case of very small separations between the membranes [2] it was shown that the rate of approach $v = -dh/dt$ $(h = H(r = 0))$ follows the Reynolds formula

$$v = V_{Re} = 2 Fh^3/3 \pi \mu R^4 \,. \tag{3}$$

R is the radius of membrane contact and F — driving force. When the separation is large, the Taylor formula can be used

$$v = V_{Ta} = Fh/6 \pi \mu R_c^2 \,, \tag{4}$$

R_c being the cell radius. An interpolation formula [12] describes the rate of approach at arbitrary separations

$$v = v_i = V_{St} V_{Ta} V_{Re}/(V_{st} + V_{Ta}) (V_{Ta} + V_{Re}) \tag{5}$$

where the Stokes rate is $V_{St} = F/6 \pi \mu R_c$.

The driving force due to dielectrophoresis is commonly given by [10]

$$F = F_p = 2 \pi R_c^3 \varepsilon_0 K_e \nabla E^2 = -\alpha \nabla E^2/2 \tag{6}$$

where ε_0 is the permitivity of the free space K_e — effective net polarizability of the cell, E — electric field intensity, and α — polarizability coefficient [12]. At close approach the dielectrophoretic force strongly increase due to the interaction between the induced dipoles. Then one must use more general formula

$$F = F_p + F_i \tag{7}$$

where F_i is the additional force due to induced electric interactions. The combination of Eqs. (5) and (7) allows the calculation of the rate of approach for arbitrary separations and, respectively, the time of approach until the establishment of the membrane contact. In order to do that we need the experimental values of the polarizability coefficients (α) and the interaction force F_i.

We constructed an assay for single cell dielectrophoresis, where a cylindrically symmetrical field is created between two concentric electrodes [12]. Recently we developed a more sophisticated method

where the electric fields are applied in a four-electrode chamber [13]. By measuring the cell velocity v as a function of the applied voltage and the separation distance one can obtain the values of the coefficients α and K_e. A typical value for pea protoplasts (radius 15 μm, in 0.5 M mannitol solution, medium viscosity — 1.41 · mPa · s, conductivity — 0.47 mS/m, temperature — 20 °C, frequency — 1 MHz) is $K_e = 110$. Polarizabilities for protoplasts and other cells at different conditions are summarized and presented in [14]. One of the basic conclusions is that in the radio frequency range most of the living cells behave as highly conductive spheres and the dielectrophoretic force F_c (Eq. (6)) can be estimated by assuming the net cell polarizability K_e of the order of the relative permittivity of the surrounding medium (for water solutions of the order of 10 to 100). This rule is also valid for determination of the interaction force F_i [15]. In this case, however, the force between two conducting spheres is given by a complicated expression, which can be evaluated only numerically [15]. Rather simple approximate formula can be written in the form

$$F_i = F_p \left(1 + Ah^{-1}\right), \tag{8}$$

where the constant A is of the order of 10 μm for pea protoplasts of 15 μm radius [15].

The expressions for the driving dielectrophoretic force also allow us to estimate the radius of membrane contact R by using a formula derived by solving Eq. (1) [2]

$$R^2 = FR_c/2 \pi T \,, \tag{9}$$

which is a modification of the well-known Deryaguin-Kussakov expression for the equilibrium radius of bubble to solid surface contact. For typical values of the electric field intensity ($E = 10^5$ V/m and $\nabla E = 10^9$ V/m [14]) and $T = 0.1$ mN/m Eq. (9) gives contact radii of the order of 1 μm for protoplasts (radii of the order of 10 μm) which is the value observed experimentally [16]. Liposomes which have internal conductivity lower than that of the external medium do not show positive dielectrophoresis and are not attracted to protoplasts by mutual dielectrophoresis [17]. Therefore, in this case dielectrophoresis is not an effective method to collect cells and liposomes for the purposes of electrofusion.

The basic conclusion from our experimental results on dielectrophoresis and this consideration is that there are frequencies (commonly in the ratio frequency range) and medium conductivities for each type of cell where the polarizability coefficients are maximal. The electrofusion should be carried out at those conditions in order to ensure minimal times of approach and maximal area of contact.

Electrically induced membrane instability

We developed theoretical models for electrical breakdown of membranes [6, 7, 18, 19] which are in agreement, at least qualitatively, with available experimental data [20–22]. The calculation of the critical thickness of rupture [6] of the film between two approaching membranes showed [23] that the dielectrophoretic force is small to induce instability. Therefore, we can expect that the membranes and the film between them are destabilized during the high voltage DC pulse. Two basic conclusions of importance for electrofusion are:

1) Formation of pores depends on the applied voltage and duration of exposure to the field. Higher voltages should be applied for shorter times to induce reversible breakdown and formation of pores. Applying the field for longer periods of time leads to irreversible breakdown and in many cases to cell death. The heat generation is proportional to the square of the applied voltage and the time of exposure. However, the time of breakdown is inversely proportional to the difference of the square of the applied voltage and its critical value. Therefore, when the applied voltages are near to the critical value, the heat generation will be large because the pulse duration must be long. Increasing the voltage to get breakdown leads to decrease of the heat generation.

2) The breakdown voltage depends on the membrane tension, respectively on the osmolarity of the medium. Decreasing the solution osmolarity leads to decrease of the breakdown voltage. Our experimental results [21] showed that decreasing the tonicity of the medium leads to a very sharp decrease of the breakdown voltage in the region near to the isotonic conditions. The further decrease of the tonicity does not change the breakdown voltage, much further.

Electrofusion

It was found that the functional dependence of the fusing voltage on the pulse duration for pea protoplasts correlates strongly to that for their electrical breakdown [22]. This helps in choosing the most appropriate parameters of the electrical field for effective fusion.

The results for membrane breakdown from the previous section can be used to make the following conclusions about how to get better fusion: 1) the applied voltage should be of the order of or higher than the critical voltage of reversible breakdown; 2) the pulse duration should be rather short, and 3) in some cases making the medium slightly hypotonic can facilitate the fusion. However, larger membrane tensions, i.e., larger osmotic forces, can decrease the contact area (see Eq. (9)) and hinder the fusion.

Recently, by electrofusion we produced hybridomas which secreted monoclonal antibodies against the Hc antigene of salmonella [24]. We also succeeded to fuse whole fragile yeast mutants with protoplasts of non-fragile yeasts [25]. In all these experiments the time of dielectrophoresis was kept to a minimum, e.g., not longer than 2 min, which corresponds to the theoretical estimations by the formula for the rate of mutual approach of cells (Eq. (5)). The fusing voltage was higher than that for breakdown. The highest yields were obtained when the number of pulses was 2 or 3; this could be due to viscous effects in the film and membrane shape relaxations, which take time.

For getting high yields with the electrofusion technique it is especially important to have the right arrangements of the electrodes and the appropriate chamber to place them in. For biotechnological purposes, i.e., for production of viable hybrids on a larger scale, the helical chamber is good [26]. We constructed several types of chambers which can also be used for fundamental research, not only of biomembrane systems, but also of colloid systems — and in some cases for the purposes of practice:

1) The cylindrical chamber [12] was used for dielectrophoretic measurements. The basic advantage of this chamber for fusion experiments is that the distribution of the electric field is known. A number of studies were performed with this chamber [27–31]; 2) sputtered metal on glass surfaces which form very thin electrodes; electrodes of different shapes were tested. The basic hope for these chambers is that they can ensure very specific cell fusion; the yield is not high and it is difficult to clean the electrodes without damaging them. 3) Plane-parallel chambers: in this case the two electrodes are flat and parallel to each other. The basic advantage of this chamber is that the field is homogeneous and the cells approach each other by mutual dielectrophoresis, i.e., the damaging field is effectively decreased; 4) the four-electrode chamber [13] is multifunctional: it can serve for electrofusion and electroporation in homogeneous and non-homogeneous fields,

for measuring simultaneously the dielectrophoretic mobilities and densities of single cells, etc.

The working chambers were connected to an electronic apparatus, Biophysical Engineering Systems and Technology (BEST-1.1). It produces AC and DC fields of different amplitude, frequency, and duration and can be used for dielectrophoresis, electroporation, and electrofusion.

Conclusion

Approach and instability of membranes are of basic importance for the kinetics of adhesion and fusion of vesicles and cells and for any other process which involves dynamic interactions between membranes, e.g., dielectrophoretically induced membrane motion and approach [32], electroporation [33], etc. Therefore, the quantitative theoretical description, based on macroscopic phenomenological concepts, such as physical chemistry of surfaces and thin films, hydrodynamics and membrane mechanics, and comparison with experimental results obtained in rigorously defined physicochemical conditions, can help in understanding the fundamental mechanisms of these processes in model and biological systems. The wide diversity and complexity of systems and conditions, especially in the living world, requires that the general concepts be applied for every concrete situation by developing a concrete theoretical model and solving the respective equations with appropriate boundary conditions.

Acknowledgement

We thank our colleagues and friends from the Department of Membrane Interactions: Drs. D. Zhelev, N. Stoicheva, M. Angelova, R. Mutafchieva, E. Bakalski, and G. Sharkov for the useful discussions and help. This work was supported by the Ministry for Culture, Science and Education through contract No. 189.

References

1. Scheludko A (1967) Adv Colloid Interface Sci 1:391-464
2. Dimitrov DS (1983) Progr Surface Sci 14:295-425
3. Dukhin SS, Rulev NN, Dimitrov DS (1986) Dynamics of thin films and coagulation. Naukova Dumka, Kiev
4. Dimitrov DS (1981) Biophys J 36:21-25
5. Dimitrov DS (1982) Colloid Polym Sci 260:1137-1142
6. Dimitrov DS, Jain RK (1984) BBA 779:437-467
7. Dimitrov DS, Zhelev DV (1985) studia biophysica 110:105-109
8. Sowers AE (ed) (1987) Cell Fusion. Plenum Press, New York
9. Berg H, Jacob H-E, Bauer E (1985) studia biophysica 110:143-149
10. Pohl AH (1978) Dielectrophoresis. Cambridge University Press, Cambridge
11. Dimitrov DS, Stoicheva N, Stefanova D (1984) J Colloid Interface Sci 98:269-271
12. Dimitrov DS, Tsoneva I, Stoicheva N, Zhelev D (1984) J Biol Phys 12:26-30
13. Doinov P, Dimitrov DS, Stoicheva N (1988) Proc Workshop on Biothermokinetics, Aberystwith, pp 42-43
14. Dimitrov DS, Zhelev D (1987) Bioelectroch Bioenerg 17:549-557
15. Zhelev DV, Kuzmin P, Dimitrov DS, Biolog Membr, submitted
16. Stoicheva N, Tsoneva I, Dimitrov DS (1985) Z Naturforsch 40c:735-739
17. Angelova M, Doinov P, Dimitrov DS, in preparation
18. Dimitrov DS (1984) J Membrane Biol 78:53-60
19. Dimitrov DS Zhelev DV, Jain RK (1985) J theor Biol 113:353-377
20. Benz R, Zimmermann U (1980) Bioelectroch Bioenerg 7:723-739
21. Zhelev DV, Tsoneva I, Dimitrov DS (1988) Bioelectroch Bioenerg 19:217-225
22. Zhelev DV, Doinov P, Dimitrov DS (1988) Bioelectroch Bioenerg 20:155-167
23. Dimitrov DS, Zhelev DV (1988) Colloid Polym Sci 76:109-112
24. Tsoneva I, Panova I, Doinov P, Strahilov D, Dimitrov DS (1986) studia biophysica 125:31-35
25. Tsoneva I, Ivanova D, Doinov P, Venkov P, Dimitrov DS (1988) FEMS Lett, submitted
26. Zimmermann U, Vienken J, Halfmann J, Emeis CC (1985) Adv Biotechn Processes 4:79-106
27. Tsoneva I, Zhelev DV, Dimitrov DS (1986) Cell Biophysics 8:89-101
28. Stoicheva N, Dimitrov DS (1986) studia biophysica 111:17-21
29. Stoicheva N, Dimitrov DS (1986) Electrophoresis 7:339-341
30. Stoicheva N, Dimitrov DS (1986) Compt rend Acad bulg Sci, 39:105-107
31. Stoicheva N, Dimitrov DS, Ivanov A (1987) Eur Biophys J 14:25-256
32. Sauer FA (1985) In: Chiabrera A, Nicolini C, Schwan HP (eds) Interactions between electromagnetic fields and cells. Plenum Press, pp 191-221
33. Neumann E, Sowers E, Jordan A (eds) (1989) Electroporation and Electrofusion in Cell Biology. Plenum Press, New York

Received October, 1988; accepted February, 1989

Authors' address:

Dr. Dimiter Stanchev Dimitrov, DSc
Central Laboratory of Biophysics
Bulgarian Academy of Sciences
Acad. G. Bonchev Str., bl. 21
Sofia 1113, Bulgaria

Progress in Colloid & Polymer Science   Progr Colloid Polym Sci 79:6–10 (1989)

Bending undulations of lipid bilayers and the red blood cell membrane: A comparative study

H.-P. Duwe, K. Zeman, and E. Sackmann

Biophysics Group, Physikdepartment E22, Technische Universität München, Garching, F.R.G.

Abstract: Biological membranes (such as envelopes of erythrocytes) and vesicles of lipid bilayers are highly flexible shells which are strongly thermally excited although in the former case the lipid/protein bilayer is coupled to a quasi-two-dimensional macromolecular network. The bending rigidity of bilayers may be even reduced to the order of kT by amphiphilic substances so that they may behave as random surfaces.

In the first part of this contribution we discuss experimental studies of thermal excitations of vesicles and cells by fast image processing techniques that allow direct measurements of the mean square amplitudes of individual modes up to the 10th order and thus enable high precision measurements of bending elastic moduli.

In the second part we discuss the role of the membrane bending rigidity and chemically induced bending moments for shape transitions of cells and models of cell plasma membranes. The latter are composed of two-dimensional macromolecular solutions and gels prepared by combining macromolecular and normal lipids.

Key words: Artificial and natural membranes, random surfaces, bending elasticity.

1. Introduction

The elastic properties of cell plasma membranes are both fascinating and biologically important. Their systematic study starts with the development of the micropipette technique by Evans and coworkers [1–3] and the Brochard-Lennon theory [4] of the flickering phenomenon. These and further studies by a variety of techniques [5–7], showed that plasma membranes exhibit exceptional elastic properties. Compared to technical materials (such as shells of polyethylene) they are orders of magnitude softer as far as the shear and bending elasticity is concerned. Thus the bending stiffness of the biological shell is by some three orders and the shear stiffness by some five orders of magnitude smaller than the corresponding values of the technical material [8]. This softness allows the cells of the blood cell-system to penetrate through narrow gaps of the capillary system or the extracellular network without loss of material. It also allows control of the cell shape by small variations of the spontaneous curvature [9]. In contrast, the stretching module of both types of materials are comparable. This low compressibility is responsible for the low ion permeability of the cell plasma membrane under stress.

The present report deals with comparative high precision measurements of the bending stiffness of lipid bilayer vesicles and erythrocytes based on the Fourier analysis of thermally excited membrane undulations. In particular, we point out the unexpected high undulation (= flickering) amplitudes (and thus low flexural rigidity) of the red blood cell membrane, which can be sustained by the cells over a wide range of osmolarities.

2. Bending elasticity of bilayer vesicles

The extension of the Brochard-Lennon theory to the shape fluctuations of quasi-spherical flaccid vesicles [10–12], paved the way for high-precision measurements of the bending stiffness by Fourier analysis of the contour of vesicles as seen by phase contrast microscopy [13]. Improved theories taking into

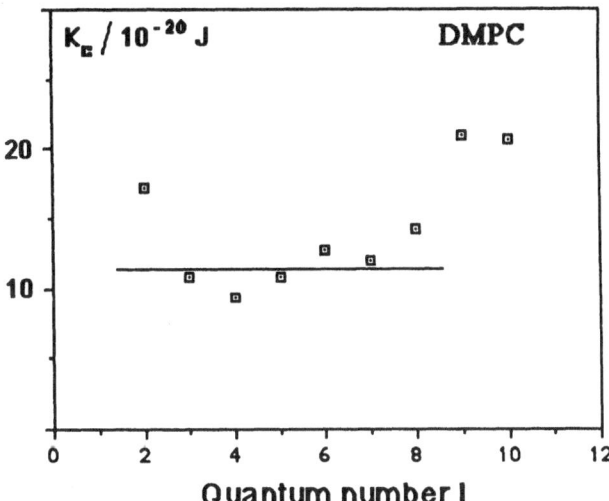

Fig. 1. Values of K_c plotted as a function of l according to Eq. (1) for vesicles of DMPC and calculated for a vanishing surface tension $y = 0$. The higher value for the second mode is due to a surface tension, owing to the constraints of fixed area and volume

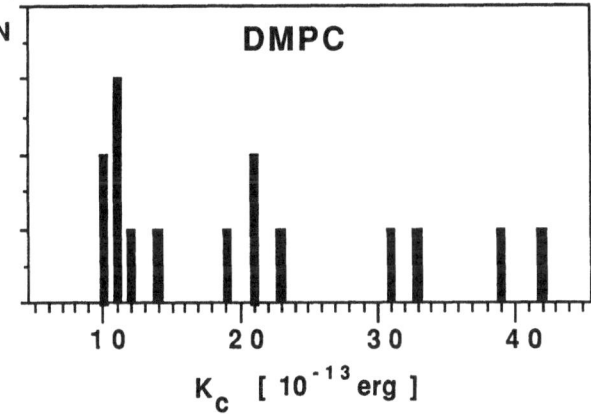

Fig. 2. Summary of bending elastic constants K_c of several vesicles. For DMPC vesicles four groups of values differing by a factor of about two are discernible. This is attributed to increasing number of bilayers

account the constraint imposed by the conditions of constant membrane area and vesicle volume lead to the following equation between the mean square amplitudes of the spherical harmonic excitation modes $U_{l,m}$ and the bending stiffness K_c [12]

$$\langle |U_{l,m}|^2 \rangle = \frac{k_B \cdot T}{K_c} \cdot \frac{1}{Z(l)} \qquad (1a)$$

$$Z(l) = (l+2)(l-1) \cdot [l(l+1) - 4\omega + 2\omega^2 - y] \qquad (1b)$$

where l is the wavenumber of excitation, ω corresponds to the spontaneous curvature of the membrane and is normally set to zero. The "Lagrange" parameter y takes into account the constant area and corresponds to a surface tension. In Fig. 1 we plot the K_c values determined for dimyristoylphosphatidylcholine (DMPC)-vesicles as a function of 1 for $y = 0$ and $\omega = 0$ according to Eq. (1). If the vesicle shape is near the spherical limit, the low order modes in particular $l = 2$ and $l = 3$ are considerably influenced by the surface tension y. The mean value of the bending constant K_c between the $l = 3$ and $l = 8$ mode is $(1.1 \pm 0.1) \cdot 10^{-19}$ J. This value holds for single bilayer vesicles. So by measuring a large number of different DMPC vesicles three groups of values differing by a factor of about two are discernible (see Fig. 2) This function is attributed to increasing number of bilayers. Mixtures of DMPC with 20–30 % cholesterol give between two-and-three-times higher values of K_c than pure DMPC.

Following Millner and Safran [12] it is possible to measure the temporal autocorrelation function of the amplitudes $U_{l,m}$.

$$\langle U_{l,m}(t) \cdot U_{l,m}(t+\tau) \rangle = \langle |U_{l,m}|^2 \rangle \cdot e - \frac{t}{\tau_l} \qquad (2a)$$

$$\tau_l = \frac{\eta R_0^3}{K_c} \cdot \frac{N(l)}{l(l+1) - (y + 4\omega - 2\omega^2)}, $$

$$N(l) = \frac{(2l+1)(2l^2 + 2l - 1)}{l(l+1)(l+2)(l-1)} \qquad (2b)$$

R_0 = radius of the corresponding sphere; η = viscosity of water.

The autocorrelation function obtained for the $l = 2$ and $l = 3$ mode of DMPC is plotted in Fig. 3. The relaxation times are $\tau_2 = 2.1$ s and $\tau_3 = 0.6$ s. The theoretical values for $K_c = 1.1 \cdot 10^{-19}$ J, $R_0 = 9$ μm and vanishing ω and y are $\tau_2 = 2.5$ s and $\tau_3 = 0.7$ s. The small discrepancies between the theoretical and experimental values are explained in terms of non-vanishing surface tension effects.

3. Flickering and apparent bending modulus of erythrocyte

One of the most striking manifestations of the high degree of dynamics of living materials is the flickering phenomenon of erythrocytes. It is generally attributed to thermally excited membrane undulations and is exploited in order to measure bending elastic constants

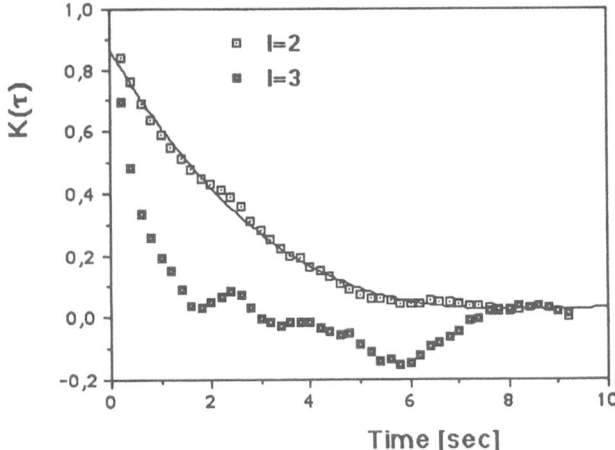

Fig. 3. Autocorrelation function of the amplitudes for the lowest bending modes with $l = 2$ and $l = 3$. The relaxation times are 2.1 s for $l = 2$ and 0.6 s for $l = 3$

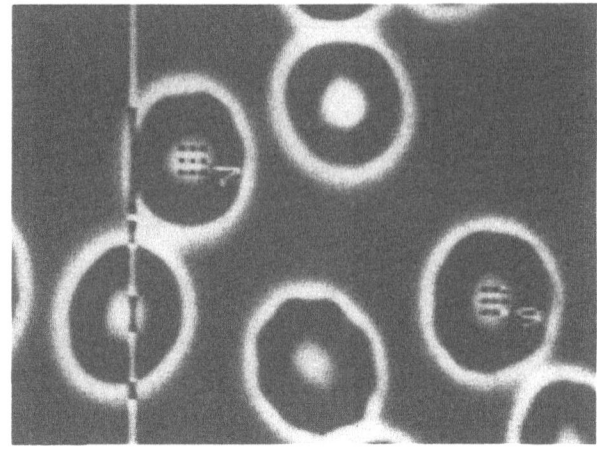

a)

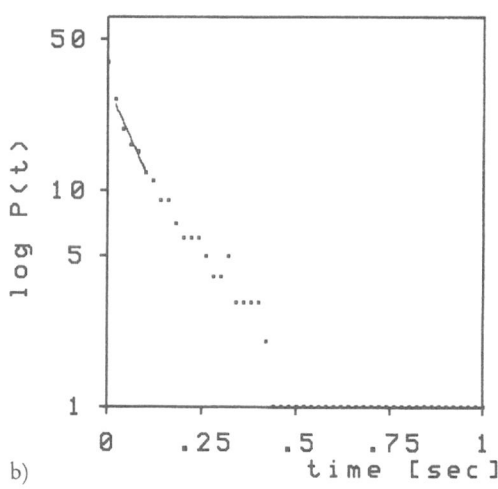

b)

Fig. 4. a) Image of a cell with the positions, r, at which $P(r, t)$ is measured, marked by dark spots; b) Intensity correlation function of cell obtained from one measurement point

[4, 6, 14]. In fact, by a new technique [14], dynamic reflection interference contrast microscopy (RICM), in combination with a fast-image processing system, we succeeded in measuring the mean square amplitudes as a function of the wave vector in the range from 10^4 cm^{-1} (1 μm) to 5×10^3 cm^{-1}. The apparent bending constant obtained by application of the equipartition theorem is about 7×10^{-20} J and is thus about an order of magnitude smaller than the K_c-value of bilayers of a 1:1 DMPC-cholesterol mixture, which would approximately correspond to the lipid composition of the biomembrane.

This provides some suggestions that the origin of the flickering is not only due to thermal excitations of the lipid-protein bilayer but may be at least partially excited within the cytoskeleton. Further evidence for this view is provided by recent systematic studies of the modification of the flickering amplitudes by variations of the external conditions. In particular we studied the effect of changes of the osmolarity of the buffer or of perturbations of the inner ion composition such as the cytoplasmic Ca^{++}-concentration.

Such studies are most effectively performed by the following modification of the classical flicker spectroscopy (cf. Fig. 4). An assembly of cells (cf. Fig. 4a) is slightly fixed to a glass plate coated with polylysine of molecular weight 3000. By using buffer containing some fetal calf serum the cells can be preserved in the discoid state for some hours. The cells are illuminated in a phase contrast microscope. The image of the cell assembly is projected onto a CCD (charge coupled

device) video camera (Aqua TV, Kempten, F.R.G.) and continuously transferred to a homemade image processing system (BAMBI). The second step consists in the calculation of the autocorrelation function

$$P(\xi, \tau) = \int_{-\infty}^{+\infty} \int_{-\infty}^{+\infty} d^2 r \, dt \, I(r, t) \, I(r + \xi, t + \tau) \quad (3)$$

of the intensity of the light passing the cell, where $\xi =$ mesh size.

With the image processing system it is possible to measure:

1) the time correlation function of the intensity of the light passing one or more of the cells, and

2) the temporal-spatial correlation function of a single cell given by Eq. (3).

The former method is applied in the present work. We average over nine points indicated in Fig. 4a. Figure 4b shows a plot of the logarithm of $P(t)$ vs. t. To a first approximation it is well represented by a single exponential. Thus the power spectrum can be characterized by an average wave vector q. From the Brochard-Lennon theory we find

$$P(t, q) \propto I_0^2 S \frac{k_B T}{q^4 K_c} \exp\left(-\frac{K_c q^3}{2\eta} t\right). \quad (4)$$

Since the average q-vector is not known we can only obtain relative values of K_c from the amplitude and of K_c/η from the decay rate of the exponential. It is very important to measure the average intensity I_0 carefully in order to account for changes of the cell shape (e.g., by osmotic swelling and shrinking).

Figure 5 shows a record of K_c and η for a single cell over a period of 50 min. Obviously these parameters are well controlled over this time period. Figure 6 shows values of I_0, K_c, and η of erythrocytes which were continuously measured while the osmolarity changed. The continuous increase of I_0 with increasing osmolarity shows the deflation of the cells. Surprisingly, the bending stiffness K_c is well preserved over a wide range of osmolarities. The increase K_c above 450 mosm is most probably caused by the interaction — in particular interpenetration — of the cytoskeletons of the opposing membranes. The viscosity η of the cytoplasm increases with increasing deflation of the cell, as

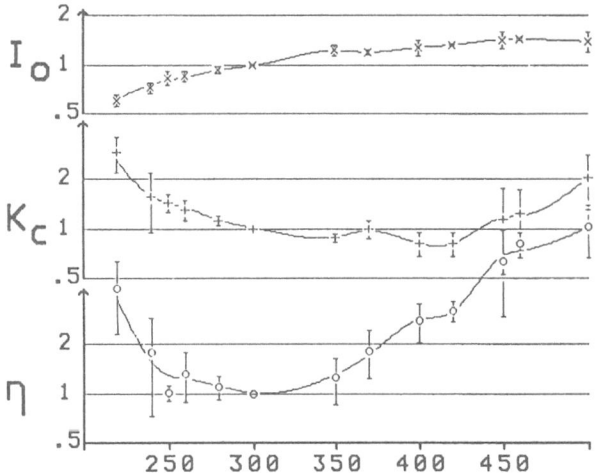

Fig. 6. Average intensity I_0 of light passing the cells (upper trace), apparent bending stiffness K_c (middle trace) and viscosity η (lower trace) of several ($n = 2$–8) erythrocytes for various osmolarities. The bars give the variations of the values obtained from different cells; the scale is logarithmic

expected. The increase of K_c and η at small osmolarities, that is when the cells assume a nearly spherical shape, can be explained in terms of the surface tension introduced by the constraint of constant volume and areas [12].

4. Is the erythrocyte flickering a thermally driven process and does it serve a biological purpose?

The low value of the apparent bending stiffness of $K_c \sim 7 \cdot 10^{-20}$ J of the red bloodcell as compared to the module of $K_c \sim 4 \cdot 10^{-19}$ J of DMPC-vesicles containing 30% of cholesterol is hard to reconcile with the assumption that the erythrocyte undulation is a purely thermally driven process. Further evidence comes from the finding shown in Fig. 6 that the apparent bending constant, K_c, does not change over a large range of osmolarities, whereas the cytoplasm viscosity, η, changes drastically. In another group of experiments (K. Zeman, unpublished results) it was found that K_c remains constant over a substantial period of time after influx of Ca^{++} into the cell via ionophore A 23187. Even when the cell shrinks, owing to the Ca^{++}-activated potassium and water efflux, K_c remains constant until the cell is strongly dehydrated and crenation takes place. When the cell-shrinking is prevented by a

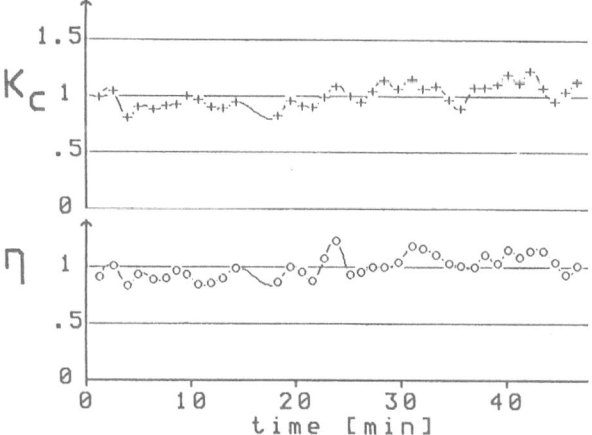

Fig. 5. Value of the apparent K_c and η of a single cell recorded over a time of 50 min. The points (O) and (+) give the averages of η and K_c, in arbitrary units, respectively

high potassium buffer [15], which inhibiting the potassium and water efflux, K_c does not alter over a prolonged period of about 150 min.

The undulations could, for instance, be driven by fluctuations of the structure of the spectrin actin network. Two mechanisms are possible:

Firstly, fluctuations in the lateral density of the network. Since it is anchored to the lipid/protein bilayer this would result in bending undulations. As described earlier relative fluctuations of the mesh size, ξ, of the network by some $5 \cdot 10^{-3}$ could account for the amplitudes of the order of 0.1 μm at an undulation wavelength of 1 μm [6]. In order to account for the observation that the mean square amplitudes scale as the fourth power of the undulation wave length (or inverse wave vector) the driving force should exhibit the same statistical character as those which drives thermal excitation. The energy could come from phosphorylation reaction of spectrin-binding proteins such as ankyrin and band 4.1 which are catalyzed by a variety of membrane-bound and cytoplasmic kinases and phosphatases. In fact, it is well established now that a large fraction of the ATP-pool is used for the turnover of the binding proteins of the cytoskeleton.

A second mechanism could be undulations of the cytoskeleton itself. These could well be driven by osmotic pressure differences between the gap separating the inner monolayer of the lipid/protein bilayer and the cytoskeleton on one side and the interface between the latter and the hemoglobin solution of the cytoplasm on the other hand. In view of the fact that the undulations are not remarkably affected by the change of the Hb-concentration (osmolarity) we consider this mechanism less likely.

Does the flickering serve a biological purpose? Two possibilities are conceivable: Firstly, it could facilitate the exchange of hemoglobin between the cortex formed by the cytoskeleton and the cytoplasm. Secondly, an even more intriguing purpose could be to prevent adhesion of the cells to the surface of the body tissue. Following Helfrich [16], the membrane undulations are expected to provide strong repulsive undulation forces.

Acknowledgement

This work was supported by the Deutsche Forschungsgemeinschaft (SA 246/13-4, SA 246/16-2), by the Fonds der chemischen Industrie, and by the Lorentz Stiftung. Enlightening discussions with E. Evans, M. D. Mitov, I. Bivas, and W. Helfrich are gratefully acknowledged.

References

1. Evans EA, Waugh R (1977) J Colloid Interface Sci 60:286
2. Evans EA, Needham D (1986) Faraday Discuss Chem Soc 81:267
3. Hochmuth RM, Buxbaum KL, Evans EA (1980) Biophys J 29:177
4. Brochard F, Lennon JF (1975) J Physique 36:1035
5. Engelhardt H, Sackmann E (1988) Biophysics J 54:495
6. Fricke K, Wirthenson K, Laxhuber R, Sackmann E (1986) Eur Biophys J 14:67
7. Peterson NO, McConnaughey WB, Elson EL (1982) Proc Natl Acad Sci 79:5327
8. Duwe H-P, Eggl P, Sackmann E (1988) In: Polymers and Biological Functions. Bad Nauheim
9. Sackmann E, Duwe H-P, Engelhardt H (1986) Faraday Discuss Chem Soc 81:281
10. Schneider MB, Jenkins JT, Webb WW (1984) J Physique 45:1457
11. Helfrich W (1986) J Physique 47:321
12. Millner ST, Safran SA (1987) Physical Review A, Vol 36
13. Engelhardt H, Duwe H-P, Sackmann (1985) J Physique lett 46:395
14. Zilker A, Engelhardt H, Sackmann E (1987) J Physique 48:2139
15. Clark MR, Mohandas N, Feo C, Jacobs MS (1981) J Clin Invest 67:531
16. Helfrich W (1978) Z Naturforsch 33a:305

Received November, 1988;
accepted February, 1989

Authors' address:

E. Sackmann
Biophysics Group
Physikdepartment E22
Technische Universität München
D-8046 Garching, F.R.G.

Progress in Colloid & Polymer Science   Progr Colloid Polym Sci 79:11–17 (1989)

Thermal fluctuations of giant vesicles and elastic properties of bilayer lipid membranes.
The role of the excess surface

J. F. Faucon[1]), P. Méléard[1]), M. D. Mitov[2]), I. Bivas[2]), and P. Bothorel[1])

[1]) Centre de Recherche Paul Pascal, C.N.R.S., Domaine Universitaire, Talence, France
[2]) Department of Liquid Crystals, Institute of Solid State Physics, Bulgarian Academy of Sciences, Sofia, Bulgaria

Abstract: Thermal fluctuations of giant unilamellar vesicles have been investigated both theoretically and experimentally. The theoretical model developed here takes the conservation of vesicle volume and membrane area explicitly into account. Using the membrane tension as a single adjustable parameter we are able to explain the experimental data. The bending elastic modulus[1]) of egg-yolk lecithin bilayers is $k_c = (0.9 \pm 0.2) \cdot 10^{-19}$ J.

Key words: Thermal fluctuations, vesicles, membranes, bending elasticity.

Introduction

Lipid membranes are one of the essential constituents of living cells. Their mechanical properties are closely related to the problem of cell stability and resistance to external influences. Soon after the beginning of research in this field the similarity of membranes to liquid crystals was realized. Applying the ideas of liquid crystal physics Helfrich [1] demonstrated that the bending elastic energy, F_c, per unit membrane area is given by the expression

$$F_c = \frac{k_c}{2} (c_1 + c_2 - c_0)^2 + \bar{k}_c c_1 c_2 \qquad (1)$$

where c_1 and c_2 are the principal curvatures of the membrane, c_0 is the spontaneous curvature, k_c and $\bar{k}_c$ are the elastic constants for cylindrical and saddle bending, respectively.

The first attempt to measure the curvature elastic modulus, k_c, was made by Servuss et al. [2]. Analyzing the thermal shape fluctuations of long tubular vesicles they obtained $k_c = (2.3 \pm 0.3) \cdot 10^{-19}$ J for egg lecithin membranes. Later, Sakurai and Kawamura [3] evaluated $k_c = 0.4 \cdot 10^{-19}$ J by bending myelin figures in a magnetic field. Measuring the time-correlation function of the shape fluctuations of long cylindrical tubes [4], as well as giant spherical vesicles [5], Schneider et al. found $k_c = (1-2) \cdot 10^{-19}$ J. Using Fourier analysis of thermally excited surface undulations of giant spherical vesicles Engelhardt et al. [6] obtained $k_c = 0.4 \cdot 10^{-19}$ J. Recently, Duwe et al. [7] reported $k_c = 1.1 \cdot 10^{-19}$ J. Introducing the angular correlation function of thermal shape fluctuations Bivas et al. [8] measured $k_c = (1.28 \pm 0.25) \cdot 10^{-19}$ J. Obviously, more precise treatment of the problem is necessary. In the present work it is shown that the bending modulus, k_c, can be accurately determined from the thermal fluctuations of giant vesicles provided that: i) all the constraints on the fluctuations are properly accounted for in the theoretical model, and ii) the experimental accuracy is improved by analyzing a large number of contours for a given vesicle.

------

[1]) A somewhat smaller value is obtained when the statistical weights of different vesicles, as well as the effect of the TV camera integration time are properly taken into account [13].

Theory

The equation of a slightly deformed spherical vesicle in polar coordinates $(\varrho, \theta, \varphi)$ is

$$\varrho(\theta, \varphi, t) = R\left[1 + u(\theta, \varphi, t)\right] \quad |u(\theta, \varphi, t)| \ll 1 \quad (2)$$

where R is the radius of a sphere enclosing the same volume as that of the vesicle and $u(\theta, \varphi, t)$ is a function describing the displacements of vesicle wall in direction (θ, φ). We can calculate the bending elastic energy of the whole vesicle, $F_c\{u\}$, integrating Eq. (1), over the vesicle surface.

Usually, the vesicles are poorly permeable to water and salts, therefore, the exchange of water through the vesicle membrane can be neglected. Moreover, to a very good approximation, the water is an uncompressible fluid. Due to the very high value of the stretching elastic modulus the vesicle membrane is practically unstretchable. So, the total vesicle volume, $V\{u\}$, and area, $S\{u\}$, do not change during the fluctuations.

We can consider the vesicle shape at a given moment, $u(\theta, \varphi, t)$, as composed of two contributions: a static part, $u_0(\theta, \varphi)$ that is the average (equilibrium) vesicle shape and a dynamic part or perturbation, $\delta u(\theta, \varphi, t)$ that gives the deviation from this equilibrium and describes the fluctuations of the vesicle. To find the equilibrium vesicle shape we have to minimize its curvature elastic energy at constant volume and area. This is a variational problem *with* constraints. The usual mathematical method to solve it is to replace the functional, $F_c\{u\}$, and the associated constraints, $V\{u\} = V_0$ and $S\{u\} = S_0$, with a new functional, $F\{u\}$ that is a linear combination of them, and to treat the last like a functional *without* constraints:

$$F\{u\} = F_c\{u\} + \sigma S\{u\} - \Delta p V\{u\}. \quad (3)$$

Here, σ and Δp are Lagrange multipliers associated with the constraints of constant membrane area and enclosed volume of the vesicle: physically, σ is the membrane tension and Δp is the difference between the internal and external hydrostatic pressures. If the function u_0 *minimizes* the functional (3), then the first variation, $\delta F\{u_0, \delta u\} = 0$ and the second, $\delta^2 F\{u_0 \, \delta u\} \geq 0$ for all small perturbations, δu, around this equi-

librium shape, u_0. The first of these conditions leads to the Euler-Lagrange equation

$$\nabla^2(\nabla^2 u_0) + (2 - \bar{\sigma})\nabla^2 u_0 + 2(\bar{\sigma} - \bar{p})u_0 = \bar{p} - 2\bar{\sigma} \quad (4)$$

where, $\bar{p} = \Delta p R^3 / k_c$ and $\bar{\sigma} = \sigma R^2 / k_c$ are dimensionless pressure and tension. We can look for a solution of Eq. (4) in a series of spherical harmonics, $Y_n^m(\theta, \varphi)$, defined in [10]:

$$u_0(\theta, \varphi) = A_0^0 Y_0^0(\theta, \varphi) + \sum_{m=-2}^{m=+2} A_2^m Y_2^m(\theta, \varphi). \quad (5)$$

Considering only vesicles fluctuating around a spherical shape we have

$$-\frac{A_0^0}{\sqrt{4\pi}} = \frac{\bar{p} - 2\bar{\sigma}}{2\bar{p} - 2\bar{\sigma}}, \quad A_2^m = 0. \quad (6)$$

We decompose the fluctuations, δu, in series of spherical harmonics as well:

$$\delta u(\theta, \varphi, t) = \sum_{n=0}^{n_{max}} \sum_{m=-n}^{m=+n} U_n^m(t) Y_n^m(\theta, \varphi). \quad (7)$$

where $U_n^m(t)$ are the time-dependent coefficients and n_{max} is a high frequency cut-off. We calculate the second variation

$$\frac{1}{2}\delta^2 F = \frac{k_c}{2}\sum_{n=2}^{n_{max}} \sum_{m=-n}^{m=+n} \{(n-1)(n+2)$$

$$\times [\bar{\sigma} + n(n+1)] + 4\bar{\sigma} - 2\bar{p}\} |U_n^m(t)|^2. \quad (8)$$

Applying the equipartition theorem to each mode of agitation we have

$$\langle |U_n^m(t)|^2 \rangle = \frac{kT}{k_c}\frac{1}{(n-1)(n+2)[\bar{\sigma} + n(n+1)] + (4\bar{\sigma} - 2\bar{p})}, \quad n \geq 2. \quad (9)$$

The equation of Euler-Lagrange (4), (6) contains two parameters, $\bar{\sigma}$ and $\bar{p}$. We need two supplementary conditions to determine them. These are the conservation of the vesicle volume and membrane area. After some algebra one can see that the last term in the denominator of Eq. (9) is very small, because $\bar{p} \approx 2\bar{\sigma}$ and therefore, it can be omitted. In that way, we come to the result of Milner and Safran [9] obtained earlier. The detailed analysis of the solutions shows that when the vesicle excess area is very small the thermally in-duced fluctuations create positive membrane tension, $\bar{\sigma} > 0$, which in turn acts on the fluctuations, decreasing their amplitudes. In this case the vesicle fluctuates around a sphere. When the excess area is increased, the membrane tension decreases and if a critical value is exceeded, the vesicle becomes elliptical ($A_2^m \neq 0$). In that case the membrane tension is negative, $\bar{\sigma} \approx -6$.

Now we have to find a relation between the amplitudes of the fluctuations, $\langle |U_n^m(t)|^2 \rangle$, and an experimentally measurable quantity. Considering the equatorial cross-section of a vesicle in the XY plane of our coordinate system, given by $\theta = \pi/2$ in Eq. (2), we write the radius of the cross-section, $\varrho_e(\varphi, t) = \varrho\left(\frac{\pi}{2}, \varphi, t\right)$. Following Bivas et al. [8] we calculate the angular autocorrelation function of the vesicle radius, $\xi(\gamma, t)$, at a given moment of time

$$\xi(\gamma, t) = \frac{1}{R^2} \left[\frac{1}{2\pi} \oint \varrho_e(\varphi + \gamma, t)\, \varrho_e^*(\varphi, t)\, d\varphi \right. $$
$$\left. - \left(\frac{1}{2\pi} \oint \varrho_e(\varphi, t)\, d\varphi \right)^2 \right]. \quad (10)$$

Taking time average of the above equation and using the series expansions (5), (7) and the expression (9) for the amplitudes we finally obtain

$$\xi(\gamma) = \langle \xi(\gamma, t) \rangle = \sum_{n=0}^{n_{max}} B_n(\bar{\sigma}, k_c)\, P_n(\cos \gamma) \quad (11)$$

$$B_n(\bar{\sigma}, k_c)$$
$$= \frac{kT}{4\pi k_c} \frac{(2n+1)}{(n-1)(n+2)[\bar{\sigma} + n(n+1)]}, \quad n \geq 2. \quad (12)$$

We see that the time-averaged autocorrelation function, $\xi(\gamma)$, is a series of Legendre polynomials, $P_n(\cos \gamma)$, with theoretical coefficients $B_n(\bar{\sigma}, k_c)$. To calculate the value of the bending elastic modulus, k_c, we decompose the experimentally measured $\xi(\gamma, t)$ into Legendre polynomials and determine the experimental coefficients, $B_n(t)$. We evaluate the mean values, $B_n = \langle B_n(t) \rangle$, and estimate the corresponding dispersions, D_n. Using $\bar{\sigma}$ and k_c as fitting parameters we minimize the function, $M(\bar{\sigma}, k_c)$:

$$M(\bar{\sigma}, k_c) = \sum_{n=2}^{n=N} \left(\frac{B_n(\bar{\sigma}, k_c) - B_n}{D_n} \right)^2 \quad (13)$$

where N is the number of amplitudes (harmonics) used for the fitting. As far as the amplitudes, B_n, are obtained as an arithmetic mean of a very large number of experimental values, $B_n(t)$, their distribution is Gaussian, and therefore, $M(\bar{\sigma}, k_c)$ obeys a χ^2-distribution with $(N - 2)$ degrees of freedom. This fact can be used to verify the quality of the fit by comparing the calculated $M(\bar{\sigma}, k_c)$ value with the tabulated χ^2 values.

Sample preparation

Egg-yolk phosphatidylcholine (EPC) was prepared according to the method of Singleton et al. [11], and dissolved in $CH_3OH/CHCl_3$ (1 : 9 v/v) at a concentration of 0.5 mg/ml. A small amount of this solution was sprayed on a microslide, then the solvent was removed under vacuum for about 1 h. The microscope cell was prepared as already described in [8] except a watertight material about 0.1 mm-thick, and made of the silicon product (CAF4 from Rhône-Poulenc, France) was used as a spacer. Both the microslide and the cover slip were previously treated with trimethylchlorosilan (Sigma, U.S.A.) to make them hydrophobic. The cell was filled with deionized water (Millipore MQ, U.S.A.), and then sealed to avoid any evaporation. Giant vesicles were formed spontaneously in the cell, and generally they were studied one day to one week after the sample preparation.

Vesicle observation and image processing

Giant vesicles were observed using an inverted phase contrast microscope IM35 (Carl Zeiss, F.R.G.; objective Ph 100 ×, NA = 1.25). The thermal fluctuations were monitored via a contrast-enhancing video camera C2400-07 (Hamamatsu Photonics, Japan) and recorded on a U-matic video tape recorder (Sony, Japan) for at least 8 min. Using a Pericolor 2001 image processing system (Numelec, France) a 512 ×512 8-bit pixel digitization of the previously recorded images was performed at regular time intervals of 1 s (10 pixels $\simeq$ 1 µm). The numerical data were finally transferred to a VAX-8600 computer (Digital, U.S.A.) and stored there for further analysis. The determination of the contour coordinates was performed as in [6] by searching for the minimum of the intensity along the vesicle radius. The number of points so determined was roughly equal to the number of pixels constituting the contours, i.e., between 500 and 1000 depending on the vesicle radius. The second step of the treatment was to eliminate the contours corresponding to noisy records, which can originate from a bad microscope focusing, for example. For this purpose, the experimental contours were roughly represented by a Fourier series limited to the first five terms. All the experimental points being farther than a chosen critical distance ($\simeq$ 2 pixels) from that representation were deleted. If the number of deleted points for a given contour exceeded some value ($\simeq$ 5 %), the whole contour itself was eliminated. Usually, less than 5 % of the total number of digitized images was lost after such manipulation. The rest were used for the determination of the bending elastic constant.

Numerical simulations

The only way to check the validity of the theoretical model is to use the function (13) with as many different harmonics, B_n, as pos-

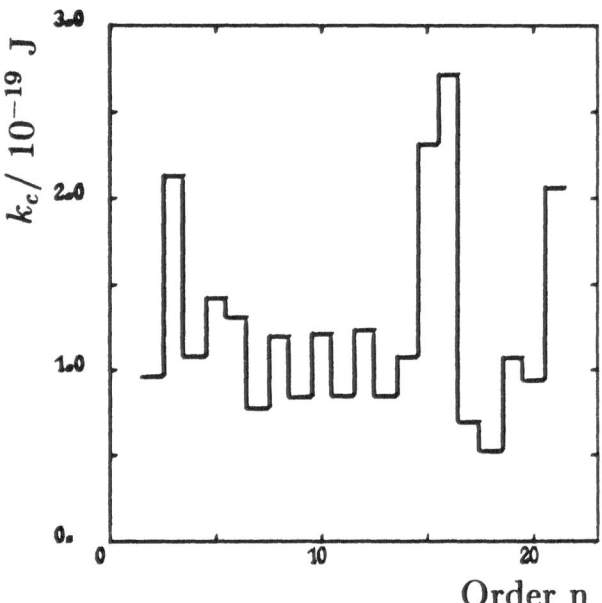

Fig. 1. Values of k_c derived from simulations vs. the order n of fluctuations. Forty contours were generated using $k_c = 10^{-19}$ J, $\bar{\sigma} = 0$, and a mean vesicle radius equal to 100 pixels. No noise was added to the calculated vesicle radii

sible. It is obvious however, that due to the experimental limitations the number, N, of the harmonics that can actually be observed and investigated is limited. The two main sources of experimental errors are: i) the inaccuracy of the contour coordinates, and ii) the limited number of analyzed images (contours). An easy way to estimate the effect of the both is to perform computer simulations.

Simulated contours were created using the theoretical values of the mean-squared Fourier amplitudes corresponding to a given set of values for k_c, $\bar{\sigma}$ and the mean vesicle radius. The squared Fourier amplitudes for a given contour were obtained by multiplying the theoretical values by random numbers obeying a χ^2-distribution. These amplitudes were then randomly decomposed into the sine and cosine components of the Fourier series, which allowed the angular dependence of the radius of the simulated contour to be computed. A Gaussian noise could then be added to the vesicle radius, with a chosen standard deviation, in order to account for the experimental errors on the contour coordinates. These simulated contours were further analyzed in the same way as the experimental ones.

An example of such a simulation is shown in Fig. 1. In this case, 40 contours have been generated, which roughly corresponds to the number used in previous works [6, 8]. It is clear that the value of k_c, obtained even in the absence of any noise on the vesicle radius, presents large variations vs. n, the order of the harmonics. However, when the number of generated contours is equal to 400, k_c remains rather constant up to $n \simeq 20$, as seen in Fig. 2, even in the presence of a Gaussian noise of 1% on the vesicle radius. Such a noise corresponds to an error of 1 pixel on the radius for a 20 μm diameter vesicle, i.e., to the expected error in normal experimental conditions.

Another interesting point is worth mentioning. It can be seen that the autocorrelation functions obtained either with or without noise are almost identical, except in one point, $\gamma = 0$. It appears that

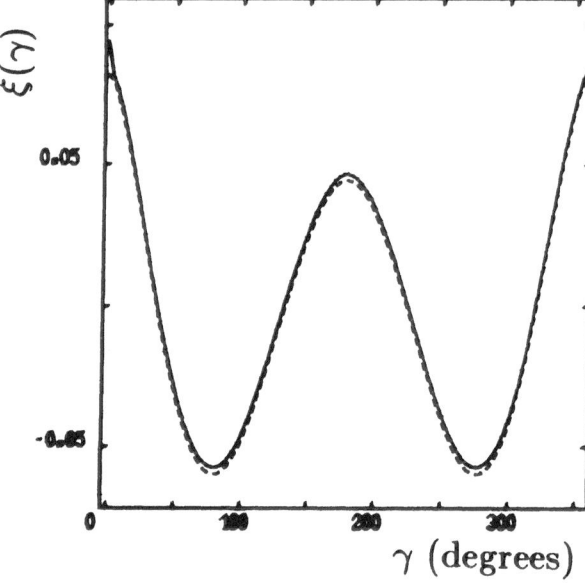

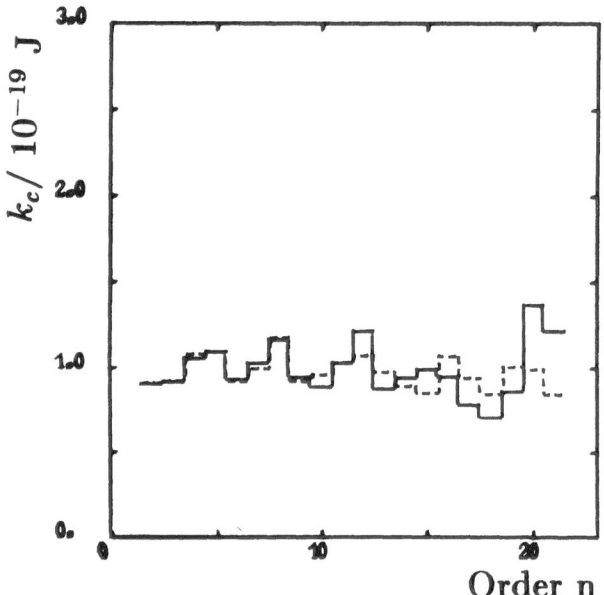

Fig. 2. Simulated data: autocorrelation functions (top figure) and dependence of k_c vs. the order n (bottom figure). Four hundred contours were generated in the presence (full line) or absence (dotted line) of a Gaussian noise equal to 1% on the vesicle radius. Entered values for simulations: $k_c = 10^{-19}$ J, $\bar{\sigma} = 0$, and mean vesicle radius = 100 pixels

the noise contributes to the autocorrelation function mainly in the form of a δ-function at $\gamma = 0$ as already pointed out in [8]. But it has practically no effect on the k_c values when the Legendre amplitudes, B_n, on the autocorrelation function are used. The same results can be obtained by direct Fourier decomposition of the contours, as in [6]. In this case, however, the noise contribution can be reduced only by subtracting a white noise from the Fourier amplitudes (data not shown).

Finally, we should mention that an accurate value of k_c can be recovered from simulated contours generated with initial values of $\bar{\sigma}$ ranging from 0 to 200, even in the presence of a 1% Gaussian noise on the vesicle radius (data not shown).

Experimental results and discussion

An example of the experimental data obtained from a giant EPC vesicle is shown in Fig. 3. It can be seen that both the autocorrelation function and the dependence of k_c vs. the order n look quite similar to those obtained by simulations. The noise contribution at $y = 0$ is clearly visible, and a nearly constant value of k_c is obtained for $n = 2$–20, provided that the value of $\bar{\sigma}$ is taken to be 83 in this case. This means that the studied vesicle has a small excess area and a spherical average-shape. On the contrary, if the experimental data are analyzed without introducing a membrane tension, as in the previous models, there is an important decrease of k_c at low orders, where the noise contribution is negligible.

We have to point out here that our theoretical model is unable to account for the behavior of *ellipsoidal* vesicles (this corresponds to negative values of $\bar{\sigma}$). In this case the fluctuations are no longer small, as supposed in the theory, and an apparent increase of k_c is observed at low orders of fluctuations (data not shown).

We checked the reproducibility of the method by studying the behavior of a large number of EPC vesicles. For this purpose, thermal fluctuations of 62 giant vesicles have been recorded and analyzed. It is worth mentioning that among them we were not able to find a vesicle that could be fitted with $\bar{\sigma} \approx 0$. This means that the occurence of a "perfect" vesicle, which has just the excess area needed to allow fluctuations to occur without apparent constraints is a very rare event.

Among all the vesicles, 14 led to negative values of $\bar{\sigma}$ and could not be well fitted. Furthermore, 15 vesicles were badly fitted even using $N < 10$ amplitudes, probably due to noisy records, and were disregarded as well. Finally, the 33 remaining "good" vesicles gave good fits for $N \geq 10$ and $7 < \bar{\sigma} < 200$ ($\bar{\sigma} < 80$ for most of them). These values of $\bar{\sigma}$ correspond to an interfacial tension $(3$–$400) \cdot 10^{-9}$ N/m.

A histogram of the k_c values obtained for these well-fitted vesicles is presented in Fig. 4. First, it is seen that k_c takes a wide range of values. However, two populations of vesicles can be easily distinguished, the first centered around $k_c = 1 \cdot 10^{-19}$ J, and the second

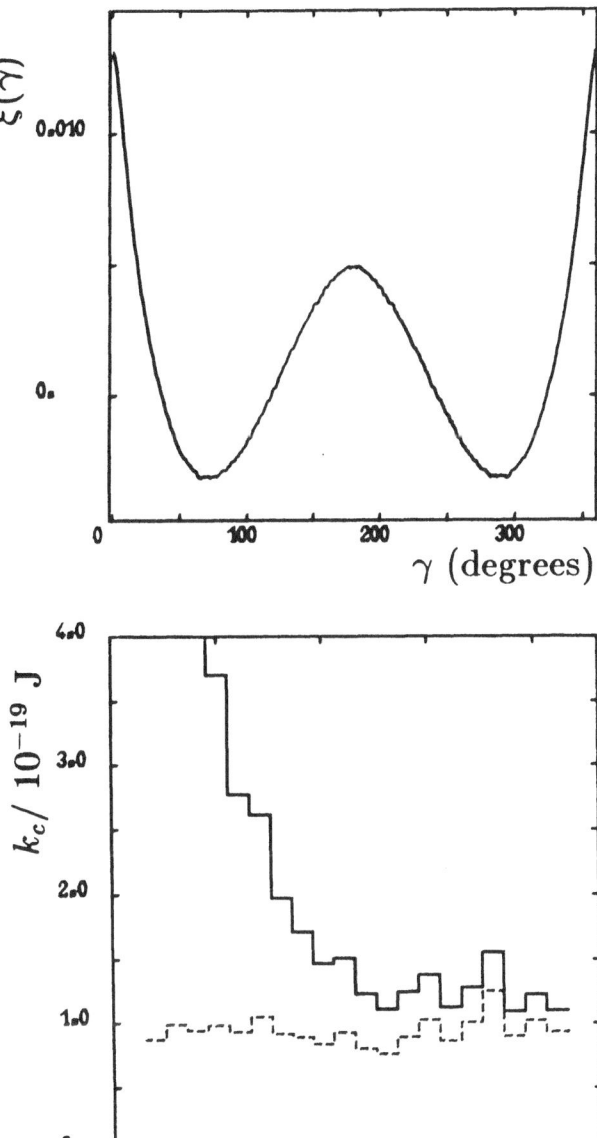

Fig. 3. Example of experimental data: autocorrelation function (top figure) and dependence of k_c vs. the order n of fluctuations (bottom figure), deduced from the analysis of 487 contours. The full line corresponds to the values obtained using $\bar{\sigma} = 0$, and the dotted line to $\bar{\sigma} = 83$. The best value of k_c for this vesicle of 25μm in diameter is $(0.93 \pm 0.04) \cdot 10^{-19}$ J

$k_c = (2$–$3) \cdot 10^{-19}$ J. A similar histogram was previously reported by Kwok and Evans [12] for the stretching modulus of EPC. Such a histogram can be interpreted assuming that the first peak corresponds to unilamellar vesicles, and the second one to bi- and/or oligolamellar

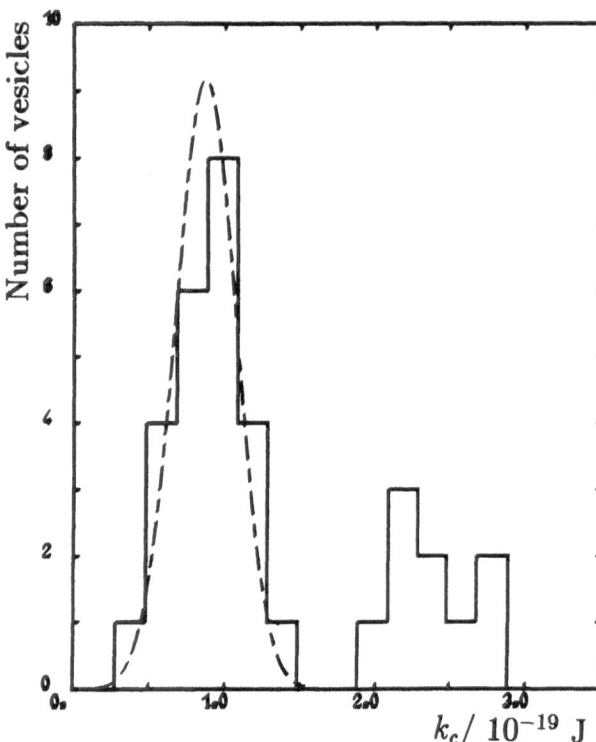

Fig. 4. Histogram of the k_c values obtained with 33 vesicles that could be fitted up to $N = 11$ or more. The first peak gives $k_c = (0.88 \pm 0.20) \cdot 10^{-19}$ J. The dotted line represents a Gaussian distribution obtained using the experimental mean value and standard deviation

vesicles. This assumption seems to be quite reasonable, due to the difficulty in assessing unambiguously the number of bilayers constituting the contours by optical microscopy. Such an explanation has already often been invoked to explain the behavior of giant spherical [6, 12] as well as tubular [2] vesicles. Thus, the mean value of k_c for egg phosphatidylcholine bilayers was derived from the first peak of the histogram[2]):

$$k_c = (0.88 \pm 0.20) \cdot 10^{-19} \text{ J} \qquad (14)$$

Conclusion

Two conclusions can be drawn — the first concerning the necessary conditions (theoretical and experimental) for an accurate determination of k_c, and the

---

[2]) A somewhat smaller value is obtained when the statistical weights of different vesicles as well as the effect of the TV camera integration time are properly taken into account [13].

second concerning the bending elastic properties of EPC bilayers.

At the experimental level it is clear that, as expected, the precision of k_c is directly related to the number of analyzed contours. But this number is obviously technically limited. In the present state of the art, several hundred images can be readily analyzed in $\simeq 1$–2 h and this allows the determination of k_c with a relatively good accuracy, as is supported by the simulations.

Another important point is the precision of the contour coordinates. The limiting factors in this case are the microscope resolution, the image digitization (512 $\times$ 512 pixels) and the video tape recorder resolution. However, the simulations clearly show that, even in the presence of a Gaussian noise of 1 pixel, the data can be fitted up to about the 20th harmonic, provided ≈ 400 contours are used. This result seems rather surprising since one can calculate that for $k_c = 10^{-19}$ J, $\bar{\sigma} = 0$, and vesicle radius 10 μm (100 pixels), the Fourier amplitude of the 10th harmonic must be about 13 nm, and it decreases down to 4.5 nm for the 20th harmonic. One could say that this is well below the resolution of the optical microscopy, but this is not the case. What is told to be the minimum resolved distance, (≈ 200 nm), is in fact the radius of the first zero-intensity interference fringe of a bright point in the object plane of the microscope. This is a good criterium when two points have to be separated and a human eye is used as a detector. But if a precise intensity measuring instrument is used instead, one can detect as low as 10% variation of the light intensity. The radius of the region where the light intensity is higher than 90% of maximum is six-fold smaller (≈ 33 nm), leading to significant precision in the position of a single point. A further 50-fold improvement is obtained because each of the 20 amplitudes are not directly measured but they are calculated as mean values of as many as 400 different autocorrelation functions, 150 points each. The final precision on the highest amplitudes is about 50–60%, against 5–7% for the lowest ones. This shows that, provided that the number of contours used is high enough, the contribution of the uncorrelated noise can be considerably reduced by using the autocorrelation function. The fluctuations can thus be investigated up to relatively high orders, which provides checks on both the validity of the theoretical model as well as the quality of the experimental data.

Finally, the time-response of the video system could also be a serious limitation, as pointed out in [6]. Indeed, one can expect to observe fluctuation harmonics only when their correlation time is much longer than the characteristic time of the video, i.e., 40 ms. In the opposite case, the fast movements should lead to an apparent decrease in the corresponding amplitudes, having as a consequence an apparent increase of k_c at higher order, n. However, such a behavior was not observed. Dynamic studies of fluctuations are now in progress to answer this important question.

From a theoretical point of view, it is now clear that one cannot consider the different modes of fluctuations to be independent, as the previous theories [5, 6, 8] did. Such an assumption clearly leads to k_c values strongly varying with the order, n, of the harmonics. The first few harmonics are in fact extremely sensitive to the excess area of the vesicle, thus when k_c is calculated from the second harmonic, for example, it can take almost any value, either very low (for a slightly ellipsoidal vesicle) or very high (for a more spherical one, Fig. 3).

It is shown here that one can readily account for both constraints exerted upon the fluctuations, i.e., the constant area and volume, by introducing two Lagrange multipliers related to the membrane tension and hydrostatic pressure difference. However, these two parameters are not independent, and only one of them, $\bar{\sigma}$, has to be considered. Hence, the final expression for k_c is almost the same as that proposed by Milner and Safran [9]. This theoretical model holds for all the vesicles having positive values of $\bar{\sigma}$, i.e., for a spherical average contour.

Among the 62 vesicles studied, 33 have been well fitted using this model, and lead to $k_c = 0.88 \cdot 10^{-19}$ J. This value, which is of the same order as those previously reported from thermal fluctuations of giant vesicles [6, 8], is strongly supported by the large number of the investigated vesicles. On the other hand, the obtained values of $\bar{\sigma}$ show that the membrane tensions involved (depending on the excess area of the vesicles) are in all cases extremely small and experimentally unmeasurable. Nevertheless, they are still important as they can considerably modify the thermal fluctuations.

Finally, despite a rigorous selection of the vesicles from both an experimental (isolated vesicle) and theoretical (goodness of the fit) basis, the standard deviation of k_c is relatively high, about 20 %. Such a deviation is of the same order as that previously obtained in the case of the stretching modulus [12]. However, it is larger than the value that can be expected from the treatment used. Moreover, no significant correlations can be observed between the obtained value of k_c for a given vesicle and any controllable parameter such as the time after the preparation, the vesicle radius or the value of $\bar{\sigma}$. So, this deviation could be attributed to some variability in the behavior from one vesicle to another. It seems unlikely that such a variability results from intrinsic differences of the bending modulus, but an alternative explanation could reside in the limited resolution of optical microscopy: indeed, it is very difficult to strictly exclude the presence of small objects close to the studied vesicle, and these could perturb the thermal fluctuations.

Acknowledgments

We thank Prof. E. Sackmann and Prof. W. Helfrich for helpful discussions during the congress and Dr. Phil Attard for his kindness in correcting the English text. One of us (M.D.M.) thanks CNRS for financial support and he thanks his colleagues for the excellent atmosphere during his stay in CRPP, (CNRS), where the work was done. This work[3]) was conducted in the frame of the scientific cooperation between CNRS and the Bulgarian Academy of Sciences.

References

1. Helfrich W (1973) Z Naturforsch 28a:693
2. Servuss RM, Harbich W, Helfrich W (1976) Biochem Biophys Acta 436:900
3. Sakurai I, Kawamura Y (1983) Biochem Biophys Acta 735:189
4. Schneider MB, Jenkins JT, Webb WW (1984) Biophys J 45:891
5. Schneider MB, Jenkins JT, Webb WW (1984) J Physique 45:1457
6. Engelhardt H, Duwe HP, Sackmann E (1985) J Phys Lett 46:L395
7. Duwe HP, Engelhardt H, Zilker A, Sackmann E (1987) Mol Cryst Liq Cryst 152:1
8. Bivas I, Hanusse P, Bothorel P, Lalanne J, Aguerre-Chariol O (1987) J Physique 48:855
9. Milner ST, Safran SA (1987) Phys Rev A 36:4371
10. Arfken G (1970) Mathematical Methods for Physicists. Academic Press, New York, San Francisco, London
11. Singleton WS, Gray MS, Blown ML, White JL (1965) J Am Oil Chem Soc 42:53
12. Kwok R, Evans E (1981) Biophys J 35:637
13. Faucon JF, Méléard P, Mitov MD, Bivas I, Bothorel P (in preparation)

Received October, 1988;
accepted February, 1989

Authors's address:

J. F. Faucon
Centre de Recherche Paul Pascal, CNRS,
Domaine Universitaire
F-33405 Talence, France

[3]) This work was supported in part by Ministère de la Recherche et de l'Enseignement Superieur (France), Grant 87C00181, by Region Aquitaine (France), and by the Bulgarian Ministry of Culture, Science and Education, Contract 587/15. July, 1987.

Progress in Colloid & Polymer Science

Progr Colloid Polym Sci 79:18–23 (1989)

Rheological properties of gelatin gels filled with phospholipids vesicles. Dynamic and uniaxial compression measurements

C. Genot, S. Guillet, and B. Metro

Laboratoire d'Etude des Interactions des Molécules Alimentaires, Institut National de la Recherche Agronomique, Nantes, France

Abstract: The effects of addition of small unilamellar vesicles (SUV) or multilamellar vesicles (MLV) of phosphatidylcholine (PC) on rheological behavior of gelatin gels (approximatively 5 % W/W) were studied. Gelation kinetics of pure gelatin gels and gelatin gels containing lipid vesicles as determined by storage modulus (G') vs. time were quite similar and no difference was observed on G' after 20 h. Apparent Young's modulus (E_0) was decreased by 15 % when 5 g/l of lipids (either MLV or SUV) were present in the gel. When the gels were submitted to three successive uniaxial deformations (respectively 25 %, 55 %, and 75 %) at a low rate (2 mm/min), a decrease in the force measured at 20 % or 50 % deformation, as a function of lipid concentration and deformation number was observed when MLV and more specially when SUV were added. Height loss of the gels after two deformations at 25 % and 55 % was the largest for gels containing SUV, and the smallest for pure gelatin gels.

Key words: Gelatin-liposome-gel rheological behavior.

Introduction

Many foods can be described as composite gels in which intrinsic properties of the polymeric network (i.e., protein or polysaccharide matrix) are modified by inclusions, such as starch granules, gas bubbles, oil droplets, or mesomorphic lipid phases. Wheat gluten, for instance, contains 5–10 % of lipids organized as vesicles, the effects of which on the rheological properties of the protein network are not yet elucidated [1]. Many dairy and delicatessen products also present a gel-like structure with inclusions. Moreover, developments of liposomes applications for food industry can be foreseen in the very near future, but the consequences of these inclusions on rheological properties of food are still not known.

In our laboratory, we studied a model system composed of gelatin gel in which phosphatidyl choline (PC) vesicles were inserted. The small and large deformation behavior of these lipid + protein gels were characterized from uniaxial compression experiments and shear dynamic measurements.

Experimental

Material

Gelatin: Powdered edible gelatin (275 Blooms, viscosity 6.67 % 60 °C: 4.55 mPa · s, pH: 5.7, isoelectric point: 7.9, ash: 0.6 %, moisture: 11 %, average molecular weight: 178 000) from swine skin was obtained from Rousselot-France.

Lipids: Frozen egg yolk phosphatidylcholine (egg PC) was purchased from Sigma (ref. P 5388). Its purity, checked by HPLC [2], was better than 90 %.

Methods

Preparation and characterization of lipid vesicles: The vesicles were prepared according to the procedure of Huang [3]. Phospholipids (10–300 mg) in $CHCP_3$ were poured into a round-bottomed glass flask. Solvent was evaporated by rotary evaporation followed by N_2 blowing. 10–50 ml of filtered and degazed NaCl 0.1 M + 0.02 % NaN_3 (Solution A) was then added above the thin lipidic film formed on the vial. Multilamellar vesicles (MLV) were obtained by vortexing the vial for 30 min after a hydratation time of 48 h. The size distribution of these vesicles is very large as shown by Counter Coulter measurements (not shown) and transmission electron microscopy (TEM) after freeze fracture (Fig. 1). Small uni-

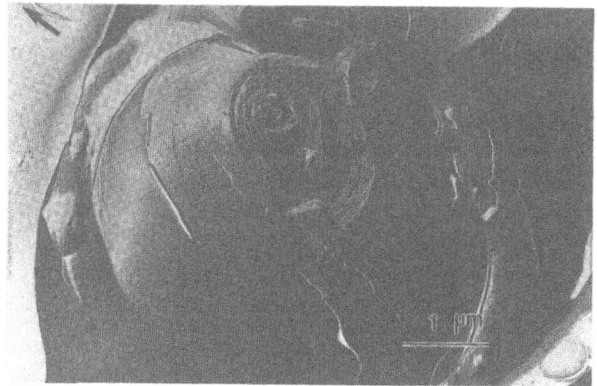

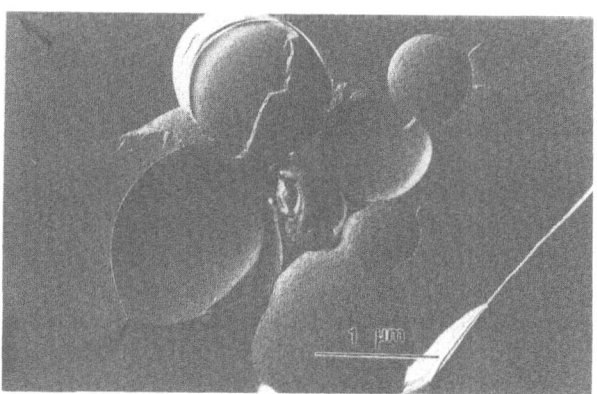

Fig. 1. Freeze-fracture electron microscopy of MLV of egg PC in NaCl 0.1 M

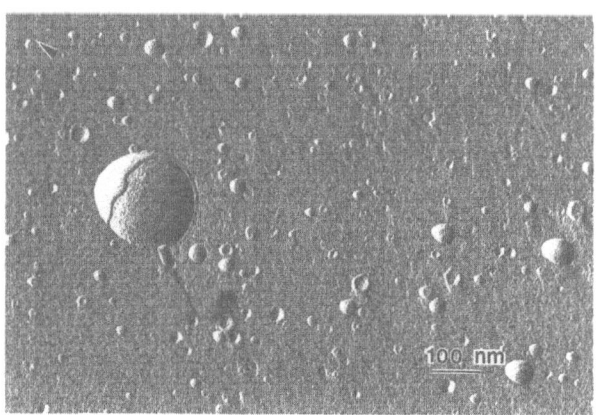

Fig. 2. Size exclusion chromatography of SUV of egg PC on Superose 6 (Pharmacia) (30 × 1 cm) equilibrated with NaCl 0.1 M. The column was previously saturated with the lipid dispersion. Flow rate: 0.4 ml/min; NaCl 0.1 M + NaNO$_3$ 0.02 %. The two peaks (1 = MLV and aggregated vesicles; 2 = SUV) were collected separately and lipid content of each fraction was estimated after phosphorus determination (4). V_0 = void volume; Vt = total volume

lamellar vesicles (SUV) were prepared from MLV by sonication for 3 h under nitrogen at 4 °C with a 500 W Sonic and Material Inc. pulse sonifier (titanium microtips; output: 4; 40 % duty cycle). Following sonication, SUV were centrifuged at 185 000 g for 6 h in order to remove titanium particles, lipid aggregates, and MLV. The translucent supernatant contained 90–95 % of SUV (expressed in lipid weight) as shown by size exclusion chromatography of liposomes on Superose 6 and phosphorus determination [4] (Fig. 2) and freeze fracture TEM (Fig. 3).

Preparation of gels: P g of dry gelatin was swollen in $V1$ ml NaCl 0.1 M, NaOH 4.5 mM + 0.02 % NaN$_3$ (Solution B) [$V1 = P(1 − 0.11)/(5 × 2)$] for 30 min at room temperature. The gelatin was then dissolved in a water bath at 55 °C for 15 min and $V1$ ml of solution A containing lipid dispersion (0.3 to 5 g lipids per gel liter) were added to obtain a 4.97 % gel (dry matter/total water) at pH 7. Some experiments were performed at 4.71 %, pH 7 gelatin gels. When dynamic measurements were carried out the progel was directly poured on the apparatus plate. For uniaxial compression measurements, the progel was poured into cylindrical molds (diameter: 30 mm; height: 30 mm) the top and bottom edges of which were raised with a sticky plastic tape. The gels which formed by cooling were allowed to mature for 3 days in controlled room temperature of 20 °C. Before measurements, the sticky tape was removed and the excess gel above and below the top and bottom edges of the mold was cut off with a thin steel wire. Thereafter the

gell was removed from the mold and covered with paraffin oil to avoid dehydration during the experiment.

Dynamic measurements: The gelatin sol introduced on the prewarmed (50 °C) horizontal plate of the Carrimed 5001 in cone-plate geometry (cone: 4°; diameter: 2 cm) was cooled to 20 °C (ap-

Fig. 3. Freeze fracture electron microscopy of SUV of egg PC

proximatively 1 min). During this time and for the next 30 h, G' and G'', the dynamic storage and loss moduli were calculated every 10 min. The oscillatory frequency was 0.5 Hz and the amplitude 3 mrad.

Uniaxial compression: Uniaxial compression experiments were carried out using the Instron universal testing instrument 1222. The cylindrical samples were compressed between flat platens at a fixed crosshead speed. The plates of the instrument were wetted with paraffin oil in order to obtain optimal slip conditions and reduce "barrelling" of the gel. Successive deformations on individual samples were 5 %, 25 %, 55 %, and 75 % using a crosshead speed of 2 mm/min. The force $F(N)$ was measured as a function of deformation. The whole experiment was performed in a controlled laboratory temperature of 20 ± 1 °C.

Small strain: determination of Young's modulus (YM)

Apparent Young's modulus E_0 $(N \cdot m^{-2})$ was defined from

$$E_0 = (F/A) \cdot (H_0/H_0 - H) \qquad (1)$$

where $F(N)$ was the strength component parallel to the direction of the deformation, A (m^2) the sample area which was considered to be constant in low deformation range and H_0 and H the heights of the gel before and after compression $(H_0 - H/H_0 \cdot 100$: deformation percentage).

E_0 was estimated from the slope of the linear part of the 5 % compression curve.

Large strain

The values of F for 20 % and 50 % deformation during the successive deformations were measured.

Results

As shown in Fig. 4, when lipid vesicles were present in gelatin gel, the intial pure, transparent gel became

turbid, the turbidity being higher when MLV rather than SUV or when increasing vesicles concentration were added.

Dynamic measurements: No significant differences were observed in gelatin gelation kinetics (Fig. 5) or in storage modulus (G') after 20 h of gelation (Table 1) when egg PC small unilamellar vesicles are present in the gels as compared to pure gelatin gels.

The loss modulus was always very low and varied between 0 and 150 Pa during the whole gelation experiments, no difference being observed between pure gelatin gels and gels containing lipid vesicles.

Uniaxial compression

Low strain: Apparent Young's modulus (E_0) of gelatin gels containing egg PC SUV and egg PC MLV (Tables 2 and 3) was decreased by 1 % to 15 % when the vesicle concentration in the gel increases from 0.12 to 5 g/l.

High strain: When three successive deformations (25 %, 55 %, and 75 %, respectively) were applied to the gels containing different lipids concentrations at a deformation rate of 2 mm/min, two phenomena were observed (Fig. 6). First, during the first compression (25 %), a decrease of the force (F) measured at 20 % deformation of the gels containing SUV or MLV as compared to pure gelatin gels occurred. This decrease was more obvious in the case of SUV addition (Figs. 6 a and 6 b). During the second compression (55 %) a similar decrease of F (measured at 20 % and 50 % deformation) after addition of lipid vesicles was observed. Second, when the third compression was

Fig. 4. Cylinders of gelatin gels with (a) and without (b) MLV (3 g/l)

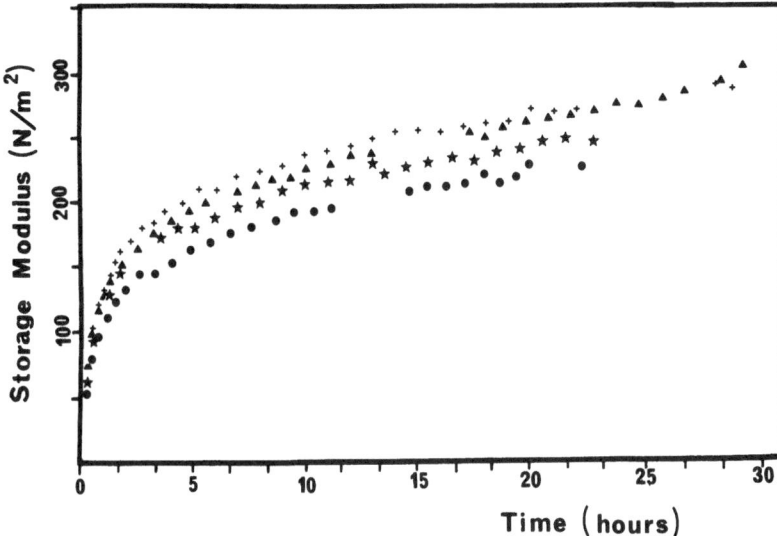

Fig. 5. Gelation kinetic of gelatin gels as shown by storage modulus vs. time. ▲, ●: gelatin 4.71%; +, ★: gelatin 4.71% + 2 g/l SUV of egg PC. (dynamic measurements – Carrimed)

Table 1. Storage modulus (G') after 20 h gelation at 20 °C of gelatin gels without or with small unilamellar vesicles (SUV) of egg phosphatidylcholine (PC) (dynamic measurements, Carrimed)

	Gelatin G' ($N \cdot m^{-2}$)	Gelatin + SUV G' ($N \cdot m^{-2}$)	lipid concentration (g/l gel)
gelatin 4.97%	3284	3443	1.18
	3499	2956	1.28
Mean value $\pm \sigma$	3391 $\pm$ 152	3199 $\pm$ 344	–
gelatin 4.71%	2239	2160	4.20
	2569	2631	4.20
	2915	2588	4.00
	2151	2356	3.15
	2196	2361	2.52
Mean value $\pm \sigma$	2414 $\pm$ 325	2419 $\pm$ 192	–

Table 2. Apparent Young's modulus (E_0) of gelatin gels (after 72 h of gelation) containing multi-lamellar vesicles (MLV) of egg PC. (Instron)

a) 4.71% gelatin gels containing different MLV concentrations prepared the same day from the same gelatin solution

Lipid concentration (g/l gel)	E_0 ($N \cdot m^{-2}$)	ΔE_0 (%)
0	7492	0
0.125	7407	0.9
0.25	7382	1.5
0.5	7229	3.5
1	7052	5.9
1.5	6834	8.8
2	7203	3.9

applied to gels containing increasing lipid concentrations (either MLV or SUV), an important decrease of F measured at 20% and 50% deformation was observed, compared to F measured on pure gelatin gels during this third compression, and to F measured on the lipid + protein gels during the first and the second compressions (Fig. 6 a–d).

Moreover, after three compressions at 5%, 25%, and 55% the height of cylinders of gelatin, gelatin + MLV, and gelatin + SUV gels was shown to be decreased by 9%, 14%, and 17% respectively.

b) 4.97% gelatin gels prepared different days (in parantheses is indicated the number of E_0 determinations for each concentration)

Lipid concentration (g/l gel)	E_0 ($N \cdot m^{-2}$)	ΔE_0 (%)
0	9700 $\pm$ 670 (5×8)	0
2.6	8720 $\pm$ 468 (6)	10.1
3.55	8386 $\pm$ 144 (6)	13.5
4.93	9047 $\pm$ 92 (4)	6.7
5.55	8230 $\pm$ 179 (6)	15.2

$\Delta E_0 = 100 \times (E_0 \text{ (no lipids)} - E_0)/E_0 \text{ (no lipids)}$

Table 3. Apparent Young's modulus (E_0) of gelatin gels (after 72 h of gelation) containing SUV of egg PC. (Instron)

a) 4.71 % gelatin gels containing different SUV concentrations prepared the same day from the same gelatin solution

Lipid concentration (g/l gel)	E_0 ($N \cdot m^{-2}$)	ΔE_0 (%)
0	7260	0
0.09	6989	3.7
0.18	7131	1.2
0.37	7015	3.4
0.75	6931	4.5
0.125	7144	1.6
1.5	6947	4.3

b) 4.97 % gelatin gels prepared different days (in parantheses is indicated the number of E_0 determinations for each concentration)

Lipid concentration (g/l gel)	E_0 ($N \cdot m^{-2}$)	ΔE_0 (%)
0	$9700 \pm 670 \ (5 \times 8)$	0
1.6	$8949 \pm 52 \ (4)$	7.7
3.75	$8364 \pm 113 \ (4)$	13.8

$$\Delta E_0 = 100 \times (E_0 \text{ (no lipids)} - E_0)/E_0 \text{ (no lipids)}$$

Discussion and conclusion

Two hypotheses can be assumed to explain the observed modifications in rheological behavior of gelatin gels after lipid vesicles addition.

The first is that gelatin-lipid interactions occurs. According to this hypothesis, previous work [5] showing an increase of gelatin solution viscosity when anionic surfactants were added was interpreted as a cooperative binding effect between surfactant micelles and gelatin molecules. In our case the phospholipid we used is zwitterionic and electrostatic interaction between PC and gelatin molecules may not be involved. Furthermore, lipid-protein interactions should occur in the sol state, when the molecules are more mobile, leading to subsequent constraints in triple-helix gelatin growth. Thus, modification in gelation kinetics, especially in the fast first stage should be observed, contrary to our observations.

The second hypothesis is a filler effect of lipid vesicles. Depending on the filler volume fraction [6, 7], the relative rigidities of the particles and the matrix [6], and on size and shape of the particles [8], a weakening or a strenghtening of the composite gel can be observed. Approximative volume fractions of phos-

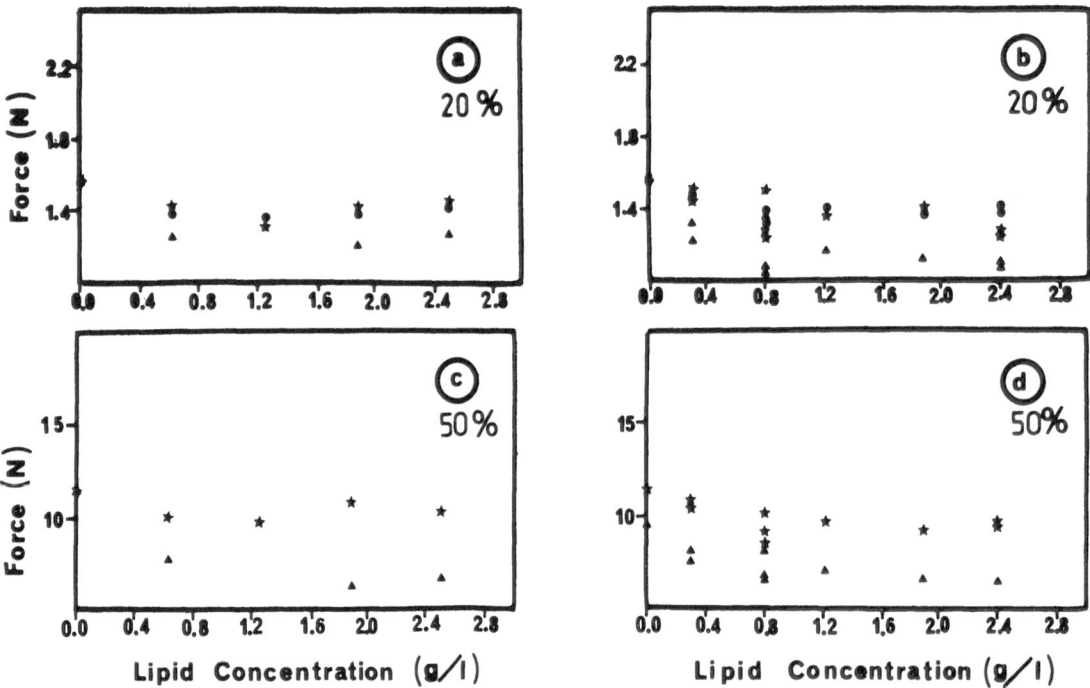

Fig. 6. Compression force (N) vs. lipid concentration (g/l) of 30 × 30 mm gelatin gels (4.97 %) cylinders containing MLV or SUV during successive deformations. (Uniaxial compression – Instron). Crosshead speed: 2 mm/min; ●: first deformation; ★: second deformation; ▲: third deformation. a: gelatin + MLV, deformation 20 %; b: gelatin + SUV, deformation 20 %; c: gelatin + MLV, deformation 50 %; d: gelatin + SUV, deformation 50 %

pholipid vesicles were calculated using the measured average vesicles diameters (MLV: 5000 nm; SUV: 25 nm), and assuming an encapsulated volume of 2.5 l/ mole for MLV and an average number of 3000 PL molecules per SUV vesicle [9]. The estimated volume fractions of vesicles are 0.25 % for 1 g of PC incorporated as MLV per liter of gelatin gel, and 1.35 % as SUV. The volume fractions of vesicles incorporated in our gels (always lower than 1.25 % for MLV and 4.2 % for SUV) are lower than the volume fractions of fillers whose effects are reported in the literature (from 5 % to 60 %). The weakening of gelatin gels we observed (lowering in E_0 and F) would indicate a very lower rigidity of lipidic vesicles as compared to gelatin gel rigidity.

Further experiments are needed to assert and quantify the relative roles of lipid-gelatin interactions and vesicle filler effect in weakening of the gelatin gels, as well as to study the influence of various parameters (size, volume fraction, affinity of lipid vesicles for gelatin matrix).

References

1. Marion D, Le Roux C, Akoka S, Tellier C, Gallant D (1987) J Cereal Sci 5:101-115
2. Stolyhwo A, Martin M, Guiochon G (1987) J Liquid Chromatogr 10(6):1237-1253
3. Huang CH (1969) Biochemistry 8:344-352
4. Barlett GR (1959) J Biol Chem 234:466-468
5. Greener J, Contestable BA, Bale MD (1987) Macromolecules 20:2490-2498
6. Brownsey GJ, Ellis HS, Ridout MJ, Ring SG (1987) J Rheol 31(8):635-649
7. Richardson RK, Robinson G, Ross-Murphy SB, Todd S (1981) Polym Bull 4:541-546
8. Ring S, Stainsby G (1982) Prog Fd Nutr Sci 6:323-329
9. Dousset N, Douste-Blazy L (1985) In: Puisieux F, Delattre J (eds) Les Liposomes. Applications thérapeutiques, Lavoisier, Paris, pp 41-72

Received October, 1988; accepted January, 1989

Authors' address:

C. Genot
Institut National de la Recherche Agronomique
B. P. 527
F-44026 Nantes Cedex 03, France

Progress in Colloid & Polymer Science   Progr Colloid Polym Sci 79:24–32 (1989)

Behavior of liposomes prepared from lung surfactant analogues and spread at the air-water interface

T. Ivanova, G. Georgiev, I. Panaiotov[1], M. Ivanova[1], M.A. Launois-Surpas[2]) J.E. Proust[2]), and F. Puisieux[2])

Pulma Laboratory, Sofia, Bulgaria
[1]) Physical Chemistry Department, University of Sofia, Sofia, Bulgaria
[2]) Physicochimie des Surfaces et Innovation en Pharmacotechnie, CNRS UA 1218, Châtenay-Malabry, France

Abstract: One exogenous surfactant therapy of respiratory distress syndrome (RDS) is to instillate into the lungs enough phospholipid to form a film at the alveolar interface because it has more similar surface properties that the natural surfactant. Liposomal suspensions are able to transport a large amount of phospholipids in situ; therefore, it is interesting to study the transformation of these vesicles into a surface film.

Lα dipalmitoylphosphatidylcholine – DPPC: *Lα* distearoylphosphatidylcholine – DSPC: soybean lecithin – SL (4:4:2) liposomes were spread at the air-physiological solution interface. The spreading kinetics were studied by means of surface pressure, surface potential, and rheological measurements. A theoretical approach to the mechanism of slow transformation of the closed bilayer structures into a surface film has been developed. The properties of surface films formed after spreading of freshly prepared or aged liposomes were compared with those of mixed DPPC:DSPC:SL(4:4:2) monolayers.

Key words: Lung surfactant, liposomes, surface films.

Introduction

Clements [1] and Scarpelli [2] have proved the important role of alveolar surfactant in respiratory mechanics and confirmed Von Neergard's idea [3]. It was then demonstrated that respiratory distress syndrome (RDS) in premature infants is due to a relative deficiency of lung surfactant [4].

As a result of these pioneering works, one possible approach to the treatment of RDS might be the introduction of exogenous surfactant into the lungs; natural surfactant, extracted from calf lung or human amniotic fluid, and artificial surfactant, made of different phospholipid mixtures, dried or in liquid suspension, have been administrated [5–15]. These trials failed to produce satisfactory results.

Another way of administrating surfactant is to instill a liposomal suspension of concentrated phospholipids directly in the lungs through the trachea [16–20]: liposomes allow administration of a large amount of phospholipid in suspension in physiologic fluid and could constitute a reserve of tensioactive material in situ, un-

dergoing a slow transformation of their closed bilayer structure into a surface film when they reach the alveolar surface. This phenomenon was shown for the first time by Pattus, Desnuelle, and Verger [21] who described the slow spreading of liposomes: unilamellar liposomes spread more reproducibly and more effectively (phospholipids remaining at surface: 75 ± 5 %) than multilamellar liposomes (phospholipids remaining at surface: 55 ± 25 %). Obladen [22] compared the spreading efficacy of large or small unilamellar and multilamellar liposomes of a dipalmitoylphosphatidylcholine (DPPC) – phosphatidylglycerol (PG) mixture. Schindler [23] has proposed a general theoretical scheme with four parameters to describe formation mechanism of phospholipid monolayer from vesicles in bulk solution. Four populations are considered: vesicles equal to those in the bulk solution, vesicles after release, vesicles after uptake, and a quantity of molecular lipids; vesicles are in molecular exchange with the monolayer and in vesicle diffusion exchange with the solution.

After these studies it was interesting to study the real mechanism of transformation of closed bilayer structures into surface films after spreading. With this objective, oligolamellar DPPC: distearoylphosphatidylcholine (DSPC): soybean lecithin (SL) 4:4:2 liposomes were prepared by the method of Bangham [24]; the liposomal suspension was then spread at the air-physiological solution interface as a model of the alveolar surface.

The aims of this work were:
— to study the kinetics of surface film formation from liposomes spread at constant surface are in relation to the spreading process, recording the evolution of superficial parameters — surface pressure Π and surface potential ΔV;
— to develop a theoretical approach to this phenomenon;
— to study the equilibrium and dynamic properties of surface films obtained after liposomes spreading and to compare them to those of monolayers spread from a solution of the same phospholipid mixture in an organic solvent.

Materials and Methods

Lipids

L-α-dipalmitoylphosphatidylcholine (DPPC) and L-α-distearoylphosphatidylcholine (DSPC) were purchased from the Sigma Chemical Company, St Louis, Missouri, USA; the synthetic crystalline DPPC and DSPC were both 99% pure.

Soybean lecithin (SL) was purchased from Lucas Meyer, Hamburg, FRG; it was made of 95–98% phosphatidylcholine, about 80% unsaturated.

Lipids were stored at $-18\,°C$ and used without further purification.

Preparation of liposomes

A chloroform solution of tocopherol acetate (6 mg in 100 ml) was used as an anti-oxydant at a concentration of 1 ml for 10 ml liposomal suspension.

Analytical grade chloroform was purchased from Merck.

The 0.15 M NaCl solution was made from triple-distilled water (second distillation from acid potassium permanganate) and Merck purest quality NaCl roasted at 700°C.

The synthetic surfactant consisted of liposomes, the preparation of which was developed in our laboratory by Bonte [25] and De Fontanges [26] using the Bangham method [24] with sonication:

The lipids DPPC, DSPC and SL (total final lipid concentration 10 mg/ml and molecular ratio 4:4:2) were dissolved in chloroform with tocopherol acetate and added to a round bottomed flask. Chloroform was then removed for 30 min by evaporation in a Rotavapor, under vacuum and above the phospholipid mixture phase transition temperature at 45–50°C (DPPC:DSPC:SL 4:4:2

phase transition zone, performed by differential scanning calorimetry with a Perkin Elmer thermal analyzer, was $30° - 45\,°C$). The thin lipid film coating the glass wall was flushed with nitrogen and then suspended at the same temperature in 0.15 M NaCl previously heated to 45°C. The milky dispersion, containing multilamellar vesicles, was then sonicated at 45°C using a probe with a Branson Sonifier cell disruptor B30 (350 W, 20 kHz) for several 5-min periods (resting time 5 min) until the solution became almost clear.

The liposomal suspension was then filtered through a 0.22 µm sterile Millex GV single-use filter unit (Millipore S.A., Molsheim, France); thus a sterile liposomal suspension can be obtained for in vitro studies as well as clinical or animal experimentation.

The suspension was then stored at $+4\,°C$.

Liposome verification

Preparations of liposomes were checked by different methods:

Total phospholipid concentration control: The total phospholipid concentration was measured immediately after sample preparation by the Takayama method [27] using a kit (Biotrol S.A., Paris, France). The results obtained were in perfect agreement with the initial phospholipid concentration 10 mg/ml.

Size measurements and structural study: Size measurements were performed with a submicron particle analyzer Coulter model N4 MD (Coulter Electronics, Hialeah, Florida, USA) — detection angle: 90°, measurement duration: 5 min.

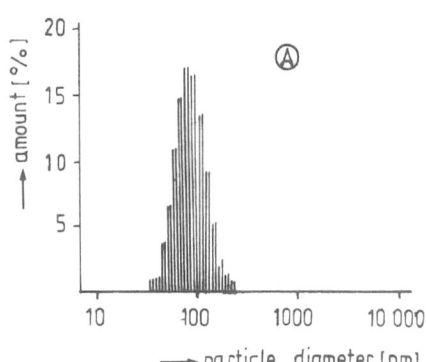

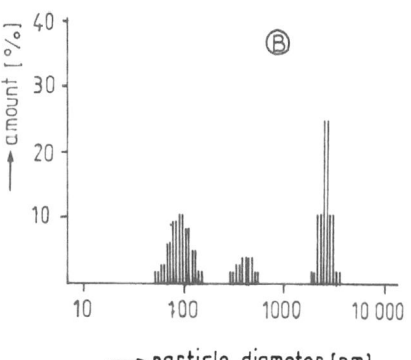

Fig. 1. Distribution of mean particle diameter for a freshly prepared liposomal suspension (A) and a 4-week stored (+4°C) liposomal suspension (B)

The structure was studied after adsorption onto carbon-coated grids and negative staining [28]; all specimens were observed at 80 kV with a Philips EM 301 electron microscope.

Immediately after sample preparation and filtration, the mean particle diameter was about 100 nm with monopeak dispersion (Fig. 1A); at the same time, electron microscope observation showed monolamellar vesicles and oligolamellar vesicles with four or five perfectly closed lamella.

After four weeks storage at + 4 °C, vesicle destruction occurs: the detected particle size increases (Fig. 1B) and electron microscope study shows large open phospholipid structures.

Measurement of superficial parameters

Surface pressure, Π, was measured by the Wilhelmy plate method using a platinum plate and an electronic balance Beckman LM600, connected to a Sefram chart recorder with a precision 0.01 dyn/cm.

Surface potential, ΔV, was measured using a gold-coated [241]Am ionizing electrode, a reference Calomel electrode, and an electrometer VA-J-51 (GDR) connected to a Sefram chart recorder with accuracy of the initial surface potential, ΔV_0 of $\pm$ 15 mV. As usual, the surface potential of the free water surface fluctuated during about 30 min; then the surface potential became constant and the spreading could be performed.

Each superficial parameter was recorded at 22–25 °C; te subphase under investigation was made of 0.15M NaCl and 0.001M $CaCl_2$ (Merck purest quality $CaCl_2$ roasted at 700 °C) in triple-distilled water.

Spreading of liposomal suspensions was performed by the classical method with an Agla syringe and different volumes were spread at an initial surface area of 186 cm².

Kinetics of surface film formation from spreading was studied by recording the evolution of Π and ΔV with time. Equilibrium isotherms were obtained by compressing surface films formed after spreading, in series of small, well defined steps; an equilibrium value was reached after each compression.

Dynamic dilatational behavior of the surface films obtained after liposome spreading was studied by the method of propagation of the longitudinal deformation along the monolayer [29–32]. This method consists of compressing the surface film ① spread at the interface ② by means of the barrier ③ moving with a constant velocity u_b (Fig. 2). As a result of the dilatational motion, the local surface pressure Π changes with time t at any point x. When the surface film behaves as a two-dimensional elastic body and when relative surface pressure and area variations $- \Delta\Pi/\Pi$ and $\Delta A/A$ — are less than 10 %, the relation between $\Delta\Pi$, t and x in the quasi steady state region $t \gg L^2 \eta/\pi^2 E h$, is found to be [29, 30]:

$$\Delta\Pi = \frac{E u_b}{L} t - \frac{4 \eta u_b}{h}\left(x - \frac{x^2}{2L} - \frac{L}{3}\right) \quad (1)$$

where E is the surface elasticity, η is the viscosity of the substrate, L is the length of the surface film, h is the thickness of the liquid subphase.

The first term in (1) describes surface film elastic behavior and the second one shows the viscous dissipation in the liquid substrate due to friction. The physical meaning of this theoretical model is simple: the propagation of a local surface concentration perturbation along an elastic surface film must be considered as instantaneous, but on account of the friction in the liquid substrate, the propagation is not instantaneous. The values of surface dilatational elasticity E can be obtained from the experimental results $\Delta\Pi(x, t)$ and the theoretical expression (1).

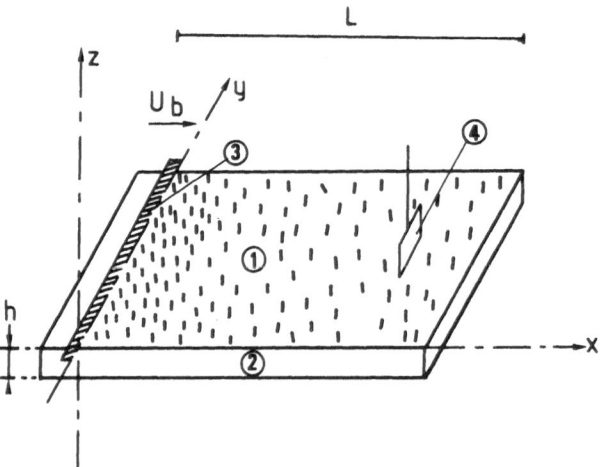

Fig. 2. Scheme of the experimental design: ① insoluble monolayer, ② Teflon trough, ③ barrier, ④ Wilhelmy plate; L is the length of the surface film; h is the thickness of the liquid subphase; u_b is the velocity of the barrier

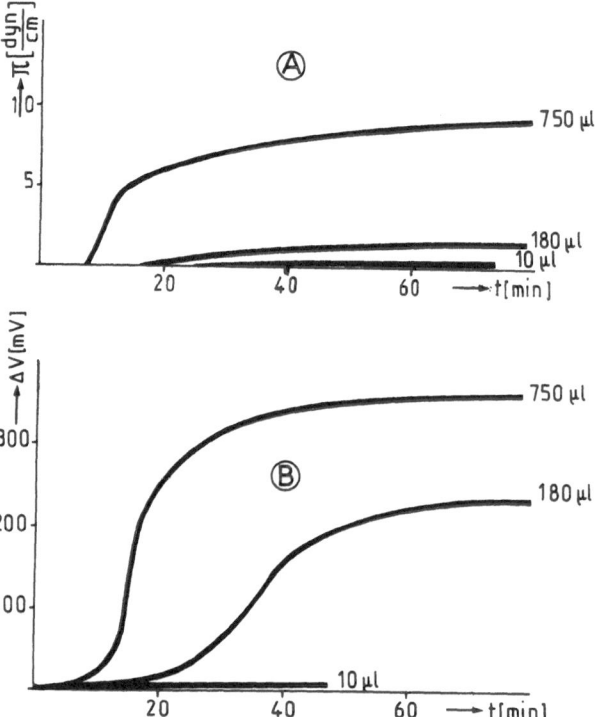

Fig. 3. Variation with time of surface pressure Π (A) and surface potential ΔV (B) for various spreading volumes of liposomal suspension (slow spreading technique)

When a compression-expansion cycle is performed, the observed surface pressure (Π) – area (A) hysteresis loop is due to the following kinetic processes:

a) simultaneous dilatational motion of the surface film and of the liquid subphase (the so called Marangoni effect);

b) interchange of molecules between the surface and the bulk phase;

c) film collapse;

d) relaxation processes in the monolayer itself, such as molecular reorganization into two-dimensional clusters, layered mesophases.

The contribution of each of those effects to Π(A) hysteresis loop can be analyzed theoretically [33].

Results and discussion

1. Liposome spreading kinetics

Figure 3 illustrates the experimental variation of Π and ΔV with time t at constant surface area, for differ-

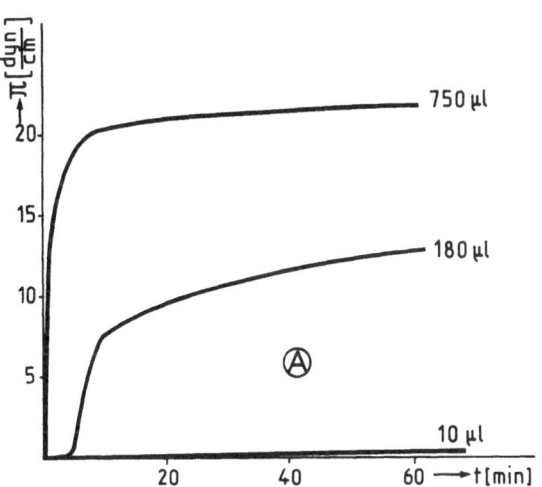

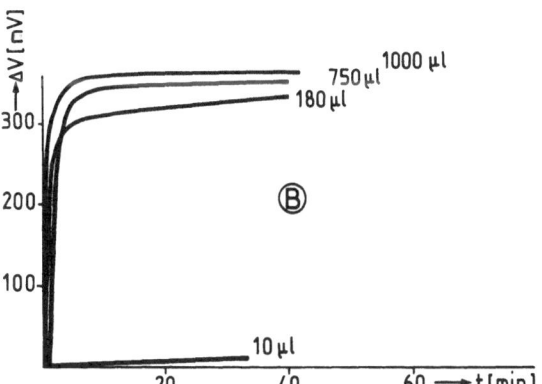

Fig. 4. Variation with time of surface pressure Π (A) and surface potential ΔV (B) for various spreading volumes of liposomal suspension (rapid spreading technique)

ent volumes of freshly prepared liposomal suspension: 10, 180, and 750 µl of 10 mg/ml phospholipid concentration suspension were slowly spread on a total area of 186 cm²; 180 and 750 µl were spread at rates of 12 µl/min and 50 µl/min, respectively (spreading duration: 15 min). As shown in Fig. 3, Π and ΔV are observed to rise to a plateau value and the rate of surface film formation tends to zero; the timecourse also depends on the volume of liposomal suspension spread.

The same experiment was performed using a rapid spreading technique: 10, 180, 750, and 1000 µl of the same suspension were spread in about 1 min; Fig. 4 shows more rapid surface film formation.

A simple theoretical approach (Fig. 5) can explain these experimental data.

For a first order approximation, we consider the presence of only two types of phospholipid structures at the interface: perfectly closed structures (type-I structures = liposomes = closed vesicles), not tensioactive and without any effect on surface potential, and open structures (type-II structures = destroyed vesicles), tensioactive and which increase the surface potential. The advantage of this simple scheme is that it allows an estimation of the rate of surface film formation using only two physical parameters: a diffusion coefficient D which can be independently calculated, and an adsorption constant K_a.

Thus, surface film formation kinetics can be described by two simultaneous processes: i) irreversible diffusion process of type-I structures to the liquid bulk phase; ii) irreversible adsorption process at the interface, changing the type-I structures into a superficial

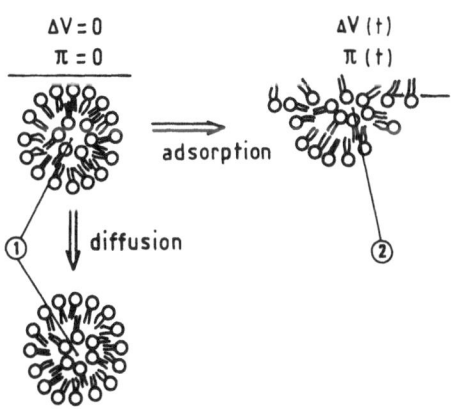

Fig. 5. Scheme describing the two processes involved in the formation of the surface film from liposomal suspension; 1) type-I structure = closed structure = intact liposome; 2) type-II structure = tensioactive open structure = destroyed liposome

film of type-II structures. Each process can be considered separately:

1.1 Kinetics governed by the diffusion process:

All the liposomal suspension initially spread at the interface is supposed to form a layer with constant thickness l (Fig. 6). The initial liposomal concentration in the bulk phase can be calculated supposing that all the phospholipids (10 mg/ml) are contained in unilamellar 150 nm diameter liposomes: $C_0 = 2.7 \cdot 10^{13}$ liposomes/cm³. The diffusion coefficient, D, can be calculated from the Einstein formula: $D = KT/6\,\pi\eta\,r = 3.10^{-8}$ cm²/s at 20 °C; $\eta = 0.01$ poise; $r =$ liposome radius = 75 nm.

To find the concentration distribution in spacial and temporal coordinates $C(x, t)$, Fick's equation must be resolved:

$$\frac{\partial C}{\partial t} = D\,\frac{\partial^2 C}{\partial x^2}. \tag{2}$$

At boundary conditions:

$$t = 0 \quad \begin{cases} 0 < x < 1 & C = C_0 \\ 1 < x < \infty & C = 0 \end{cases} \tag{3a}$$

$$t > 0 \quad x = 0 \quad \left(\frac{\partial C}{\partial x}\right)_{x=0} = 0 \tag{3b}$$

$$t > 0 \quad x = \infty \quad \left(\frac{\partial C}{\partial x}\right)_{x=\infty} = 0. \tag{3c}$$

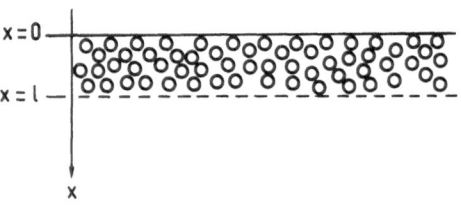

Fig. 6. Initial distribution of liposomes (type-I structures) in the subsurface after spreading at the interface; l is the thickness of the liposome layer

The solution is given by [29]:

$$C(x, t) = C_0 \left[1 - \frac{1}{2} \sum_{n=1}^{\infty} \operatorname{erfc}\frac{(2n-1)\,l - x}{2\sqrt{Dt}} \right.$$
$$\left. + \operatorname{erfc}\frac{(2n-1)\,l + x}{2\sqrt{Dt}} \right]. \tag{4}$$

An expression for evaluating the liposome concentration $C(0, t)$ for $x = 0$ in the subsurface first layer (solution zone underlying the interface), can also be obtained:

$$\frac{C(0, t)}{C_0} = 1 - \sum_{n=1}^{\infty} \operatorname{erfc}\frac{(2n-1)\,l}{2\sqrt{Dt}}. \tag{5}$$

The numerical result obtained from (5) for three different thicknesses, l, (Fig. 6) corresponding to 10, 180, and 750 µl of spread liposomal suspension is given in Fig. 7.

For 10 µl, 1 s is almost sufficient for all the liposomes in the first layer to irreversibly desorb into the bulk; this result is in perfect agreement with kinetics shown

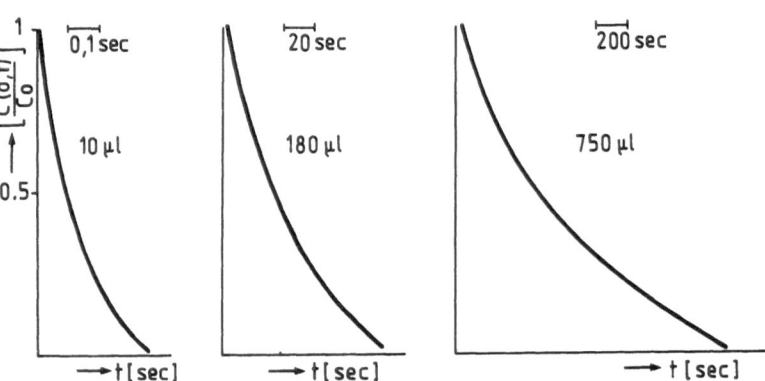

Fig. 7. Variation with time of the relative superficial concentration of spread liposomes calculated for a pure diffusion process

in Figs. 3 and 4. However, all the type-I structures (10 µl) dip into the subphase after crossing the interface and no tensioactive material (type-II structures) appears at the interface.

In contrast, when 750 µl of suspension is spread the time for irreversible diffusion of type-I structures in the bulk (2 000 s) is greater than characteristic time for adsorption and the second process becomes preponderant.

1.2 Kinetics governed by the adsorption process:

An irreversible adsorption process can be described by the kinetic equation of Langmuir, neglecting the desorption term:

$$\frac{dn^*}{dt} = K_a \, (dC_0 - n^*) \left(1 - \frac{n^*}{n_\infty^*}\right). \tag{6}$$

$n^*(t)$ is the number of type-II structures adsorbed on a 1 cm^2 area at time t,

dC_0 is the initial number of liposomes (type-I structures) which are able to adsorb on 1 cm^2,

n_∞^* is the maximal number of type-II structures adsorbed in a close-packed layer,

$1 - [n^*(t)/n_\infty^*]$ is the available surface area at time t,

K_a is the adsorption constant.

Solving (6) with $n^* = 0$ at $t = 0$ leads to an undimensional parameter:

$$\xi = \frac{\xi_0 \left(e^{(\xi_0 - 1)\tau} - 1\right)}{\xi_0 \, e^{(\xi_0 - 1)\tau} - 1} \tag{7}$$

with $\xi = \frac{n^*}{C_0}$, $\xi_0 = \frac{n_\infty}{dC_0}$, $\tau = K_a \frac{dC_0}{n_\infty} \, t$.

For unilamellar liposomes with external diameter $d = 150$ nm and for $C_0 = 2.7 \cdot 10^{13}$ liposomes/cm^3, $\xi_0 = 10.9$.

To compare this theoretical expression (7) describing the adsorption process with the experimental data in Figs. 3 and 4, the following strategy was adopted. The surface potential, ΔV, directly depends on the number of type-I structures transformed into type-II structures; thus,

$$\Delta V = 4 \, \pi \frac{\mu}{\varepsilon} \, n^* \tag{8}$$

where ε represents dielectric permitivity and where the vertical component of dipole moment μ does not change during the process.

Saturation might correspond to the transformation of all type-I structures which are able to adsorb (dC_0) into type-II structures:

$$\Delta V_{MAX} = 4 \, \pi \frac{\mu}{\varepsilon} \, dC_0. \tag{9}$$

From (8) and (9), we can obtain the following expression:

$$\frac{\Delta V}{\Delta V_{MAX}} = \frac{n^*}{dC_0} = \xi(t). \tag{10}$$

Experimental results $\xi(t)$, obtained from $\Delta V(t)$ curves (Figs. 3B and 4B) and expression (10), were compared to the theoretical curve $\xi(t)$. As shown in Figs. 8A (slow spreading) and 8B (rapid spreading), experimental curves obtained from 750 and 1 000 µl of liposomal suspension spread are similar to the theoretical curve $\xi(t)$: in this case, adsorption process is more important.

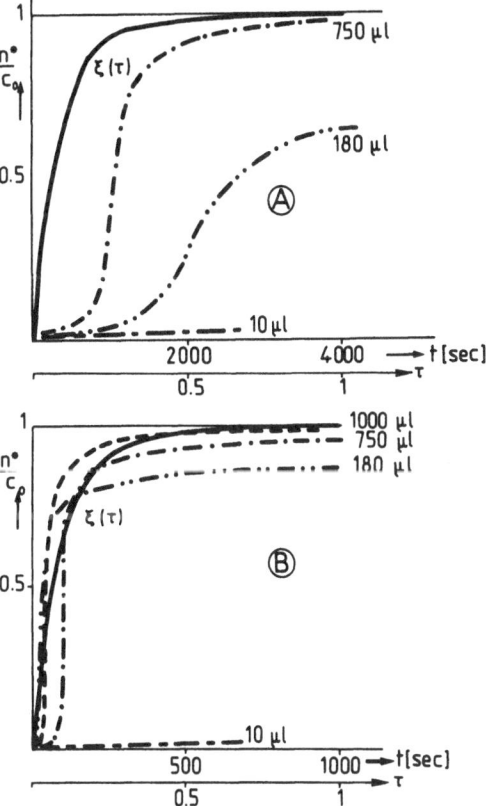

Fig. 8. Variation with time of the relative superficial concentration of type-II structures after slow spreading technique (A) and rapid spreading technique (B); continuous lines: calculated from (10) and (7); dashed lines: experimental results; τ is a reduced time

In conclusion, the kinetics of surface film formation from liposomal suspension volumes more than 1000 µl are totally explained by the adsorption process and by the irreversible transformation of type-I structures (liposomes = closed vesicles) into type-II structures (superficial film); in such cases, an experimental value of K_a can be calculated: $K_a = 10^{-2} \, s^{-1}$. Surface film formation kinetics for liposomal suspension volumes less than 10 µl are determined by irreversible vesicle diffusion into the bulk phase. Between 10 and 1000 µl the kinetics are governed by both processes and must be described by solving (2) with boundary conditions (3a), (3c), and (6) instead of (3b).

Figures 9A and 9B show the experimental results obtained by spreading of liposomal suspensions after 4 weeks storage; they contain large, open phospholipid structures (see § liposome verification above). The surface film is almost instantaneously formed after spreading the suspension at the interface: a spreading volume of less than 10 µl can already form a surface film with high surface pressure and potential.

2. Equilibrium and dynamic properties of surface films formed by spreading of phospholipid

Figure 10A shows the equilibrium isotherms of superficial films obtained after spreading freshly prepared (180 and 750 µl) and 4-week stored (4 and 9 µl) liposomal suspensions; the equilibrium isotherm of a monolayer with the same phospholipid composition spread from a solution in solvent is also represented (M). The mean molecular area value, A_M, corresponds to the monomolecular layer; A_1 and A_5 mean molecular area values are calculated supposing all the 150 nm diameter liposomes make a superficial film of unilamellar (A_1) or five lamellar (A_5) vesicles.

Equilibrium isotherms of 4-week stored liposomal suspensions are displaced to area values lower than the monomolecular layer corresponding area A_M: the monomolecular layer formation seems more difficult

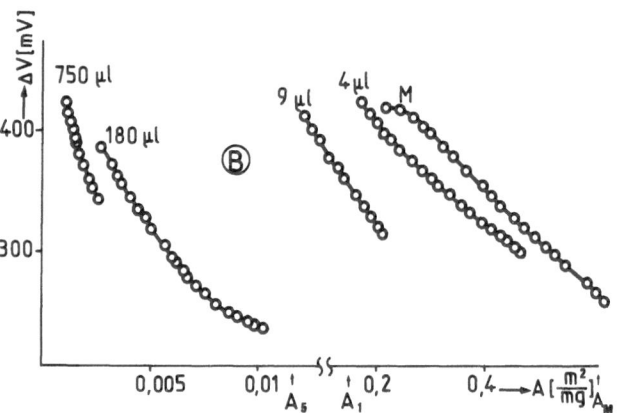

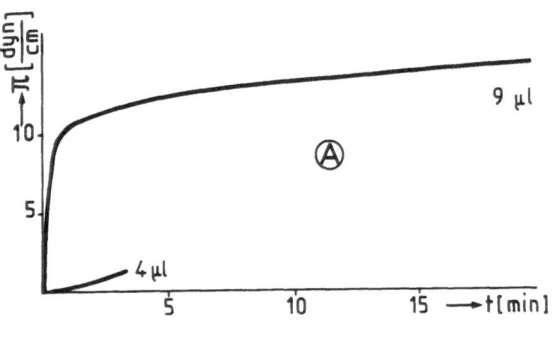

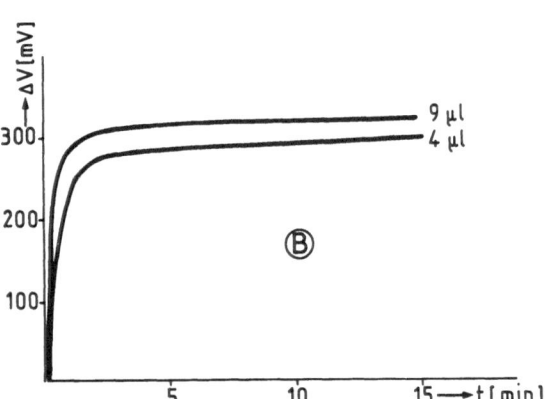

Fig. 9. Variation with time of surface pressure Π (A) and surface potential ΔV (B) for various spreading volumes of 4-week stored (+4 °C) liposomal suspension

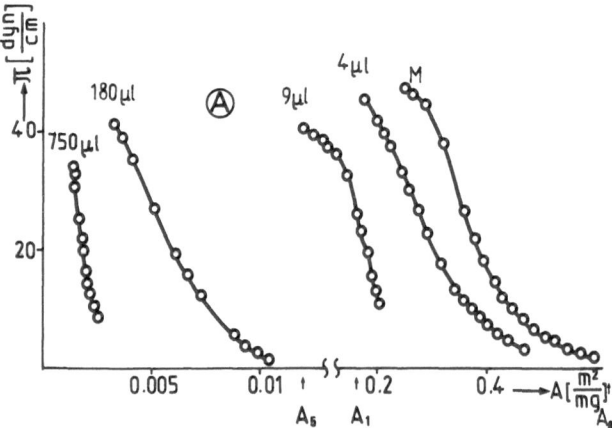

Fig. 10. Quasi-static compression isotherms — surface pressure Π (A) and surface potential ΔV (B) — after spreading of a phospholipid mixture in organic solution (M), liposomes = closed structures (180 and 750 µl) and 4-week stored liposomal suspension = large open structures (4 and 9 µl)

and the surface film built is partially structured into bilayered mesophases. The formation of the mono-molecular layer from freshly prepared liposomal suspension is much less effective and the equilibrium isotherms are displaced to area values lower than A_M, A_1, and A_5. These results confirm that, during spreading, some liposomes go to the bulk phase through the interface as in the theoretical scheme presented above.

Experimental data obtained for $\Delta V(A)$ (Fig. 10B), also confirm the theoretical scheme.

In Fig. 11A, surface potential, ΔV, vs. surface pressure, Π, is compared for the different superficial layers formed after spreading of 180 µl freshly prepared liposomal suspension, 9 µl of 4-week stored liposomal suspension, and the monomolecular layer (M). These results show that surface film structure is not exactly the same: aqueous phospholipid suspension does not form monomolecular layers but probably inhomogeneous surface films, as shown in Fig. 5. A study of dynamic dilatational properties could give more information about the structure of the three superficial films, since rheological properties are linked to interfacial structure. The surface dilatational elasticity values, E, of the three superficial films are presented in Fig. 11B. The rheological behavior of all three films is about the same up to a Π value of 15 dyn/cm. Above 15 dyn/cm, the elasticities become different and for the surface film obtained from the 4-week stored liposomal suspension, relaxations occur at surface pressures above 20 dyn/cm; this result was confirmed by the hysteresis study. For example, Fig. 12 shows the $\Pi(A)$ hysteresis obtained at initial surface pressures, Π_0, of 8 and 37.5 dyn/cm. Close to $\Pi_0 = 8$ dyn/cm (Fig. 12A) the hysteresis loop only depends on the Marangoni effect [33], whereas near $\Pi_0 = 37.5$ dyn/cm (Fig. 12B) the hysteresis loop shows a relaxation process in the surface film itself, in addition to the Marangoni effect.

The complete dynamic behavior of surface films built up from liposomes should be described in order to explain the structure of these films.

Conclusion

The kinetics of surface film formation from phospholipid suspensions depends on the volume spread and whether the liposomes are closed or open structures. A simple theoretical approach can describe these kinetics by two processes: irreversible diffusion into the bulk phase through the interface (concerning closed vesicles) and irreversible adsorption; this latter

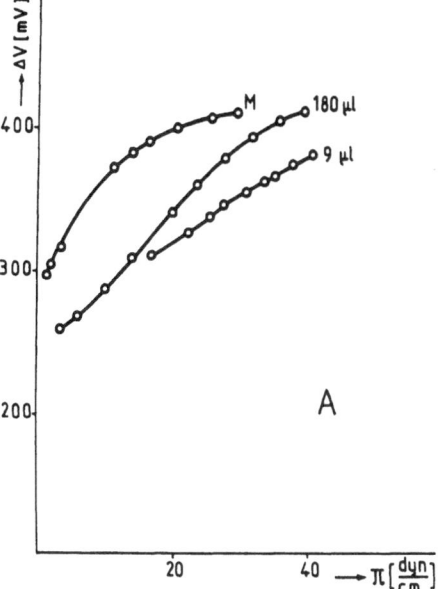

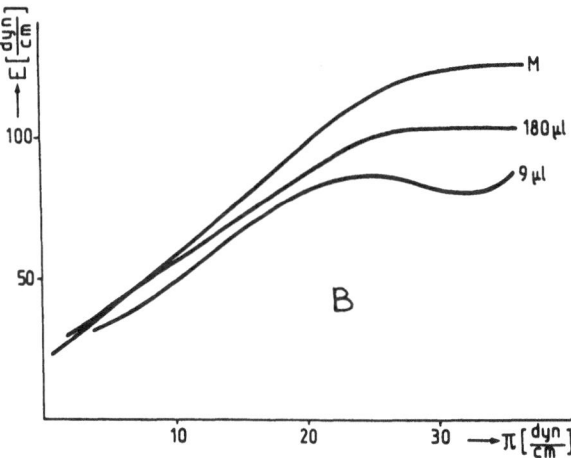

Fig. 11. Surface potential ΔV (A) and superficial elasticity modulus E (B) vs. surface pressure Π for three surface films: monomolecular film from an organic solution of the phospholipid mixture (M), surface film from liposome spreading (180 µl), surface film from large open structure spreading (9 µl)

process represents an irreversible transformation of closed vesicles (type-I structures) into destroyed vesicles (type-II structures) which form the surface film.

As shown by surface potential measurements and dilatational surface properties, the structure of the surface films obtained by spreading phospholipid suspensions is not the same as that of those obtained by spreading solutions in solvent; these former surface films are not real phospholipid monolayers.

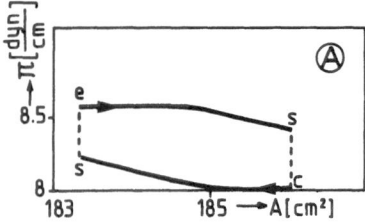

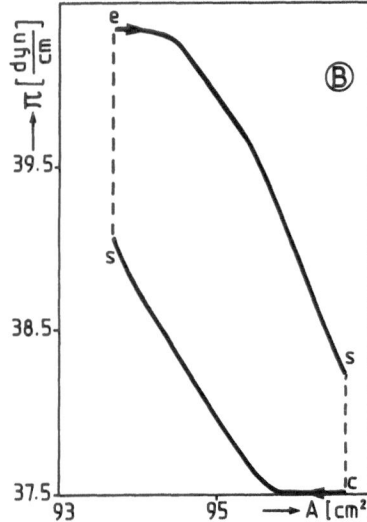

Fig. 12. Hysteresis loops (surface pressure Π vs. total area A) of surface film obtained after spreading of 4-week stored (+ 4 °C) liposomal suspension; at 8 dyn/cm (A) and 37.5 dyn/cm (B)

Acknowledgements

This work was partly-supported by the Bulgarian Ministry of Sciences.

References

1. Clements JA, Hustead F, Johnson RP, Gribetz I (1961) J Appl Physiol 16:44
2. Scarpelli EM (1972) The surfactant system of the lung. Lea and Febiger, Philadelphia
3. Von Neergard KZ (1929) Ges Exp Med 66:373
4. Avery ME, Mead J (1959) Am J Dis Child 97:517
5. Robillard E, Alarie Y, Dagenais-Perusse P, Baril E, Guilbeault A (1964) Can Med Ass J, 90:55
6. Chu JS, Clements JA, Cotton EK, Klaus MH, Sweet AY, Tooley WH (1967) Pediatrics 40:709
7. Obladen M, Brendlein F, Krempien B (1979) Eur J Pediatr 131:219
8. Fujiwara T, Tanaka Y, Takei T (1979) IRCS Med Sci 7:311
9. Fujiwara T, Maeta H, Chida S, Morita T (1979) IRCS Med Sci 7:313
10. Fujiwara T, Maeta H, Chida S, Morita T, Watabe Y, Abe T (1980) Lancet 1:55
11. Morley CJ, Robertson B, Miller N, Lachmann B, Nilsson R, Bangham A, Grossmann G (1980) Arch Dis Child, 55(10):758
12. Morley CJ, Bangham AD, Miller N, Davis JA (1981) Lancet 1:64
13. Morley CJ, Bangham AD, Johnson P, Thorburn GD, Jenkin G (1978) Nature 271:162
14. Smyth JA, Metcalfe IL, Duffty P, Possmayer F, Bryan MH, Enhorning G (1983) Pediatrics 71(6):913
15. Hallman M, Merritt TA, Schneider H, Epstein BL, Mannino F, Edwards DK, Gluck L (1983) Pediatrics 71(4):473
16. De Fontanges A, Bonte F, Taupin C, Ober R (1984) J Colloid Interface Sci 101(2):301
17. Bonte F, Dehan M, De Fontanges A, Lindenbaum A, Perret L, Puisieux F (1985) Arch Fr Pediatr 42:397
18. Puisieux F (1978) Labo-Pharma 281:899
19. Delattre J (1979) J Fr Biophys Med Nucl 3(1):3
20. Puisieux F, Delattre J (1985) Les liposomes, applications thérapeutiques. Ed Technique et Documentation, Lavoisier, Paris
21. Pattus F, Desnuelle P, Verger R (1978) Biochim Biophys Acta 507:62
22. Obladen M, Popp D, Scholl C, Schwartz H, Jahnig F (1983) Biochim Biophys Acta 735:215
23. Schindler H (1979) Biochim Biophys Acta 555:316
24. Bangham AD, Standish MM, Watkins JC (1965) J Mol Biol 13:238
25. Bonte F (1985) Thèse, Paris Sud
26. De Fontanges A (1985) Thèse, Paris Sud
27. Takayama M, Itoh S, Nagasaki T, Tanimizu I (1977) Clin Chim Acta 79:93
28. Bangham AD, Horne RW (1964) J Mol Biol 8:660
29. Dimitrov DS, Panaiotov I (1975-76) Ann Univ Sofia, Fac Chim 70:103
30. Panaiotov I, Dimitrov DS, Georgiev G (1980) Studia Biophys 78:95
31. Dimitrov DS, Panaiotov I, Richmond P, Ter-Minassian-Saraga L (1978) J Colloid Interface Sci 65(3):483
32. Panaiotov I, Dimitrov DS, Ter-Minassian-Saraga L (1979) J Colloid Interface Sci 72(1):49
33. Panaiotov I (1986) Thesis, Sofia
34. Carslaw HS, Jaeger JC (1959) Conduction of heat in solids. Oxford Univ Press, Clarendon, Oxford

Received October, 1988;
accepted February, 1989

Authors' address:

M. A. Launois-Surpas, J. E. Proust, F. Puisieux
Physicochimie des Surfaces et Innovation en Pharmacotechnie
CNRS UA 1218
5 rue J. B. Clément
F-92290 Châtenay-Malabry, France

Progress in Colloid & Polymer Science   Progr Colloid Polym Sci 79:33–37 (1989)

Laser-induced temperature jump and time-resolved x-ray powder diffraction on phospholipid phase transitions

P. Laggner[1]), M. Kriechbaum[1]), A. Hermetter[2]), F. Paltauf[2]), J. Hendrix[3]), and G. Rapp[4])

[1]) Institut für Röntgenfeinstrukturforschung der Österreichischen Akademie der Wissenschaften und der Forschungsgesellschaft Joanneum, Graz, Austria
[2]) Institut für Biochemie und Lebensmittelchemie der Technischen Universität Graz, Graz, Austria
[3]) European Molecular Biology Laboratory (EMBL) Hamburg Outstation at DESY, Hamburg, F.R.G.
[4]) Max-Planck-Institut für Medizinische Forschung, Heidelberg, F.R.G.

Abstract: Temperature-jump x-ray small-angle powder diffraction was used to investigate the kinetics and structural mechanism of the thermotropic phase transitions of a phosphatidylethanolamine (1-stearoyl-2-oleoyl-sn-3-phosphatidylethanolamine)/water system. An Erbium laser ($\lambda = 1.5\,\mu$m) with a pulse energy of 2 J delivered in 2 ms was used for rapid heating, leading to a T-jump of about 12 °C. The small-angle diffraction pattern was detected with 1 ms resolution during the relaxation following the T-jump. The results showed that the $L_\beta - L_\alpha$ transition (~ 29 °C) is at least as fast as the heating time (2 ms). The lamellar-to-hexagonal ($L_\alpha - H_{II}$) transition was found to be considerably slower with a half-time in the order of 100 ms. Both transitions were found to be two-state, with the phase structures coexisting during the transition and without detectable nonlamellar intermediates.

Key words: Phospholipids, structure, phase transition, kinetics, synchrotron radiation, laser-induced T-jump.

Introduction

In view of their important role in biological membranes and lipoprotein complexes, the various structures exhibited by phospholipids in aqueous dispersion have been studied extensively by virtually all physico-chemical techniques [1]. Thus, a solid body of knowledge exists on the equilibrium properties of various liquid-crystalline phases and on the thermodynamics of their transitions. However, very little is known about the kinetics and the structural mechanisms of their interconversion.

In recent years several attempts have been made to use the high x-ray flux offered by synchrotron radiation sources to study the process of phospholipid phase transitions in real-time by time-resolved x-ray diffraction [2–6]. All these studies involved temperature jumps effected by external heating and passive heat conduction through the sample container (glass-capillary). By this approach, time-resolution in the

order of 1–2 s were achieved which, for many transitions, is too slow to reveal the intrinsic kinetics of the system. The limiting factor in reaching higher time resolutions is the heat conductivity of the various materials involved (metal, glass, water). This makes it impossible to reach a satisfactory T-jump situation for a transition of a few degrees width with relaxation times on the millisecond time-scale [4].

In order to overcome this problem of passive heat diffusion we have chosen to use absorptive heating by illumination of the sample with a short, powerful infrared laser-pulse (Er-laser). Holzwarth et al. have used a similar approach with an iodine-laser for optical T-jump studies on various systems, including phospholipids [7]. In the present work we combined for the first time this powerful heating method which yields jump rates in the order of 10 K/ms with fast time-resolved x-ray powder diffraction in order to analyze the structural pathways of phospholipid phase transitions. We report on the results obtained with 1-stearoyl-2-oleoyl-sn-3-phosphatidylethanolamine (SOPE) in excess water, which undergoes a gel-to-liquid-crystal $(L_\beta - L_\alpha)$ transition around 29 °C and a lamellar-to-inverted hexagonal $L_\alpha - H_{II})$ transition around 57 °C.

Materials and methods

SOPE was synthesized according to procedures published elsewhere [8, 9]. The material was found to be better than 98 % pure by GLC analysis of fatty acid methyl esters. Aqueous suspensions were prepared by mixing weighed amounts of lipid and H_2O and vortexing.

The thermotropic behavior of the system ($c = 0.2 \%$, w/w) was characterized by differential scanning calorimetry (DSC) in a Privalov DASM-4 high sensitivity calorimeter (Mashpriborintorg, Moscow, USSR).

Static x-ray powder diffraction experiments were performed in a Kratky compact camera (A. Paar, Graz, Austria) equipped with a position sensitive electronic x-ray detector (M. Braun, Munich, F.R.G.). For time-resolved x-ray measurements the X-33 small-angle camera at the EMBL Outstation at DESY (Hamburg) was used in combination with a linear position sensitive detector and a suitable time-slicing data collection system, allowing data accumulation in a time-scale down to 0.5 ms [13].

The principle of the experimental arrangement is shown schematically in Fig. 1. The sample contained within a 1 mm i.d. glass capillary (Marck capillary, Hilgenberg, F.R.G.) of 0.01 mm wall thickness was placed in a thermoelectrically controlled cuvette (A. Paar, Graz, Austria). The laser beam (oval cross-section of approximately 0.6 × 0.4 cm) was aligned with the help of a low power He-Ne laser (0.5 mW) vertically to the direction of the x-ray beam (cross-section approximately 0.1 × 0.5 cm) onto the capillary from above.

The design of the Erbium laser is described elsewhere [10]. It produces a pulse with $\lambda = 1.5 \mu m$ of 2 ms duration and 1–2 J energy. The absorption coefficient of water at this wavelength is 6.5 cm^{-1}. To optimize infrared energy absorption and to reduce thermal gradients within the exposed sample volume of about 5 μl, a gold-plated glass mirror was put underneath the sample so that the laser beam passes the sample twice during one pulse.

Results and discussion

The isothermal phase behavior of SOPE was characterized by static x-ray small-angle powder diffraction (X-SAPD) and differential scanning calorimetry (DSC). The results are shown in Fig. 2. Between 28° and 30 °C, a major endothermic transition occurs in which the lamellar repeat period changes from 65 Å to 57 Å. Wide-angle diffraction experiments (data not shown) have confirmed that this transition refers to the chain melting transition $L_\beta \to L_\alpha$. It is noteworthy that the static SAPD data (Fig. 2 B) show a coexistence of the two structures in the transition range even under isothermal conditions. Since these experiments involved a 30 min equilibration time between temperature steps of 1 °C, and in view of the rapid transition kinetics (see below), it is justified to refer to these data as equilibrium states.

The lamellar-to-inverted hexagonal phase $(L_\alpha - H_{II})$ change occurs in the temperature range between 52° and 60 °C. Here, too, the SAPD results reveal a coexistence of the two structures over the temperature range of the transition (Fig. 2 B).

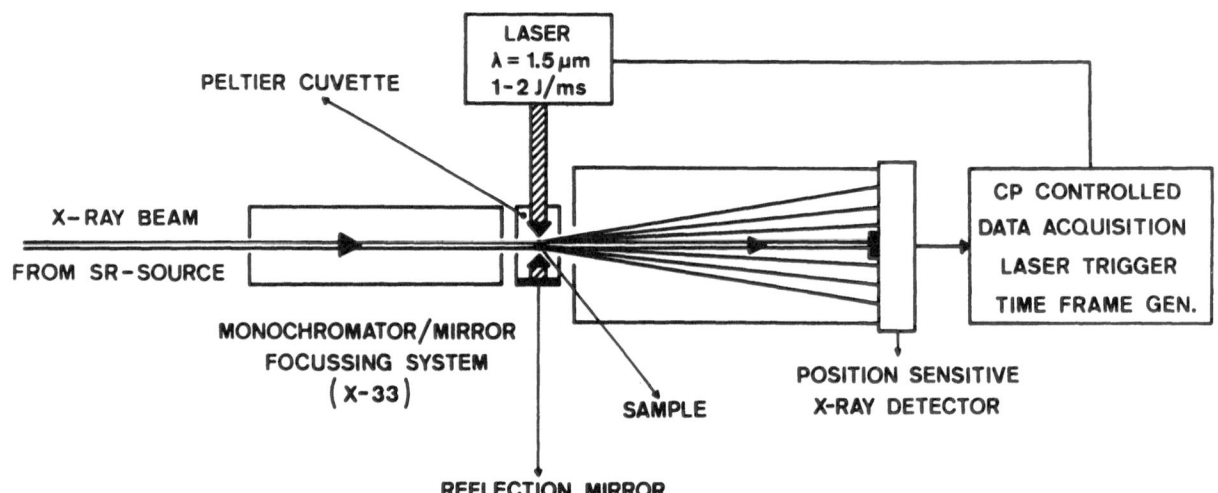

Fig. 1. Schematic drawing of the arrangement used for time-resolved small-angle powder diffraction in combination with an infra-red laser-induced temperature jump

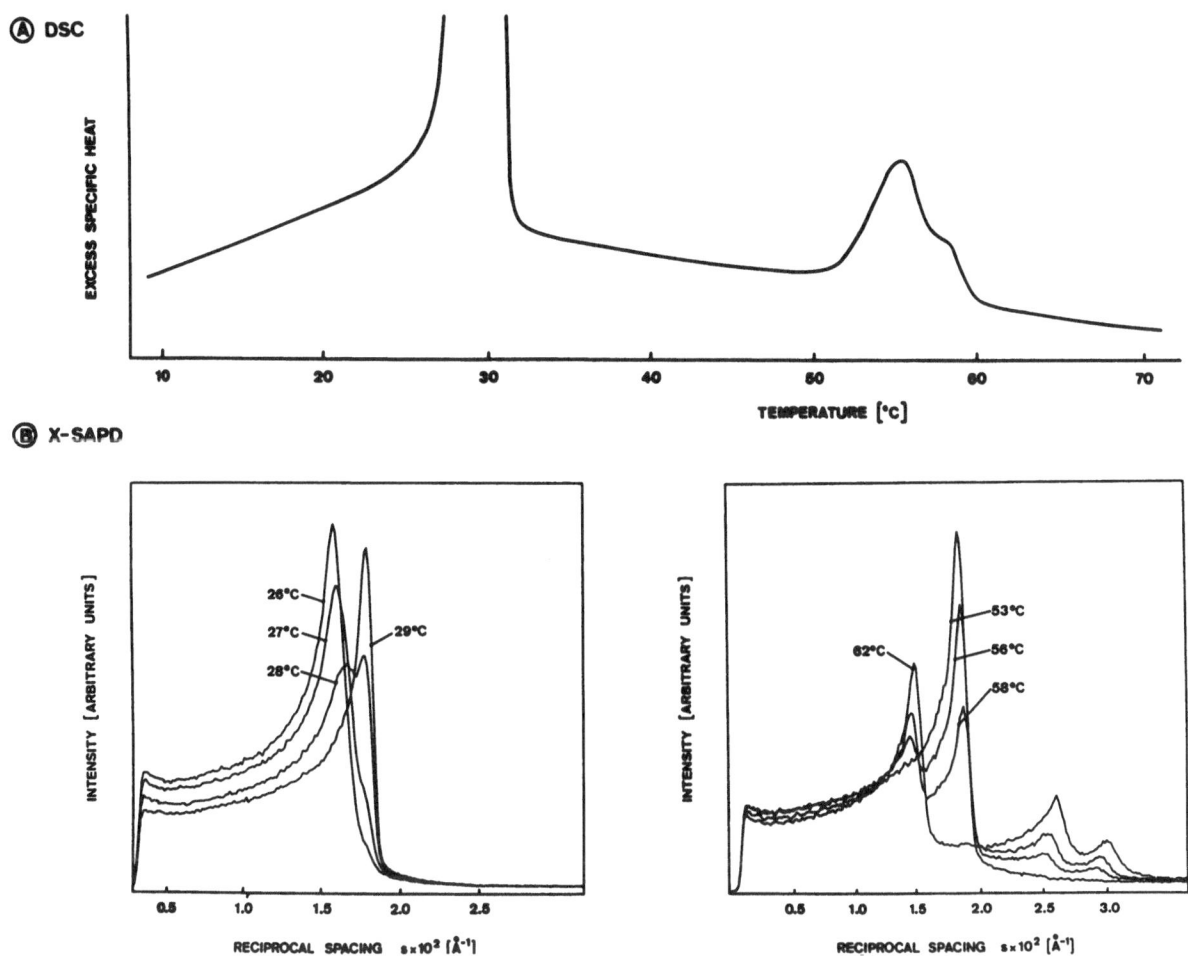

Fig. 2. Thermotropic mesomorphism of aqueous dispersions of 1-stearoyl-2-oleoyl-sn-3-phosphatidylethanolamine (SOPE). A) Differential scanning calorimetry trace of a 0.2 % (w/w) aqueous suspension; heating rate 0.25 °C/min; B) x-ray small-angle powder diffraction of an SOPE suspension (20 %, w/w) in the region of the $L_\beta - L_\alpha$ transition (left) and the $L_\alpha - H_{II}$ transition (right); $s = (2 \sin \theta)/\lambda$ 2θ ... diffraction angle, λ ... wavelength

The results of a T-jump experiment through the $L_\beta - L_\alpha$ transition are shown in Fig. 3. The starting temperature has been chosen at 26 °C. The three-dimensional representation of the powder diffraction pattern in the region of the lamellar first-order peak shows that the L_β structure transforms into the L_α state within a period of less than 5 ms. An evaluation of the peak intensities reveals a complementary growth-and-decay process for the two structures (Fig. 3 B). This time characteristic corresponds closely to that of the laser pulse structure, and therefore we do not consider this as a lower limit of the transformation rate. No intermediate disorder or transition state other than the coexistence of the two lamellar periods of L_β and L_α can be detected. Pending a quantitative modelling of

the data, which shall also involve experiments with a faster heating rate using a Q-switched laser (under construction), this suggests that the transformation proceeds via strain-free coherent interfaces between the two phases with different lattices [11]. Thus, the present results confirm the notion that the $L_\beta - L_\alpha$ transition in phospholipids is a relatively fast process [12] and that the transition times (1–2 s) reported from previous time-resolved x-ray experiments indeed reflect only the limits of the experimental time resolution [4, 5] in experiments involving passive heat conduction.

The performance of the laser-heating arrangement was tested in a series of experiments where the starting temperature was successively decreased below the

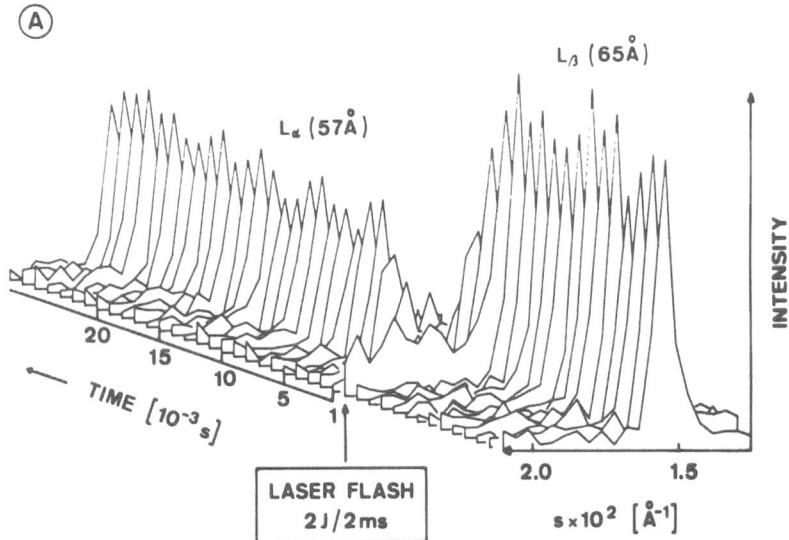

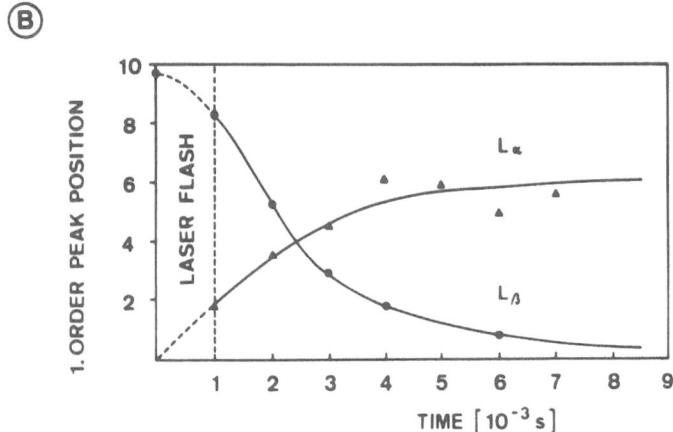

Fig. 3. Time-resolved x-ray powder diffraction of SOPE (20% aqueous suspension) in the angular range of the first-order lamellar repeat spacing in the $L_\beta - L_\alpha$ transition. A) Three-dimensional plot of 13, 1 ms time frames before the laser flash and 27 frames (1 ms) after the flash; B) Intensities in the peak positions of the first-order peaks of the L_β and L_α phases, respectively after the temperature jump

$L_\beta \rightarrow L_\alpha$ transition. The lowest starting temperature from which the laser-pulse jump took the system into the transition range was 17 °C. Thus, the amplitude of the T-jump reached by this technique is approximately 12 °C.

The $L_\alpha \rightarrow H_{II}$ transition has been investigated in a similar way starting from a temperature of 50 °C. In this case the rate of the transformation was found to be considerably slower: Figure 4 shows that 50% transformation was reached only after approximately 200 ms. This confirms our previous notion [5, 6] that the $L_\alpha \rightarrow H_{II}$ transition, which is associated by a relatively low positive enthalpy (Fig. 2 A), is strongly kinetically hindered, and that the process involves no detectable non-lamellar intermediates or disorder, (see

appendix). Even though the T-jump has reached a final temperature above the isothermal transition range (Fig. 2 A) a complete interconversion could not be observed even 10 s after the laser pulse. Since at this time the passive heat dissipation into the unheated sample domains has already become effective, the data cannot be used for a complete kinetic description of the long-time relaxation process. This drawback could be overcome by using a feed-back system which allows to maintain the temperature after the T-jump. The initial (below 50 ms) decay and growth rates, however, can be taken as nucleation rates with a half-life in the order of 0.1 s. For the (temperature-dependent) growth our previous, slow T-jump experiments [5, 6] have indicated a half-life of approximately 5 s.

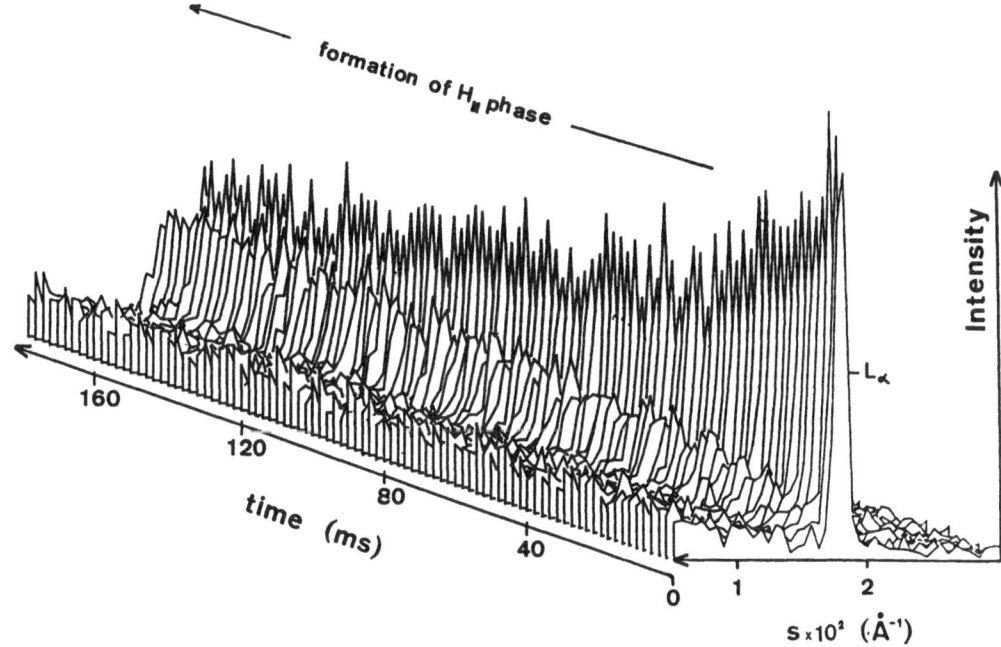

Fig. 4. Time-resolved x-ray powder diffraction of SOPE (20% w/w) in the initial stages of the $L_\alpha \to H_{II}$ transition (2 ms/time frame)

Appendix

Closer inspection of the initial relaxation in the $L_\alpha - H_\parallel$ transition shows a rapid shrinkage of the lamellar lattice period from 56 to 51 Å within the first 5 ms.

Acknowledgements

The present work has been supported by grant no. P6287C from the Österreichischer Fonds zur Förderung der Wissenschaftlichen Forschung and a matching grant from the EEC.

References

1. Small DM (1986) In: Small DM (ed) Handbook of Lipid Research, Vol 4, The Physical Chemistry of Lipids. Plenum Press, New York, pp 475-522
2. Laggner P (1986) In: Bartunik HD, Chance B (eds) Structural Biological Applications of X-Ray Absorption, Scattering and Diffraction. Symp Bristo 1984, Academic Press, New York, pp 171-182
3. Ranck JL, Letellier L, Shechter E, Krop B, Pernod P, Tardieu A (1984) Biochemistry 23:4955-4961
4. Caffrey M (1985) Biochemistry 24:4826-4844
5. Laggner P, Lohner K, Müller K (1987) Mol Cryst Liq Cryst 151:373-388
6. Laggner P (1988) Top Curr Chem 145:173-202
7. Holzwarth JF, Eck V, Gentz A (1984) In: Bayley P, Dale R (eds) Spectroscopy and Dynamics of Biological Systems. Academic Press, London, pp 354-377
8. Hermetter A, Stütz H, Franzmeier R, Paltauf F (1988) Proc Natl Acad Sci USA, submitted for publication
9. Comfurius P, Zwaal RFA (1977) Biochim Biophys Acta 488:36-42
10. Rapp G, Poole KJV, Maeda Y, Kaplan JH, McCray JA, Goody RS (1989) Ber Bunsenges Phys Chem 93
11. Porter DA, Easterling KE (1981) Phase Transformations in Metals and Alloys. Von Nostrand Reinhold Co, New York
12. Tsong TY, Kanehisa MI (1977) Biochemistry 12:2674-2680
13. Hendrix J, Fürst H, Hartfield B, Dainton D (1982) Nucl Instrum and Methods 201:139-144

Received October, 1988;
accepted February, 1989

Authors' address:

P. Laggner
Institut für Röntgenfeinstrukturforschung der
Österreichischen Akademie der Wissenschaften
und der Forschungsgesellschaft Joanneum
Steyrergasse 17
A-8010 Graz, Austria

Progress in Colloid & Polymer Science

Progr Colloid Polym Sci 79:38–42 (1989)

Modifications of the structure and dynamics of dimyristoylphosphatidic acid model membranes by calcium ions and poly-L-lysines. A Raman and deuterium NMR study

G. Laroche[1]), M. Pézolet[1]), J. Dufourcq[2]), and E. J. Dufourc[2])

[1]) Centre de Recherche en Science et en Ingénérie des Macromolécules, Département de Chimie, Université Laval, Québec, Canada
[2]) Centre de Recherche Paul Pascal C.N.R.S., Domaine Universitaire, Talence, France

Abstract: Interactions of calcium ions (Ca^{2+}) or poly-L-lysines (PLL) of different molecular weight with negatively charged model membranes (dimyristoylphosphatidic acid (DMPA) dispersions) have been followed by Raman and deuterium (^2H) NMR spectroscopies.

Results show that both short (m.w. = 3 300 to 4 000) and long (m.w. = 180 000 to 260 000) PLL increase the degree of ordering of the lipid acyl chains either in gel or fluid phases. This effect is attributed to a neutralizing action of the PLL positive charges on the phospholipid head group negative charges, which leads to a better chain packing. In addition, a 20 °C upshift in the temperature, T_c, of the gel-to-fluid phase transition is promoted by long PLL, whereas no change in T_c is observed with short PLL. These observations are correlated to temperature-induced conformational changes of the polypeptides. As detected by Raman amide I bands, long PLL retain their β-sheet conformation over all the temperature range of the study, whereas short PLL undergo at about T_c, a transition from β-sheet ($T < T_c$) to random coil ($T > T_c$).

Effect of calcium leads to an increase in ordering such that lipids are in their gel state up to 75 °C. Above this temperature a structural modification occurs which is not yet identified but which cannot be attributed to a conventional gel-to-fluid phase transition.

Key words: DMPA, model membranes, calcium, poly-L-lysines, Raman, ^2H-NMR.

Introduction

Multivalent cations and polycationic species are known to induce dramatic structural changes in model membranes composed of acidic phospholipids. Among these species, calcium has been extensively studied [1–4] because it induces membrane fusion. The interaction of poly-L-lysine (PLL) with phospholipids has also been the subject of several investigations [5–10] because this polypeptide is a good model to mimic extrinsic proteins. In this study we have thus investigated the effect of calcium ions and poly-L-lysines of different molecular weight on the anionic lipid dimyristoylphosphatidic acid (DMPA) by Raman and deuterium nuclear magnetic resonance (^2H-NMR) spectroscopies.

These techniques are particularly useful to follow the thermotropism of phospholipids and to monitor the structure and dynamics of bilayers. Moreover, Raman spectroscopy gives information about the secondary structure of polypeptides bound to phospholipid, whereas ^2H-NMR spectroscopy is very sensitive to the dynamics and conformational properties of membranes.

Materials and methods

Materials

The disodium salts of dimyristoylphosphatidic acid and of dimyristoylphosphatidic acid with a perdeuterated *sn*-2 chain were obtained from Avanti Polar Lipids (USA). The hydrobromide salts

salts of PLL of molecular weight 3300–4000 (short PLL) and 180000–260000 were purchased from Sigma (St. Louis, Missouri, USA). All materials were used without further purification.

Raman measurements

Dispersions of DMPA were prepared by mixing appropriate amounts of lipid in a 150 mM NaCl solution containing 10 mM EDTA at pH 6.5. Samples (10 % weight in total lipid) were then heated to approximatively 65 °C for 10 min, shaken with a vortex mixer and cooled below the gel-to-fluid phase transition temperature. This cycle was repeated at least three times. The pH of the dispersion was measured and ajusted to 6.5, if necessary. Samples were then transferred in glass capillary tubes and centrifuged in an hematocrit centrifuge. Samples containing poly-L-lysine were prepared similary except that the lipid concentration was 1%. Appropriate amounts of 3 % PLL were added to the dispersions, and the incubation cycle was repeated once again. The lipid-to-lysine residue incubation molar ratio (R_i) was 1. Samples containing calcium were prepared in a 70 mM $CaCl_2$ solution in order to have R_i = 0.33. The Raman spectrophotometer used in this work is described elsewhere [11, 12].

^{2}H-NMR experiments

Samples were prepared at the same incubation molar ratio as for Raman experiments in order to have a final lipid concentration of 50 mg in 1 ml of solution. NMR signals were obtained on a Bruker MSL-200 spectrometer operating at 30.7 MHz by means of the quadrupolar echo sequence [29]. Quadrature detection was utilized and temperature regulated to ± 1 °C. Samples were allowed to equilibrate at a given temperature for at least 30 min prior to recording the NMR signal. Typical experimental parameters are: $\pi/2$ pulse length of 5.25 µs, spacing of the two $\pi/2$ pulses in the quadrupolar echo sequence of 25 µs, spectral window of 500 kHz and recycle time of 1.5 s. Data treatment was performed on a VAX/VMS 8600 computer.

Results

Raman spectroscopy

To study phospholipid conformation and thermotropism, the signal arising in the C-H stretching mode region (2750–3100 cm^{-1}) is the most interesting. The assignment of the features of this spectral region is now well established [13–16].

Two peak height intensity ratios are particularly well suited to follow the thermotropic behavior of phospholipid bilayers [13, 17, 18]. It has been previously shown that the h_{2880}/h_{2850} and h_{2930}/h_{2880} ratios monitor essentially the overall disorder of the lipid acyl chain matrix; moreover, the former ratio is also sensitive to the vibrational coupling between the acyl chain [9]. Figures 1a and 1b illustrate how these ratios are affected when a dispersion of DMPA in water undergoes the gel-to-fluid phase transition. The transi-

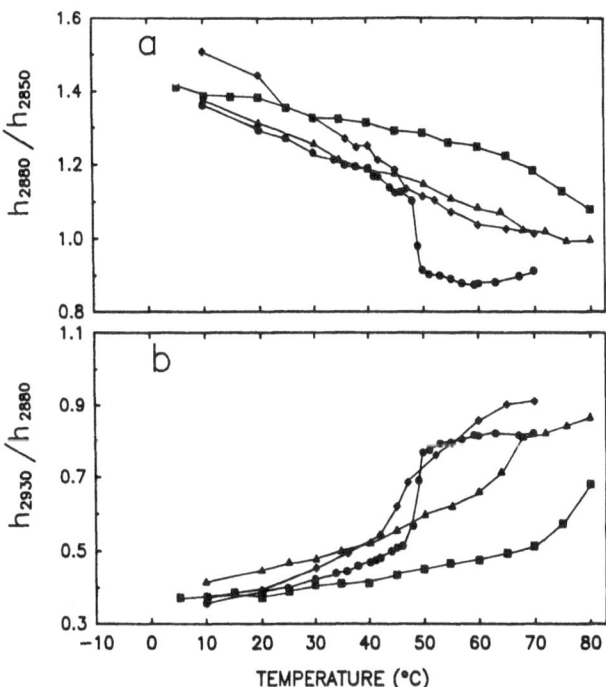

Fig. 1. Temperature profiles of DMPA in the absence (●) and in the presence of long PLL (▲, R_i = 1), short PLL (◆, R_i = 1), or of calcium ions (■, R_i = 0.33) derived from the Raman h_{2930}/h_{2880} (a) and h_{2880}/h_{2845} (b) peak intensity ratios

tion occurs at 49 °C and is highly cooperative, which is in good agreement with the results of Graham et al. [19] and Mushayakarara and Levin [20]. When long PLL is added to DMPA at R_i = 1, the transition is shifted towards higher temperatures and is considerably broader. It is clear from these results that long PLL increases the packing as well as the order of the lipid acyl chains, specially at high temperature.

The effect of short PLL on DMPA is quite different. In this case, the transition temperature is almost not affected by the addition of the polypeptide. In order to understand this difference in behavior between long and short PLL, we have also determined the conformation of the bound polypeptides from their amide I bands (Fig. 2). For the DMPA/long PLL complex, the spectrum of the amide I region is almost superimposable with that of β-sheet PLL over the whole range of temperature investigated. Figures 2b and 2c reveal that the conformation adopted by short PLL bound to DMPA is also β-sheet at low temperature, but the polypeptide undergoes a conformational change to the random coil structure at high temperatures. In order to follow this change of conformation, we have meas-

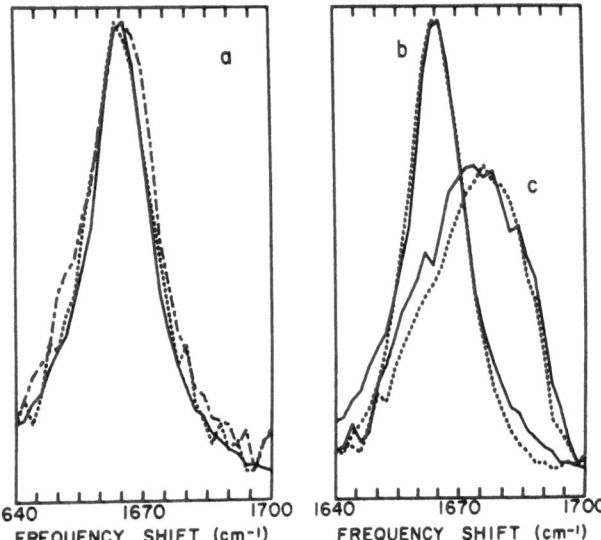

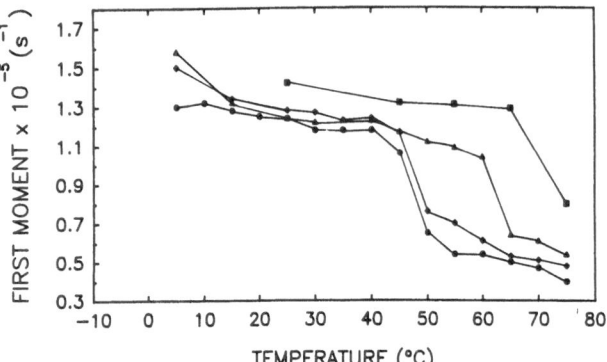

Fig. 4. Temperature variation of the first moment of the ^{2}H-NMR spectra for pure DMPA (●), DMPA/long PLL (▲), DMPA/short PLL (◆) and DMPA/Ca^{2+} complexes (■)

Fig. 2. a) Raman spectra of the amide-I-region of polylysine in the β- sheet conformation (solid line) at pH 11.5 and 60 °C (4 scans at 2 s/2 cm^{-1}); DMPA/long PLL complex at $R_i = 1$ at pH 6.5 and 20 °C (dotted line) (10 scans at 2 s/2 cm^{-1}); b and c) Raman spectra of the amide-I-region of polylysine in the β-sheet (b, solid line) and in the random coil (c, solid line) conformations (4 scans at 2 s/2 cm^{-1}); DMPA/short PLL complex with $R_i = 1$ at 20 °C (b, dotted line) and at 70 °C (c, dotted line) (25 scans at 2 s/2 cm^{-1}). These spectra have been corrected for the water spectral contribution

ured the ratio of the peak height intensity of the lipid carbonyl band at 1740 cm^{-1} relative to the amide I band of PLL. Figure 3 illustrates these results in comparison with the h_{2930}/h_{2880} ratio of the lipid. As it can be seen, the break around 45 °C on the temperature profile of the DMPA/short complex nearly coincides with the onset of the conformational transition of PLL.

^{2}H-NMR spectroscopy

In order to support our Raman results, we carried out ^{2}H-NMR experiments. The calculation of the moments of the ^{2}H-NMR spectra [21, 22] is known to be an efficient and quantitative method to follow the order and dynamics of the lipid acyl chains [23, 30, 31]. The results of such calculations performed on pure DMPA and DMPA/PLL complexes as a function of temperature are presented in Fig. 4. As can be seen, these results are in very good agreement with those obtained by Raman spectroscopy. Long PLL produces an important shift of the gel-to-fluid phase transition to high temperature while short PLL has almost no effect on this transition. The value of the spectral moments in the fluid phase shows that long PLL has a large ordering effect on the DMPA acyl chain matrix. Short PLL produces a similar but less important effect. Therefore, both the neutralization of the DMPA negative charges by the positively charged lysine groups and the conformation adopted by PLL seem to affect the packing of the lipid acyl chains. In order to establish the ordering effect of PLL along the acyl chains of DMPA, we have dePaked [24, 25] the spectra of pure DMPA and DMPA/PLL complexes taken at 7 °C

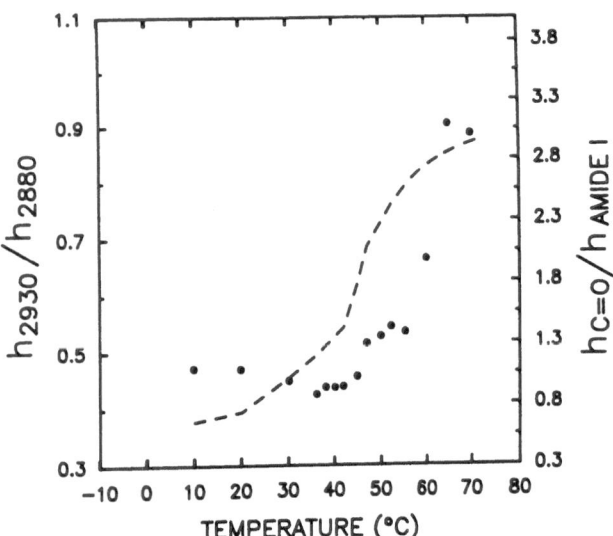

Fig. 3. Temperature profile of DMPA (dashed line) and of short PLL (●) in the DMPA/short PLL complexes ($R_i = 1$) derived from the h_{2930}/h_{2880} and $h_{C=0}/h_{\text{amide I}}$ peak height intensity ratios, respectively

above the phase transition (results not shown). This technique allows the determination of the quadrupolar splitting for each carbon position of the lipid acyl chains. These data show that both long and short PLL have an ordering effect due to the charge neutralization as mentioned above. From these data it is clear that this effect is most likely felt at the "plateau" region, e.g. from carbon 3 to 10. At the center of the bilayer, all systems seem to have approximatively the same ordering.

Complexes of Ca^{2+} with DMPA appear to be quite different from those made of DMPA and PLL. The thermotropic behavior followed by the Raman h_{2930}/h_{2880} and h_{2880}/h_{2850} intensity ratios, or by the first moment of the ^2H-NMR signal indicates that this complex undergoes a structural or dynamical modification near 75 °C. However, this cannot be identified to a gel-to-fluid phase transition since the ^2H-NMR spectrum DMPA/Ca^{2+} complex taken at 75 °C displays an isotropic peak (results not shown). Moreover, this change is reversible since after cooling the system to 25 °C the original spectrum is obtained.

Discussion

The results presented here shed more light on the nature of the interactions between acidic phospholipids and positively charged species. It is clear from these data that two effects have to be taken into account when dealing with lipid/poly-L-lysine interactions.

Firstly, both long and short PLL seem to be as effective to neutralize the negative charges of DMPA. This neutralization reduces the electrostatic repulsion between the DMPA head groups and leads to a better packing of the lipid acyl chains as evidenced by an increase in T_c and in the chain local order in the fluid phase. The increase in ordering is more important at the "plateau" positions than towards the center of the bilayer. This effect can be paralleled to that of acidic neutralization of the PA charges which leads to a comparable 20 °C increase of the phase transition temperature [26]. It must be emphasized that this is the first time that an electrostatic interaction at the head group level is shown to promote a very important change in ordering of the acyl chains.

Secondly, the conformation adopted by the bound polypeptide seems to affect significantly the temperature of the phase transition of the lipid and the order of the lipid acyl chains in the fluid phase. The Raman data reveals that long PLL bound to DMPA adopts the

β-sheet conformation even at high temperatures whereas short PLL undergoes a change in conformation from β-sheet to random coil near 45 °C. Since long PLL is still in the β-sheet structure when the lipid is in the fluid phase, this leads to a greater ordering of the lipid acyl chains for the DMPA/long PLL complex as revealed by the first moments of ^2H-NMR spectra. This conformational effect was first proposed by Carrier and Pézolet [9] on the DPPG/PLL system where long PLL adopts the α-helical conformation and produces a 3 °C-shift of the gel-to-fluid phase transition, while short PLL remains unordered, giving rise to a lowering of 3 °C of the transition temperature [32].

The effect of calcium ions is somewhat different from that of polylysines. Raman and ^2H-NMR results show that addition of calcium promotes an increase of the packing and order of the DMPA acyl chains. It has been shown [27, 28] that the DMPA/Ca^{2+} system forms an ordered cochleate phase. Our results are thus in good agreement with this hypothesis. However, both Raman and NMR spectroscopies show that the complex undergoes a transition near 75 °C. At this temperature the NMR data show that the DMPA/Ca^{2+} complex gives rise to an isotropic peak (data not shown) which can be due to a very fast motion of the lipid acyl chains or to a change of symmetry of the system. More experiments are necessary to further characterize this system.

In summary, we have shown that the DMPA/PLL interaction is governed by electrostatic interactions and the conformation adopted by PLL. Both long and short PLL increase the order of the lipid acyl chains by neutralizing the DMPA negative charges. However, this ordering effect is much more important for the DMPA/long PLL complex since this polypeptide retains the β-sheet structure over the whole range of temperature investigated whereas short PLL undergoes a change of conformation from β-sheet to random coil at 45 °C.

Acknowledgements

Prof. J. H. Davis (Guelph, Canada) is thanked for making his "de-Pake-ing" program available.

References

1. Onishi SI, Ito T (1974) Biochemistry 13:881-887
2. Jacobson K, Papahadjopoulos D (1975) Biochemistry 14:152-161
3. Dluhy RA, Cameron DG, Mantsch HH, Mendelsohn R (1983) Biochemistry 22:6318-6325

4. Kouaouci R, Silvius JR, Graham I, Pézolet M (1985) Biochemistry 24:7132-7140
5. Galla HJ, Sackmann E (1975) Biochim Biophys Acta 401:509-529
6. Galla HJ, Sackmann E (1975) J Am Chem Soc 97:4114-4120
7. Hartmann W, Galla HJ (1978) Biochim Biophys Acta 509:470-474
8. Hartmann W, Galla HJ, Sackmann M (1977) FEBS Lett 78:169-172
9. Carrier D, Pézolet M (1984) Biophys J 46:497-506
10. Laroche G, Carrier D, Pézolet M (1988) Biochemistry 27:6220-6228
11. Savoie R, Boulé B, Genest G, Pézolet M (1979) Can J Spectrosc 24:112-117
12. Pézolet M, Boulé B, Bourque D (1983) Rev Sci Instrum 54:1364-1367
13. Gaber BP, Peticolas WL (1977) Biochim Biophys Acta 465:260-274
14. Spiker RCJr, Levin IW (1975) Biochim Biophys Acta 339:361-363
15. Bunow M, Levin IW (1977) Biochim Biophys Acta 487:388-394
16. Hill IR, Levin IW (1979) J Chem Phys 70:842-851
17. Faucon JF, Dufourcq J, Bernard F, Duchesneau L, Pézolet M (1983) Biochemistry 22:2179-2185
18. Tarashi T, Mendelsohn R (1980) Proc Natl Acad Sci USA 77:214-219
19. Graham I, Gagné J, Silvius JR (1985) Biochemistry 24:7123-7131
20. Mushayakarara E, Levin IW (1984) Biochim Biophys Acta 776:185-189
21. Davis JH (1979) Biophys J 27:339-358
22. Bloom M, Davis JH, Dahlquist FW (1978) In: Lipporan E, Saluvere T (eds) 20th Ampere Congress, Tallin, Estonia, Springer, Berlin p 551
23. Dufourc EJ (1986) Thèse d'Etat, Université de Bordeaux I, France
24. Bloom M, Davis JH, McKay AL (1981) Chem Phys Lett 80:198-202
25. Sternin E (1982) M Sc Thesis, Vancouver, Canada
26. Trauble H, Eibl H (1974) Proc Natl Acad Sci USA 512:84-96
27. Liao MJ, Prestegard JH (1981) Biochim Biophys Acta 645:149-156
28. Van Dijck PWM, de Kruijff B, Verkleij AJ, Van Deenen LLM, de Gier J (1978) Biochim Biophys Acta 512:84-96
29. Davis JH, Jeffrey KR, Bloom M, Valic MI, Higgs TP (1976) Chem Phys Lett 42:390
30. Davis JH (1983) Biochim Biophys Acta 737:117-171
31. Dufourc EJ (1988) In: Op den Kamp (ed) Membrane Biogenesis. Springer Verlag, Berlin Heidelberg, pp 141-176
32. Carrier D, Dufourcq J, Faucon JF, Pézolet M (1985) Biochem Biophys Acta 830:131-139

Received November, 1988;
accepted January, 1989

Authors' address:

E. J. Dufourc
CRPP/CNRS
Domaine Universitaire
F-33405 Talence Cedex, France

Progress in Colloid & Polymer Science   Progr Colloid Polym Sci 79:43–48 (1989)

Lipid mixtures in monolayer and BLM

P. Lo Nostro, A. Niccolai and G. Gabrielli

Dipartimento die Chimica, Università degli Studi di Firenze, Italy

Abstract: The relationship between spreading monolayers at the water/air interface and planar lipid bilayers (BLM) of 1-monooleoyl-glycerol MON), 1-monostearoyl-glycerol (MOS), galactocerebrosides (GAL), and MOS/MON, GAL/MON mixtures has been studied.

Spreading isotherms were obtained and the monomolecular surface phases of these substances were examined and described through state equations. Molecular orientations of MON, MOS, and GAL hydrocarbon chains were discussed according to the surface phase rule. The two-dimensional miscibility of MOS/MON and GAL/MON mixtures was also studied.

The MON, MOS/MON, and GAL/MON planar bilayers were obtained and their electrical capacitance and resistance were measured.

Some relations between miscibility in monolayer and BLM formation were established:

1) substances giving liquid films at the w/a interface produce stable BLMs;

2) components miscible in monolayers at the w/a interface give stable planar bilayers.

Key-words: Monolayer, planar lipid bilayer (BLM), water/air interface, 1-monooleoyl-glycerol, 1-monostearoyl-glycerol, galactocerebrosides.

Introduction

Monolayers, BLMs, and vesicles are useful membrane models. Usually these systems give only particular information that allows the investigation of some features of the biological membranes.

In this study are reported possible relationships between the physico-chemical behavior of monolayers and that of BLMs. Information about the thermodynamic properties of galactocerebrosides and some derivatives of glycerol are obtained from the spreading isotherms of these substances at the water/air interface and are correlated with the BLMs properties.

Substances which constitute biomembranes, such as galactocerebrosides, or which are similar to biomembrane components such as 1-monooleoyl-glycerol and 1-monostearoyl-glycerol were used. In fact, glycosphingolipids act in many cell-surface phenomena: cell-cell interaction, cell growth and oncogenic transformations, recognition of external ligands and antigenic reactions [1]; and monoolein and monostearin have the same glyceric skeleton as do the biomembrane phospholipids.

Experimental

1-monooleoyl-glycerol (MON), 1-monostearoyl-glycerol (MOS), galactocerebrosides (GAL) from bovine brain and squalene were obtained from Sigma Chemie GmbH, West Germany, and they were used without further purification (purity > 99 %).

a) Monolayers

Benzene (purity > 99 %) supplied by Merck, Darmstadt, FRG, was used as the spreading solvent of MON, MOS, and MOS/MON monolayers. The subphase was bidistilled water purified of colloidal impurities by a Milli-Q system (Millipore). Chloroform (purity > 99 %), supplied by Farmitalia Carlo Erba S.p.A., Milan, Italy, was used as spreading solvent for GAL and GAL/MON monolayers. GAL and GAL/MON solutions were obtained by gently heating. NaCl (purity > 99.8 %), obtained from Farmitalia Carlo Erba S.p.A., Milan, Italy, 0.1 M was used as subphase.

The surface pressure π (mN/m) vs. the molecular area A (m²/mg) measurements were carried out following Wilhelmy's method. The film was compressed discontinuously at a speed of 3.8 cm²/min: this speed was necessary to ensure the equilibrium conditions. Measurements were performed at the w/a interface and at three different temperatures: 15°, 25° and 30°C.

The surface pressure, the area, and the temperature were determined with an accuracy of ± 0.016 mN/m, ± 0.01 m²/mg and ± 0.3°C, respectively.

b) BLMs

Membranes were obtained by gently heated dispersions prepared from 10 mg of substance in 1 ml of squalene. They were studied at room temperture ($24° \pm 1°C$). The 1.3/1 MOS/MON system was studied at about 60°C.

BLM preparation followed the Mueller et al. technique [2], using a NaCl 0.1 M solution. The membrane area was determined by means of an ocular micrometer. Resistance and capacitance measurements were made with a square wave with a linewidth varying between 20 and 240 mV, and a frequency of 300 mHz; all the experiments were performed in "voltage clamp". Membrane capacitance was measured at 30 Hz with a 2 % precision ac impedance bridge.

Results and discussion

1. Monolayers

1.2 Pure components

Figure 1 shows the isotherms of MON, MOS and GAL as a function of temperatures.

The MON spreading isotherms were in good agreement with the literature data [3] and with our previous data obtained by a continous compression of the film, where the Langmuir method was used to measure the surface pressure [4].

According to the phase classification [5,6], the compressibility modulus K reported in Table 1 showed MON to be in a liquid-intermediate state for all the areas and temperatures considered. K values were obtained from the spreading isotherms by using the following relationship:

$$K = - A \left(\partial\pi/\partial A\right)_{P,T} \qquad (1)$$

where π is the surface pressure, A is the molecular area, and P and T are, respectively, the external pressure and the temperature. The monolayer was considered to be a unitary fluid [7] and the following relation was applied to the liquid intermediate phase:

$$\pi A = kT \left(1 + a_1/A + a_2 A^2 + \ldots\right) \qquad (2)$$

where k is the Boltzmann constant, a_1 and a_2 are, respectively, the first and the second virial coefficients.

Table 1 reports the first virial coefficient values a_1. These were positive with small absolute values. Therefore we could conclude the existence of small repulsive interactions among the molecules at the interface. In fact, a liquid-intermediate phase is characterized by hydrophobic random chains and weak attractive or repulsive interactions [7,8].

Figure 1 and Table 2 show that MOS had both a liquid condensed and a liquid intermediate phase at all temperatures considered. The liquid condensed phase corresponds to the linear part of the monolayer isotherm at highest surface pressures, before the collapse point, and it has a high slope. The following equation was used for the liquid condensed phase:

$$\pi = a + bA \qquad (3)$$

where the a and b parameters depend on the substance and temperature. Limiting areas were determined by

Table 1

	A (m²/mg)	π(mN/m)	K(mN/m)	a_1(Å²/mol)
15°	1.45–0.42	0.10–36.80	1.45–69.00	38.82
25°	1.15–0.50	1.28–35.22	4.85–65.21	46.52
30°	1.16–0.34	0.23–41.01	0.00–48.96	21.25

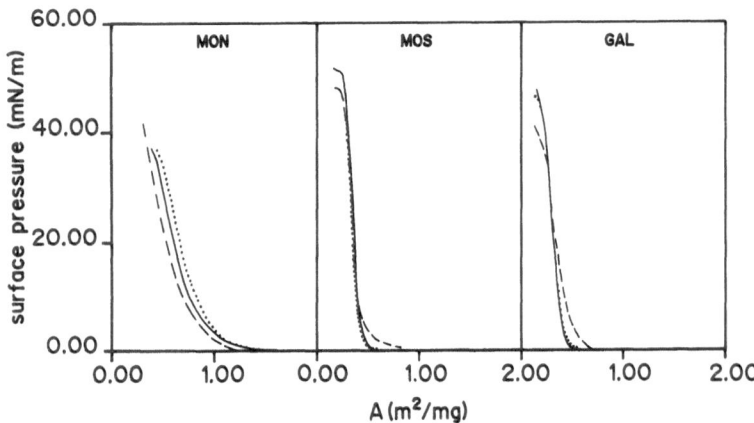

Fig. 1. 1 MON, MOS, and GAL spreading isotherms at (——) 15°, (···) 25° (20° for MOS), (- - -) 30°C

Table 2

	$A_{\lim}$(Å²/mol)	K(mN/m)	equation
Liquid-condensed phase			
15°	48.51	134.2	$\pi = 135.14 - 2.77A$
25°	51.55	113.8	$\pi = 114.79 - 2.20A$

	A(m²/mg)	π(mN/m)	K(mN/m)	a_1(Å²/mol)
Liquid-intermediate phase				
15°	0.56−0.30	0.03−26.00	0.84−68.55	48.31
25°	0.50−0.35	0.30−14.44	3.00−60.34	12.60
30°	0.70−0.26	0.02−35.23	0.07−60.45	52.69

extrapolating to $\pi = 0$ the straight lines corresponding to the liquid condensed phases; for MOS films these values were very close to the cross-sectional area of hydrocarbon chains with a vertical orientation at the w/a interface. In fact, this phase is characterized by strongly interacting hydrophobic chains with an almost vertical orientation in respect to the w/a interface [7]. The liquid intermediate phase corresponds to the first part of the spreading isotherms and it was de-

Table 3

	A(m²/mg)	π(mN/m)	K(mN/m)
15 °C			
MON	0.42−1.45	0.10−36.80	1.45− 69.60
GAL/MON 1/4	0.43−0.99	1.00−29.20	5.22−100.94
GAL/MON 1/2	0.32−1.12	0.10−36.95	5.65− 72.03
GAL/MON 1/1	0.72−0.26	0.07−44.55	2.52− 78.05
GAL/MON 2/1	0.20−0.66	0.07−45.65	2.30− 83.67
GAL/MON 5/1	0.20−0.55	0.02−44.18	0.22− 85.98
GAL	0.15−0.70	0.00−47.40	0.84−134.20
25 °C			
MON	0.50−1.15	1.28−35.22	4.85− 65.21
GAL/MON 1/4	1.30−0.55	0.36−33.61	1.27− 78.72
GAL/MON 1/2	0.42−0.96	0.44−33.87	8.44− 95.04
GAL/MON 1/1	0.37−0.71	0.09−39.39	2.13−102.92
GAL/MON 2/1	0.33−0.75	0.11−39.20	1.65− 83.38
GAL/MON 5/1	0.30−0.65	0.15−38.00	1.95−115.20
GAL	0.25−0.50	0.30−42.68	3.00−113.80
30 °C			
MON	0.34−1.16	0.23−41.01	0.00−48.96
GAL/MON 1/4	0.33−0.96	0.36−33.79	6.91−47.40
GAL/MON 1/2	0.41−1.03	0.40−37.10	20.60−78.00
GAL/MON 1/1	0.26−0.84	0.01−42.60	0.42−82.25
GAL/MON 2/1	0.23−0.61	0.39−39.67	6.34−82.17
GAL/MON 5/1	0.28−0.66	0.10−41.60	3.30−77.00
GAL	0.26−0.70	0.02−35.23	0.07−60.45

cribed by the relation (2). The a_1 coefficient was negative and had a small absolute value indicating the presence of weak interactions among the chains and the interface.

GAL is a commercial mixture of galactocerebrosides with different hydrophobic chains but with the same polar head group (i.e., the galactose moiety). However, the heterogeneity of the hydrophobic chains makes a negligible contribution with respect to the polar glycidic head group [1]. The GAL monolayers (Fig. 1) showed − in addition to the liquid intermediate phase − a liquid condensed one at 15 °C and 25 °C. This phase is described by Eq. (3). Table 3 shows the parameters of the GAL liquid condensed and liquid intermediate phases. For GAL we assumed a mean molecular weight of 789.76, obtained from GC-analysis [1]. The values of a_1 confirmed the previous deductions obtained for the liquid intermediate phase in the case of MON, i.e., the presence of weakly interacting flexible chains.

1.2 MOS/MON and GAL/MON mixtures

Figure 2a gives the molecular areas vs. the molar ratios at constant surface pressure at three different temperatures for the MOS/MON system. Small deviations from the additivity rule are observed.

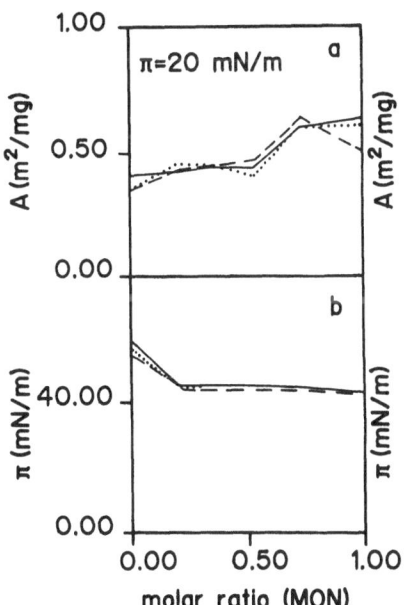

Fig. 2. a) MOS-MON molecular area vs. molar ratio at (——) 15°, (····) 20°, (−−−) 30 °C; b) MOS-MON π_{coll} vs. molar ratio at (——) 15°, (····) 20°, (−−−) 30 °C

Therefore, no definitive conclusion regarding the bidimensional miscibility of MON and MOS could be made. However, according to the surface phase rule [9], the invariance of collapse surface values as a function of the molar ratios (Fig. 2b) are a strong indication of the immiscibility between the two components.

For GAL/MON system, the compressibility moduli of all considered mixtures are reported in Table 4. The K values suggested that the mixtures gave films more fluid than the GAL film; in fact, the liquid condensed phase appeared only at 25 °C for the 5/1 GAL/MON mixture.

Figure 3a shows the mixture area at $\pi = 20$ mN/m vs. the weight ratios. At 15 °C we observed negative deviations from the additivity for all mixture compositions suggesting the complete miscibility and attractive interactions between GAL and MON hydrocarbon chains in monolayer at this temperature.

As reported above for the MOS/MON system, only small deviations from the area's additivity were found; therefore no definitive conclusion regarding GAL/MON miscibility can be made.

In Fig. 3b the collapse pressures of these mixture films, π_{coll} (mN/m), are plotted vs. weight ratios at 15 °, 25 °, and 30 °C. The bidimensional state phase rule [9] was applied to the monolayers of GAL/MON mixtures: the weight ratio-dependence of the collapse pressure indicates the total miscibility of these components.

The ΔG_{mix}, ΔH_{mix}, and $T\Delta S_{mix}$ values (see Fig. 4) were determined according to Bacon and Barnes [10]. At 15 ° and 30 °C, $\Delta G_{mix} < 0$; therefore the GAL/MON monolayers wer more stable with respect to the films of pure components. At 25 °C $\Delta G_{mix} < 0$ for 1/1 and 1/2 GAL/MON mixtures, whereas $\Delta G_{mix} \geq 0$ in the other cases (i. e., 1/4, 2/1 and 5/1 GAL/MON mixtures). Thermodynamic parameters indicate that miscibility is favored by the enthalpic factor.

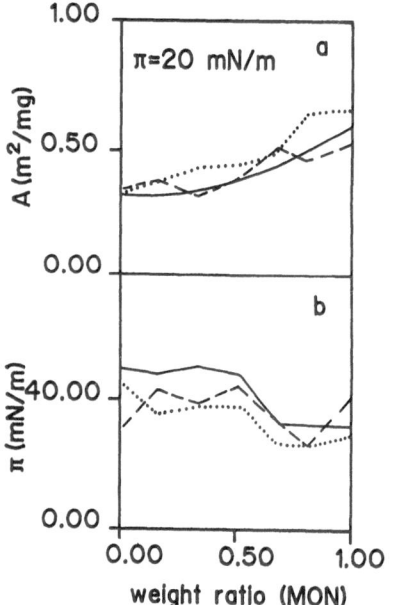

Fig. 3. a) GAL-MON molecular area vs. weight ratio at (—) 15 °, (····) 25 °, (– – –) 30 °C; b) GAL-MON π_{col} vs. weight ratio at (—) 15 °, (····) 25 °, (– – –) 30 °C

2) BLMs

Table 4 shows the membrane capacitance of BLMs prepared from squalene dispersion. MON bilayers presented a high specific capacitance value, in agreement with the literature data for solvent-free mem-

Table 4

components	$C(\mu F/cm^2)$	$\alpha(mV^{-2})$	
MON	0.79 ± 0.01	$4.4 \cdot 10^{-6}$	
MON/MOS 5.1/1	0.69 ± 0.01	$1.8 \cdot 10^{-6}$	
MON/MOS 1.3/1	0.49 ± 0.01	$1.8 \cdot 10^{-4}$	$(T = 60°)$
GAL/MON 1/17	0.67 ± 0.01	$1.7 \cdot 10^{-5}$	
GAL/MON 1/15	0.66 ± 0.01	$1.8 \cdot 10^{-6}$	
GAL/MON 1/10	0.63 ± 0.01	$1.5 \cdot 10^{-6}$	
GAL/MON 1/5	0.70 ± 0.01	$4.3 \cdot 10^{-6}$	
GAL/MON 1/2.6	0.70 ± 0.01	$2.1 \cdot 10^{-6}$	

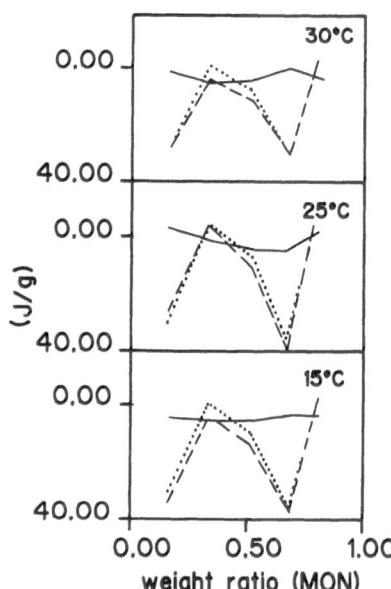

Fig. 4. (—) ΔG_{mix}, (····) ΔH_{mix}, (– – –) $T\Delta S_{mix}$, at 15 °, 25 °, 30 °C

branes [11]. The following equation was used for MON BLMs:

$$C = C(O)(1 + \alpha V^2) \qquad (4)$$

whre C = specific capacitance at an applied potential V;

$$C(0) = \text{specific capacitance at } V = 0.$$

MON bilayers showed a small α value. This confirmed that the BLMs were solvent-free. Therefore, the area A per MON hydrocarbon chain value in the bilayer has been obtained using the following equation [11]:

$$A = 2 C V_{mol}/\varepsilon_o \varepsilon_B \qquad (5)$$

where: $\varepsilon_o = 8.87 \cdot 10^{-8}$ µF/cm², ε_B is the dielectric coefficient of the hydrocarbon core of the MON acylic chain, and V_{mol} is the single hydrocarbon chain volume. According to Requena and Haydon [12], $\varepsilon_B = 2.20$ and $V_{mol} = 4.75 \cdot 10^{-22}$ cm³ at 20 °C, therefore $A = 38 \pm 2$ Å²/molec. In monolayer at the w/a interface, this area value corresponds to a liquid intermediate phase; probably also in the planar bilayer the MON molecules present a liquid state.

Very instable MOS BLMs were obtained by increasing the temperature to about 60 °C, in agreement with the literature data [13]. Planar bilayers were obtained from 5.1/1 weight ratio mixture in squalene at room temperature, and from 1.3/1 mixture at 60 °C; all the BLMs were rigid and instable. The BLMs resistance values was always about 10^7 Ω · cm². In mono-

layer at the w/a interface, MOS produces condensed films [4].

The galactocerebrosides in squalene did not give BLMs at room temperature, but it was possible to form lipid bilayers from GAL/MON systems in squalene.

The GAL/MON system specific resistance was almost ohmic, and its value was about 10^6–10^7 Ω · cm². Figure 5 shows the best fit lines of the membrane current I (nA/cm²) vs. applied potential V (mV) for the BLMs with different GAL/MON weight ratios; the increasing of I as a function of potential V is probably due to the channels formation across the bilayer membrane.

The specific capacitance of the GAL/MON bilayers was always smaller in respect to that of MON BLMs (Table 4). The very small α coefficient suggested that the BLMs were solvent-free. Further studies must be performed to see if the smaller C value depends on a decrease of the hydrocarbon-core dielectric coefficient or on an increase of the BLM thickness.

Conclusions

From monolayer measurements GAL and MOS give condensed films, whereas MON produces liquid films; MOS and MON are immiscible, whereas GAL and MON are completely miscible at the w/a interface at 15°, 25°, and 30 °C. The MON BLMs are very stable, whereas GAL or MOS planar bilayers are very rigid and instable, at least at room temperature. The addition of GAL or MOS to MON produces membranes more rigid and more sensitive to mechanical vibrations: this is in agreement with observations reported in literature [1]. The instability of GAL and MOS BLMs is probably due to the high molecular rigidity and to the increased hydrocarbon chain length.

From our results concerning BLM and monolayer properties we can finally conclude that:

1) at a fixed temperature stable BLMs are obtained from those substances which produce liquid monolayers;

2) amphiphilic mixtures of components completely miscible at the w/a interface give stable BLMs.

This statement allows us to establish the stability of planar bilayers from the monolayer phase knowledge, and vice-versa. These considerations confirmed what had been deduced for other components [14].

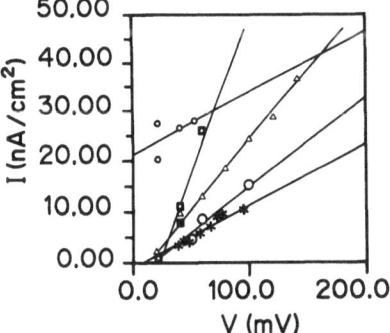

Fig. 5. Membrane current I vs. potential V for MON/GAL systems BLMs: MON (✳), GAL-MON 1/17 (○), GAL-MON 1/15 (□), GAL-MON 1/5 (△), GAL-MON 1/3 (⊕)

Acknowledgement

M.P.I. and C.N.R. are greatefully acknowledged for partial financial support. P. L. N. thanks ENIRICERCHE for a fellowship.

References

1. Gambale F, Robello M, Usai C, Marchetti C (1982) Biochim Biophys Acta 693:165
2. Mueller P, Rudin DO, Ti Tien H, Wescott WC (1962) Circulation 26:1167
3. Lundberg B, Ekman S (1979) Acta Chem Scand B 33:395
4. Niccolai A, Baglioni P, Dei L, Gabrielli G (1989) Colloid Polym Sci 267:262
5. Davies J, Rideal E (1963) Interfacial Phenomena. Academic Press, New York London, p 265
6. Adamson AW (1976) Physical Chemistry of Surfaces. John Wiley & Sons, New York, p 138–139
7. Gabrielli G, Niccolai A, Dei L (1986) Colloid Polym Sci 264:972
8. Pethica BA, Glasser ML (1981) J Colloid Interface Sci 81:41
9. Crisp DJ (1949) Surface Chemistry Suppl, Research, London, pp 17–23
10. Bacon KJ, Barnes GT (1978) J Colloid Interface Sci 67:70
11. White SH (1978) Biophys J 23:337
12. Requena J, Haydon DA (1975) Proc Roy Soc Lond A Math Phys Sci 47:161
13. Pagano E, Cherry RJ, Chapman D (1973) Science 181:557
14. Gabrielli G, Gliozzi A, Sanguinetti A, D'Agata A (1989) Colloid Interface 35:262

Received October, 1988;
accepted April 1989

Authors' address:

Prof. G. Gabrielli
Università degli Studie
Dipartimento di Chimica
via Gino Capponi, 9
I-50121 Firenze, Italy

Characterization of small lipid vesicles prepared by microfluidization

G. Masson

Nestlé Research Centre, Nestec Ltd., Vers-chez-les-Blanc, Lausanne, Switzerland

Abstract: The recent technique of microfluidization has been used for the production of small unilamellar lipid vesicles.

The particular properties of the vesicles produced depend very highly on the number of passages of the preparation through the Microfluidizer M 110.

The small unilamellar vesicles have been characterized in shape and size by electron photomicrography, and their size distribution has been measured by photon correlation spectroscopy.

Key words: Vesicles, properties, influence of the preparation, size distribution.

Introduction

Liposomes have been used for a long time for various research applications at the laboratory scale. However, up to the present, and because of the absence of a suitable technology, the production of liposomes has been limited.

This constraint seems to have been overcome by a recent microfluidization technique which will be briefly described. This method allows large-scale preparation of small lipid vesicles of uniform size distribution with satisfactory entrapment efficiency [1–4].

In addition to classical laboratory applications of liposomes as model membranes and as active substance carriers, some new uses have been reported in the food industry for immobilization of enzymes. Other fields of potential commercial interest in cosmetics, dermatology and medicine have been recently discovered.

Materials and methods

23 g of purified egg lecithin (Ovothin 170, Lucas Meyer. Hamburg, FRG) was dispersed in 200 ml of buffer phosphate solution 0.025 M at pH 6.88. The solution corresponded to about 150 µmol phospholipids per ml of solution. The dispersion was heated to 30 °C and stirred for 1 h under a light stream of nitrogen. The hy-

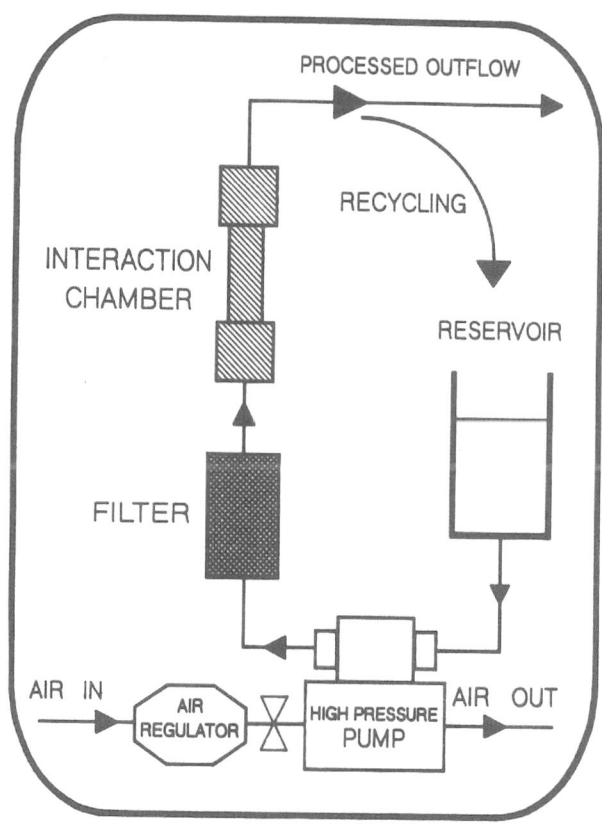

Fig. 1. Schematic representation of the Microfluidizer M110

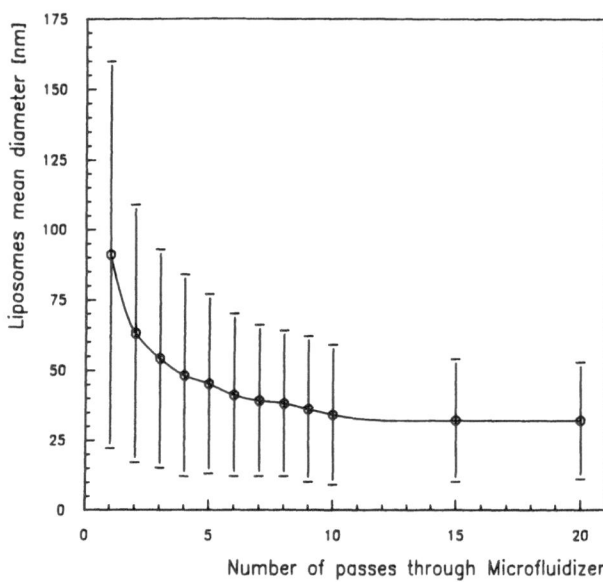

Fig. 2. Size reduction of liposomes

drated lipids were then passed through the Microfluidizer. Figure 1 shows a schematic representation of the Microfluidizer M110 which was used (Microfluidics. Newton, Massachusetts, USA).

The crude suspension of phospholipids was placed in the reservoir and the air regulator adjusted to the selected operating pressure (860 bars). With such a setting, when the air valve is open, the liquid dispersion flows through a filter (5 μm) into the interaction chamber where it is separated into two streams which interact at extremely high velocities in dimensionally defined microchannels. The suspension can be recycled through the machine and, in this eventuality, the suspension must be cooled because of the temperature increase in the interaction chamber at high operating pressure. Flow rates in the order of 100 ml/min were employed, and volumes of 200 ml were processed.

A Malvern photon correlation spectrometer (Malvern Instruments), was used to measure the mean size and the size distribution of the liposomes by light scattering. The spectrometer was equipped with a 64-log-channels Malvern Autocorrelator 7032 and a Spectra-Physics 15 mW He-Ne laser (wavelength 632.8 nm).

Freeze-fracture electron microscope photographs of liposomes were obtained by use of a Cryofract (Reichert-Jung, France) and an electron microscope EM 300 (Philips, Holland).

Characterization of liposomes

As already mentioned by Mayhew [2], vesicle formation can be modified by varying the pressure or the number of passes through the interaction chamber. Figure 2 shows the reduction of the mean size of liposomes during successive passes through the Microfluidizer.

In conjunction with the size reduction, a progressive decrease of the spread of the size distribution which corrresponds — in Fig. 2 — to the contraction of

the bar magnitude as a function of the number of passes, was observed.

Increasing the number of microfluidization cycles beyond 20–25 did not result in any further reduction in liposome size.

Figure 3 shows a typical liposome size distribution obtained after the 20th pass in the Microfluidizer. This distribution was obtained from light scattering data. The mean size of the particles was 31 nm and about 82 % of the liposomes population was situated in the range 10 nm – 50 nm.

A common example of electron micrographs of unilamellar liposomes is shown in Fig. 4. These images illustrate the smooth spherical shape of the liposomes obtained by microfluidization, and the trapped volume compartment. The smooth fracture faces are representative of the entire field, and the absence of multistep surfaces unambiguously demonstrates the unilamellar nature of the liposomes. Moreover, the vesicles are seen to be fairly homogeneous in size. This confirms the results obtained by light scattering. The few larger vesicles, and also the material which does not produce vesicles, can easily be removed by centrifugation or by size-exclusion gel chromatography.

Discussion

Microfluidization provides a practical and convenient means of preparing research or commercial quan-

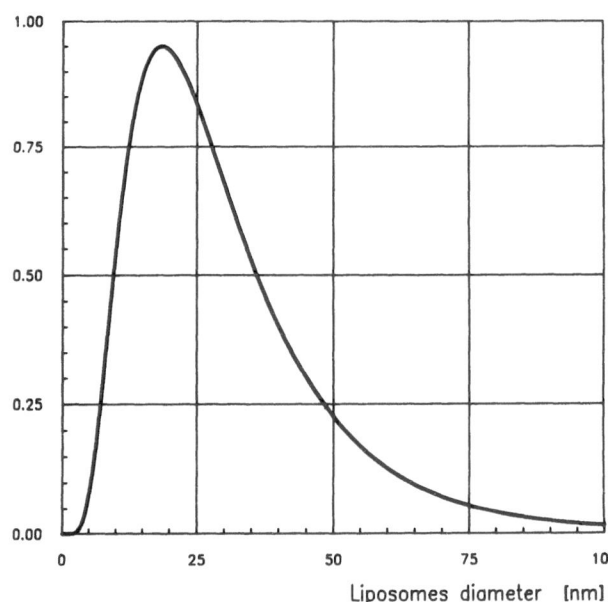

Fig. 3. Liposome size distribution

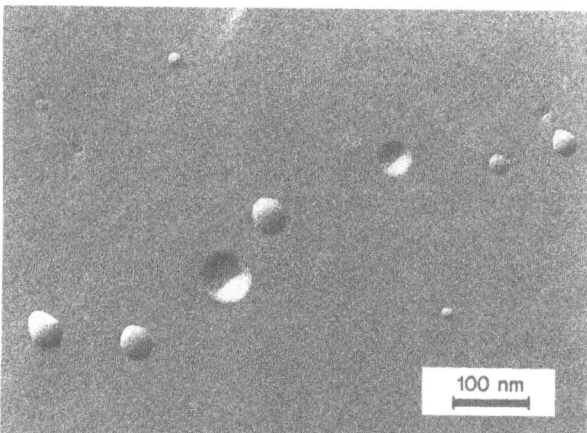

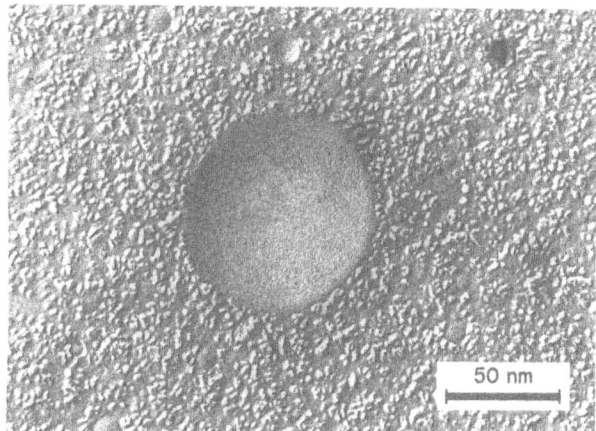

Fig. 4. Freeze-fracture electron microscope photos. An example of electron micrographs of unilamellar liposomes is shown in Fig. 4. These images illustrate the smooth spherical shape of the liposomes obtained by microfluidization, and the trapped volume compartment. The smooth fracture faces are representative of the entire field, and the absence of multistep surfaces unambiguously demonstrates the unilamellar nature of the liposomes. Moreover, the vesicles are seen to be fairly homogeneous in size. This confirms the results obtained by light scattering.

tities of small unilamellar vesicles. According to our own experience, the best characterized unilamellar liposomes were obtained by microfluidization of aqueous suspensions of egg phospholipids. As con-

firmed by light scattering measurements, the main advantages of the use of the Microfluidizer technique are the uniformity of the size distribution and that the liposomes formed are smaller than the smallest easily prepared vesicles.

The composition of lipids may have great importance on the size of the unilamellar vesicles that can be produced. However, the Microfluidizer can operate at considerably higher lipid concentrations than are possible with other techniques.

The Microfluidizer process does not involve the use of organic solvents and the liposomes can be prepared by a continuous process rather than a batch process normally required for other preparations.

Electron microscope pictures of the vesicles obtained by microfluidization show smooth spherical shapes and give clear evidence of the unilamellarity of the liposomes.

References

1. Korstvedt H, Bates R, King J, Siciliano A (1984) Drug and Cosmetic Industry, November
2. Mayhew E, Lazo R, Wail WJ, King J, Green AM (1984) Biochimica & Biophysica Acta 775:169–174
3. Mayhew E, Nikolopoulos G, King J, Siciliano A (1985) Pharmaceutical Manufacturing, August 1985, 18–20
4. Koide K, Karel M (1987) International Journal of Food Science and Technology 22:707–723

Received November, 1988;
accepted February, 1989

Author's address:

G. Masson
Nestle Research Centre
Nestec Ltd.
Vers-chez-le-Blanc
1000 Lausanne 26, Switzerland

Progress in Colloid & Polymer Science   Progr Colloid Polym Sci 79:52–58 (1989)

Mixed monolayers of two polypeptides at the water/air interface

M. Puggelli, G. Gabrielli, and C. Domini

Department of Chemistry, University of Florence, Florence, Italy

Abstract: Mixed monolayers of two polypeptides, poly-γ-glutamate (PγMG) and poly-L-alanine (PLA), in the same α-helix surface comformations were studied at the water-air interface.

Their interphasal arrangement and their α-helix conformation were deduced by determining the spreading π-A isotherms and comparing them with Huggins' theory, by infrared spectra obtained with multiple internal reflection (MIR) of the two polypeptides, and by ellipsometrical measurements carried out on transferred monolayers.

The study of mixed monolayers allows the deduction that the two polypeptides are completely miscible at the water-air interphase, and that they both keep their α-helix conformation in mixtures.

The hydrophobic nature of the attractive interactions in the PLA and PγMG mixtures was demonstrated by applying Joos' theory to the collapse pressures determined at equilibrium conditions.

The collapse of the two polypeptides under equilibrium conditions was also studied using the scanning electron microscope.

The kinetic study of the collapse mechanism, both of the two separate components and for their mixtures, allows conclusions that the process consists essentially in nucleation and growth of a tridimensional phase and that in the collapsed phase, the two polypeptides are also compatible.

Key words: Monolayers, polymer mixtures, collapse of monolayers.

Introduction

In previous works [1] mixed monolayers of non-ionizable synthetic polymers at the liquid/air and the liquid/liquid interfaces have been studied. These studies show that a favorable reciprocal orientation of the hydrophobic chains is the prevailing criterion in determining bidimensional miscibility [2].

Nevertheless, it is known [3] that, depending on the spreading solvent used, it is possible to obtain stable monolayers of synthetic polypeptides at the water/air interface in different macromolecular conformations. It is also known that the collapse studies of mixed monolayers, in either static or dynamic conditions, give useful information concerning the reciprocal miscibility of components both in monolayers and in the collapsed phase [4].

The aim of the present work was to determine the orientation and the conformation of synthetic polypeptides at the water/air interface and to study their reciprocal miscibility in monolayer and in the collapsed phase in order to define the importance of the macromolecular conformation in bidimensional miscibility.

The polypeptides selected for this study were: poly-γ-methyl-L-glutamate (PγMG) and poly-L-alanine (PLA). The choice was justified because these polypeptides are stable in monolayers in the same conformation (α-helix), obtained from the same spreading solvent [4, 5]. Furthermore, the study appeared particularly interesting from a biological point of view because polypeptide mixtures constitute a simple and useful protein model, and these same monolayers are considered to be a useful model of biological membranes.

Experimental

PyMG and PLA, previously studied [6], were the substances used. Their molecular weights were, respectively, 46 000 and 63 000.

Chloroform and dichloroacetic acid (12 %) were the spreading solvents used to obtain the α-helix monolayers, and conditions have already been described by us [6] and by others [7]. Bidistilled water, further purified using the Milli-Q water system apparatus (Organex system, supplied by Millipore), was used as the support. Film compressions were carried out discontinuously; the rate of area compression was $1.7 \cdot 10^{-4}$ to $3.3 \cdot 10^{-4}$ m²/mg s. The following compression modalities were used to study collapse: readings were taken every 0.03 m²/mg, and, as previously described [4], the time interval between two successive readings was gradually increased from 5 to 40 min in order to obtain a thermodynamic equilibrium before each new compression [6].

The spreading isotherms were obtained with the Langmuir method by a computer-controlled Lauda Filmwaage balance, as already described [8]. The collapse kinetics were determined on a Cahn balance [4].

Temperature, controlled at the beginning and the end of each measurement, was maintained thermostatically constant within ± 0.05 °C. Accuracy was within ± 0.01 m²/mg for the surface area and 0.01 mN/m for the surface pressure.

Ellipsometric measurements on transferred films were taken using a Rudolf-research model 437-02 ellipsometer and a program described elsewhere [9].

Transfer from a liquid to a solid support (a layer of chromium evaporated onto a glass plate) was obtained with a Joyce-Loebl Model 4 apparatus using the Blodgett-Langmuir method. The pressure chosen for the transfer corresponds to the most condensed phase possible before collapse in order that the transfer be quantitative and not alter the surface conformations.

To determine the MIR spectra, the monolayers were transferred onto Germanium plates which were inserted into a Perkin-Elmer Model 983 spectrometer.

The scanning electron micrographs were taken with a JSM-13 electronic microscope on collapsed films as previously described [4].

Results and discussion

The results obtained for the two polymers and their mixtures will be considered separately.

Pure polymers

In order to determine the conformation of PyMG and of PLA at the water/air interface their spreading isotherms were determined at 15°, 20°, 25°, and 30 °C. Figure 1 shows, as an example, the PyMG and PLA isotherms at 25 °C. As can be seen, around 20 mN/m the PyMG isotherm showed a net "plateau" of

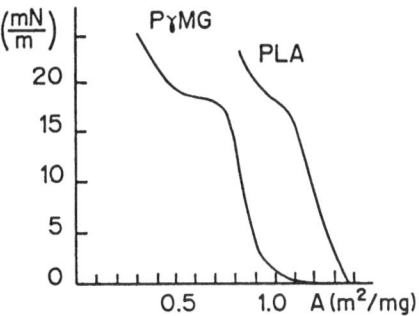

Fig. 1. Spreading isotherms at 25 °C of PyMG and PLA

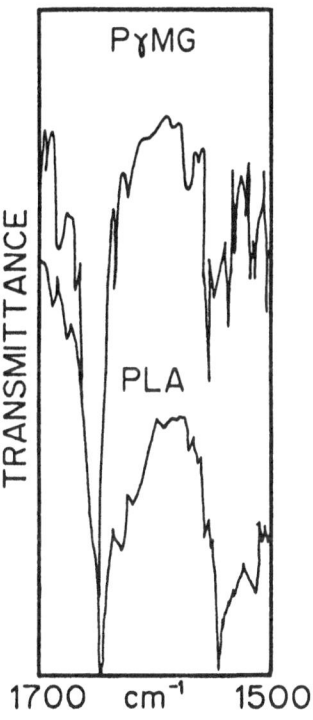

Fig. 2. MIR spectra of PyMG and PLA

Table 1. Huggins' parameters at 25 °C

	PyMG low pressure	high pressure	PLA low pressure	high pressure
ψ	-3.2	-24.5	-3.9	-14.3
K'	4.40	9.80	3.80	5.80
K	0.48	0.29	0.51	0.41
ε_Δ	40 000	16 500	600	3 700

ε_Δ expressed in ergs 10^8

π, whereas the PLA isotherm had an inflection. Both the "plateau" and the inflection are attributable to the α-helix conformation as already observed [3, 4].

To obtain further information concerning the distribution and orientation of the macromolecules at the interface, the experimental isotherms were compared to those calculated using Huggins equation [10]. As described elsewhere [11], to obtain a good fit it is necessary to consider each isotherm as having two separate parts. Table 1 reports the parameters for PyMG and for PLA at 25 °C, both at high and at low pressures. Comparison of the single parameters has been amply discussed elsewhere [6] and there is general agreement

that the two polymers exist in a rigid form on the surface; this allows their transfer from a liquid support to a solid one using the Blodgett-Langmuir technique and their later examination with MIR spectroscopy. Figure 2 shows the two peaks characteristic for α-helices at about 1654 cm^{-1} and 1550 cm^{-1}.

Further information concerning the surface conformation of the two polymers was obtained by studying the collapse of their monolayers under both static and dynamic conditions. The MIR spectra were performed on films of PyMG and PLA transferred at a surface pressure above the "plateau" pressure or the inflection, and they were identical to those obtained at

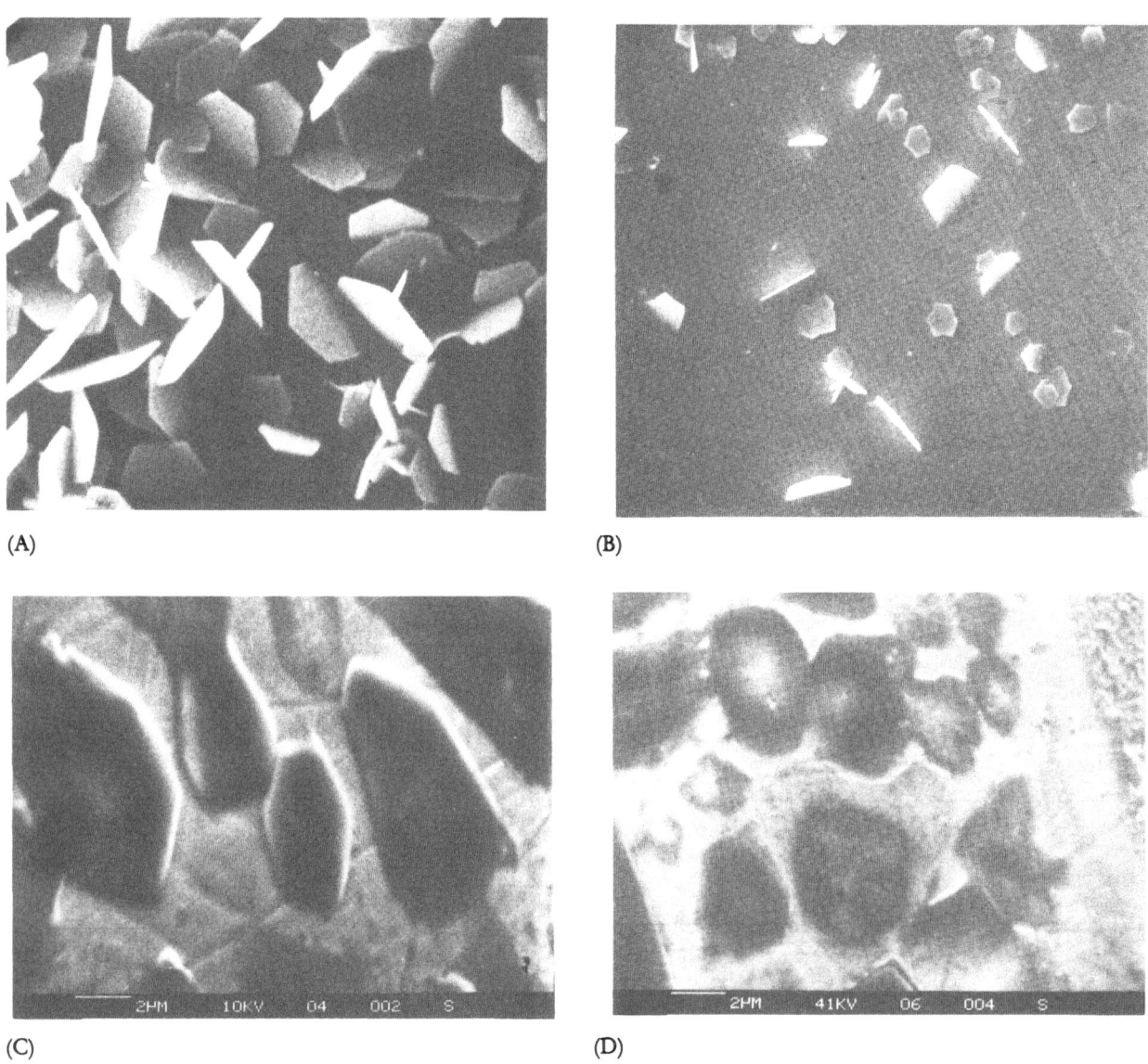

(A)

(B)

(C)

(D)

Fig. 3. Electron scanning micrographs of collapsed monolayers of PyMG (A and B) and PLA (C and D)

Table 2. Activation energies (*E* Kcal/mol), free energies (*ΔG** Kcal/mol), enthalpies (*ΔH** Kcal/mol), entropies (*ΔS** cal/K mol)

Area	PLA k				E	ΔG*	ΔH*	ΔS*	PyMG E	ΔG*	ΔH*	ΔS*
	15 °C	20 °C	25 °C	30 °C								
First	0.34	0.38	0.47	0.50	1.99	17.84	4.17	−45.87	4.0	18.1	3.5	−49.1
Second	0.31	0.37	0.44	0.44	1.58	16.10	3.67	−41.65	4.0	18.2	3.4	−49.5
Third	0.32	0.36	0.40	0.40	0.44	17.90	3.18	−49.41	3.8	18.2	3.5	−49.6

lower pressures and particularly below that of the "plateau". This confirmed that the "plateau" could not be attributed to a conformational transition from an α-helix to a β-helix. The collapsed monolayers, that is those with pressures definitely above 20 mN/m, were also observed using scanning electron microscopy (SEM) and the microscopies are reported in Fig. 3. As can be seen, both PyMG and PLA showed the same hexagonal structure.

The ellipsometrical measurements of the film thickness before and after transition for PyMG [11] (10 and 19 Å) and for PLA [12] (4–5 and 9–10 Å) allowed us to confirm a horizontal conformation at the water/air interface characteristic of the α-helix and to assign a "plateau" or inflection to a particular type of collapse: that is, the passage from a mono- to a bilayer. This is characteristic of the α-helices and has been previously found by us [4] and by others [5]. The diversity of the formal aspect of the isotherms at collapse has already been found by others [13] and is attributed to the different flexibility or length of the side chain, as was the case in our situation.

The study of the collapse kinetics for both polymers was performed by determining the surface pressure in relation to time at a constant area. As done previously [14], the collapse was considered to be the chemical reaction of a compound A (bidimensional A → tridimensional A) whose kinetic trend could be evaluated by the surface pressure decrease (proportional to the amount of substance that is separated) in relation to time. At three separate surface areas and at 15°, 20°, 25°, and 30°C the kinetic trend for PLA and for the first part (about 10–15 min) of PyMG agreed with the Prout-Tompkins law [14] and indicated a process of nucleation and growth of the nucleus formed with interference between the polymer chains during growth. However, for the second part of PyMG the relation which best agreed with the experimental data was a zero or second-order relation [11] which repre-

sented the formation and sliding on a preceding monolayer followed by a mechanism of nucleation and growth of a tridimensional phase. The second stage of the collapse process did not appear in PLA: this could be attributed to the diverse flexibility and length of the side chain which can prevent the second layer from sliding on the first. Table 2 reports the "*k*" rate constants for both polymers which are represented by the slopes of the relation $\pi = f(t)$. As can be easily seen, at all temperatures studied, the *k*-rate constants had the same order of magnitude but varied with temperature variations: this demonstrated that the collapse process was activated.

The ΔH^*, ΔS^*, and ΔG^* values for PLA were also deduced from the activation energies by applying the transition state theory. These are reported together with the previously deduced PyMG values [11] in Table 2. As can be seen from the table, these values were also the same order of magnitude for the two polymers in the α-helix form.

Furthermore, for PLA and for the first stage of PyMG the value of ΔG^* remained more or less constant as the area chosen for the kinetic decreased. This was due to maximum entropic and minimum enthalpic contributions in the lowest area caused by the obviously more ordered arrangement present in this area.

Mixtures

The MIR spectra of the PLA and the PyMG mixtures obtained, (as for the pure components) from films transferred onto solid supports showed the same bands typical of the α-helix conformations.

Figure 4 shows the plots of the π-A isotherms of the same mixtures in different molar ratios previously determined [6]. The trend of the areas in relation to the molar ratios between moles of monomeric units at 15°, 20°, 25°, and 30°C and at 10 mN/m of pressure was obtained. Negative deviations from additivity

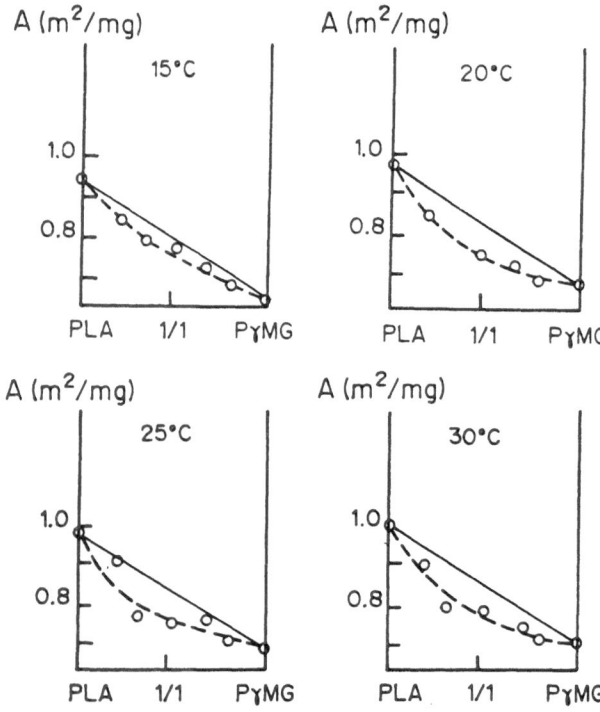

Fig. 4. Plot of surface area A (m²/mg) against molar ratios of PLA/PγMG mixture at 15°, 20°, 25°, and 30°C

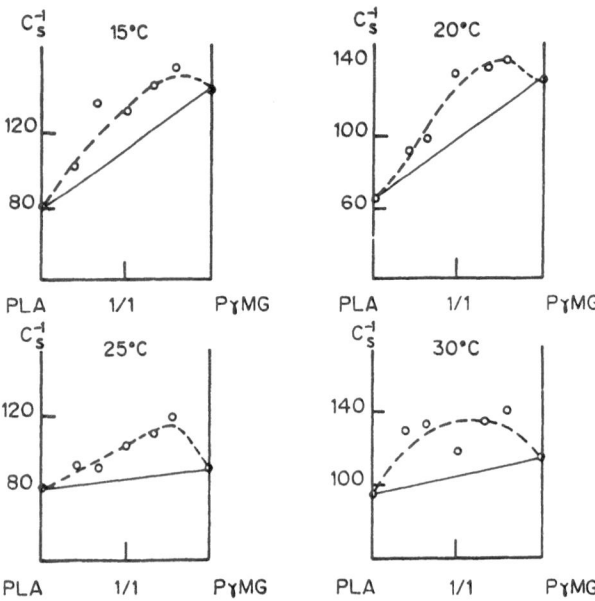

Fig. 5. Plot of surface compressional modulus C_s^{-1} (dys/cm) as a function of molar ratios at 15°, 20°, 25°, and 30°C

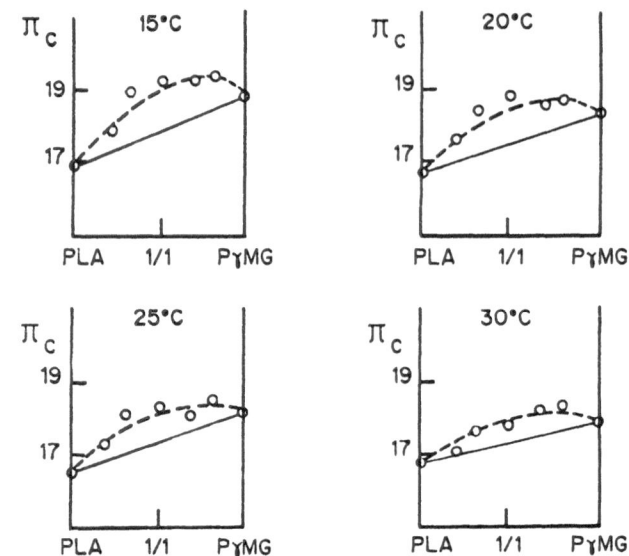

Fig. 6. Plot of collapse pressure π_c against molar ratios at 15°, 20°, 25°, and 30°C

were noted at all temperatures and this constituted an initial indication that the two components were miscible with prevailing attractive interactions.

To confirm this initial result, the surface compressional moduli, defined as $C_s^{-1} = -A\left(\dfrac{\partial \pi}{\partial A}\right)_T$, for mixtures having different molar ratios were calculated, and the results are shown in Fig. 5. The positive deviations from additivity indicated the presence of more condensed phases for the mixtures than for the pure components and, therefore, the possibility of attractive interactions. Since the two components presented all the miscibility characteristics, the free surface energy of mixing ΔG_m, was computed by graphic integration of the isotherms, as done in previous works [2, 6]. The values of ΔG_m were negative at all temperatures for all mixtures, demonstrating that the mixtures were more stable than the pure components.

Furthermore, since ΔG_m was almost the same at all temperatures examined, the entropic contribution of mixtures should be considered negligible and, therefore, the only determinable contribution was enthalpic [6].

As for the pure components, further information concerning the mixtures was also obtained by studying the collapse of the respective monolayers under both static and dynamic conditions. Figure 6 reports the π_c values as a function of the molar ratios at 15°, 20°, 25°, and 30°C. As can be seen these varied as the molar ratios varied. On the other hand, the isotherms of the mixtures in the same ratios presented only one collapse pressure value, and this ulteriorly confirmed

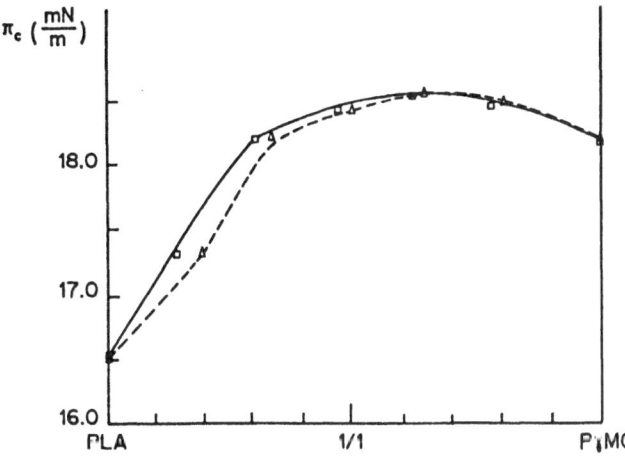

Fig. 7. Plot of collapse pressure π_c against molar ratios in monolayer ($\triangle$---$\triangle$) and molar ratios in the collapsed phase (O———O) at 25 °C

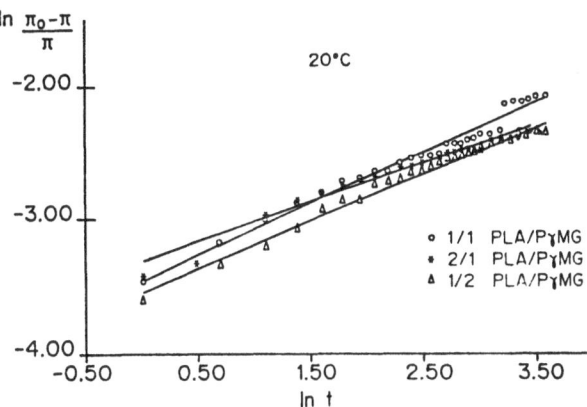

Fig. 8. Agreement of the experimental data with the Prout-Tompkins equation (straight line) at 20 °C of three mixtures in different molar ratios

the total miscibility of PLA and PyMG. Furthermore, the π_c values, which were higher for all the mixtures than for the pure components, demonstrated that the mixtures of the two polymers were more stable than the pure components. To have more quantitative information about the increased stability of the mixtures, the attraction energies between the macromolecular segments in the zone where this is maximal, (that is, at the collapse pressure) was calculated using Joos's theory [15] as previously described [6].

The values of these energies are determined from the parameter ξ of the Joos equation [15] considering z, the bidimensional coordination number, (usually considered equal to 6 for low molecular weight substances) considered equal to 2 from a comparison of Huggins' theory with Singer's relationship which considers monolayers to be pseudo-lattice and, therefore, characterized by a coordination number. The value of 2 on the other hand, agreed with the presence on the surface of rigid macromolecular segments such as the α-helices. Furthermore, since z was nearly identical for both polymers, it was also considered invariable for all mixtures. The values obtained for these energies were all quite small — about 0.1 cal/mole, in agreement with the existence of only hydrophobic interactions between the two polypeptides, as reported elsewhere [6]. Knowledge of these parameters and of the π_c for the single components and for their mixtures allowed calculation of the respective compositions of the two polypeptides in the collapsed phase. Figure 7 shows the trend of the collapse pressure in relation to the molar ratios in the monolayer and in the collapsed

phase. The plot was typical of complete miscibility of the two phases and demonstrated that there was always separation in the collapsed phase of the mixtures richest in PLA: that is, the component with the lowest collapse pressure.

Furthermore, for a ratio between moles of monomeric units PyMG/PLA = 2/1 there are mixtures in monolayer and in the collapsed phase of the same composition, such an azeotrope, as already revealed by Joos for other compounds [15].

As for the pure components, the kinetics of the collapse process was also determined for the mixtures. As for pure PLA, a kinetic law of the Prout-Tompkins-type best represented the experimental data. Figure 8 shows as an example the agreement between the experimental data and the Prout-Tompkins relation for the three mixtures at 20 °C. As can be observed, agreement was acceptable and therefore it could be assumed that even for the mixtures the collapse mechanism was made up of a process of nucleation and growth. This unique and equal mechanism for all the mixtures further confirmed miscibility in the collapsed phase and in monolayer. Since the formal constants of rate of the process varied with temperature variations, it was possible to obtain both the activation energies and the values of ΔH^*, ΔS^*, and ΔG^*, as done for the pure PLA. The activation energies and the values of ΔH^*, ΔS^*, and ΔG^* for all mixtures were constant for the whole process and the same order of magnitude: from this it could be affirmed that their kinetic behavior was that of a single component. This could be considered an indication of the presence of a single phase, or rather, of the miscibility between components.

Conclusions

From the reported results, the following conclusions can be drawn:

1) The two polypeptides considered, PLA and PγMG, could be obtained at the water/air interface in the α-helix form, and this form was maintained in the bidimensional mixtures;

2) The study of the mixtures indicated miscibility between the two components both in monolayers and in the collapsed phase;

3) The interactions that determined miscibility were attractive and prevalently hydrophobic due to the similar interphasal arrangement of the hydrophobic chains;

4) The collapse mechnism indicated a process of nucleation and growth, usually activated for both the pure components and for their mixtures;

5) The information reported in the preceding points indicates the possibility of additionally defining conformations and miscibility between components that constitute simple and useful protein models.

References

1. Gabrielli G, Puggelli M, Ferroni E (1972) Berichte Int Kongress Grenzflakenaktive Stoffe Band II C, Verlag H München; (1974) J Colloid Interface Sci 47:145; Gabrielli G, Maddii A (1978) J Colloid Interface Sci 64:19
2. Gabrielli G, Puggelli M, Baglioni P (1982) J Coll Interface Sci 86:485
3. Gabrielli G, Puggelli M (1975) Adv Chem Ser 144:347; Malcolm BR (1968) Proc Roy Soc London A305:363; (1973) Proc Surface Membrane Sci, Academic Press 7:183-229; (1985) Thin Solid Films 134:201
4. Gabrielli G, Baglioni P, Fabbrini A (1981) Coll Surface 3:147; Gabrielli G, Baglioni P (1981) J Coll Interface Sci 83:221; Baglioni P, Dei L, Ferroni E, Gabrielli G (1985) J Colloid Interface Sci 109:109
5. Malcolm BR (1962) Nature 195:901; (1966) Polymer 7:595; Loeb GL (1968) J Colloid Interface Sci 26:223; Albert A, Cordoba J (1984) Colloid Polym Sci 262:811
6. Gabrielli G, Puggelli M, Dei L, Domini C (1988) Colloid Polym Sci 266:429
7. Loeb GI, Baier RE (1968) J Colloid Interface Sci 27:38; Malcolm BR (1971) J Polymer Sci C34:87
8. Baglioni P, Carlà M, Dei L, Martini E (1987) J Phys Chem 91:1460
9. McCracking FL (1960) Technical Note US 479
10. Huggins ML (1973) Kolloid Z Z Polym 251:449
11. Baglioni P, Dei L, Gabrielli G (1983) J Colloid Interface Sci 93:402
12. Puggelli M, Gabrielli G, Caminati G (1989) Colloid Polymer Sci 267:65
13. Malcolm BR (1975) In: Applied Chemistry at Protein Interfaces. Adv Chem Series 145:338
14. Gabrielli G, Baglioni P, Ferroni E (1981) J Colloid Interface Sci 81:139
15. Joos P, Demel RA (1969) Biochim Biophys Acta 183:447; Joos P (1969) Bull Soc Chim Belges 78:207

Reveived October, 1988;
accepted February, 1989

Authors' address:

G. Gabrielli
Dipartimento di Chimica
Via Gino Capponi 9
I-50121-Firenze, Italy

Progress in Colloid & Polymer Science   Progr Colloid Polym Sci 79:59–63 (1989)

Enzyme activity and cation exchange as tools for the study of the conformation of proteins adsorbed on mineral surfaces

H. Quiquampoix[1][2]), P. Chassin[1]), and R. G. Ratcliffe[2])

[1]) Station de Science du Sol, I.N.R.A., Versailles, France
[2]) Department of Plant Sciences, University of Oxford, Oxford, U.K.

Abstract: The modification of the conformation of proteins adsorbed on different mineral surfaces was followed indirectly using two methods. Firstly, the effect of the pH on enzyme activity of two β-D-glucosidases on different mineral surfaces was studied; a diminished activity would indicate a departure from the active conformation. The mineral surfaces were chosen for their electrical charge and hydrophobicity/hydrophilicity (montmorillonite, talc, goethite). Secondly, the exchange of paramagnetic charge compensating cations on adsorption of bovine serum albumin was studied by an NMR method and the specific interfacial area occupied by the protein on the clay surface was deduced.

The results with montmorillonite showed that below the i.e.p. the proteins were adsorbed up to the saturation level of the surface and adsorbed proteins unfolded; above the i.e.p. the surface coverage decreased and there was no major change in the conformation. In addition to these electrostatic interactions, the comparison between the results obtained with goethite and talc showed that hydrophobic interactions can also induce conformational changes.

Key words: Protein adsorption, protein conformation, electronegative surfaces, hydrophobic surfaces, nuclear magnetic resonance.

Introduction

The adsorption of a protein on a negatively charged surface often shows a maximum near the isoelectric point (i.e.p.) of the macromolecule [1–4]. Several explanations of this phenomenon have been proposed, including i) competition between the protein and protons in the solution for the negative sites on the surface below the i.e.p., and a coulombic repulsion of the protein above the i.e.p. [1]; ii) an increase in the lateral coulombic repulsions between adsorbed proteins above and below the i.e.p., decreasing the surface coverage [2]; and iii) unfolding of the adsorbed protein above and below the i.e.p. as a result of internal electrostatic repulsions [3, 4]. Recent work on the adsorption of a sweet almond β-D-glucosidase suggests that the pH dependence of the adsorption may be best understood in terms of an unfolding of the protein below the i.e.p., as a result of electrostatic interactions between the protein and the surface, and an electrostatic repulsion of the protein above the i.e.p. [5, 6]. The generality of this fourth explanation is the subject of this paper.

It is not possible to determine the exact three-dimensional structure of a protein adsorbed on a solid surface and, in order to assess the conformational changes that occur on adsorption, we have chosen to follow two parameters that are more or less directly related to the protein structure. The first parameter is enzyme activity, a decrease of which indicates a departure from the active conformation of the protein. The second parameter is the amount of charge-compensating cations released from the clay surface on protein adsorption. This last parameter, divided by the quantity of adsorbed protein, gives an estimate of the specific area occupied by a single protein molecule on the surface and allows conformational changes to be detected.

Materials and methods

Proteins

Bovine serum albumin (BSA; A-7638), sweet almond β-D-glucosidase (G-8625, type II) and *Aspergillus niger* β-D-glucosidase, which is actually an impurity contained in a commercial preparation of cellulase (C-7377, type I), were all obtained from Sigma.

Minerals

A Wyoming montmorillonite sample with a size fraction < 2 μm was saturated with sodium by the method described previously [5]. A synthetic goethite was prepared by the method of Schwertmann et al. [7]. A chlorite-free talc was obtained by collecting the white parts from a talc sample from Luzenac (France). The absence of chlorite was checked by IR spectroscopy.

Enzyme activity measurements

The experimental procedures have been described elsewhere [5] and it is only necessary to summarize them here. Procedure A is a measurement of the enzyme activity without adsorbent surfaces and acts as a control. Procedure B measures the enzyme activity in the presence of adsorbent surfaces, but all the enzyme is not necessarily adsorbed. Procedure C measures the activity in the supernatant after centrifugation and represents the contribution of non-adsorbed enzyme to the overall activity measured by procedure B. The concentrations were 8.35 μg·ml⁻¹ for sweet almond β-D-glucosidase, 385 μg·ml⁻¹ for *A. niger* β-D-glucosidase and 3 mg·ml⁻¹ for montmorillonite, goethite, and talc.

NMR measurement of manganese displaced from montmorillonite by BSA

The adsorption of proteins on smectites displaces charge compensating cations [1]. The release of manganese from montmorillonite during BSA adsorption was measured by ^{31}P nuclear magnetic resonance (NMR) spectroscopy. The montmorillonite was pre-equilibrated with Mn^{2+} and the release of the paramagnetic ion was detected by its line-broadening effect on the signal from orthophosphate in the suspending medium. This method, which is sensitive and does not require a centrifugation step, will be fully described elsewhere [8]. The NMR spectra were recorded at 121.49 MHz on a Bruker CXP 300 NMR spectrometer.

The amount of BSA adsorbed on the montmorillonite was measured by centrifuging the suspension and measuring the UV adsorption of the supernatant at 279 nm.

Results

Adsorption and activity of the two β-D-glucosidases on montmorillonite

Figure 1 shows the results obtained with the procedures A, B, and C on *A. niger* β-D-glucosidase. A comparison of curves A and B below pH 5 shows a reduction in the catalytic activity of the adsorbed enzyme; whereas a comparison of the curves A, B, and C at a pH higher than 6 shows that no enzyme was adsorbed on montmorillonite in this pH range.

In order to ease the comparison and quantification of these effects Fig. 2 plots the relative activity of the adsorbed enzyme R and the proportion of enzyme not adsorbed F against pH. R is the ratio of the activity due to the fraction adsorbed (B–C) and that of an equal quantity of enzyme in solution (A–C), namely $R = 100 (B–C)/(A–C)$. The non-adsorbed proportion of enzyme is $F = 100 \, C/A$.

For *A. niger* β-D-glucosidase, Fig. 2 reveals a linear increase of R from pH 2 to pH 5 and an increase in F

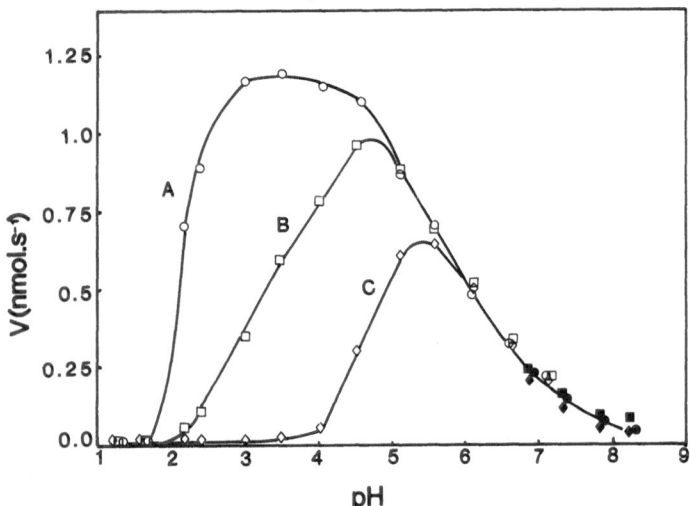

Fig. 1. Effect of montmorillonite on the activity of *A. niger* β-D-glucosidase against pH. O: control (A); □: in the presence of montmorillonite (B); ◇: supernatant (C). Open symbols for citrate buffer; closed symbols for phosphate buffer

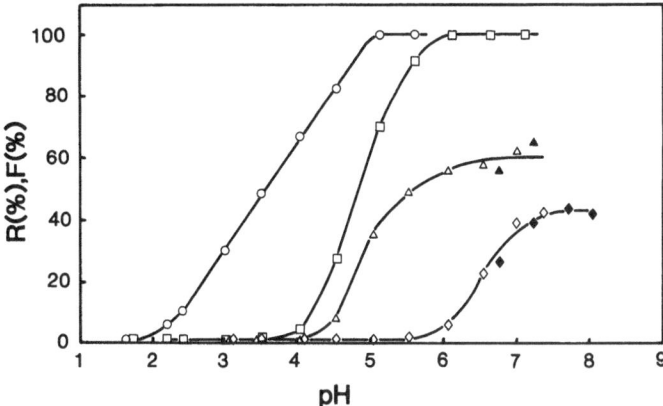

Fig. 2. Effect of montmorillonite on the relative activity (R) of adsorbed β-D-glucosidase and on the proportion of non-adsorbed β-D-glucosidase (F) against pH. *A. niger*: R (O), F ($\square$); sweet almond: R ($\triangle$), F ($\diamond$). Open symbols for citrate buffer; closed symbols for phosphate buffer

from pH 4 to pH 6. The maximum value obtained was 100% for both parameters. For sweet almond β-D-glucosidase, Fig. 2 shows a similar pH dependence for the evolution of R and F, but R and F reach a plateau lower than 100% at higher pH.

Activity of sweet almond β-D-glucosidase on talc and goethite

Figure 3 shows the activity of sweet almond β-D-glucosidase in solution and adsorbed on talc, goethite, or montmorillonite. The parameter A × $(B-C)/(A-C)$ for adsorbed enzyme and the parameter A for

enzyme in solution are plotted in order to compare equal amounts of enzyme. It can be seen that the inhibition is higher with talc than with goethite, but that talc and goethite lack the very marked inhibition observed with montmorillonite at acid pH.

Adsorption of BSA on montmorillonite and displacement of charge compensating cations

Figure 4 shows that the maximun amount adsorbed Γ_{max} vs. pH presents a maximum near the i.e.p. of the BSA at pH 4.5. The fraction of Mn^{2+} displaced from the clay surface by the protein shows a different pat-

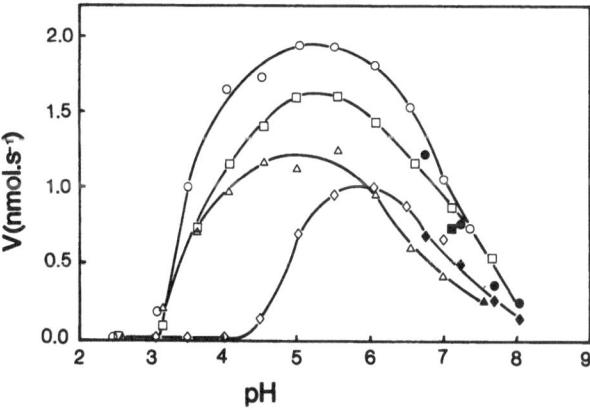

Fig. 3. Effect of goethite, talc, and montmorillonite on the activity of sweet almond β-D-glucosidase against pH. Activities of equal amounts of enzyme in the solution and in the adsorbed state are plotted. Control in solution (O), adsorbed on goethite ($\square$), on talc ($\triangle$), and on montmorillonite ($\diamond$). Open symbols for citrate buffer; closed symbols for phosphate buffer

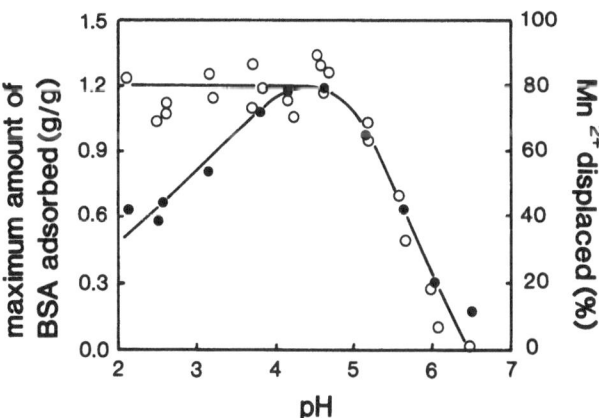

Fig. 4. Maximum amount of bovine serum albumin adsorbed on montmorillonite ($\bullet$) and fraction of Mn^{2+} displaced by bovine serum albumin from the montmorillonite surface (O) against pH

tern: below pH 4.5 we observe a plateau at 80% and above pH 4.5 the displaced fraction decreases until 0%. Interestingly, this decrease follows the alkaline branch of the curve Γ_{max} vs. pH if the plateau of the exchange curve and the maximum of the Γ_{max} curve are normalized to the same value.

Discussion

1. Relative importance of electrostatic and hydrophobic interactions on the conformation of adsorbed enzymes

1.1. Electrostatic interaction

It has been shown previously [5] that the decrease in the activity of an enzyme adsorbed on montmorillonite was due to a modification of the protein conformation. Montmorillonite is a 2 : 1 phyllosilicate with a high specific area (800 $m^2 \cdot g^{-1}$) and the major part of this area is constituted from basal surfaces bearing a permanent electronegative charge due to isomorphic substitutions in the octahedral alumina layer. This electric charge is independent of the pH, in contrast to the pH-dependent charge of a protein and this simplifies the interpretation of the electrostatic interactions between the surface and the macromolecule.

The maximum quantity of protein adsorbed on montmorillonite depends on the nature of the protein, the pH and the ionic strength, but it is typically between 0.5 and 2 $g \cdot g^{-1}$ [1]. The quantities of protein used (2.78· 10^{-3} $g \cdot g^{-1}$ for the sweet almond β-D-glucosidase and 0.128 $g \cdot g^{-1}$ for the *A. niger* β-D-glucosidase) are too low to saturate the montmorillonite surfaces and the lateral interactions between adsorbed macromolecules are thus kept to a minimum.

The *A. niger* β-D-glucosidase appears to be a good example of a protein for which electrostatic interactions with the surface are most important. Below the i.e.p., which is at pH 4.0 [9], all the enzyme is adsorbed (Fig. 2). Above the i.e.p., the free fraction F of the protein increased progressively until the point at which no enzyme was adsorbed. In this case the protein is electronegative like the mineral surface and the electrostatic interaction becomes repulsive, decreasing the amount adsorbed. The decrease in the relative activity R of the enzyme with decreasing pH (Fig. 2) indicates a progressive change in the conformation of the adsorbed protein. When the pH falls below the i.e.p., the positive charge on the enzyme increases and the attractive coulombic interaction with the electronegative surface causes an unfolding of the protein.

The sweet almond β-D-glucosidase shows qualita-

tively the same pattern, confirming the important role of the coulombic interactions. However, an exact analysis is more complex than with the *A. niger* β-D-glucosidase because the sweet almond β-D-glucosidase is composed of two isoenzymes with approximatively the same molecular weight [10] and the same optimal pH [11], but with different isoelectric points: 4.4 and 6.4 [10]. Nevertheless, as shown in Fig. 2, below pH 4, when both isoenzymes bear a positive charge, $R = 0$, and above pH 6 when they begin to bear a negative charge, F increases. In contrast to the *A. niger* β-D-glucosidase parameters R and F did not reach 100%. The maximum values of R and F were 60% and 45%, respectively, indicating a significant involvment of forces other than electrostatic.

1.2. Hydrophobic interaction

The hydrophilic residues in a protein are generally found on the surface of the molecule, whereas the hydrophobic residues are buried inside, protected from contact with the water. A conformational change that brings the hydrophobic residues from the core of the protein into contact with a hydrophobic surface may well be energetically favored by an increase in entropy [4, 12, 13]. The hydrophobic residues in a native protein are often found in α-helices and the reduction in the α-helix content in the adsorbed protein will increase the rotational freedom of the protein chain.

The results in Fig. 3 for sweet almond β-D-glucosidase confirm the existence of such hydrophobic interactions. Talc is a hydrophobic solid bearing no electrical charge on its basal surfaces. Its surface free energy consists only in its dispersive component which originates from the surface oxygen of the tetrahedral sheet [14]. Goethite is hydrophilic and with variable charge, but in the presence of citrate buffer the net charge is low in the pH range employed here because of the complexation of the oxyanions with surface hydroxyls [15]. Two deductions can be drawn from the data in Fig. 3. First, the lower enzyme activity on the talc than on the goethite shows the existence of a hydrophobic interaction. Second, the fact that the talc did not provoke the high decrease of activity at acid pH induced by the electronegative montmorillonite confirms the importance of electrostatic interactions. We can conclude that the two types of interaction act together in the modification of the conformation of the adsorbed sweet almond β-D-glucosidase, but that the relative importance varies with both the pH and the ionic strength of the suspending medium.

2. Interpretation of the maximum of adsorption of proteins on electronegative surfaces near their i.e.p.

2.1. Phenomenon occuring below the i.e.p.

Figure 4 shows that below the i.e.p. of the BSA (~ pH 4.5) the maximum amount adsorbed Γ_{max} decreases progressively, but the amount of Mn^{2+} displaced by the protein from the clay surface remains constant. If the amount of displaced cation is taken as a measure of the occupied area of the surface, then it can be seen that as the pH decreases below the i.e.p., the specific interfacial area of the protein-clay contact increases, showing that the coulombic interactions cause an unfolding of the protein. This independent method confirms the interpretation obtained from the study of the activity of the adsorbed enzymes.

2.2. Phenomenon occuring above the i.e.p.

Figure 4 shows that above the i.e.p., Γ_{max} and the amount of displaced Mn^{2+} decreased in the same proportion. We interpret this as a reduction of the surface coverage as the pH increases above the i.e.p., without extensive modification of conformation of the BSA, as far as we can judge from the fact that the specific interfacial area remains approximatively constant. Here too, this new approach confirms the results obtained from the proportion of free enzyme F (Fig. 2) at high pH when the ratio of protein: clay is such that the amount of protein is far too low to saturate the clay surface.

Conclusion

When a protein interacts with a highly electronegative surface like montmorillonite, electrostatic forces play a leading role. For two of the three proteins studied here (*A. niger* β-D-glucosidase and BSA) these electrostatic forces dominate, but in the case of the sweet almond β-D-glucosidase we need to suppose that both electrostatic and hydrophobic interactions are important.

When the pH and ionic strength are such that the forces acting between the surface and the protein are highly attractive, the protein unfolds and spreads over the surface. If it is an enzyme, it loses its catalytic activity. When the conditions are such that the forces are repulsive, the coverage of the surface decreases. These two mechanism acting in opposite directions explain the maximum in protein adsorption that occurs on electronegative (or symetrically electropositive) surfaces near the isoelectric point of the protein.

References

1. Mc Laren AD, Peterson GH, Barshad I (1958) Soil Sci Soc Am Proc 22:239-244
2. Eirich FR (1977) J Colloid Interface Sci 58:423-436
3. Norde W, Lyklema J (1978) J Colloid Interface Sci 66:257-265
4. Norde W (1986) Adv Colloid Interface Sci 25:267-340
5. Quiquampoix H (1987) Biochimie 69:753-763
6. Quiquampoix H (1987) Biochimie 69:765-771
7. Schwertmann U, Cambier P, Murad E (1985) Clays Clay Miner 33:369-378
8. Quiquampoix H, Ratcliffe RG, Manuscript in preparation
9. Mc Cleary BV, Harrington J (1988) Methods Enzymol 160:575-583
10. Jdanov YuA, Kessler RM, Yakubova NR, Cherstnev KB (1977) Biokhimiya 42:26-33
11. Nanasi P, Kandra-Berti L, Vuong P (1976) Acta Biol Debrecina 13:11-15
12. Norde W, Lyklema J (1978) J Colloid Interface Sci 66:295-302
13. Norde W, Lyklema J (1979) J Colloid Interface Sci 71:350-366
14. Chassin P, Jouany C, Quiquampoix H (1986) Clay Miner 21:899-907
15. Bowden JW, Nagarajah S, Barrow NJ, Posner AM, Quirk JP (1980) Aust J Soil Res 18:49-60

Received November, 1988; accepted February, 1989

Authors' address:

Hervé Quiquampoix
Station de Science du Sol, I.N.R.A.
Route de Saint-Cyr
F-78026 Versailles Cedex, France

Progress in Colloid & Polymer Science   Progr Colloid Polym Sci 79:64–69 (1989)

Electrokinetic and stability properties of cholesterol in aqueous NaCl and NaCl + bile salt solutions

J. Salcedo, A. Delgado, and F. González-Caballero

Departamento de Física Aplicada, Facultad de Ciencias, Universidad de Granada, Granada, Spain

Abstract: Experimental data on the electrophoretic mobility of cholesterol particles in aqueous NaCl and NaCl + bile salt solutions are employed for the estimation of the zeta potential (ζ) of the particles in these media. The results obtained through the use of the well known Smoluchowski formula are compared with those deduced from a more elaborate treatment [11] so as to study the relative influence of the relaxation and electrophoretic effects for these systems. The use of the parameter $\zeta^2 \varkappa^{-1}$ ($\varkappa^{-1}$ being the double layer thickness) allows us to correlate the electrokinetic properties of cholesterol with its tendency to coagulate in suspension. The influence of pH and NaCl concentration on this parameter is first discussed. Furthermore, the zeta potential of cholesterol in the presence of a series of bile salts is used to study the stability conditions of such suspensions, the results being discussed in terms of the characteristics of the different bile salts in solution.

Key words: Electrophoresis, cholesterol, bile salt, stability.

Introduction

Cholesterol is an important lipid not only because it is an essential component of the biological structure, but also because of the important functions it carries out in the human organism. Due to the relationship of cholesterol to many diseases such as atherosclerosis or gallstone formation, most investigations on this substance have focussed on biological or clinical aspects. Physicochemical studies of a more fundamental character are considerably more scarce.

Cholelithiasis is the term used to describe a number of processes involved in the formation of stones or calculi from the bile components, and their subsequent deposit in the bile ducts. From the point of view of interest for this work, it is worth noting that the primary step in the formation of a cholesterol calculus is the existence of an unstable bile with a high cholesterol content. The growing process is not precisely determined, and, although some theories have been elaborated, it can be considered a matter under investigation. The important point in the context of the present work is the presence of solid material immersed in an aqueous solution of high complexity. The surface of the solid particles consists mainly of cholesterol and their growing will depend primarily on the characteristics of the solid/solution interface. The aggregation process must be connected with the electrical structure of that interface. Hence, it seems of interest to study what conditions determine the alteration of the electrical state of cholesterol particles in such a complex medium as bile is.

In order to ensure stability, every colloidal particle must retain its individuality without joining other particles. This can be achieved, for example, via adsorption of electrolyte or polyelectrolyte molecules or ions on the colloid. Such an adsorption can favor the existence of important repulsion (mainly electrostatic) between the particles to prevent their approach to short distances where attractive van der Waals forces are intense.

It is well known that electrokinetic phenomena are an important source of experimental information on the electrical properties of interfaces. In this paper, electrophoresis is used to estimate the electric potential

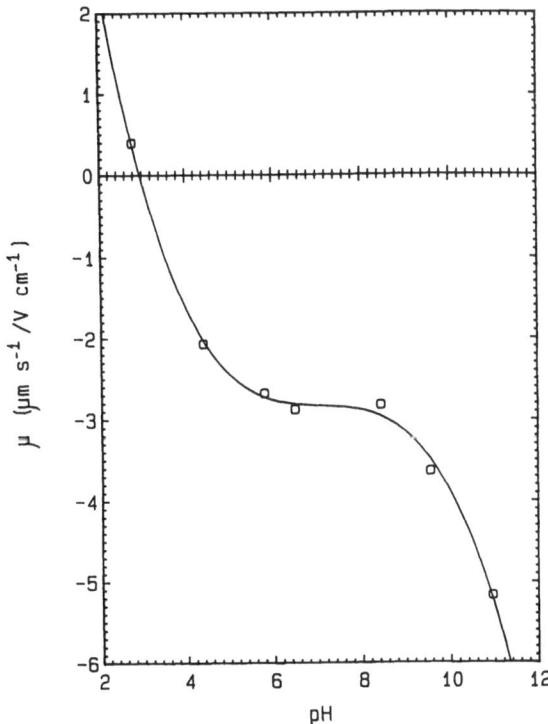

Fig. 1. Electrophoretic mobility of cholesterol as a function of pH in the presence of 5×10^{-6} M NaCl

(ζ) at the cholesterol surface for different pH values and different concentrations of bile salt in the dispersion medium. A rapid information of the stability of such systems can be obtained by means of the index proposed by Eilers and Korff [1]. This index is given by the product $\zeta^2 \varkappa^{-1}$ ($\varkappa^{-1}$ being the double layer thickness). We will discuss the stability conditions of the cholesterol suspensions under study by means of that index.

Experimental

The anhydrous cholesterol sample employed was obtained from Carlo Erba (Italy) with "USP-Cod. Franc." quality. Sodium glycocholate (NaGC), glycodeoxycholate (NaGDC), glycochenodeoxycholate (NaGCDC), taurocholate (NaTC), taurolithocholate (NaTLC), taurodeoxycholate (NaTDC) and taurochenodeoxycholate (NaTCDC) were from Sigma (USA). NaOH and HCl used for adjusting the pH of the suspensions were from Merck (Germany) and Carlo Erba, respectively. Their quality, like that of NaCl (Carlo Erba) was the highest available. All these chemicals were used as received, without further purification. Water used in all the experiments was doubly distilled in an all-Pyrex still and filtered through 0.2 µm membranes. Its specific conductivity was always lower than 10^{-6} Scm^{-1}.

Cholesterol dispersions for microelectrophoresis determination were prepared by dilution of concentrated stock dispersions. These were obtained by dispersing 500 mg of cholesterol in one liter of water, and siphoning off the 10 upper cm of the suspension (not including the surface) after 7 h sedimentation. Particles with a spherical equivalent radius lower than 5.4 µm were selected, according to Stokes equation, for the experiments.

A commercial microelectrophoresis apparatus (Zeta-Meter, USA) was used for the determination of the electrophoretic mobility of cholesterol. The details of the method followed have been given elsewhere [2]. About 30 particles were timed for each direction of the applied electric field, the electrophoretic mobility finally assigned to each system being the average of at least three experiments carried out on different days and using always fresh sample. The cylindrical electrophoresis cell was immersed in a thermostated cell maintained at 25.0 ± 0.1°C by means of a Haake F3K (F.R.G.) circulating bath.

Results and discussion

Mobility

Figure 1 shows the variation of the electrophoretic mobility, μ, of cholesterol with the pH of the dispersion medium at a constant NaCl concentration (similar results, not shown here, were obtained with NaCl concentrations up to 5×10^{-3} M). The symbols correspond to experimental data, while the line has been obtained by least squares fitting of the data, according to the equation:

$$\mu = \sum_{n=0}^{3} a_n \, (\text{pH})^n . \tag{1}$$

A simple approach to the charge generation mechanism in cholesterol/water systems can be given. It is known [3] that cholesterol molecules have a dipole moment of approximately 1.99 debye. In the crystal structure at 25 °C, they are arranged in the form of alternately hydrophobic and hydrophilic bilayers [4, 5]. Thus, when immersed in water, cholesterol particles lead water dipoles to be oriented in the vicinity of their surfaces [6]. The combined effect of adsorbed water dipoles and ions coming from water dissociation gives rise to an electric double layer cholesterol/water, responsible for the electrophoretic mobility of the particles.

At neutral pH, the electrophoretic mobility of cholesterol is negative (Fig. 1), this indicating that the main contribution to the electric double layer generation is the preferred orientation of adsorbed water dipoles.

Results in Fig. 1 show that μ is strongly dependent on pH. Increasing H$^+$ concentration in cholesterol suspensions originates a decreasing effect of $|\mu|$, and

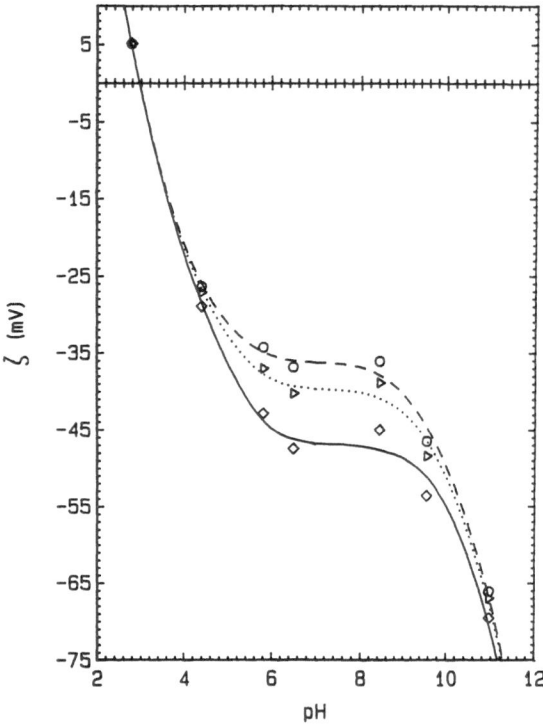

Fig. 2. Zeta potential of cholesterol as a function of pH in the presence of 5×10^{-6} M NaCl. (O, - - -): Smoluchowski; ($\triangle$, $\cdots$): O'Brien and White, $r = 5.4$ μm; ($\diamond$, ——): O'Brien and White, $r = 1.4$ μm

even reverses its sign at very low pH's. This points to the potential determining character of H^+ ions for cholesterol particles. The pH at which μ becomes zero (isoelectric point) was found to be around pH 3. Similarly, OH^- ions can be considered as potential determining ions because of their effect on μ shown in Fig. 1. Similar results were found by different authors [7–10], although the experimental conditions were different. The very weak effect of NaCl on the mobility found in this work, suggests that it behaves as an indifferent electrolyte for the cholesterol/solution interface.

Zeta potential

Two different theoretical approaches have been used for the calculation of the zeta potential of cholesterol i) the Smoluchowski formula, valid in the limit of plane interfaces:

$$\mu = (\varepsilon/\eta)\,\zeta, \tag{2}$$

where ε is the dielectric permittivity of the liquid and η its viscosity and ii): the theory elaborated by O'Brien and White [11], which can be applied to spherical par-

ticles of arbitrary radius, and that, under certain conditions, reproduces the results of Smoluchowski model. The application of O'Brien and White theory requires a previous estimate of the particle radius of the cholesterol employed. As already mentioned, the Stokes equivalent diameter was 5.4 μm. On the other hand, by comparison between the experimental mobility of cholesterol in different dispersion media [12] and the maximum mobilities predicted by O'Brien and White theory [11], a minimum radius of 1.4 μm is deduced. This value is close to the minimum size obtained from scanning electron microscope pictures, 1.0 ± 0.4 μm [12]. The model will be used in this paper assuming both hypotheses about the equivalent spherical radius of cholesterol, viz., 5.4 and 1.4 μm. Thus, the zeta potential of cholesterol corresponding to each pH, NaCl concentration and model can be calculated, as shown in Fig. 2 for 5×10^{-6} M NaCl concentration. In the positive mobility region, there is practically no difference between the models. This fact can be explained by the important ionic strength of the medium for that pH interval, and the consequent small thickness of the diffuse double layer, which changes between 30 Å at pH = 2 and 170 Å at pH = 3.5. Both figures are several orders of magnitude below the estimated particle radius, and therefore correspond to the validity interval of Smoluchowski's equation. Thus, the relationship between electrophoretic mobility and zeta potential is practically independent of pH in such conditions, and hence, the observed variation of μ with pH must be essentially due to changes in the activity of potential determining ions in the inner part of the double layer.

In the acid pH and negative mobility region, the differences between the models begin to be noticeable and increase with pH. The double layer thickness now increases from 75 Å at pH = 2.8 to 1344 Å at pH 6.9. That increase in the relation double-layer thickness/ particle radius provokes an increase in the importance of the relaxation effect. It is well known that such effect gives rise to a retardation in the particle motion, and hence, in our case to a higher absolute value of zeta potential for a given mobility. A different effect not considered in Smoluchowski's formula is the so-called electrophoretic retardation, which depends on the concentration, charge, and mobility of the ions present. Such effect also implies an increase in $|\zeta|$. The separation between the curves corresponding to Smoluchowski and O'Brien and White models in Fig. 2 gives an idea of the relative importance of these retardation effects.

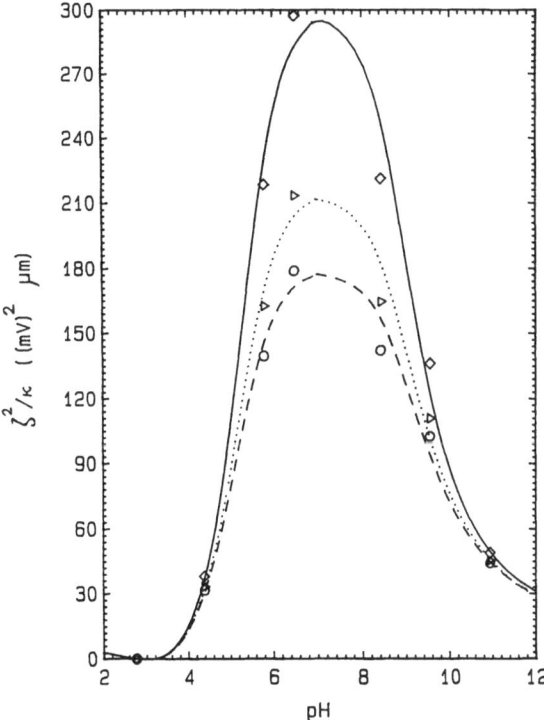

Fig. 3. Stability index of cholesterol suspensions as a function of pH in the presence of 5×10^{-6} M NaCl. O and - - - : Smoluchowski; $\triangle$ and $\cdots$: O'Brien and White, $r = 5.4$ μm; □ and —— : O'Brien and White, $r = 1.4$ μm

by changing the concentration or type of electrolyte the stability index will change due to the modifications in ζ and $\varkappa$. Eilers and Korff [1] found that the onset of instability can be associated to values of $\zeta^2 \varkappa^{-1}$ between 10 and 100 mV² μm.

Figure 3 shows the variation of the index with pH for different NaCl concentrations. Note that for the lowest concentration, 5×10^{-6} M, the maximum stability of the cholesterol suspension corresponds to neutral pH (the bile pH is slightly lower). This was to be expected since one has the lowest ionic strength (and hence the largest double-layer thickness) at this pH. Furthermore, the O'Brien and White model, with radius 1.4 μm, predicts the maximum value of $|\zeta|$, i.e., the maximum electrostatic repulsion between the particles. At lower pH values, the stability index decreases sharply, due to the decrease in both $\varkappa^{-1}$ and $|\zeta|$ (see Fig. 2) upon acid addition to lower the pH. At high pH values the stability index shows a decrease (less sharp than in the acid range) with pH: the system is again unstable even though ζ is relatively high in absolute value. The decrease in double-layer thickness appears now as the essential factor responsible for the lack of stability of cholesterol suspensions.

Two significant facts are worth noting when the concentration of NaCl in the system is increased (not shown here for brevity). Firstly, the maximum stability index decreases with concentration due to the increase in ionic strength. Secondly, the maximum shifts to higher pH values, due to the fact that the highest $|\zeta|$ values are estimated in the basic pH range. This behavior agrees with the experimental results reported by Saad and Higuchi [15]. These authors found that at pH > 6.5 the rates of growth and dissolution of cholesterol at constant ionic strength were markedly retarded. They also reported that at pH > 8 the processes were practically inhibited for at least 72 h, even though the supersaturation of cholesterol was 100-times higher than its solubility limit.

In the basic pH region, the situation is reversed. Thus, the maximum differences between the models appears for the lower basic pHs. Such differences become less important as pH is raised. As before, the decrease in double-layer thickness from 1344 Å at pH 7.1 to 30 Å at pH 12 will explain that behavior. The hypothesis of an increasing adsorption of OH⁻ ions when the pH is higher seems to be confirmed by the calculations of zeta potential.

Stability of the suspensions

An empirical relationship between ζ and the coagulation behavior of the suspension was proposed by Eilers and Korff [1], who showed that the onset of instability was associated with a rapid decrease in the value of the quantity $\zeta^2 \varkappa^{-1}$. The theoretical justification of this rule is found in the DLVO theory of colloidal stability. In particular, Derjaguin [13, 14] showed that there is important instability when $\zeta^2 \varkappa^{-1} < A\varepsilon^{-1}$, A being the Hamaker constant. For a given system, the values of A and ε are practically constant. Therefore,

Effect of bile salt addition on the zeta potential of cholesterol

Figures 4 and 5 show the variation of the zeta potential of cholesterol as a function of the concentration of different taurine- and glycine-conjugated bile salts. A constant NaCl concentration of 10^{-2} M was added to all the systems to maintain approximately constant the ionic strength of the suspensions. In this way, zeta potential changes will be essentially due to the different affinity of the bile salts for cholesterol surface. Note

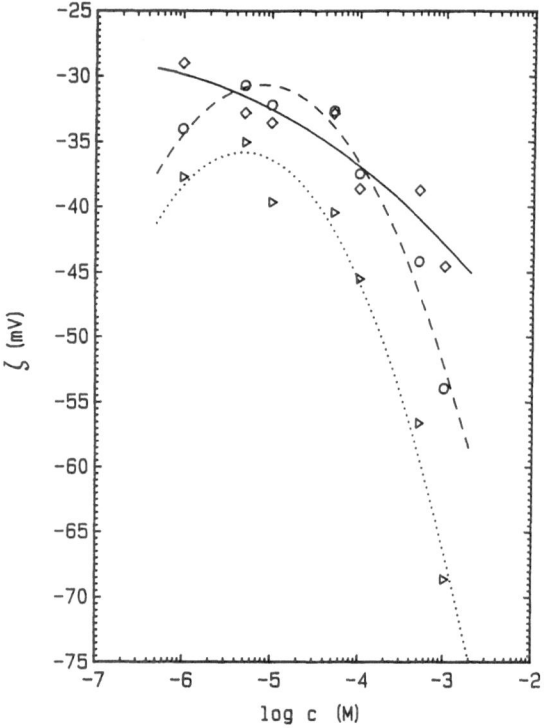

Fig. 4. Zeta potential of cholesterol in aqueous solution 10^{-2} M NaCl as a function of the concentration of different bile salts. (O, ---): NaGCDC; (Δ, ···): NaGDC; (◇, —): NaGC

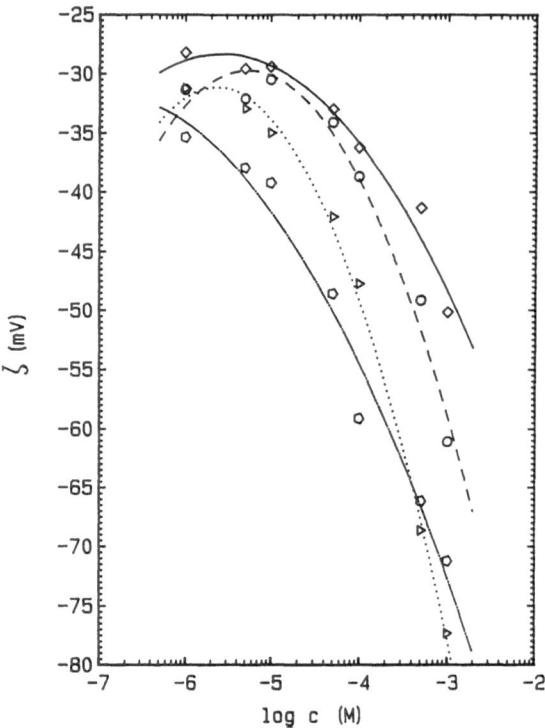

Fig. 5. Zeta potential of cholesterol in aqueous solution 10^{-2} M NaCl as a function of the concentration of different bile salts. (O, ---): NaTCDC; (Δ, ···): NaTDC; (◇, —): NaTC; (O, —·—·—·): NaTLC

too that an increase in $|\zeta|$ will automatically mean an increase in the value of the stability index and hence a better stability of the system. Furthermore, such ionic strength is high enough for the differences between the three calculations of zeta potential described above to become negligible. For this reason, only ζ values obtained from Smoluchowski formula have been included in Figs. 4 and 5.

The comparison between the results obtained with the different salts shows that ζ increases in absolute value in most cases according to the sequence:

$$\text{NaGC} < \text{NaTC} < \text{NaGCDC} < \text{NaTCDC}$$

$$< \text{NaGDC} < \text{NaTLC} < \text{NaTDC} .$$

This will also be the sequence of increasing stability of cholesterol in the presence of these bile salts. However, the trends of variation of ζ with concentration are very similar for all the salts studied. The reason for this lies probably in the fact that the hydrophobic parts of their molecules are practically identical. The

differences must be due to the different structures of their lateral chains and different number of hydroxyl groups. The following points are worth noting

a) Each bile salt conjugated with taurine (polar head SO_3^-) gives a higher $|\zeta|$ value than the corresponding bile salt conjugated with glycine (polar head COO^-).

b) In salts of the CDC type, the hydroxyl groups in positions 3α and 7α, are closer to each other than they are in DC salts (positions 3α and 12α, see [16]). This makes CDC salts more hydrophobic than DC salts: the former are less adsorbed than the latter and hence they give lower $|\zeta|$ values, as observed for glycocholates (Fig. 4) and taurocholates (Fig. 5).

c) It can be seen in Fig. 5 that, at low concentrations, $|\zeta|$ increases according to the sequence:

$$\text{NaTC} < \text{NaTDC} < \text{NaTLC}$$

and at high concentrations the order is

$$\text{NaTC} < \text{NaTLC} < \text{NaTDC} .$$

These bile salts differ precisely in their number of hydroxyl groups. It can be expected that the physical properties in which these groups are involved will be ordered in the same way for conjugated and ordinary bile salts. The dipole moment is one of such properties. For cholic acid (HC; three OH groups) the moment is 3.84 debye; for deoxycholic acid (HDC; two groups), 3.22 debye, and for lithocholic acid (HLC; one OH group) the dipole moment is 2.5 debye [17]. Hence, the hydrophobic character of these acids will increase in the order

HLC < HDC < HC

and in consequence the capacity of these acids to adsorb onto cholesterol surface will be ordered as follows:

HC < HDC < HLC

in agreement with the observed behavior of conjugated salts at low concentrations. At high concentrations that ordering partly disappears since $|\zeta|$ is higher for NaTDC than for NaTLC. This happens precisely for concentrations higher than 10^{-4} M, a value close to the critical micelle concentration of these bile salts. The above result would indicate that NaTDC micelles are better adsorbed on cholesterol surface than NaTLC micelles.

Acknowledgement

Financial support by CAICYT, Spain, under project No. 1228/84 is gratefully acknowledged.

References

1. Eilers H, Korff J (1940) Trans Faraday Soc 36:229
2. Delgado A, González-Caballero F, Salcedo J, Cabrerizo MA (1988) Mater Chem Phys 19:327
3. Kumler WD (1945) J Amer Chem Soc 67:1901
4. Craven BM (1976) Nature 260:727
5. Shieh HS, Hoard LG, Nordmand CE (1977) Nature 267:287
6. Shabd R, Upadhyay BM (1981) Carbohydrate Research 93:191
7. Douglas HW, Shaw DJ (1957) Trans Faraday Soc 53:512
8. Whikehart DR, Lees MB (1971) Biochem Biophys Acta 231:561
9. Hollingshead S, Johnson GA, Pethica BA (1965) Trans Faraday Soc 61:577
10. Seaman (1958) Thesis, Cambridge
11. O'Brien RW, White LR (1978) J Chem Soc Faraday Trans II 74:1607
12. Salcedo J (1988) Ph D Thesis, University of Granada, Spain
13. Derjaguin BV (1940) Trans Faraday Soc 36:203
14. Derjaguin BV (1940) Trans Faraday Soc 36:730
15. Saad HY, Higuchi WI (1965) J Pharm Sci 54:1303
16. O'Connor CJ, Wallace RG (1985) Adv Colloid Interface Sci 22:1
17. Kumler WD, Halverstadt IF (1942) J Amer Chem Soc 64:1941

Received October, 1988;
accepted February, 1989

Authors' address:

Dr. F. González-Caballero
Departamento de Física Aplicada
Facultad de Ciencias
Universidad de Granada
E-18071 Granada, Spain

Progress in Colloid & Polymer Science

Progr Colloid Polym Sci 79:70–75 (1989)

Phase behavior of cationic lipoaminoacid surfactant systems

C. Solans, R. Infante, N. Azemar, and T. Wärnheim

Instituto de Tecnología Química y Textil (C.S.I.C.), Barcelona, Spain
Institute for Surface Chemistry, Stockholm, Sweden

Abstract: Alkyl esters of long chain basic aminoacids are known as cationic surfactants which have a very good solubility in water; many of them possess antimicrobial properties and are generally considered milder and less irritant than other surfactants.

Long chain N^{α}-acyl-L-α-amino-ω-guanidine alkyl acid derivatives have recently been synthesized. Physico-chemical and antimicrobial studies of these compounds as a function of the alkyl ester or sodium salt (R), the straight chain length of the fatty acid residue (x) and the number of carbons between the ω-guanidine and α-carboxyl group (n) were carried out. Among the different aminoacid surfactant derivatives synthesized, the methyl ester of N^{α}-lauroyl arginine (LAM) showed higher activity of both surface and antimicrobial properties.

In this study, some fundamental studies on LAM phase behavior in binary and multicomponent systems have been undertaken. The phase equilibria has been determined in the binary LAM/water and ternary LAM/water/alkanol systems. Solubilization of nonpolar compounds such as hydrocarbons has also been investigated.

Key words: Lipoaminoacid, surfactants, liquid crystals, phase diagrams, micellar solutions, microemulsions.

1. Introduction

The increasing concern about environmental protection in recent years has prompted the development of surfactants with a higher degree of biocompatible properties [1]. In this respect lipoaminoacid surfactants constitute an interesting class of surfactants because, due to their chemical structure, it can be expected that they are milder and much less irritant than other commonly used surfactants [2].

Although different kinds of lipoaminoacid surfactants have been synthesized [3, 4] and some of them are already used in practical applications, mainly in cosmetic and food technologies [4], only a few studies of their properties have been reported [3, 5]. In particular there is no information concerning their phase behavior in binary H_2O/lipoaminoacid surfactant and ternary H_2O/lipoaminoacid surfactant/organic solvent systems. Because of their peculiar chemical character a different phase behavior from that of ordinary surfactants is to be expected.

Lipoaminoacid surfactants have been synthesized by our team for the last few years [6]. Among the different aminoacid surfactant derivatives, a cationic one, the methyl ester of N^{α}-lauroyl arginine (LAM) (I) showed higher activity on both surface and antimicrobial properties.

$$CH_3\text{-}(CH_2)_{10}\text{-}CO\text{-}NH\text{-}\overset{\overset{\textstyle COOCH_3}{|}}{CH}\text{-}(CH_2)_3\text{-}NH\text{-}\overset{\overset{\textstyle \diagup NH_2}{}}{\underset{\diagdown NH_2}{C\oplus}} \cdot Cl^-$$

As a part of a systematic study on phase behavior of lipoaminoacid surfactant systems, preliminary results concerning LAM systems are reported.

2. Experimental

Materials

N^{α}-Lauroyl Arginine methyl ester hydrochloride salt (LAM): was synthesized in our laboratories according to the method previously described [6]. n-Pentanol (Fluka) and toluene (Merck) were purist grade, and water was twice distilled.

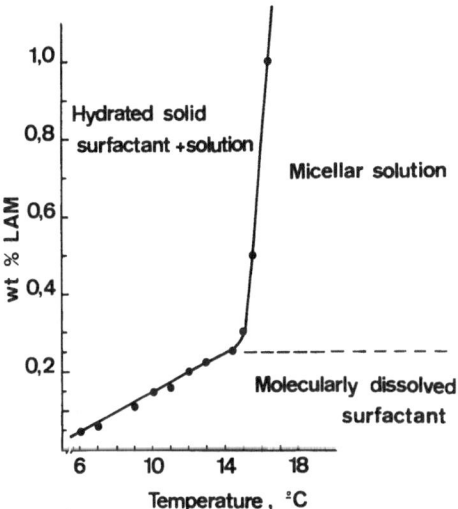

Fig. 1. Solubility curve for the LAM/water system as a function of temperature

Methods

– *Phase Diagrams* were determined by weighing all components in ampules which were sealed, homogenized by stirring, and placed under thermostated conditions. Presence of liquid crystalline phases was determined by observing the samples under polarized light.

– *Optical microscopy* was carried out by means of a polarizing microscope equipped with a hot stage.

– *Self-diffusion measurements:* Self-diffusion measurements were performed according to the FT-NMR pulsed gradient spin echo method [7] on a JEOL FX-100, operating at 100 MHz for 1 h, and at the ambient probe temperature 23° C. Experimental conditions and parameters have been described previously [8]. The accuracy of the measurements is estimated to 5 %.

3. Results

Binary LAM/H₂O system

The solution behavior of LAM in water as a function of temperature is shown in Fig. 1. The melting point of the hydrated solid surfactant, or Kraft Point, is 14.5° C. The shape of the curve in Fig. 1, which represents the solubility curve, is similar to that of typical ionic surfactants [9]. Aqueous solutions of LAM experience supercooling phenomenon. When cooled at temperatures lower than the melting point of the hydrated solid surfactant they remain unchanged for several days before the surfactant solidifies. Analogous behavior has been described for other cationic [10] and anionic [11] surfactant solutions.

In order to know the number and type of mesophases formed in the LAM/H₂O system, optical microscopy examinations were performed according to the "flooding" or "penetration method" of Lawrence [12, 13]. In a "flooding" experiment water is allowed to diffuse into anhydrous surfactant placed between a slide and a cover-slip. After a short time, gradients in composition are produced and the different meso-

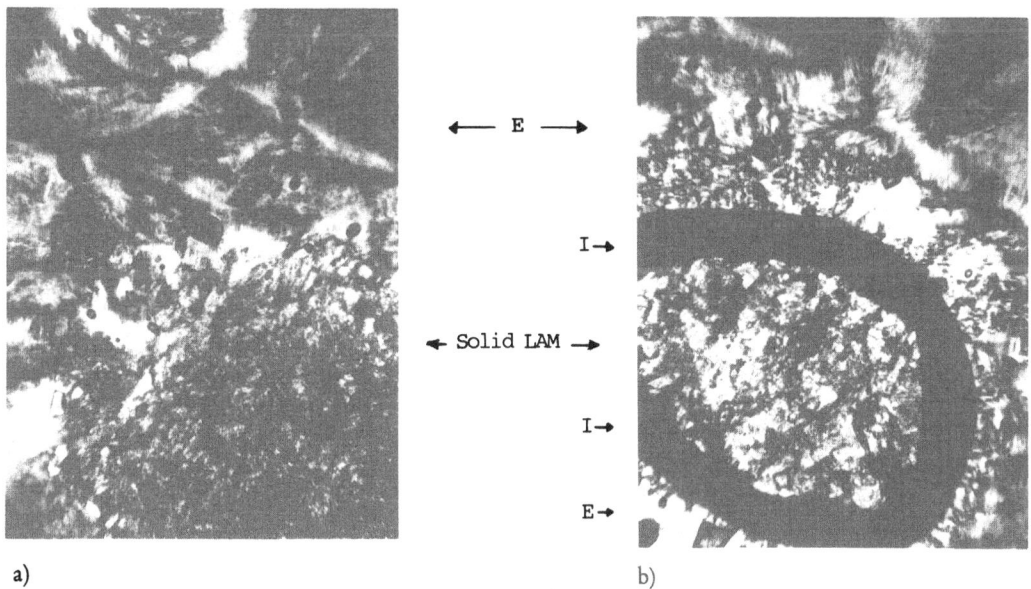

Fig. 2. Optical micrograph illustrating the phases formed by penetration of water into crystalline LAM under polarized light: a) at 30°C and b) at 40 °C; (magnification: *x* 60; *E*: hexagonal liquid crystal; *I*: isotropic viscous (cubic) liquid crystal)

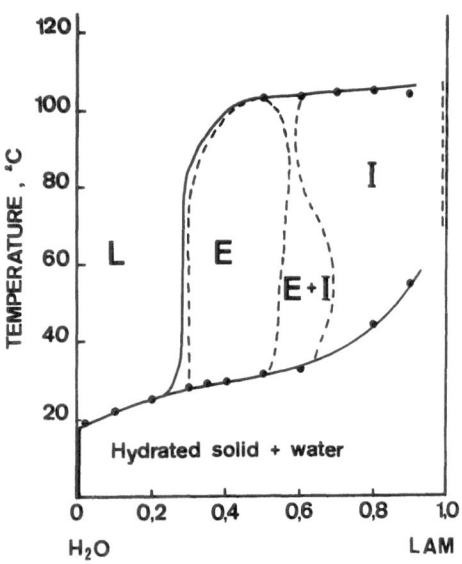

Fig. 3. Phase diagram for the system LAM/H$_2$O. *L*: micellar solution; *E*: hexagonal liquid crystalline phase; *I*: isotropic viscous (cubic) phase

phases develop as separate rings around crystalline surfactant. At 20 °C only solid LAM and water were observed. By increasing the temperature, two different bands were observed, each one forming at a definite temperature. The first one, in order of increasing surfactant concentration, developed at 28.5 °C and showed typical hexagonal liquid crystalline pattern. The second one, which occurred at 38 °C, was isotrop-

ic and very viscous as revealed by pressing the coverslip. Consequently it corresponds to a cubic phase. Figure 2a shows that a 30 °C only hexagonal liquid crystalline phase and solid surfactant are present while at 40 °C (Fig. 2b) an isotropic phase has developed between the hexagonal phase and solid surfactant.

The phase behavior of LAM/H$_2$O mixtures as a function of temperature, is indicated by the partial phase diagram of Fig. 3. The main features of this diagram are: a) the presence of an isotropic solution (*L*) over a wide range of temperature (20 °C and 100 °C) at low concentration of surfactant; b) the occurrence of a hexagonal liquid crystalline phase (*E*) between approximately 30 % and 55 % LAM; and c) the existence of a viscous isotropic (cubic) liquid crystalline region (*I*) above 68 % LAM.

The exact boundaries of the liquid crystalline phases as well as the two phase regions between them and between *L* and *E* regions have not been determined yet. The melting temperature of the hexagonal and cubic liquid crystalline phases was in the range 103 °–105 °C, giving rise to an isotropic liquid phase. Phase behavior of LAM at temperatures higher than 125 °C cannot be determined due to hydrolysis of LAM in solution.

Ternary and quaternary microemulsion systems

The phase diagrams for the ternary LAM/water/pentanol-system at two different temperatures, 25 °C

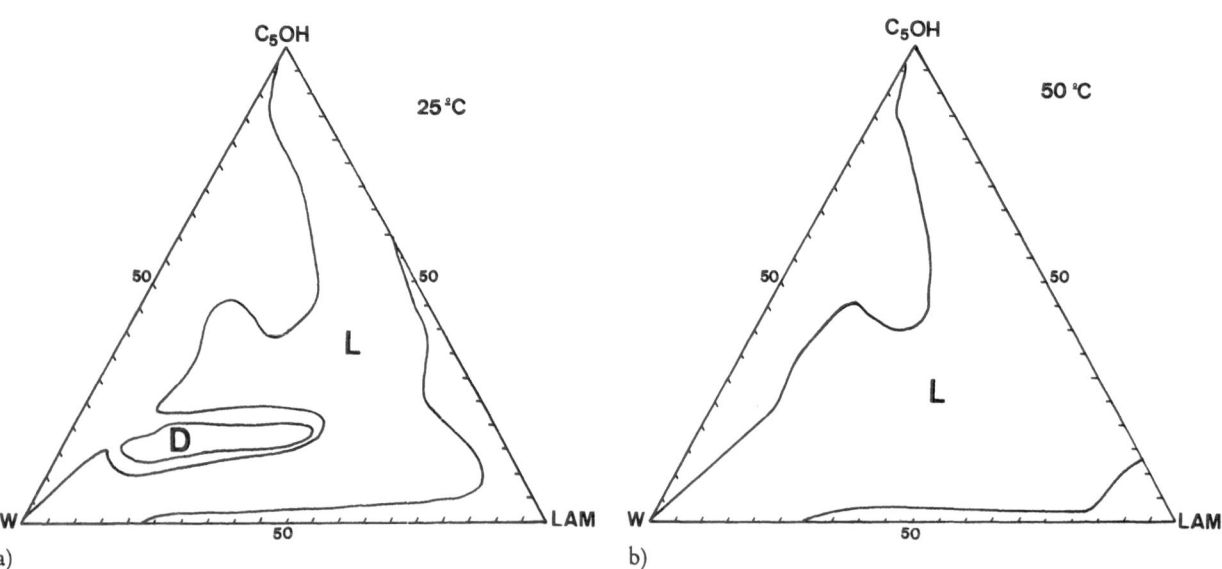

Fig. 4. Phase diagrams of the ternary LAM/H$_2$O/pentanol system. a) at 25 °C and b) at 50 °C. *L*: isotropic liquid solution; *D*: lamellar liquid crystalline phase

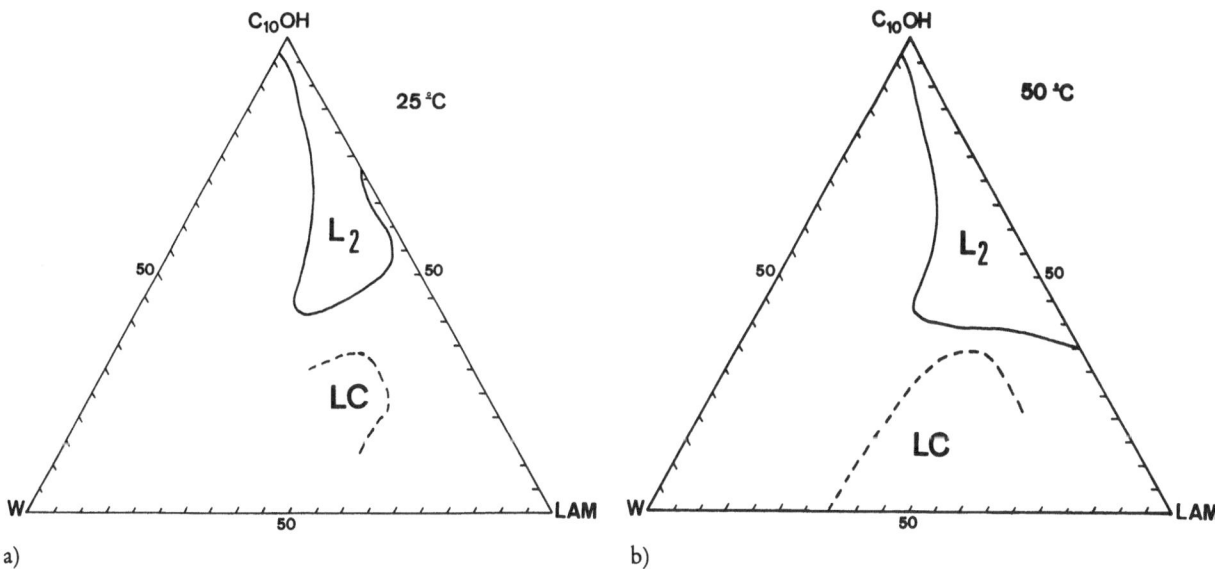

Fig. 5. Phase diagrams of the ternary LAM/H$_2$O/decanol system: a) at 25 °C and b) at 50 °C. L_2: inverse micellar solution; *LC*: liquid crystalline phases

and. 50 °C, is show in Figs. 4a and b. At 25 °C, the phase diagram is characterized by a very large solution phase region with a small lamellar liquid crystalline phase in the water-rich region. No liquid crystalline phases are formed on the binary water/surfactant axis at that temperature (cf. binary phase diagram). At 50 °C, the lamellar liquid crystalline phase has melted into an isotropic solution phase, while the hexagonal and viscous isotropic (cubic) phase, extending from the water/surfactant axis, is able to incorporate only some wt % pentanol. The partial phase diagrams for the ternary LAM/water/decanol-system at 25 °C and 50 °C is shown in Figs. 5a and b. Only a small solution phase L_2 is formed with decanol as cosurfactant, while the phase diagram is dominated by one- and multiphase regions with liquid crystalline phases.

The general behavior of the ternary system does to some extent conform to previously examined systems [14]. The influence of the alcohol cosurfactant chain length has been studied in great detail, and the growth of the liquid crystalline phases at expense of the solution phases with increasing chain length is anticipated for any ionic surfactant, and indeed found here as well. However, one point which by no means is expected is very large extension of the solution phase in the pentanol system. The normal behavior is that two separate solution phases, extending from the water (L_1-phase) and the alcohol corner (L_2-phase), respectively, are formed. The lamellar liquid crystalline phase found in these systems is usually stable over a wide concentration and temperature interval. The most evident explanation for this behavior can be based on simple packing considerations for the surfactant and the change in head-group/head-group interactions upon adding an uncharged but polar cosurfactant [15]. While the repulsion prohibits a closer packing of the surfactant, the hydrophobic interaction will minimize the water/hydrocarbon contact at the aggregate interface. The addition of a cosurfactant mediates the electrostatic interaction and will decrease the optimal surface area. This matter has been treated in great detail by Jönsson and Wennerström [16], who has performed a successful calculation of a complete ternary phase diagram with a surfactant and a cosurfactant of comparable chain-length. This well-understood behavior contrasts with our obtained phase-equilibria, in particular the pentanol-system in Fig. 4. The most obvious reason for this is of course the bulkier head-group of the lipoaminoacid, making ordered structure with a low-surface area, normally stable at higher amounts of cosurfactant, energetically disfavored.

In view of the large solution phase region obtained for the ternary system, the formation of microemulsions and their structure compared with normal type surfactant poses an interesting problem. Microemulsions are formed with aliphatic and aromatic hydro-

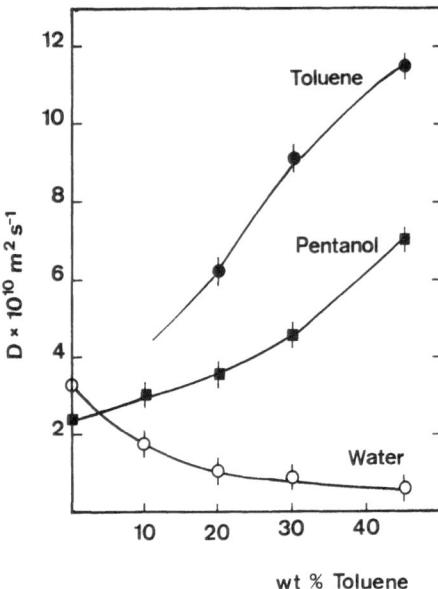

Fig. 6. Self-diffusion coefficients derived from NMR-measurements on LAM/water/pentanol/toluene samples, with a fixed weight fraction LAM/water/pentanol 40/30/30, respectively. (O) water, (■) pentanol, (●) toluene

carbons, and the structure of the system LAM/water pentanol/toluene has been examined in some detail. Multi-component self-diffusion measurements have proved to be an efficient tool in providing structural data in these systems [17]. A large body of structural data from this and other methods has been accumulated, and the different types of structures encountered are by now comparatively well-characterized. Ternary systems containing medium-chain alcohol such as pentanol normally exhibit a bicontinuous type structure, with rapid diffusion of both water and cosurfactant, sufficiently rapid to exclude the presence of closed droplet aggregates [17]. Upon adding hydrocarbon, there is a gradual transition from this bicontinuous structure to more ordered reverse micelles. The behavior has been studied in detail in, e.g., the water/pentanol/alkali oleate systems [18,19]. The results from the LAM-system are qualitatively similar to these and other systems (Fig. 6): an initially rapid water self-diffusion decreases gradually with the addition of toluene: at 45 wt % toluene there is more than an order of magnitude difference in the self-diffusion between the water and the toluene. The behavior is completely analogous to, e. g., the oleate systems [18,19].

Conclusions

— Phase behavior of the binary LAM/H$_2$O system has revealed that LAM assemble in structures (micelles, liquid crystalline phases) similarly to conventional surfactants.

— The general behavior of ternary LAM/H$_2$O/*n*-alkanol systems does to some extent conform to the behavior of ionic surfactant/H$_2$O/*n*-alkanol systems: as the alkanol chain length increases the liquid crystalline phases grow at expense of the solution phase. However, two unexpected features are found in the LAM/H$_2$O/*n*-C$_5$OH system: a) instead of two separate solutions a very large solution phase (*L*) is formed, and b) the lamellar liquid crystalline phase is not stable over a wide temperature interval.

Microemulsion formation and structure in LAM/H$_2$O/C$_5$OH/toluene system is analogous to typical ionic surfactant systems.

Acknowledgements

The authors acknowledge Ms. A. Vilchez and Ms. I. Carrera for their assistance in the experimental work. They also thank CICYT, for financial support.

References

1. Richtler HJ, Knaut J (1988) Proc 2nd World Surfactant Congress, Paris, Vol 1, p 3
2. Than P (1985) In: Rieger MM (ed) Surfactants in Cosmetics. Marcel Dekker, pp 361
3. Gallot B, Haj Hassan H (1989) In: El-Nokaly M (ed) Polymer Association Structures: Microemulsions and Liquid Crystals. ACS Symposiums series 384
4. Takehara M (1988) Proc 6th International Conference on Surface and Colloid Science, p 77
5. Sanson A, Egret-Charlier M, Boulousse O, Maget-Dana R, Charles M, Itab M (1986) In: Mittal KL, Bothorel P (eds) Surfactants in Solution. Vol 5, p 793
6. Infante R, García Domínguez J, Erra P, Juliá R, Prats M (1984) Int J of Cosm Sci 6:275
7. Stilbs P (1987) Progr NMR Spectrosc 19:1
8. Wärnheim T (1986) Thesis, Royal Institute of Technology, Stockholm
9. Shinoda K (1978) Principles of Solution and Solubility. Marcel Dekker, Inc
10. Kunieda H, Shinoda K (1978) J Physs Chem 82:1710–1714
11. Leigh ID, McDonald MP, Wood RM, Tiddy GJT, Trevethan MA (1981) J Chem Soc Faraday Trans I 77:2867
12. Stevenson DG (1961) In: Durham (ed) Surface Activity and Detergency. MacMillan, NY
13. Rendall K, Tiddy GJT, Trevethan MA (1983) J Chem Soc Faraday Trans I 79:637
14. Ekwall P (1975) Adv Liq Cryst 1:1
15. Mitchell DJ, Ninham BW (1981) J Chem Soc, Faraday Trans 2, 77:601

16. Jönsson B, Wennerström H (1987) J Phys Chem 91:338
17. Lindman B, Stilbs P (1987) In: Friberg S, Bothorel P (eds) Microemulsions. CRC Press, p 119
18. Sjöblom E, Henriksson U (1984) In: Mittal KL, Lindman B (eds) Surfactants in Solution. Plenum, New York, Vol 3, p 1867
19. Wärnheim T, Sjöblom E, Henriksson U, Stilbs P (1984) J Phys Chem 88:5420

Authors' address:

Dr. C. Solans
Instituto de Tecnología Química y Textil (C.S.I.C.)
C/Jorge Girona Salgado, 18–26
E-08034 Barcelona, Spain

Received November 1988;
accepted February, 1989

Progress in Colloid & Polymer Science

Progr Colloid Polym Sci 79:76–80 (1989)

Cytochrome c monolayer and mixed surfactant-cytochrome c monolayer

M. Saint-Pierre-Chazalet[1]), F. Billoudet[1]), and M. P. Pileni[1, 2])

[1]) Université P. et M. Curie, S.R.I., Batiment de Chimie-Physique, Paris, France
[2]) C.E.N. Saclay D.P.C., S.C.M., Gif sur Yvette, France

Abstract: In this paper, preliminary results show the formation of cytochrome c monolayer. At low pressure, cytochrome c molecules remain at the interface, whereas at high pressure part of the cytochrome c is redissolved into the bulk phase. A surfactant monolayer formed at the air/water interface is strongly disturbed by injecting cytochrome c into the subphase. These changes are mainly attributed to the adsorption and penetration of cytochrome c at the water/air interface, which is promoted by electrostatic interactions between cytochrome c and the anionic surfactant.

Key words: Cytocrome c, lipid-protein interactions, monolayer.

Introduction

Cytochrome c is a well-known protein active in several biological and photosynthesis processes [1, 2]. It acts as an electron carrier. Cytochrome c is, at neutral pH, a positively charged hemoprotein. Many important biological processes depend on transfer (electron energy) from a bilayer membrane to the adjacent water phase. To understand these processes it is therefore of the greatest importance to obtain a very detailed picture of the protein and membrane/water interface. Protein-lipid interactions are of fundamental importance to the function of biological membranes. The investigation of intermolecular interactions in monolayers at the air/water interface provides the advantage that this model system and its environmental parameters can be varied to a large extent. This concerns chemical composition, temperature, surface pressure, phospholipid phase state and surface electrostatic interactions. Several groups are studying phospholipid-protein interactions by the Langmuir technique [3–5]. A few have studied the interaction of cytochrome b_5 or c [6–8] with various phospholipids. Cytochrome c is a highly water-soluble protein. Contrary to what was expected, it has been demonstrated by two different techniques [9–11] that cytochrome c acts as a surfactant in reverse micellar solution. From that it could be expected that cytochrome c can form a monolayer.

In the present paper it is shown that cytochrome c alone forms monolayers. When a monolayer is made of a charged lipid surfactant, cytochrome c penetrates the monolayer or is adsorbed on it. Different charged surfactants are used.

Materials and methods

Materials

Stearic acid (SA) elementary analysis standard supplied by Merck.

Dipalmitoylphosphatidyl acid (DPPA) purum and dipalmitoylphosphatidyl-choline (DPPC) puriss was purchased from Fluka. Egg-phosphatidylcholine (EPC) was prepared from egg yolk according to the method of Patel and Sparrow [12].

Horse heart cytochrome c was obtained from Fluka.

Preparation of monolayers

Monolayers were prepared in a Langmuir trough (LAUDA balance) of small volume (20 × 15 × 1 cm) to obtain relatively high protein concentration.

Cytochrome c (concentration: 5×10^{-4} mol/l^{-1}) was injected with a microsyringe at numerous points underneath the monolayer and stirred.

All monolayer experiments were carried out at $t = 21 \pm 1$ °C and at pH = 7.

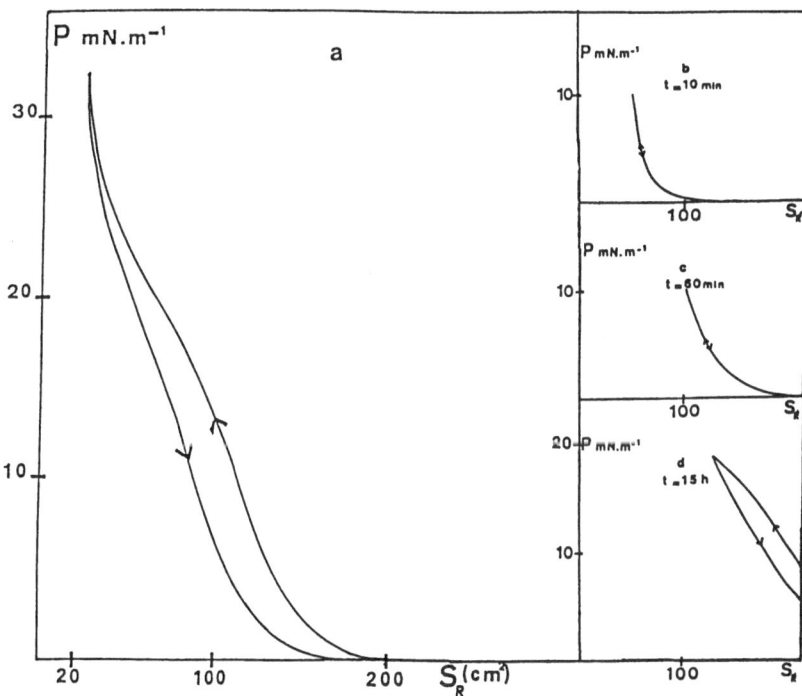

Fig. 1. a) Pressure-area isotherm of cytochrome c monolayer obtained 1 h after injection of cytochrome c into the subphase at $C = 1.25 \times 10^{-6}$ mol l^{-1}. S_r = relative surface area. b,c,d) Pressure-area isotherms of cytochrome c monolayer obtained at various times, in the same conditions

Results

1. Monolayer of cytochrome c

A 5×10^{-4} mol/l^{-1} of cytochrome c concentration in water is injected into the aqueous subphase to obtain after dispersion a final concentration equal to 1.25×10^{-6} mol/l^{-1}. Then, despite the solubility of the protein, a film of protein becomes established at the surface resulting in the isotherm shown in Fig. 1a. At a surface pressure equal to 16 mN · m^{-1} a plateau appears which could be attributed either to a change in orientation of the cytochrome c at the interface or to redissolving of the protein in the trough. The area per molecule for the protein cannot be calculated directly from the curve because the partition coefficient of protein between bulk water and the air/water interface is unknown. Therefore, the monolayer surface areas given in all the figures are relative values (S_r). However, the monolayer area increases with time (Figs. 1b, c, d), and we observed that below 10 mN · m^{-1}, the isotherm remains the same during a rapid compression and decompression. Figure 1d shows that 15 h after injection of cytochrome c, a saturation pressure of 10 mN · m^{-1} is reached, corresponding to the maximum of absorbed molecules at the interface. Equilibrium will be reached for this pressure. The increase in the monolayer area vs. time for concentrations varying between

2.5×10^{-7} and 1.75×10^{-6} mol/l^{-1} is shown in Fig. 2a at $P = 8$ mN · m^{-1}. An immediate effect is observed after addition of cytochrome c and the area increases to reach a plateau which corresponds to an equilibrium monolayer area independent of the cytochrome c concentration injected into the subphase at relatively high cytochrome c concentration (up to 5×10^{-7} M). At fixed surface area (Fig. 2b), just after injection of cytochrome c, the pressure increases with time until 10 mN · m^{-1}, then decreases above this pressure. We can conclude that the equilibrium pressure is 10 mN · m^{-1}.

These results indicate that a water-soluble protein such as cytochrome c is able to form a monolayer as a surfactant. However, the equilibrium is reached after a long period of time (15 h), and the stability of the monolayer is highly dependent on the surface pressure. At high pressure (16 mN · m^{-1}), part of the cytochrome c could be redissolved or reorganized.

2. Penetration of cytochrome c in monolayers

The lipids used are either a negatively charged surfactant such as stearic acid (SA) or dipalmitoylphosphatidylacid (DPPA) or zwitterionic surfactant such as dipalmitoylphosphatidylcholine (DPPC) or egg-phosphatidylcholine (EPC). Monolayers of these vari-

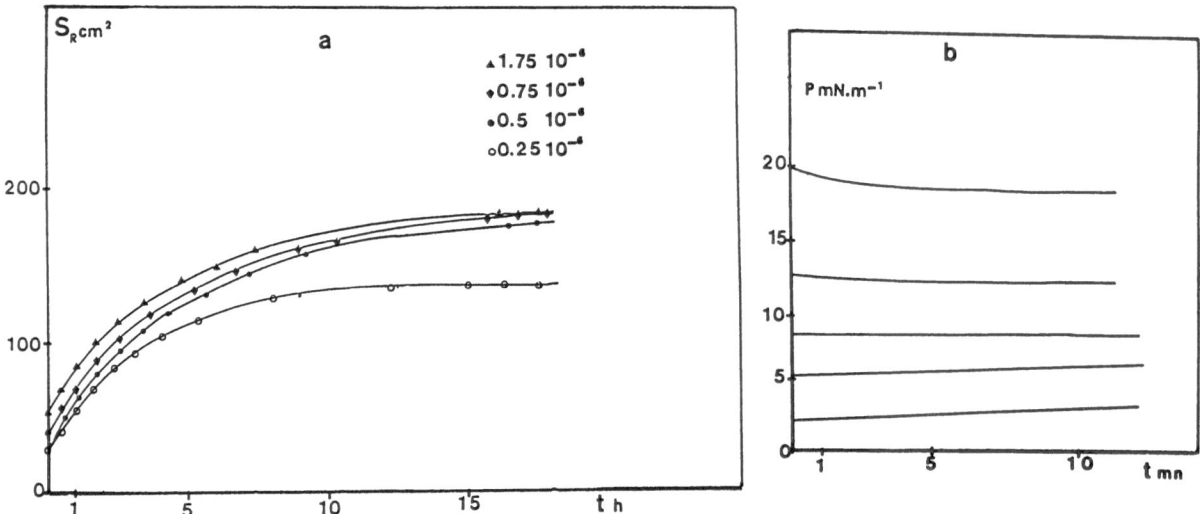

Fig. 2. a) Increase of monolayer area of cytochrome *c* monolayer vs. time at various concentration of cytochrome *c* and at $P = 8 \, \text{mN} \cdot \text{m}^{-1}$.
b) Increase of pressure of cytochrome *c* monolayer vs. time just after injection of cytochrome *c* into the subphase at various initial pressures

ous lipids, differing in the charge of the polar head group, are compressed to an initial surface pressure and then cytochrome *c* is injected into the subphase. These experiments were carried out either by holding the monolayer area constant at its initial value and recording changes in film pressure or by allowing the area to expand while holding the pressure constant at its initial value. We observed that with time the surface area or the surface pressure of the lipid monolayer increases as a result of cytochrome *c* uptake. The presence of cytochrome *c* in the surfactant monolayer induces an isotherm pressure-area very different from that obtained in the absence of cytochrome *c* (Fig. 3). It should be noted that the pressure ($16 \, \text{mN} \cdot \text{m}^{-1}$) cor-

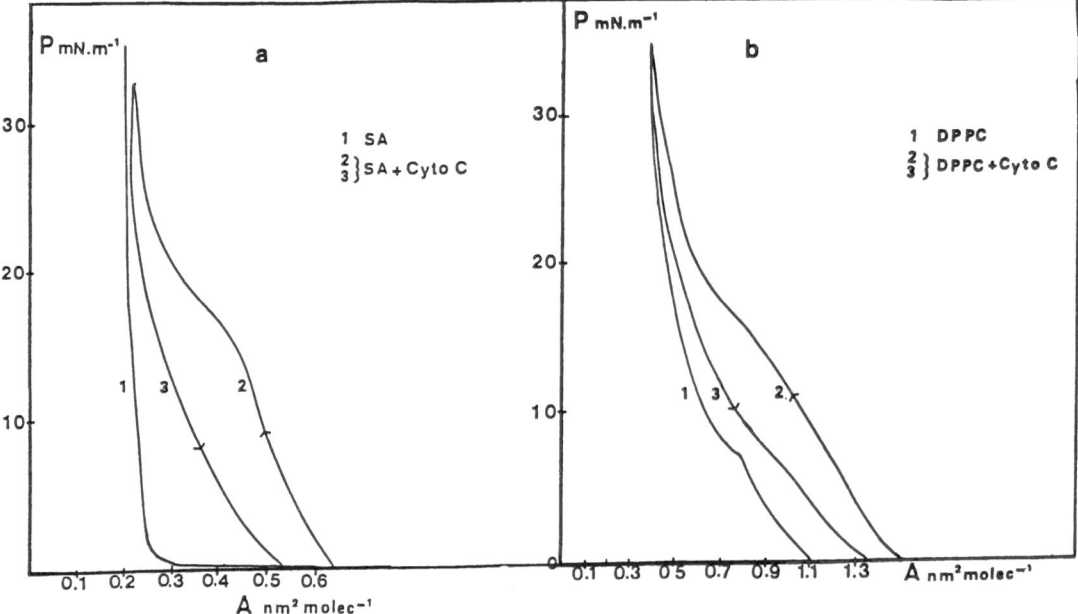

Fig. 3. a) Pressure-area isotherms for stearic acid before and after incorporation of cytochrome *c*. b) Pressure-area isotherms for DPPC before and after incorporation of cytochrome *c*. In these experiments cytochrome *c* was added by injection into the subphase at multiple sites beneath a lipid film maintained at $P = 0.5 \, \text{mN} \cdot \text{m}^{-1}$

responding either to a change in the cytochrome c conformation or to its redissolution is similar both in the presence of lipid monolayer and in its absence.

Measurements at constant surface area

When the surface area of the previously mentioned lipids is constant, the addition of cytochrome c induces an increase in pressure with time which reaches a plateau after 5 min (Fig. 4). This shows that the surfactant monolayer promotes the stabilization of cytochrome c at the interface and that equilibrium is reached faster than in the absence of surfactant. The increase of surface pressure vs. time is stronger for DPPA (dianionic surfactant) than that observed with SA (monoanionic surfactant). From such data it can be deduced that the penetration of cytochrome c is promoted when a dianionic surfactant is used. With DPPC and EPC (zwitterionic surfactants), the addition of cytochrome c induces in the monolayer area a small increase by comparison of the data obtained with the other surfactants which are negatively charged. The difference observed with DPPC (gelliquid phase transition at $P = 6$ mN $\cdot$ m^{-1}) and EPC (liquid phase) could be due to the difference of compressibility of the monolayer. While surface pressure changes at constant surface area are dependent upon the compressibility of the lipid, changes in film area at constant surface pressure are a more quantitative measurement of protein insertion into monolayer surfactants. Therefore, we used this second method.

Measurements at constant surface pressure

Fig. 5a shows the increase in the monolayer area of SA vs. time at constant pressures. At low initial surface pressure of SA (5 mN $\cdot$ m^{-1}), a major increase in the monolayer area is obtained by adding the protein, whereas at high initial surface pressure (14 mN $\cdot$ m^{-1}), the increase in the monolayer area is much smaller. This is probably due to the fact that at low pressure more space is available to allow the migration of cytochrome c at the interface. A comparison with the different lipids at $P = 8$ mN $\cdot$ m^{-1} and for a given cytochrome c concentration (2.5×10^{-6} M) is shown in Fig. 5b. The major increase in surface area observed with DPPA and SA may be due to the ability of cytochrome c not only to be absorbed but also to penetrate the lipid monolayer.

It is well known that cytochrome c is highly positively charged, so that attractive electrostatic interactions promote migration of cytochrome c at the interface. This is confirmed by the fact that when using SA (monoanionic surfactant) instead of DPPA (dianionic surfactant), the attractive electrostatic interaction decreases and less cytochrome c is located at the air/water interface. It should be noted in Figs. 4 and 5b that even with DPPC and EPC (zwitterionic surfactants), a small increase in the monolayer area or surface pressure is obtained. This indicates that electrostatic interactions are not the only cause of the increase in surface pressure. Penetration of cytochrome c in the lipid monolayer could be due to a hydrophobic characteristic of the protein, as has been observed in

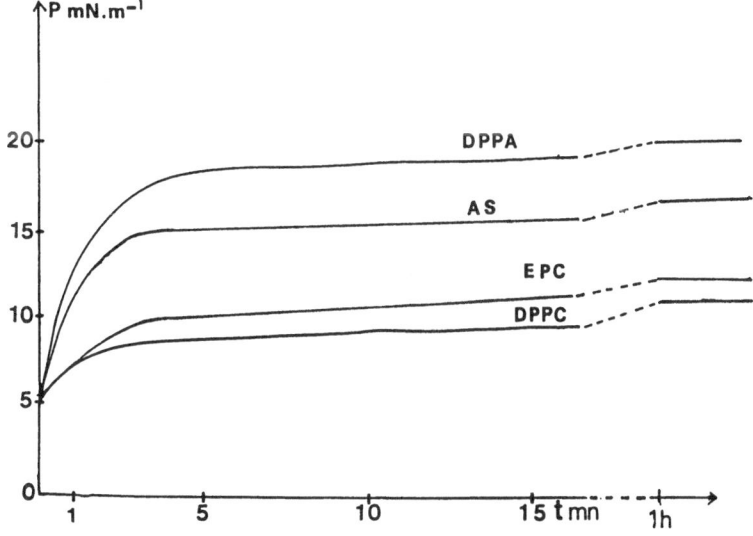

Fig. 4. Increase of pressure with time for various lipids in monolayer compressed until $P = 5$ mN $\cdot$ m^{-1} following injection of cytochrome c into the subphase

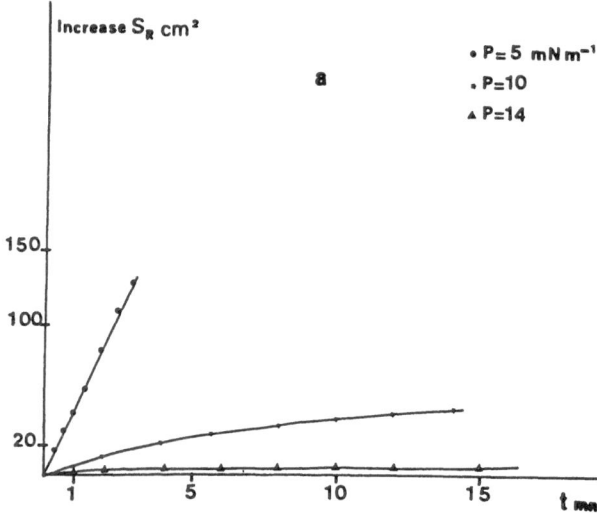

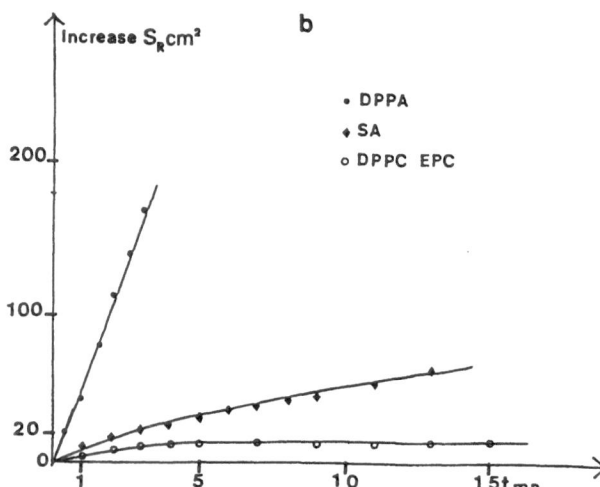

Fig. 5. a) Increase of monolayer area vs. time for a stearic acid film after injection of cytochrome c beneath a film maintained at various constant pressures. b) Increase of monolayer area vs. time for various lipid films after injection of cytochrome c beneath a film maintained at $P = 8$ mN $\cdot$ m^{-1}

reverse micelles [9–11]. It should be noted that the monolayers formed with various lipids differ by their rigidity (gel or cristal liquid) and by the viscosity of the monolayer.

Conclusion

In this paper, preliminary results show that pure cytochrome c forms a monolayer characterized by a specific isotherm. This result was not expected because of the fact that cytochrome c is not reported to be a membrane protein. This could be due to "polymerization" of cytochrome c. The addition of cytochrome c to a monolayer surfactant assembly induces an increase in the monolayer area which could be attributed to adsorption and penetration of cytochrome c into the surfactant monolayer. Such a penetration process is strongly promoted by electrostatic interaction between cytochrome c and an anionic surfactant. A previously published paper [7] showed interactions between cytochrome and phospholipids in monolayer. Most results in that study were obtained using cytochrome b_5 instead of c. Concerning the latter protein, the major difference between the two papers is that in our case small interactions between cytochrome c-DPPC is observed, whereas in the earlier study there were no interactions. This difference could be due to the method of preparing the protein-lipid monolayer.

References

1. Dickerson RE, Timkovich R (1975) In: Boyer PD (ed) The enzymes. Academic Press, New York, 11:7
2. Nicholls P (1974) Biochim Biophys Acta 346:261-271
3. Verger R, Pattus F (1982) Chem Phys Lipids 30:189-199
4. Schnindle H (1979) Biochim Biophys Acta 555:316-325
5. Saint-Pierre-Chazalet M, Thomas C, Dupeyrat M, Gary Bobo CM (1988) Biochim Biophys Acta 944:477-486
6. Wilkinson MC, Zaba BN, Taylor DM, Laidman DL, Lewis TJ (1985) Biochim Biophys Acta 857:189-197
7. Heckl WM, Zaba BN, Möhwald H (1987) Biochim Biophys Acta 903:166-176
8. Pilon M, Jordy W, De Kruijfft B, Demel RA (1987) Biochim Biophys Acta 902:207-216
9. Pileni MP, Zemb T, Petit C (1985) Chem Phys Lett 118:414-420
10. Brochette P, Petit C, Pileni MP (1988) J Phys Chem 92:3505-3511
11. Petit C, Brochette P, Pileni MP (1987) J Phys Chem 90:6517-6524
12. Patel KM, Sparrow JT (1979) J Chromatogr 150:542-547

Received October, 1988;
accepted March, 1989

Authors' address:

M. P. Pileni
C.E.N. Saclay
DPC SCM
F-91191 Gif sur Yvette, France

Orientation of rods formed by aggregated surfactants in organic media: the steroid-cyclohexane physical gel case

P. Terech

Institut Laue-Langevin, Grenoble, France

Abstract: We study the physical gels obtained by aggregation of a steroid derivative in cyclohexane. The amphiphilic steroid molecule can give infinite rod-like aggregates by a multiple hydrogen bonding process. At a critical filamentary concentration viscoelastic gels are obtained through a sharp sol-gel threshold. We present here some preliminary results concerning rod orientation effects in this system.

Long-range orientation order is observed in concentrated gel samples as a consequence of excluded volume effects. Typical Schlieren optical textures characterize nematic-like domains in the samples.

The use of strong magnetic fields during the steroid aggregation kinetics can give strikingly higher order parameters. The initial state of these experiments is a diluted sol phase. A slow steroid aggregation reaction gives rise to a slow viscosity increase of the sol phase and allows the reorientational motion of the rod-like aggregates.

An oriented xerogel is obtained from demixion of the unstable gel phase. This behavior is in qualitative agreement with the theoretical predictions concerning rigid rod-like polymers in solution. Polarizing optical microscopy, magnetically induced optical birefringence and two-dimensional small-angle neutron scattering patterns are used to detect the anisotropic states.

Key words: Gel, rod orientation, neutron scattering, x-ray diffraction, surfactant, organic media.

1. Introduction

Gel systems are important in medicine, biology, chemistry, and polymer science and find numerous applications in, for example, the photographic, cosmetic, food and petroleum industries. The increasing practical importance of gels has stimulated recent structure-oriented studies using modern spectroscopic methods, neutron and x-ray scattering and electron microscopy. A gel phase is a solid-like three-dimensional network of interconnected molecular aggregates embedded in a liquid dispersing medium. Chemical gels involving covalent reticulating bonds are distinguished from physical gels which involve low-energy reticulating bonds.

As examples, DNA and gelatin are two natural macromolecules which give physical gels in water. But a large number of low molecular weight molecules are also known to form physical gels when put in solution. Sodium deoxycholate in water [1] and 12-hydroxyoctadecanoic acid in carbon tetrachloride [2] have received increased attention in recent years.

The system described here belongs to a family of physical gels, the two components of which are an organic apolar solvent (hydrocarbon) and a low molecular weight (see formulae) steroid derivative ($MW = 375$). A simple and well defined two-component system of this type is not only of interest in its own right, but it may serve as a model for more complex gel systems. Under conditions defined by the phase diagram [3] a steroid aggregation process takes place to give growing rod-like aggregates. When a critical rod-like aggregate concentration is reached a transition solution-gel occurs. To a great extent it is the solid network which determines the physical properties of a gel, which therefore can only be adequately understood if the structural features of the network are known.

Properties below and just above the gelation threshold are respectively that of large clusters (sol phase) and infinite cluster (gel phase) in solution. In a percolation model the related quantities such as viscosity and modulus of elasticity follow power laws with universal exponents [4, 5]. This universality characteristic originates from a geometrical invariancy property valid within a given length range. As a consequence, the gel network in conventional experimental preparation procedures is random. Numerous gelling systems show such a regular random network. Among large molecules, gelation [6] gels reveal these typical networks. This is also the case for steroid gels in cyclohexane [7] obtained when the constitutive fibers are allowed to grow in the absence of any external orientational force (mechanical, electrical, or magnetic).

We report in this paper the orientation effects obtained with a gelling steroid solution from two distinct procedures. First, rod orientation is observed without any external forces in concentrated systems. Secondly, when strong magnetic fields are used, rod alignment is obtained.

hours delay reaction. The driving force of aggregation kinetics is strongly dependent upon the steroid concentration ratio C/C^* where C^* refers to the critical solution/gel threshold at a given temperature ($C^* = 1.1$ 10^{-2} M at 20 °C). A detailed structural study of the gel network in cyclohexane by electron microscopy (freeze etching method [7]) and small-angle neutron scattering (SANS) experiments [8] provided the following picture of the native gel network: it is constituted by an assembly of very long entangled 9.9 nm diameter chiral fibers made up of two coiled finer filaments. From stereoscopic views an average mesh size of about 300 nm is found in the case of gels formed at room temperature from a solution of initial concentration about 3×10^{-2} M. Fibers are at least 5 μm long. A statistical analysis shows the most frequently occurring fiber diameter to be 9.1 nm and also reveals the presence of finer filaments some 4.6 nm in diameter. We deduced that at least two of these finer filaments are required to form the 9.9-nm-diameter fibers. The connections between the filaments were found to consist of fusion zones produced by exchange of filaments, filaments bodily entwining one about the other, or of juxtaposed fibers.

Further neutron scattering experiments give an estimation of the mass per unit length of fiber $n_L = 70$ molecules/nm. The following expression gives the molecular mass M_R of a rod aggregate as a function of its length L

$$M_R = 70 \, L M_o / N \tag{1}$$

where M_o is the steroid molecular mass and N is Avogadros's number.

2. Description of the steroid/cyclohexane system

Gelling solutions are prepared by dissolving the steroid in hot cyclohexane under vigorous shaking. Gels are obtained when the temperature is decreased. Stability of the gel phases is very good in cyclohexane where steroid crystallization is never observed.

Consistent structural, kinetic and thermodynamics informations have already been obtained concerning the gel state and the related sol-gel transition. In the following we summarize the results of use in the present paper. We have shown that the kinetic of gelation [3] could be varied in a range of characteristic times spreading from quite instant phenomenon to a several

3. Experimental

For small-angle neutron scattering experiments, samples were held in 1-mm-path QS quartz cells which are standard in SANS. Experiments were performed on spectrometers D 11 and D 17 of the Institute Laue-Langevin with the neutron momentum tranfer Q ranging between 10^{-3} and 2×10^{-1} Å^{-1}. Absolute intensity measurements were done by comparison with a 1-mm-thick water sample. Corrections for detector efficiency, sample holder scattering and background were performed. For orientation experiments in magnetic fields, the steroid amine derivative concentration is adjusted to get long gelation times at 20 °C ($C \sim 2.1 \times 10^{-2}$ M; $t_G \geq$ 60 min). A 4.2 T or 7 T magnetic field is used for the experiment from a Brucker spectrometer or water-cooled Bitter-type solenoid, respectively. An Olympus BH-2 microscope equipped with crossed polarizers is used for optical characterization. An Elliot rotating anode generator GX 20 was used for the x-ray studies.

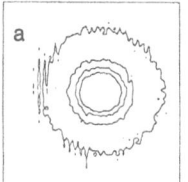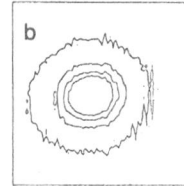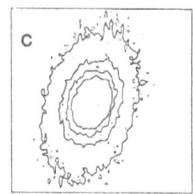

Fig. 1. Neutron scattering patterns $\lambda = 10$ Å; approximate Q range: 0.002 Å^{-1} − 0.15 Å^{-1}. This missing intensity in the middle of the contour maps is due to the shadow thrown by the beam stop. a) Isotropic scattering, $C_o = 1.3 \times 10^{-2}$ M; b) anisotropic scattering, concentration effect, $C_o = 5.0 \times 10^{-2}$ M; c) anisotropic scattering, magnetic orientation, $C_o = 1.1 \times 10^{-2}$ M

4. The reference dilute system

The neutron scattering appeared to be isotropic (see Fig. 1a). The intensities are averaged over all possible orientations of the scattering planes for a given value of the momentum transfer modulus Q.

$$Q = \frac{4\pi}{\lambda} \sin \theta/2 \qquad (2)$$

with the full scattering angle θ and the wavelength λ of the neutrons. Here the rod-like aggregates are non-interacting particles and the scattered intensity is only function of the form factor and number of particles. No birefingence can be detected from this system when observed by optical microscopy. Figure 2 shows the scattering curve and the typical Q^{-1} divergence of

the intensity at low Q. The rod scattering function (expression 3) is adequate to describe the intensity decrease:

$$I(Q) \propto \frac{\pi}{QL} \left[\frac{2J_1(QR)}{QR} \right]^2. \qquad (3)$$

Figure 3 shows a distribution of the growing rod-like steroid aggregates during the sol-gel transition observed by electron microscopy.

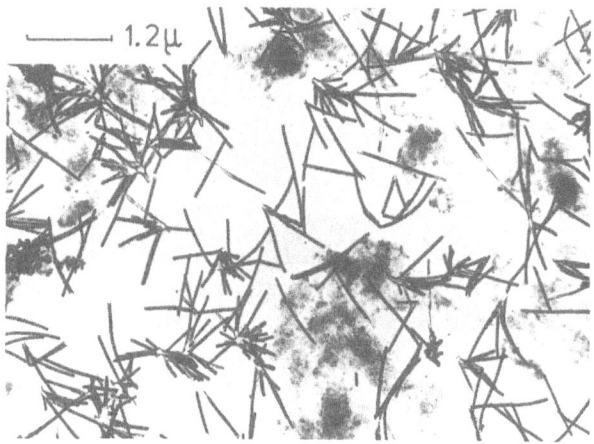

Fig. 3. Electron microscopy replica of a dried gelling solution showing the finite length fibers. Darkened areas between the fibers are a consequence of the EM procedure. They are produced by the fast evaporation of the solvent, which deposits the remainder of non-aggregated steroid species

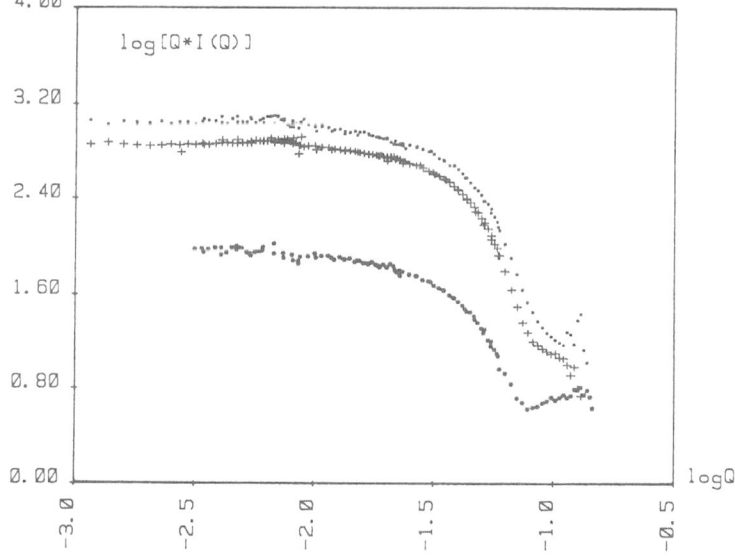

Fig. 2. Scattering curves for three gel samples of different initial concentration at room temperature. The specific log-log representation $Q\,I(Q)$ vs. Q is used. (✳) $C_o = 1 \times 10^{-2}$ M; (+) $C_o = 2 \times 10^{-2}$ M; (■) $C_o = 3 \times 10^{-2}$ M

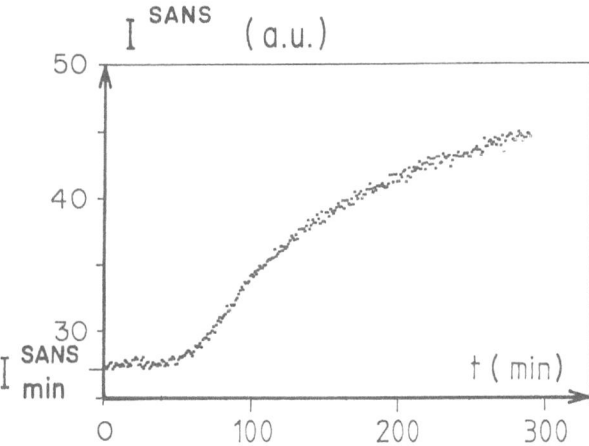

Fig. 4. Typical integrated neutron scattered intensity I^{SANS} during the gelation of solution $C_o = 2.4 \times 10^{-2}$ M at 19 °C

For the fluid phase, no Q-dependent small-angle scattering signal exists. By contrast, a strong scattered intensity defined by Eq. (3) is growing during the gelation process. Since the shape of the scattering curve does not change with time, as confirmed by SANS experiments [3], the integrated intensity is proportional to the total amount of scattering objects.

Figure 4 shows a typical variation of the intensity $I^{SANS}(t)$, integrated between 0.008 and 0.2 Å^{-1}, vs. time. $(I^{SANS}(t) - I^{SANS}_{min})$ values are directly proportional to the amount of aggregated steroid species; I^{SANS}_{min} is a background level mainly due to incoherent scattering.

5. Rod orientation in concentrated steroid gels

A) Results

The filamentary organization of the gel network is strongly affected by the volume reduction of the sample [9] during solvent evaporation and mesophases are suspected, as is the case for some aqueous systems of surfactants [10] or polymeric solutions [11].

When a hot droplet of a gelling steroid solution is deposited on a glass, the successive steps of gelation, steroid concentration by cyclohexane evaporation and xerogel formation are observed by polarizing optical microscopy. As cyclohexane evaporates, birefringence is enhanced and extinction crosses with black brushes and dark threads of typical Schlieren textures appear and spray over the preparation (Fig. 5).

Furthermore, the neutron scattering can be used to detect an ordering of the rod-like aggregates in the sample (see Fig. 1b). Despite that the contours do not follow a smooth line, due to statistical deviations, it can be seen that the scattering becomes elliptic. This anisotropic scattering increasing with the steroid concentration together with the typical related birefringent optical patterns argue for a description of these solutions as nematic-like lyotropic phases.

B) Discussion

In a model case of rigid rods in solution, the concentration limit C^{**} above which excluded volume effects give rise to a nematic liquid phase is

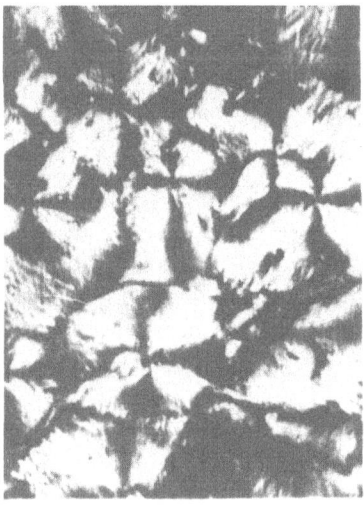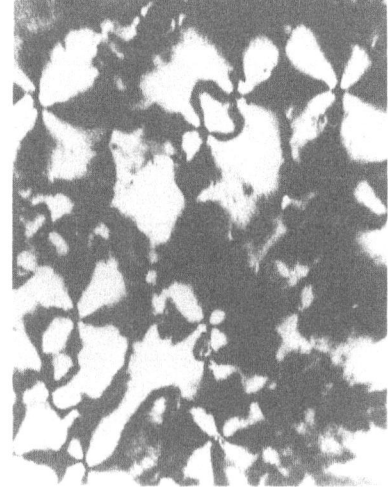

Fig. 5. Optical textures observed in concentrated gels prepared by solvent evaporation   ├───┤ 100 μm

$$C^{**} = M/dL^2 \qquad (4)$$

where M, d and L are the rod mass, diameter, and length, respectively. The situation of an entangled rods solution is that of the sol phase before the gelation threshold. Previous TEM results [9] on quickly dried gelling solutions have shown that steroid aggregates could be approximated to finite rigid rods.

For a mean L value of 500 nm frequently encountered in the electron micrographs and a 9.5 nm rod diameter an indicative C^{**} value of 2.5×10^{-2} M is obtained from Eq. (1). The $C^{**} + C^*$ value of 3.6×10^{-2} M is thus not so different from the experimental microscopic observation of enhanced birefringence around 10×10^{-2} M, taking into account that a gel is made or randomly connected filaments. This explains that anisotropic domains may exist in such concentrated systems.

Furthermore, it is remarkable that the mesomorphic state characterized by the optical textures during solvent evaporation remains present in the solid xerogel [12] where it can be analyzed by a calorimetric study. The related molecular structure and filament distribution in the xerogel is a characteristic feature of the lyotropic system which cannot exist reversibly in the solid state.

6. Rod orientation under magnetic fields

A) Results

When the sol-gel transition is performed under strong magnetic fields a large magnetic optical birefringence jump ($\Delta n = 10^{-6}$) is observed. In addition, the SANS pattern of the obtained gel phase is anisotropic (see Fig. 1c). Some samples gelled within the magnetic fields are mechanically softer and can give, after a very variable time delay of several days, a solid collapsed bundle of fibers in a liquid cyclohexane phase. This behavior is statistically very surprising in the steroid system for which several hundred gel samples were prepared without the use of magnetic fields and have always been very stable even over several years.

When observed under crossed polars, the fibers appear highly birefingent and oriented (see Fig. 6).

Furthermore, x-ray diffraction photographs (Fig. 7) show, on the one hand, a set of uniform rings, and on the other hand, a set of nonuniform rings typical of orientational effects. Among the nonuniform rings we can distinguish a first class of reflections with a 3.1 nm

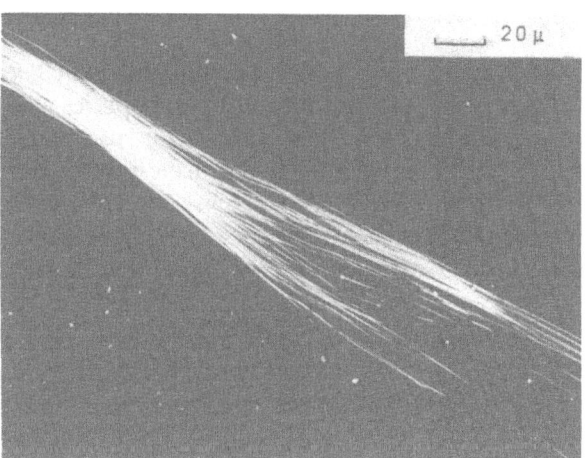

Fig. 6. Polarizing optical microscopy of the phase-separated gel network showing highly birefringent oriented steroid fibers

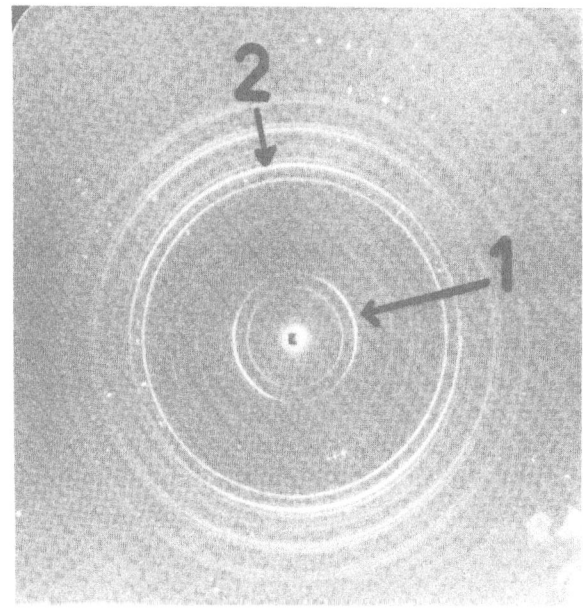

Fig. 7. X-ray diffraction photograph of the dried, magnetically oriented steroid gel network; Copper $K\alpha$ radiation. The arrows point to the two main nonuniform rings. The corresponding spacings are 1.52 and 0.55 nm, respectively. The diffraction pattern is that of a poor alignment of filaments within the fiber bundles which indicate an important random angular orientation about the fiber axis in the individual diffracting units

periodicity over four orders of diffraction, the first one being weak, and a second class of anisotropic diffractions in an orthogonal direction with an intense reflection at 0.55 nm.

B) Discussion

Usually ordering of organic compounds in magnetic fields is only possible for anisometric species with a strong anisotropy of the diamagnetic susceptibility tensor ($\Delta\chi$). This is not the case with the individual steroid molecules. In a gel network any orientation effect is impossible because of the entanglement of the constitutive fibers. By contrast, we expect that the intermediary situation may be different during the kinetics of steroid aggregation from the "monomeric" steroid species to the rod-like aggregates (see Fig. 4).

In a static experiment with rod-like particle orientation in a magnetic field, the degree of orientation ϕ depends on a balance between the stabilization orientational energy W_{OR} and thermal agitation. The energy W_{OR} is proportional to the resultant diamagnetic anisotropy of the rod-like aggregate susceptibility tensor which possesses an axis of rotational symmetry. This anisotropy is an additive property of the molecular quantity $\Delta\chi_{mol}$ ($\Delta\chi = \chi_{\parallel} - \chi_{\perp}$). The energy balance expression is thus defined by Eq. (5), where H is the magnetic field, K is a form factor of the rod-like

$$\Phi \propto K \Delta\chi_{mol} (\cos^2 \theta) H^2/2k_B T \qquad (5)$$

aggregate ($K = 1$ for a linear aggregate, for instance), k_B is Boltzmann's constant, T is the absolute temperature, and θ is the angle between the particle symmetry and the field direction.

To rotate the rod-like particles, the driving magnetic torque has to compete with a viscous torque characterized by the rotational diffusion coefficient D_R. The Brownian motion of the rods follows various models for different rod concentration ranges. Here we assume that the rod concentration C (g/cm³) is such that $C^* \ll C \ll C^{**}$, where $C^* = M_R/L^3$ and $C^{**} = M_R/dL^2$, where L is the rod length, d is the rod diameter, and M_R is the molecular mass of the rod. These are the conditions of the "semi-dilute entangled regime" where the rotational diffusion coefficient dependence on the rod length is given by [13]:

$$D_R \propto D_{RO} C^{-2} L^{-7} \qquad (6)$$

where D_{RO} is the rotational diffusion coefficient at zero concentration and C is the mass rod concentration.

In the steroid system, the orientation experiments are carried out under these conditions ($C_o \sim 8 \times 10^{-3}$ g/cm³ $\sim 2.0 \times 10^{-2}$ M). Indeed, an indicative value of

$C^{**} = 14 \times 10^{-3}$ g/cm³ (3.5×10^{-2} M) can be calculated following the specific expressions:

$$C_{ster}^{**} = C_o + C^{**} \qquad (7)$$

where $C_o \sim 4 \times 10^{-3}$ g/cm³ ($\sim 10^{-2}$ M) at 20 °C is a limit concentration below which no long filaments can exist [3, 8].

By contrast, in a dynamic experiment of steroid aggregation, the rod length is a kinetic function of the initial steroid concentration. This is the actual situation of the magnetic orientation experiments for which the initial state is always the unstable sol phase. The kinetic function is deduced from Avrami plots of SANS kinetic data [3] during the first stages of the unidirectional aggregation reaction and can be approximated by:

$$L(t)/L_o = 1 - \exp(-Kt) \qquad (8)$$

where L_o is the mean rod length equilibrium value and K is a kinetic constant (sec⁻¹). Electron microscopy experiments frequently show a mean L_o value of 10^4 nm. Various aggregation kinetic data [3] indicate $K \sim 0.1$ mn⁻¹ $= 2 \times 10^{-3}$ s⁻¹ in this range of low initial steroid supersaturation. The number of steroid rods n per cubic centimeter of solution is found to be equal to $C/M_R = 10^{-13}/$cm³ for a 10^{-2} M ($\sim 4 \times 10^{-3}$ g/cm³) steroid concentration involved in the rod structure.

We assume n is constant during the growth step. For various time values of the kinetic aggregation reaction. Table 1 indicates the length $L(t)$ (Eq. (8)) and the ratio $D_{RO}/D_R = L^9(t) (5 \times 10^{-20} n)^2$ for different numbers of

Table 1. Reduced rotational diffusion coefficient D_{RO}/D_R for three different distributions of n growing rods (equilibrium length $L_o = 10^4$ nm) at various kinetic steroid aggregation times

t, s	$L(t)$, nm	D_{RO}/D_R $n=10^{13}$ $d=100$ nm	$n=10^{10}$ $d=3163$ nm	$n=10^9$ $d=10000$ nm
2	40	65[b]	0	0
5	99	2×10^{15}	0.2	0.002
10	198	∞	117[b]	1.1
20	392	∞	5×10^4	546[b]
50	952	∞	∞	2×10^6
1000	8647	∞	∞	∞

[a]) d is the corresponding gel network mesh size at equilibrium;
[b]) is critical time and length values for which the high D_{RO}/D_R values prevent any rod orientation.

rods n or germination sites. The first case ($n = 10^{13}$) corresponds to the statistical case where all the germination sites are growing at the same time. The second case ($n = 10^{10}$) concerns a situation where the germination sites are growing by domains, which is more realistic in this range of low steroid supersaturation as demonstrated by previous kinetic experiments. The third case ($n = 10^9$) corresponds to the limit case of the minimum number of rods (C^*) of length L_0 to gelify the system. The corresponding mesh-size values of the porous network are calculated from a simple model in which rods of a finite range L are placed in cubes of size d^3. We get the mean distance between the rods of radius from:

$$d = R\pi^{-1/2}\phi^{1/2} \qquad (9)$$

where ϕ is the volume fraction of the rods in solution. The first case ($n = 10^{13}$) which implies a mass conservation relation at equilibrium gives a mean size of 100 nm which is experimentally observed by electron microscopy experiments.

It is clearly seen that whatever the effective situation, any rotation becomes impossible in a few seconds (or minutes), the system being completely "frozen" by excluded volume effects. This is why we can only expect to build oriented rods of finite lengths (198 nm in the third case) which act as oriented precursors for the consequential growth steps. The oriented xerogel-phase separation confirms that the rod-like aggregates of the sol phase have been oriented and have guided the subsequently growing fibers. Similar behavior has been reported with collagen in which small assembly of molecules are precursors of fibrils [14]. A comparable situation is found in the polymerization reaction of fibrinogen [15] where the reaction has to take place slowly to give highly oriented fibrin gels. Additionally, collagen above a certain denaturation pH assumes a sufficient rigidity of the triple-helical structure for a spontaneous formation of an isotropic dilute phase and concentrated anisotropic ordered phase. The observed phase separation in the steroid case is in agreement with the Flory lattice model for rigid rods in solution which predicts an extension of the heterogeneous biphasic domain when the ordered phase is aligned [16, 17].

The orientation effect becomes inefficient as the length and concentration increase during the steroid aggregation kinetics and specially when entanglements occur in the filament dispersion. This explains the relative poor local ordering observed in x-ray photographs. Nevertheless, orientation effects are clearly seen on the two above-mentioned sets of reflections, where it may be noticed that the corresponding Bragg distances of 1.52 and 0.55 nm can be correlated on one hand with the steroid molecular length [18] in a bilayer structure and, on the other hand, with a lateral molecular spacing in such mesomorphic states.

Magnetic field orientation experiments have underlined some analogies with polymeric systems. Complementary structural studies of orientation in these low molecular weight gelling compounds are now in progress using stronger magnetic fields and also mechanical shear stresses.

Acknowledgements

The author is grateful to Drs. C. Berthet, R. Ramasseul, J, Torbet, F. Volino and R. H. Wade for their continuous help in this work.

References

1. Conte G, Di Blasi R, Giglio E, Paretta A, Pavel NV (1984) J Phys Chem 88:5720
2. Tachibana T, Mori T, Hori K (1980) Bull Chem Soc Japan 53:1714; ibid 54:73
3. Terech P (1985) J Colloid Interface Sci 107:244
4. De Gennes PG (1980) In: Scaling Concepts in Polymer Physics. Cornell University Press, Ithaca, pp 128
5. Guyon E, Roux S (1987) La Recherche 191:1050
6. Djabourov M, Leblond J, Papon P (1988) J Physique, Paris 49:319
7. Wade RH, Terech P, Hewat EA, Ramasseul R, Volino F (1986) J Colloid Interface Sci 114:442
8. Terech P, Volino F, Ramasseul R (1985) J Physique, Paris 46:895
9. Terech P, Wade RH (1988) J Colloid Interface Sci 125:542
10. Luzzati V, Hustachi H, Skoulios A, Husson F (1960) Acta Cryst 13:660
11. Frost HM, Cohen Y, Thomas EL (1987) In: Russo PS (ed) Reversible Polymeric Gels and Related Systems. ACS Symposium Series No 350
12. Terech P (1989) Mol Cryst Liq Cryst 166:29
13. Doi M (1981) J Polym Sci 19:229
14. Murthy NS (1982) IUPAC international Symposium, Amherst Proceedings, p 833
15. Torbet J, Freyssinet JM, Hudry-Clergeon G (1981) Nature, London 289:91
16. Flory PJ (1956) Proc R Soc London A 234:73
17. Miller WG, Wu CC, Wee EL, Santee GL, Rai JH, Goebel KG (1974) Pure Appl Chem 38:37
18. Terech P, Berthet C (1988) J Phys Chem 92:4272

Received October, 1988; accepted February, 1989

Author's address:

P. Terech
Insitut Laue-Langevin
156 X
F-38042 Grenoble Cedex, France

Progress in Colloid & Polymer Science   Progr Colloid Polym Sci 79:88–93 (1989)

Use of microemulsion systems as media for heterogeneous enzymic catalysis

A. Xenakis, T. P. Valis, and F. N. Kolisis

The National Hellenic Research Foundation, Institute of Biological Research, Athens, Greece

Abstract: Lipases catalyze specifically the cleavage of triglycerides to free fatty acids and glycerol. Depending on the reaction media, these enzymes can also be used to catalyze the synthesis or the transesterification of some specific triglycerides. Since these reactions are heterogeneous, taking place at interfaces, they can be performed in water-in-oil microemulsion systems under mild conditions. This approach can give information about the interactions between protein molecules and the surfactant membrane separating the oil/water phases. These interactions can have dramatic influence on the enzymic activity.

The nature of the microemulsion used is crucial, i) in controlling the equilibrium of the reaction from hydrolysis to condensation, depending on the water content; ii) for differences of the solubilization site and catalytic action of the enzyme hosted in the dispersed aqueous phase, the structure of which depends on the surfactant used.

In the present work we studied the catalytic behavior of lipase from *Rhizopus delemar* in the direction of hydrolysis by using AOT anionic, CTAB cationic, and $C_{12}EO_4$ nonionic microemulsion systems. Various parameters of the enzymic reaction (such as K_m, pH optimum, T optimum, water content) were studied in these different systems. The results show that in the AOT system the activity of the enzyme is higher in comparison to the other systems.

Key words: Microemulsions, AOT, CTAB, $C_{12}EO_4$, lipase.

Introduction

Lipase (triacylglycerol acylhydrolase, EC 3.1.1.3.) is the enzyme which hydrolyzes the long-chain aliphatic esters to glycerol and free fatty acids [1]. This catalytic process is heterogeneous. The lipophilic substrates are catalyzed by the hydrophilic enzyme molecule at the interface [2].

It is of great technological interest to increase the interface, since this results in the increase of the number of available to react with substrate molecules. This can be accomplished by using microemulsions [3–6].

The enzyme molecules can be entrapped in the reversed micelles avoiding direct contact with the unfavorable organic medium and, thus, protected against denaturation. The interior of the reversed micelles acts like a microreactor, which provides a favorable aqueous microenvironment for enzyme activity.

The size, shape and solubilizing ability of the reversed micelles, which affect the enzyme activity, depends strongly on the nature of the amphiphile used.

The aim of this investigation was to examine the catalytic behavior of lipase from *Rhizopus delemar* in different types of microemulsions:

i) Water reversed micelles stabilized by the anionic AOT (bis-(2-ethylhexyl)sulfosuccinate sodium salt) in isooctane;

ii) microemulsions formulated with the cationic CTAB (cetyltrimethylammonium bromide) in isooctane, using pentanol-1 as cosurfactant, and

iii) nonionic systems consisting of $C_{12}EO_4$ (tetraethyleneglycoldodecylether)/*n*-decane/water.

Different parameters such as pH, temperature, and water content, affecting the enzyme activity were studied in the above mentioned systems.

Table 1. Compositions of the microemulsion samples studied. Values are in w/w%

Microemulsion type	Organic solvent		water (buffer)	Surfactant	Cosurfactant	R
Anionic	Isooctane	Triolein		AOT	—	
	72.88	23.72	0.52	2.88		4.5
	72.49	23.60	1.04	2.87		9.0
	72.43	23.59	1.12	2.86		9.6
	72.40	23.57	1.17	2.86		10.1
	72.36	23.56	1.22	2.86		10.5
	71.91	23.42	1.82	2.85		15.8
	71.55	23.29	2.33	2.83		20.4
	71.17	23.17	2.85	2.81		25.0
	71.78	23.04	3.36	2.80		29.7
Cationic	Isooctane	Triolein		CTAB	Pentanol	
	54.40	20.00	2.00	8.60	15.00	4.7
	54.40	20.00	2.50	8.60	14.50	5.9
	53.95	20.00	3.00	8.55	14.50	7.0
	53.50	20.00	4.00	8.50	14.00	9.5
	52.61	20.00	5.00	8.39	14.00	11.4
	49.47	20.00	10.00	8.03	13.50	25.2
	44.99	20.00	15.00	7.51	12.50	40.4
Nonionic	Decane	Triolein		$C_{12}EO_4$	—	
	71.46	19.05	4.73	4.76		20.0
	63.68	18.19	9.03	9.10		20.0
	62.57	4.17	16.57	16.69		20.0
	58.40	8.34	16.57	16.69		20.0
	52.22	17.41	12.96	17.41		15.0

Experimental

Materials

Rhizopus delemar lipase purchased from Fluka, was further purified by ammonium sulfate fractionation and Sephadex G-100 chromatography. The final enzyme preparation with a specific activity of 980 units/mg, was 10-times improved in purity, giving two distinct bands in SDS polyacrylamide gel electrophoresis.

AOT (bis-(2-ethylhexyl)sulfosuccinate sodium salt) was obtained from Serva and purified as described by Martin and Magid [7]. Moisture was removed by distillation in toluene and lyophilization and checked periodically by Karl Fischer titrations. CTAB (cetyltrimethylammonium bromide) from Serva was recrystallized from methanol/ether. $C_{12}EO_4$ (tetraethyleneglycoldodecylether) was obtained from Nikko Chemicals Co.Ltd. isooctane (2,2,4-trimethylpentane) and 1-pentanol were products of Ferak, Berlin, *n*-decane was obtained from Merck, Darmstadt. All other compounds and reagents used were of the higher commercially available purity, while double-distilled water was used.

Methods

Preparation of microemulsions

In the case of AOT/isooctane/water microemulsions the compositions of the systems used were chosen to be in the monophasic area of the phase diagram [8]. A stock solution of 0.1 M AOT in isooctane was used, to which the appropriate amounts of water and of a concentrated stock lipase solution in buffer (a few microliters) were added. Various quantities of triolein were added and the final concentration of AOT was adjusted to 50 mM by adding isooctane. Solubilization was achieved within less than 1 min by gentle shaking. The total amount of water was calculated to give the desired value of the molar ratio $R = [H_2O]/[AOT]$.

A similar procedure was followed for the cationic and nonionic systems, considering in all cases the substrate (triolein) as part of the oil phase, assuming that substrate consumption during hydrolysis did not alter the overall composition of the microemulsions.

The compositions of the different microemulsions used for this study are presented in Table 1.

Lipase assay

Hydrolysis of triolein by the enzyme in AOT microemulsion systems was carried out according to Han and Rhee [9]. Free fatty acids determination was made as described by Lowry [10], by following the absorbance at 715 nm. A calibration curve was plotted, using oleic acid as an internal standard.

This method was extended for the measurements of lipase activity in CTAB and $C_{12}EO_4$ systems, with the addition of a centrifugation step (2500 rpm for 4 min), after the reaction was stopped, giving satisfactory results.

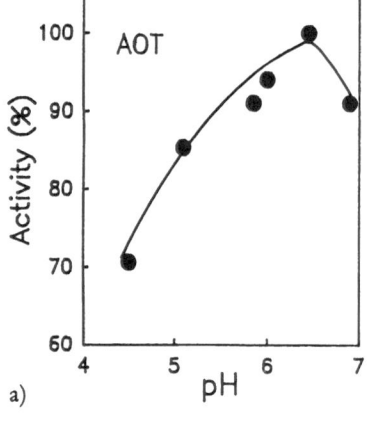

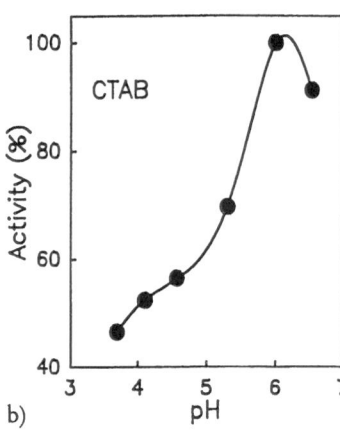

Fig. 1. Lipase activity in microemulsions (a) of AOT (50 mM in isooctane) at $R = 9$ and (b) in CTAB (0.2 M in isooctane) at $R = 7$, at various pH values (pH values are those of the stock enzyme solutions in acetate buffer)

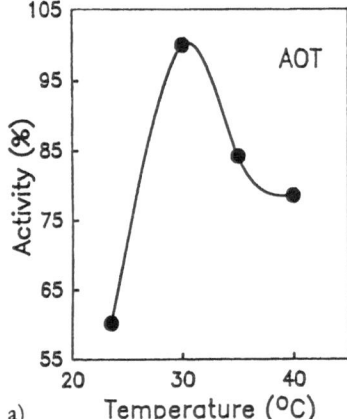

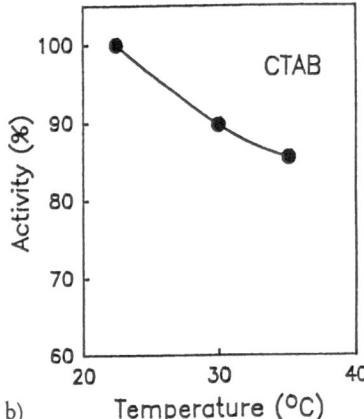

Fig. 2. Effect of the temperature on lipase activity in microemulsions of (a) AOT (50 mM in isooctane) at $R = 9$; pH of the stock enzyme solution was 6.5. (b) CTAB (0.2 M in isooctane) at $R = 7$; pH of the stock enzyme solution was 6.0

One unit of enzyme was defined as the amount of lipase liberating 1 μmol of fatty acids/min under the assay conditions.

Results and discussion

Anionic systems

Activity tests

The effect of various pH values in microemulsions on lipase activity is shown in Fig. 1a. The optimum pH was found to be 6.45. As pH values we considered the values of the buffered enzyme solutions used for the preparation of the various microemulsion systems. These values do not necessarily correspond to the pH of the dispersed aqueous phase [11].

Enzyme activity was also studied as a function of various temperatures. As can be seen in Fig. 2a the optimum temperature for lipase activity was 30 °C. We assumed that the stability of this type of micro-emulsion systems is not affected in the range of studied temperatures.

The release of fatty acids as a function of reaction time at different enzyme concentrations and fixed substrate concentration (20 %, v/v, where $[S] \gg K_m$) was studied. Fatty acids formation was proportional to the enzyme concentration. The Kcat of the reaction was calculated to be 1.10×10^2 s^{-1}. Hydrolysis of triolein by lipase in microemulsions followed Michaelis-Menten kinetics. The apparent K_m was determined from a double-reciprocal plot and was found K_m, app = 155 mM (Fig. 3).

Effect of water content on the enzyme activity

The R ($R = [H_2O]/[AOT]$) values were used to express water content in microemulsions. Figure 4a represents the dependence of hydrolytic enzymic activity on R. As one see there is an increase of the

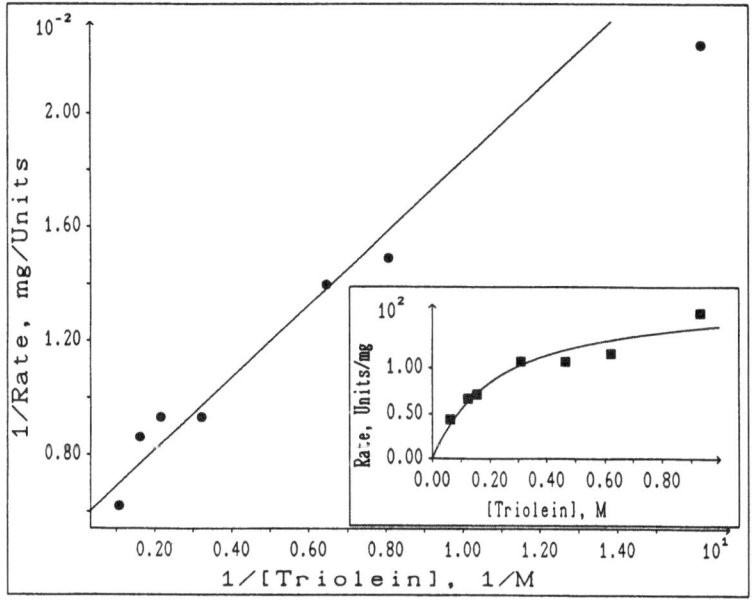

Fig. 3. The apparent K_m of lipase in AOT microemulsion system for triolein, calculated by Lineweaver-Burk double reciprocal plots. Enzyme activity was measured as function of various triolein concentration at $R = 9$, (AOT) = 50 mM and pH of stock enzyme solution 6.5 (inset shows the dependence of initial velocity on triolein concentrations)

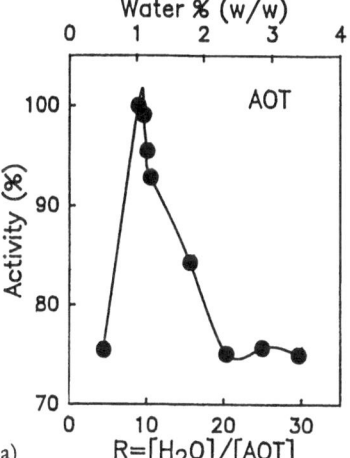

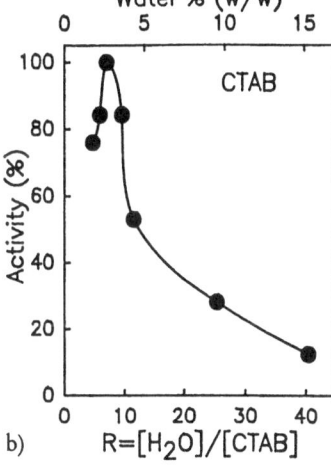

Fig. 4. Effect of R on lipase activity in microemulsions of (a) AOT (50 mM in isooctane); total enzyme concentration was in all cases 2 units/ml; pH of the stock enzyme solution was 6.5. (b) CTAB (0.2 M in isooctane); in this case the enzyme concentration was 4.5 units/ml; pH of the stock enzyme solution was 6.0

enzyme activity with increasing R up to $R = 9$. After this R value a decrease of the enzyme activity is observed. The above R value is almost the same with other R optimum values reported for other enzymes studied in the same microemulsion systems [12].

Structural studies on the AOT based microemulsion systems have shown that the water dispersed phase constitutes spherical droplets, formulated by a monolayer of the surfactant molecules [13]. Since their size depends on the water content, R can be a useful parameter to express the behavior of the enclosed enzyme. Thus, in small droplets (low R values) the restricted enzyme molecule cannot perform its maximum activity. In addition the hydrolytic reaction catalyzed by lipase is not favored, since the few water molecules are rather bound to the surfactant layer. At R values around 9 it seems that the enzyme can have its optimal conformation, resulting in a maximum activity. At higher R values the size of the droplet is such that a quantity of free water appears in the water core of the droplet. This excess amount of water is intercalated between the enzyme molecule and part of the

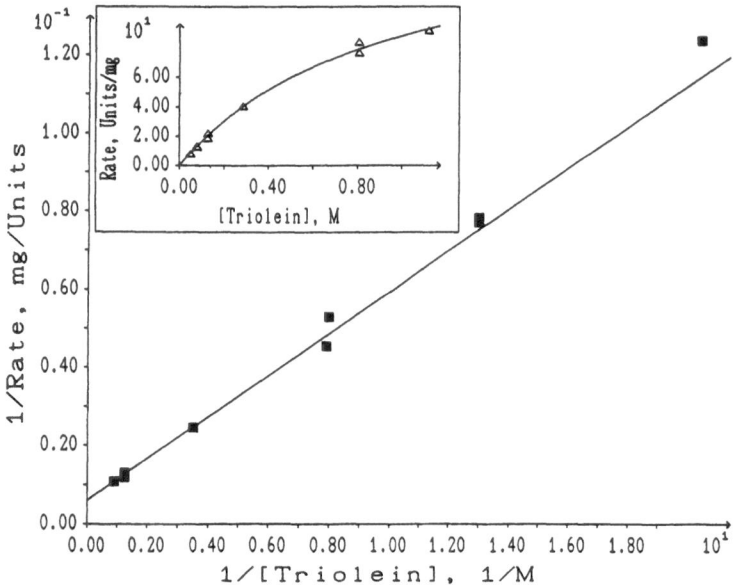

Fig. 5. The apparent K_m of lipase in CTAB micro-emulsion system for triolein, calculated by Lineweaver-Burk double reciprocal plots. Enzyme activity was measured, at (CTAB) = 0.2 M, $R = 7$ and pH of stock enzyme solution 6.0 (inset shows the dependence of initial velocity on triolein concentrations)

surfactant membrane of the droplet, lowering the interface through which the substrate contacts the enzyme.

Cationic systems

In order to elucidate the role of surfactants on enzymic activity, apart from the anionic type of AOT, we used in our studies a cationic type of surfactant. A typical cationic amphiphile is the cetyltrimethylammonium bromide (CTAB), which can form microemulsions in the presence of a cosurfactant. In our studies we chose to use as cosurfactant 1-pentanol, which is poorly soluble in water, while its five-carbon chain is small enough not to interfere in the reaction catalyzed by lipase.

Activity tests

Hydrolysis of triolein, used as substrate, was proportional to the enzyme concentration. The Kcat of the reaction was found to be Kcat = $1.167 \times 10^2 \text{ s}^{-1}$ and the K_m, app was calculated to be K_m, app = 876 mM, according to the Lineweaver-Burk double reciprocal plot (Fig. 5). In all cases a normal Michaelis-Menten behavior can be observed.

It is interesting to note that although the Kcat values are almost identical in both systems, the apparent K_m

value of lipase for triolein in CTAB is considerable higher compared to the K_m in the AOT systems, which means a lower affinity of the enzyme for this substrate. This is possibly due to the linear hexadecyl chain of the CTAB molecule, which can be compared to the length of the aliphatic chain of the oil substrates of the enzyme. This is not the case for AOT, where the hydrophobic moiety is a double-branched chain of medium length.

Optimum pH was found to be 5.8 (Fig. 1B). As one can observe, there is a slight shift towards lower values, compared to the optimum pH in AOT systems. This difference is possibly due to the nature of the ions present in the water pools of the two different systems.

The effect of the temperature on the reaction rate was also studied (Fig. 2B). It can be noticed that the activity decreases with increasing temperature over 22.5 °C. The stability of these microemulsions limit the number of temperature values that could be tested.

The dependence of the enzyme activity on the water content was also studied and the same pattern with the AOT systems was observed (Fig. 4B). The optimum enzyme activity was found at $R = 7$. In this case, however, R values may not reflect the sizes of the droplets because of the presence of the cosurfactant in the interface, as well as the continuous isooctane phase.

Nonionic systems

We used tetraethyleneglycoldodecylether ($C_{12}EO_4$)/ decane/water systems to study the activity of lipase. Hydrolytic activity was observed after an initial delay of about 1 h. Then, the appearence of free fatty acids increased linearly with time, up to a limit, when the stability of the microemulsion was perturbed. Similar results were obtained for microemulsions prepared with various quantities of detergent (5–20 %), or water (all the range of the monophasic area of the phase diagram [14]).

This unexpected behavior of the enzyme can be due to the presence of the small amounts of free fatty acids released during the reaction, affecting the thermodynamic stability of these very sensitive systems. It seems that the particular structure of these microemulsions is not suitable for hydrolytic assays of lipase, since very few water molecules are available [15].

It is possible that the low water activity may favor the reaction to the opposite direction [16]. Further work is in progress in this area.

In conclusion, among the different systems studied, the anionic one seems to be more suitable for the hydrolytic activity of lipase. Cationic surfactants can also be used, although a lower activity is observed, while nonionic systems do not seem adequate for this type of reactions.

Acknowledgments

This work was supported by an EEC grant (Contract N° BAP-0051-GR).

References

1. Mattson FH, Beck LW (1955) J Biol Chem 214:115
2. Desnuelle P (1961) Adv Enzymol 23:129
3. Martinek K, Levashov AV, Klyachko NL, Khmelnitski Y, Berezin IV (1986) Eur J Biochem 155:453
4. Luisi PL, Magid L (1986) Crit Rev Biochem 20:409
5. Luisi PL, Giomini M, Pileni MP, Robinson BH (1988) Biochim Biophys Acta 947:209
6. Fletcher PDI, Freedman RB, Robinson BH, Rees GD, Schoemacker P (1987) Biochim Biophys Acta 912:278
7. Martin CA, Magid L (1981) J Phys Chem 85:3985
8. Eicke HF (1982) Chimia 36:241
9. Han D, Rhee S (1986) Biotech Bioeng 28:1250
10. Lowry RR, Tiusley IJ (1976) J Am Oil Chem Soc 53:470
11. Smith RE, Luisi PL (1982) Helv Chem Acta 63:2302
12. Martinek K, Levashov AV, Klyachko NL, Pantin VI, Berezin IV (1981) Biochim Biophys Acta 657:277
13. Zulauf M, Eicke HF (1979) J Phys Chem 83:480
14. Ravey JC (1987) Prog Colloid Polym Sci 73:107
15. Ravey JC, Buzier M, Oberthur R (1987) Prog Colloid Polym Sci 73:113
16. Macrae AR (1983) J Am Oil Chem 60:281

Received November, 1988;
accepted January, 1989

Authors' address:

A. Xenakis
National Hellenic Research Foundation
Institute of Biological Research
48, Vas. Constantinou Ave.
GR-116 35 Athens, Greece

Progress in Colloid & Polymer Science   Progr Colloid Polym Sci 79:94–100 (1989)

B. Colloids of industriel interest

Preparation of monodisperse, magnetizable, composite metal/polymer microspheres

D. Charmot

Rhône-Poulenc Recherches, Aubervilliers, France

Abstract: Colloidal microparticles are being increasingly used as microcarriers in bioengineering, immunodiagnostics, and also for "in vivo" uses, for example, in drug targeting or as contrast aids in imaging systems. The advantages of microspheres over traditional supports are mainly related to the high specific surface and the absence of porosity which enhance reaction kinetics. Magnetic particles are even more attractive since they can be collected very simply using a single magnet. Among the different characteristics of the particles, the magnetic properties, internal morphologies, and size distribution are the most important.

In this paper we describe the preparation of monosized magnetic particles made up of a crosslinked polystyrene that entraps finely divided cobalt crystallites. The metal is introduced within performed polymer particles by thermolysis of metal carbonyl complexes. It is shown that nucleophilic pyridine units present in the polymer promote the thermolysis selectively inside the particles. The distribution of cobalt is discussed in relation with the chemical composition and crosslinked level of the microspheres.

Key words: Magnetic microsphere, monodisperse, metal encapsulation, dicobaltoctacarbonyl, thermolysis.

1. Introduction

Latex particles have long been recognized as a means of detecting antibody-antigen reactions in agglutination diagnostic tests [1]. They show various advantages over traditional fixation supports (coarse beads, tubes, plates), such as: large specific surface area, ease of chemical bonding, and high kinetics due to absence of porosity. These colloidal particles must, however, be collected and re-dispersed simply, rapidly, and reversibly (without agglomerating).

In this respect magnetic particles have some quite remarkable features in that they can be recovered even from heterogeneous media (fermentation broth, culture medium). Various methods have been proposed to prepare magnetic particles by incorporating a metal oxide within polymer particles [3–11]. In the processes described, the best conditions have been sought to obtain an ideal magnetic carrier, in terms of particle morphology (sphericity, magnetic oxide homogeneity in the polymer matrix), particle size distribution and magnetic properties. Indeed, actual user conditions of these magnetic lattices require:

1) superparamagnetic materials with zero magnetization in a zero field so as to avoid any irreversible agglomeration;

2) a minimum of interaction between the embedded magnetite and the surrounding biological fluids;

3) absence of internal porosity;

4) a narrow particle-size distribution, which, although not necessarily indispensable, allows for a better control of the surface and particle recovery efficiency.

The procedure described in this paper makes it possible to prepare magnetizable monodisperse particles in the micronic size range. These microspheres are made up mainly of a non-porous cross-linked styrene/divinylbenzene skeleton, and they enclose a finely divided metal (Fe, Co), deposited by thermal decomposition of metal carbonyls.

Table 1. Conditions of preparation of the particles

Latex ref.	PS 1	PS 2	MM 10/0	MM 10/5	MM 10/15	MM 10/25	MM 5/5	MM 5/15	MM 5/25
latex PS 1 (g)		52							
latex PS 2 (g)			100	100	100	100	100	100	100
K2S2O8 (g)	1	1							
S.D.S. (g)	9	4.5	1.1	1.1	1.1	1.1	1.1	1.1	1.1
PVP K30 (g)			0.82	0.82	0.82	0.82	0.82	0.82	0.82
water (g)	2100	2065	530	530	530	530	530	530	530
styrene (g)	900	885	108	102	90	78	108	96	84
D.V.B. (g)			12	12	12	12	6	6	6
4-VP (g)				6	18	30	6	18	30
Benzoylperoxide (g)			0.72	0.72	0.72	0.72	0.72	0.72	0.72
diameter (microns)	0.35	1.35	2.20	2.20	2.20	2.27	2.10	2.27	2.20

2. Experimental section

Materials

Styrene and divinylbenzene (actually a 55% pure grade of ortho-, meta-, para-isomers) were supplied by Norsolor (France); 4-vinylpyridine was purchased from Raschig (F.R.G.); sodium lauryl sulfate (SDS), polyvinylpyrrolidone (PVP K 30), and potassium peroxodisulfate were obtained respectively from Sidobre-Sinnova (France), GAF Products (U.S.A.), and Prolabo (France). All ingredients were used without further purification except the styrene, which was re-distilled and stored at $-15\,°C$ before use. Toluene was dried on a molecular sieve (3 Å). The water was deionized. Dicobaltoctacarbonyl $Co_2(CO)_8$ was provided by Ventron Alpha Products (F.R.G.) and ironpentacarbonyl $Fe(CO)_5$ by Aldrich (U.S.A.). Both metal carbonyls were titrated with bromine by measuring the carbon monoxide released. Purity was better than 99% in every case.

Polymerization

The polymer particles were synthesized using a standard emulsion polymerization technique: first, a polystyrene seed (PS1) was prepared by batch polymerization using SDS and potassium persulfate as emulsifier and initiator, respectively. Then PS1 (0.35 µm) was used in a seeded polymerization to increase the particle size up to 1.35 µm (PS2).

The particles used in this study were grown from latex PS2 by suspension polymerization using an oil-soluble initiator and the proper choice of monomer composition (styrene, DVB, 4-VP). A typical suspension recipe consisted of SDS (1.1 g), PVP K 30 (0.82 g), seed latex PS2 (100 g at 30% solid content), monomer mixture (120 g), and benzoylperoxide (0.72 g). The SDS and PVP were dissolved in water and mixed with the seed latex, which was allowed to swell with the monomer/initiator mixture at 40 °C for 6 h under vigorous stirring (400 rpm). Then the temperature was raised to 80 °C and held constant for 12 h. Monomer completion was achieved by raising the temperature to 90 °C for another 2 h. Finally, the particles were steam-stripped to remove unreacted monomer. All of the microspheres obtained were monodisperse with an average diameter in the range of 2.1–2.3 microns. (See Table 1 for the characteristics of the particles).

Metal impregnation

The particles prepared according to the procedure described in the last section were spray-dried. Two grams of the powder thus obtained and 18 g of toluene were loaded into a 50 ml, three-necked, round-bottom flask equipped with a thermometer, a reflux condenser, a magnetic stirrer, a gas inlet, and a serum cap to introduce the metal carbonyl. The condenser was connected to a gas burette. The entire system was heated in an oil bath to 110 °C to remove residual water by azeotropic distillation, purged overnight with dry nitrogen, and then cooled to room temperature. The metal carbonyl (4.1 ml of a solution in toluene at 12.5 wgt. %) was introduced with a syringe. Spontaneous CO release was monitored by displacement of water from the gas burette. The oil-bath temperature was then increased to 120 °–130 °C until the reaction was completed. Finally, the particles were collected magnetically, washed in ethanol, redispersed in water, and steam-stripped to remove the solvent.

Particle characterization

The particles were examined with a JEOL 100 CX transmission electron microscope and a JEOL JSM 35 scanning electron microscope. Internal morphologies were observed on ultrathin sections using Epon 601 as an embedding resin and a Reichert-Jung ultramicrotome. Absence of porosity was checked by the B.E.T. method. Metal content was determined by atomic absorption after mineralization. Magnetic properties were measured on a Hyste 5000 (Rhône-Poulenc Systèmes, France) B-H meter.

3. Results

A first experiment was carried out to highlight the role of vinylpyridine by comparing particles MM 10/0 and MM 10/5 impregnated with cobalt. The cross-sections shown in Figs. 1a and b make it clear that without pyridine functions the cobalt is deposited only on the surface of the beads. For particles MM 10/5 the metal grains are visible within the spheres with a slightly higher concentration near the surface. It is to be

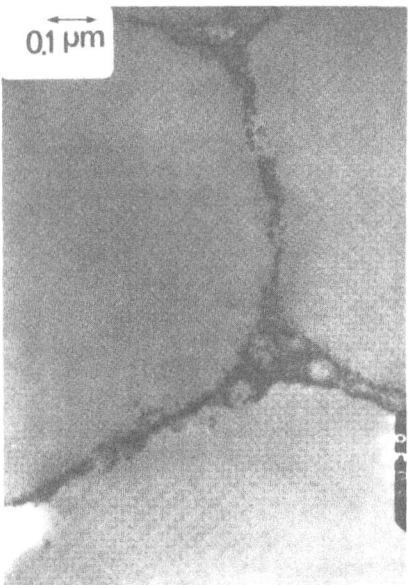

Fig. 1a. TEM views of ultrathin sections of particles MM 10/0 impregnated with cobalt. Cobalt/polymer: 0.09. The metal is localized outside the particles

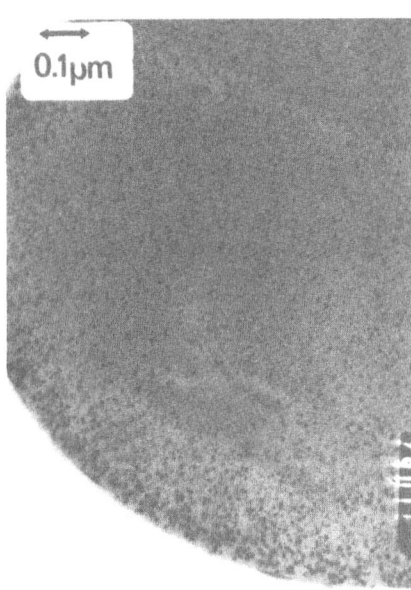

Fig. 1b. TEM views of ultrathin sections of particles MM 10/5 impregnated with cobalt. Cobalt/polymer: 0.09. The metal is seen inside the particles with a higher concentration at the outer shell of the beads

observed that in these experiments and those described further on, the metal is deposited exclusively on or in the polymer, and the dicobaltoctacarbonyl is totally consumed during thermolysis. This in indicated by the absence of any infrared signal at 2000 cm^{-1} in the

Fig. 2. SEM views of particles MM 10/5 impregnated with cobalt. Cobalt/polymer: 0.09. The polymer surface has been altered during the thermolysis

toluene phase. It is also seen from the scanning microscope views (Fig. 2) that the thermal decomposition process of $Co_2(CO)_8$ within the solvent-swollen beads is accompanied by a kind of blistering of the particles, which were originally smooth.

In a second set of experiments, the proportion of 4-VP in the terpolymer was varied, keeping the cross-linking level constant at 10% DVB. Once the carbonyl complex is added to the dispersion at 20°C, carbon monoxide is freed. The CO volume increases as the 4-VP content is augmented from 5 to 25% (Fig. 3). On the other hand, after thermal decomposition the tendency is reversed. Thermolysis efficiency (as gauged by the ratio of measured CO to theoretical CO) is close to 100% for MM 10/5 but only 70% for MM 10/15 and MM 10/25 (Fig. 4), with a slower reaction kinetic. The cross-sections (not shown) reveal a fine, homogeneous metal distribution in the polymer. When the pyridine content increases from 5% to 25%, the size of the cobalt crystallites decreases from 40–100 Å to 20–40 Å.

In another series of experiments the pyridine content was varied and the rate of crosslinking kept at a lower level (DVB: 5%). The CO release kinetics are given in Fig. 5, and the tendencies already observed show up again: the speed and efficiency of thermolysis decreases as the level of 4-VP is increased. The decomposition curves all show a common feature — a fast expulsion of about 25% of the CO in the first minutes of the reaction.

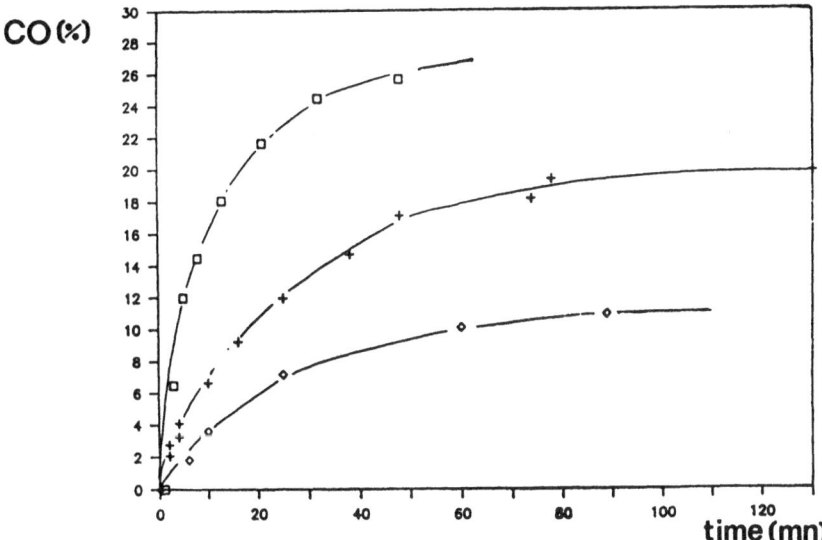

Fig. 3. Kinetics of carbon monoxide release after addition of $Co_2(CO)_8$ to the particle dispersion. Temperature: 20 °C. MM 10/5: –◇–◇–◇, MM 10/15: –+–+–+, MM 10/25: –□–□–□–

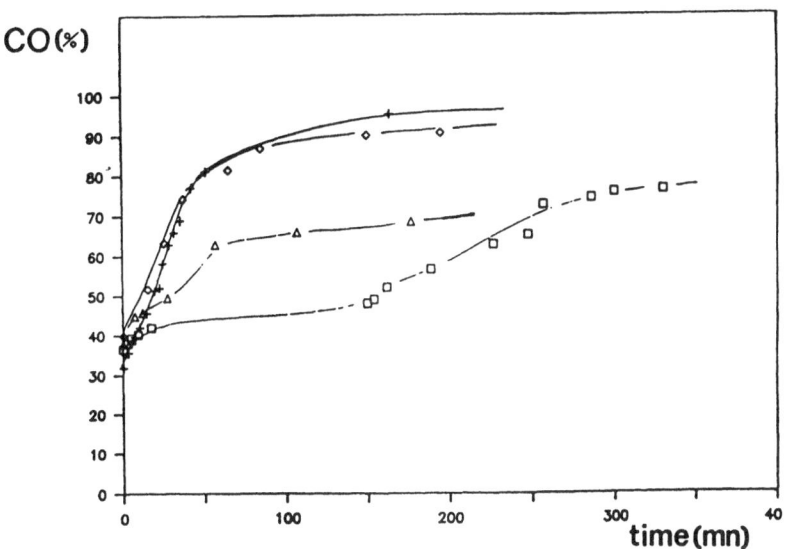

Fig. 4. Kinetics of carbon monoxide release after addition of $Co_2(CO)_8$ to the particle dispersion. Temperature: 110 °C. MM 10/5: –◇–◇–◇, MM 10/15: –□–□–□–, MM 10/25: –△–△–△

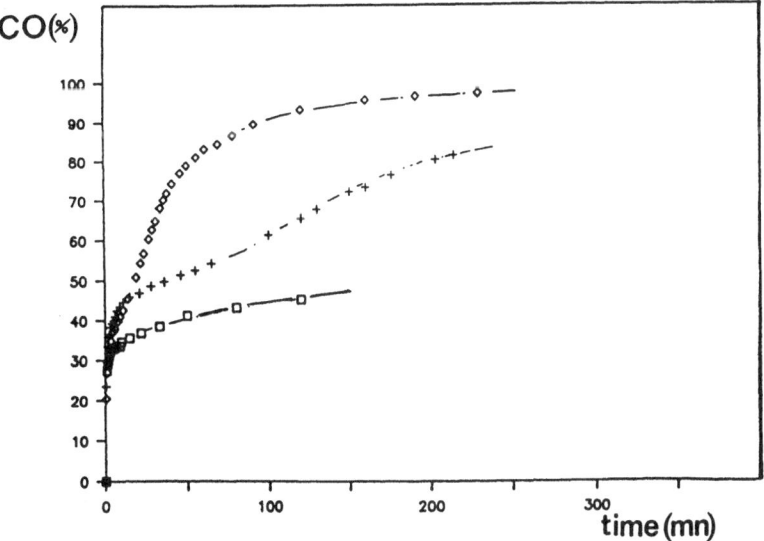

Fig. 5. Kinetics of carbon monoxide release after addition of $Co_2(CO)_8$ to the particle dispersion. Temperature: 110 °C. MM 5/5: –◇–◇–◇, MM 5/15: –+–+–+, MM 5/25: –□–□–□–

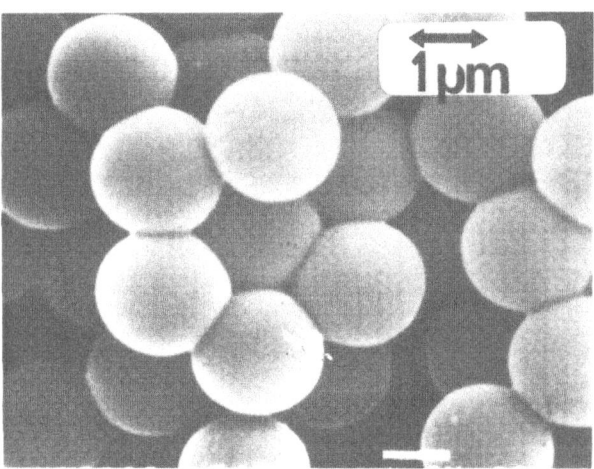

Fig. 6. SEM views of particles MM 5/5 impregnated with cobalt. Cobalt/polymer: 0.09. Magnification: 10 000

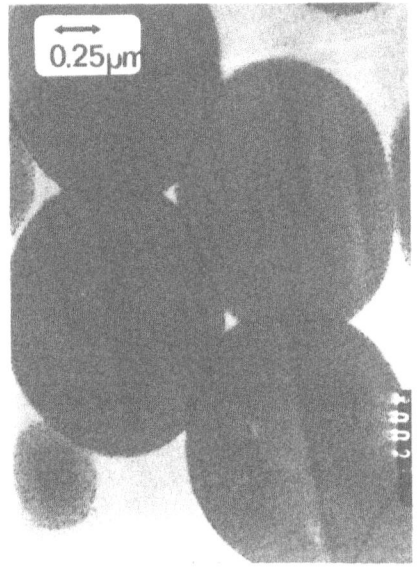

Fig. 7. TEM views of ultrathin sections of particles MM 5/5 impregnated with cobalt. Cobalt/polymer: 0.09

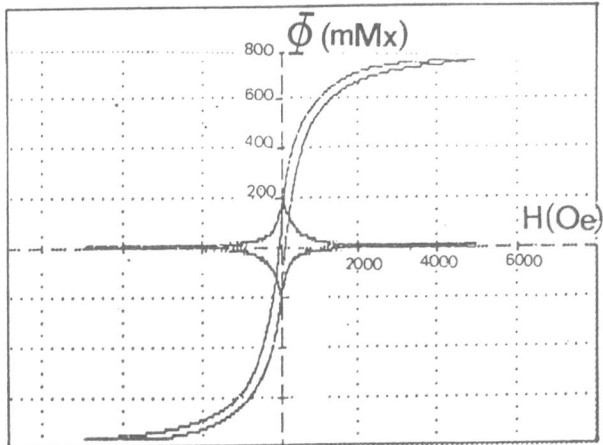

Fig. 8. Aimantation curve of particles MM 10/5 impregnated with cobalt. Cobalt/polymer: 0.09. The quasi-absence of hysteresis demonstrates the super paramagnetic behavior of the microspheres. Aimantation: at saturation 3.8 e.m.u./g; and residual 0.7 e.m.u./g

For the product with low pyridine group substitution (MM 5/5) a clear acceleration of the reaction is observed beyond 45 % CO release, reaching 100 % efficiency in 60 min. For the particles with a higher 4-VP level, this acceleration begins late (MM 5/15) or not at all (MM 5/25). Here again the size of the cobalt grains decreases as a function of the pyridine group level. Accordingly the magnetism of the microspheres falls off sharply. Under the scanning microscope (Fig. 6) it

is seen that the decrease in crosslinking from 10 % to 5 % restores a perfectly smooth surface to the spheres.

A decrease in cross-linking improves the homogeneity of metal distribution even more, as shown in Fig. 7, which represents the cross-sections of the MM 5/5 particles impregnated with 9 % cobalt

The magnetization curve of this product, given in Fig. 8, is characteristic of superparamagnetic materials (no hysteresis). The value of saturation magnetization of the entrapped cobalt is 42 e.m.u./g; this value is lower than that of saturation magnetization of bulk cobalt: 160 e.m.u./g [12].

The tests carried out with ironpentacarbonyl $Fe(CO)_5$ are summarized in Table 2. As $Fe(CO)_5$ is thermally more stable, solvents with higher boiling points are used, and thermolysis times are prolonged. The products obtained do not show any magnetic properties. Under examination of particle cross-sections the metal grains cannot be distinguished (at a microscope resolution of around 20 Å)

4. Discussion

The thermal decomposition of labile metal complexes, such as carbonyl complexes, when initiated within polymer particles, is a good way to produce mixed metal/polymer particles directly. The solvent used to dissolve the metal complex swells the particles so as to ensure a homogenous distribution of the

Table 2. Thermolysis of Fe(CO)$_5$

latex ref.	Fe/polymere (%)	Temp. (°C)	solvent	time (h)	Co evolved (%)	residual Fe(CO)$_5$[a]) (%)
MM 10/15	13.3	130	chlorobenzene	22	100	0.1
MM 10/15	13.0	135	xylene	14	93	n.d.

[a]) measured by IR spectroscopy at $2\,000$ cm^{-1}

metallic precursor in the polymer. It is clear that cobalt distribution in the particles is controlled by the presence of pyridine groups, which initiate metal precipitation. In the absence of pyridine functions, it is the polar units on the particle surface (sulfate groups, polyvinylpyrrolidone) that play this role, and the metal is deposited on the surface.

Previous works on ferrofluid synthesis by thermolysis of metal carbonyls [15–17] have shown that strongly polar groups catalyze the decomposition of metal carbonyls and induce nucleation of the metal grains. In the experiments performed at 20 °C (see Fig. 3), the extent of carbon monoxide released vs. the 4-VP content indicates that a ligand exchange between CO and the pyridine sites probably takes place in the Co$_2$(CO)$_8$ coordination sphere. When the temperature is raised about 25 % of the carbon monoxide is evolved almost immediately, which corresponds to a possible formation of the Co$_2$(CO)$_8$ dimer complex (Eq. (1)), or to a ligand substitution between a CO molecule and a pyridine group (Eq. (2)).

$$2\ Co_2(CO)_8 \longrightarrow Co_4(CO)_{12} + 4\ CO \qquad (1)$$

$$Co_2(CO)_8 + 2\ Pyr$$
$$\longrightarrow [Co_2(CO)_6(Pyr)_2] + 2CO. \qquad (2)$$

The CO release kinetics at 120 °C show that intermediate complexes are formed all along the reaction. Depending on the respective proportions of Co$_2$(CO)$_8$ and 4-VP present, two kinds of complexes are likely to form: Co$_4$(CO)$_{12}$ and Co$_2$(CO)$_{7-n}$ (Pyr)$_n$, respectively. As Co$_4$(CO)$_{12}$ is only slightly soluble in toluene, it is most likely to precipitate immediately in the polymer phase. These intermediate complexes are progressively thermolyzed into metal-rich clusters which give the nuclei. The acceleration observed at 45 % conversion is due to the catalytic effect of the growing nuclei on the decomposition reaction. The increase in pyridine sites multiplies the

nucleation sites and consequently diminishes the size of the crystallites.

It is not clearly understood why the reaction stops beyond 70 % for 4-VP-rich particles. The absence of dicobalt-octacarbonyl and cobalt-tricarbonyl at the end of the process indicates that the residual carbon monoxide is associated with the metallic clusters inside the microspheres. These mixed Co/Pyr/CO complexes, enriched in pyridine ligands, are probably not prone to thermolysis and do not give metallic cobalt. This could explain, at least partially, the drastic decrease in aimantation level.

For the particles treated with Fe(CO)$_5$, the absence of magnetic properties may be linked to the small size of the metal crystallites, since the crystal defects induced on the surface become very great and oxidation progresses more rapidly [13]. The formation of an amorphous Fe-C alloy during the decomposition of Fe(CO)$_5$ may also be a cause [14].

The surface blistering is most probably linked to CO release during the impregnation step. When crosslinking is low, the particles swell up considerably in the solvent, and their internal viscosity is low enough to allow the gas to escape without deforming the spheres. (It must be realized that 1 gram of particle gives off nearly 150 cm^3 of CO!). This is not the case when the DVB content is higher, leading to severe deformation of the beads after thermolysis. Moreover, during the reaction, the strongly cross-linked polymer hinders precursor diffusion in the particles and leads to a high peripheral concentration of cobalt. On the other hand, crosslinking at less than 5 % (expressed in percent DVB) gives gels that are difficult to handle. Finally, styrene/DVB/4-VP terpolymer particles (MM 5/5) offer the best compromise in terms of metal distribution and surface aspect of the beads.

5. Conclusions

It has been demonstrated that mixed cobalt/polymer monodisperse particles, nonporous and with the metal perfectly distributed within the microspheres, can be produced. The particles are made up of a cross-linked styrene/divinyl benzene/4-vinylpyridine terpolymer and are synthesized by emulsion polymerization with two successive seeding steps. After redispersion in an organic medium in the presence of dicobaltoctacarbonyl, a thermolysis reaction forms cobalt nuclei within the particles. Size and distribution of the cobalt crystals are controlled by the pyridine group density in the polymer. The CO release kinetic curves tend to indicate the existence of intermediate complexes (clusters) which form within the particles in the course of the reaction. These clusters lead to the formation of metallic nuclei, which in turn catalyze the decomposition reaction. It is also shown that the rate of crosslinking affects the surface morphology of the beads and, to a lesser extent, the gradient of metal concentration within the particles. These observations are analyzed in terms of internal bead viscosity, which resists carbon monoxide release and metal precursor diffusion in the swollen particles.

This encapsulation process is easy to use and can theoretically be applied to any kind of cross-linked polymer particle that has polar groups, using thermally decomposable complexes.

The particles exhibit superparamagnetic behavior, which render them particularly suitable for biomedical applications as fixation supports in diagnostic tests or affinity chromatography.

Acknowledgment

The author thanks N. Pecate, who performed most of the experimental work, and L. Mennecier for her contribution in electron microscopy, and Rhône-Poulenc for permitting the publication.

References

1. Singer J (1956) Am J Med 21:888
2. Munro PA, Lilly MD, Dunnill P (1977) Biotechnology and Bioengeneering 19:101-124
3. Margel S, Zisblatt S, Rembaum H (1979) J Immunol Meth 28:341-353
4. Margel S et al (1982) German Pat Appl No DE 3224484 A1
5. Margel S et al (1982) J Cell Sci 56:157-175
6. Rembaum H et al (1980) US Pat 4224359
7. Rembaum H et al (1979) US Pat 4157323
8. Rembaum H et al (1980) US Pat 4197220
9. Rembaum H et al (1978) Chemtech 182
10. Ugelstad et al (1985) J Polym Sci Polym Symp 72:225
11. Daniel JC, Schuppiser JL, Tricot M (1981) Rhône-Poulenc, Eur Patent 38730
12. Cullity BD (1972) In: Introduction to magnetic materials. Addison-Wesley Pub, p 386
13. Griffiths CH, O'Horo MP, Smith TW (1979) J Appl Phys 50(11):7108
14. Van Wonterghem J et al (1988) J Colloid Interface Sci 121(2)
15. Hess PH, Parker PH (1965) J Appl Polym Sci 10:1915-1927
16. Thomas JR (1966) J Appl Phys 37:2914
17. Smith TW, Wychick D (1980) J Phys Chem 84: 1621-1629

Received November, 1988;
accepted January, 1989

Author's address:

D. Charmot
Rhône-Poulenc Recherches
52, Rue de la Haie-Coq
F-93308 Aubervilliers Cedex, France

Progress in Colloid & Polymer Science   Progr Colloid Polym Sci 79:101–105 (1989)

Fractal structure of Portland cement paste during age hardening analyzed by small-angle x-ray scattering

M. Kriechbaum[1]), G. Degovics[1]), J. Tritthart[2]), and P. Laggner[1])

[1]) Institut für Röntgenfeinstrukturforschung der Österreichischen Akademie der Wissenschaften und der Forschungsgesellschaft Joanneum, Graz, Austria
[2]) Institut für Werkstoffkunde, Festigkeitslehre und Materialprüfung der Technischen Universität Graz, Graz, Austria

Abstract: The changes in inner structure of hydrating Portland cement paste during the process of hardening were studied by small-angle x-ray scattering. The scattering curves taken at different times between 1 and 28 days after preparation showed two different decay exponents as demonstrated by the linear slopes in their log-log plots, related to the mass and surface fractal structure, respectively, of Portland cement. The time-course of the mass fractal dimension changes from an initial value of 1.9 (day 1) to 2.8 (day 28); that of the surface fractal dimension remains almost constant around 2.8. This may indicate the development from a very ramified and porous structure to a more compact and homogeneous one by interlinking and space-filling the loose initial gel-network with solid calcium silicate hydrate and calcium hydroxide.

Key words: Portland cement, small-angle x-ray scattering, age hardening, fractals, inner surface.

Introduction

The knowledge of material property and behavior of Portland cement, a major raw building material used world-wide is of great importance for its technical use. Therefore, characterization of its static nature or of dynamic processes during and after hardening may be helpful to specify or to further improve its technical-mechanical qualities.

The main components of anhydrous Portland cement are calcium silicate, calcium aluminate, and ferrite solid solution of variable stoichiometric compositions, with the major and most important component being tricalcium silicate [1]. Complex chemical hydration reactions take place upon addition of water, resulting in the solidification of hardened cement in a slow process (age hardening). Among these reactions the formation of the calcium silicate hydrate (CSH) gel around the initial grain particles of the anhydrous powder and the formation of calcium hydroxide crystals play a major role during age hardening.

Ultrastructural investigations of Portland cement have been focused on the determination of inner surface area (specific surface) by means of vapor sorption techniques [2], small-angle x-ray scattering (SAXS) [3–5] and small-angle neutron scattering (SANS) [6, 7], which all gave different results. Therefore, it seems difficult to characterize the inner surface area of this porous material and the values reported depend quite strongly on physical methods used.

Because of its disordered and irregular inner structure, and due to the material growth processes involved in age hardening, Portland cement may exhibit fractal properties. The conception of fractal dimension was introduced by Mandelbrot [8] and can be applied to the internal structure of many materials such as rocks, clay, coal, and silica [9]. Theoretical models involving fractal geometries have been successfully proposed for diffusion limited aggregation and growth mechanisms [10]. In two recent works, one with SANS [7] and one with NMR [11], the possible significance of fractal geometry to the structural description of Portland cement was taken into account.

The goal of our present work was to investigate by SAXS the slow dynamic structural changes of Portland cement structure in the process of age hardening, and to discuss the results in terms of the concept of fractal geometry.

Methods and materials

Cement paste samples of Portland cement type PZ 375 were prepared with a water/cement ratio of 0.5. For the SAXS measurements the freshly prepared sample was filled in a punched oval hole (13 × 6 mm) of a 0.4 mm thin stainless steel sample holder (20 × 10 mm) and sealed on both sides with 0.15 mm thin cover slides.

SAXS experiments were carried out with a Kratky compact camera (A. Paar, Graz, Austria) with slit-collimation geometry, using Ni-filtered CuK$_\alpha$ radiation (wavelength $\lambda = 1.54$ Å) from a conventional x-ray tube (Philips, Eindhoven, The Netherlands) with copper anode, operated at 50 kV and 40 mA. Scattering curves of the cement sample were recorded in acquisition times of 1 000 s with a one-dimensional position-sensitive x-ray detector (M. Braun, Munich, F.R.G.) with energy discrimination, at different intervals between 1 and 28 days after preparation. The data were evaluated after subtraction of blank and instrumental background without correction for primary beam geometry (desmearing).

Determination of fractal dimension by SAXS

The typical property of an ideal inifinite sized fractal object is its scale-invariant selfsimilarity, which means that its structure (volume or surface) has the same appearance at all length scales. Scattering techniques like SAXS can determine the scaling properties in the submicroscopic region, since they probe the density-density correlation function on length scales covering the range between one and several thousand Å. From the decay of the scattered intensity I towards larger scattering angles 2θ (θ is related to the reciprocal scattering vector $h = (2\pi/\lambda)\sin 2\theta$, with λ = wavelength of the x-rays) one can directly obtain the fractal dimension D by the relationship:

$$I(h) \sim h^p . \tag{1}$$

For objects whose volume or mass is fractal (clusters, aggregates) the exponent p reflects directly the mass fractal dimension D_m by:

$$p = -D_m, \text{ with } 1 < |p| < 3 \text{ and } 1 < D_m < 3 .$$

For objects whose surface is fractal (powders with porous, irregular surface), the exponent p is related to the surface fractal dimension D_s by:

$$p = -(6 - D_s), \text{ with } 3 < |p| < 4 \text{ and } 2 < D_s < 3 .$$

These relationships hold only for scattering curves measured with point-collimation geometry of the incident x-ray beam. In case of slit-collimation geometry the absolute value of p will be exactly 1 less (in the ideal case of infinitely long slits) than would be

obtained with point-collimation and has to be corrected to $|p| + 1$ [12].

For a uniform particle with a homogeneous core and a smooth surface, which is non-fractal ($D_m = 3$ and $D_s = 2$), the well known Porod's law [13] is valid:

$$I(h) = k \cdot h^{-4} . \tag{2}$$

With the Porod-constant k, evaluated, for example, from a $I(h) \cdot h^4$ vs. h^4 plot, the specific inner surface S_i of a two-phase system with the volume fractions φ_1, φ_2 can be calculated by the relationship:

$$S_i = \pi k \cdot \varphi_1 \cdot \varphi_2 / \int_0^\infty I(h) \, h^2 dh . \tag{3}$$

This approach, however, fails if the structure has fractal properties, i.e., if the exponent in Eq. (2) is less than 4.

To illustrate the determination of D_m of a fractal object from its scattering curve, the so-called Menger's-sponge [8] — a fractal with $D_m = 2.72$ (Fig. 1) was approximated by modelling its geometric shape, placing 8 000 spheres of radius $r = 0.62$ Å on specified lattice coordinates in a cubic 27 × 27 × 27 Å lattice with 1 Å distance spacing. This approximates the Menger's-sponge of the third generation. The scattering curve $I(h)$ of this finite-sized fractal was then calculated analytically, using the Debye-formula [14]. From the log I vs. log h plot a slope of $p = -2.7$ can be obtained (Fig. 1), in good agreement with the fractal dimension according to Eq. (1).

Results

The SAXS curves of cement paste, measured at different times of age-hardening after preparation are shown in Fig. 2 in double-logarithmic plots. They exhibit two linear regions over the h-range 0.015 to 0.07 Å^{-1} and 0.07 to 0.18 Å^{-1} with different slopes p_1 and p_2, respectively. These two slopes might be interpreted as mass and surface fractal dimensions, D_m and D_s, respectively, of the cement structure. The values of D_m and D_s calculated from Eq. (1) — assuming infinite slit length — are compiled in Table 1.

In order to check the influence of the beam slit geometry on the value of the slope p, the theoretical scattering curve of a statistical fractal with $D_m = 1.7$ (Fig. 3), approximated by 149 spheres (with the same specification as for the Menger's-sponge) was smeared with various beam-profiles. These simulations were performed according to the method of Glatter [15], choosing a constant slit-width profile and different slit-length profiles, (the transformation factor between the h-scale and the slit-width and slit-length scale in this simulation is 1). Figure 3 shows the unsmeared (curve 1) and the beam-profile (shown in the insert) smeared curves (curves 2–5) in a log-log plot and the corresponding values of the slopes p in the linear range between $h = 1/R$ (R is the radius of gyration of the

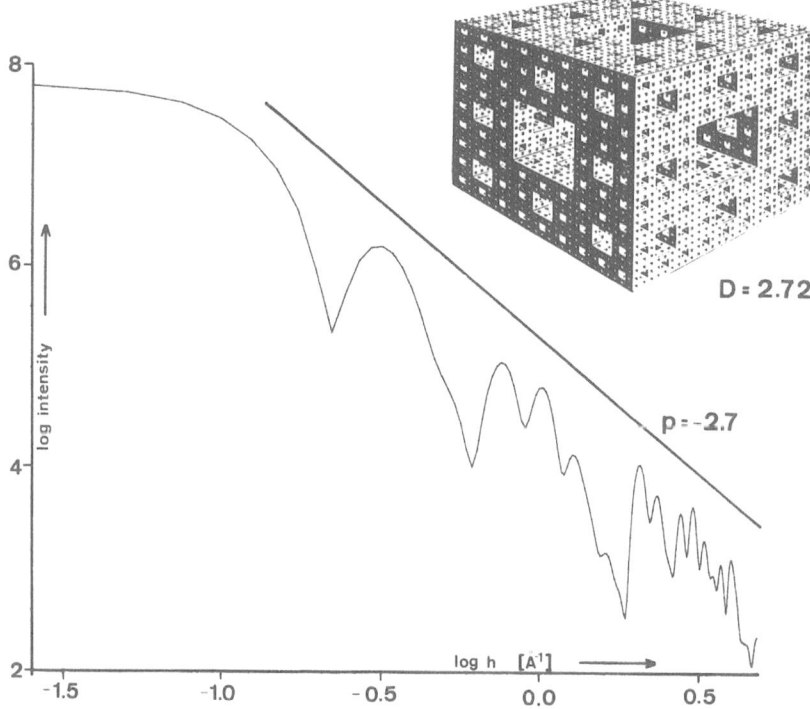

Fig. 1. The theoretical scattering curve of a Menger's-sponge structure approximated by spheres in places of the mass-filled subcubes (see text). Insert shows a Menger's-sponge of the fourth generation, i.e., containing four steps of scale-invariant selfsimilarity in its structure

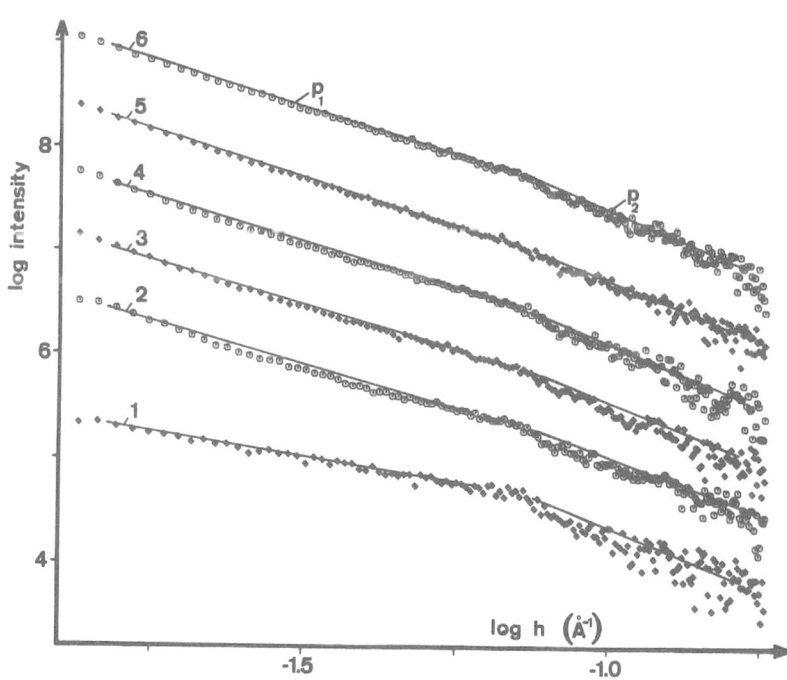

Fig. 2. Double logarithmic SAXS-curves of Portland cement paste at different stages of age hardening, increasing from curve 1 to curve 6. The age in days and the numerical values of the slopes p_1 and p_2, obtained by weighted least-squares fits are listed in Table 1. The curves are vertically displaced for clarity and the log intensity scale is in relative units

Table 1. Fractal dimensions D_m and D_s of Portland cement paste at different time intervals of age hardening after preparation. The two slopes p_1 and p_2 for the calculation of $D_m = |p_1| + 1$ and $D_s = 6 - (|p_2| + 1)$, assuming infinitely long slits, are obtained from the linear parts of the log-log plots of the scattering curves (Fig. 2) in the h-range 0.015–0.07 Å^{-1} and 0.07–0.18 Å^{-1}, respectively

curve no. in Fig. 2	days of age hardening	slope p_1	$D_m \pm 0.1$	slope p_2	$D_s \pm 0.1$
1	1	−0.9	1.9	−2.1	2.9
2	2	−1.6	2.6	−2.1	2.9
3	4	−1.7	2.7	−2.3	2.7
4	8	−1.7	2.7	−2.3	2.7
5	14	−1.8	2.8	−2.2	2.8
6	28	−1.8	2.8	−2.3	2.7

entire particle) and $h = 1/R_u$ (R of the structural subunits, used for the approximation of the entire particle).

Discussion

The SAXS results (Fig. 2 and Table 1) reveal that the structure of Portland cement can be considered as having both mass fractal (slope p_1) as well as a surface fractal (slope p_2) properties. During age-hardening, D_m changes significantly from an initial value of 1.9 (after 1 day), to 2.8 after 28 days, whereas D_s changes only slightly from 2.9 to 2.7 in the same time period. It is

noteworthy that the decay exponents p in the measured h-range do not reach the value −4 (−3 for infinitely long slit-length) as expected from Porod's law for ideally smooth inner surfaces, even after 28 days of hardening. Thus, no definable specific inner surface can be calculated in this h-range for the hardened cement sample as well as for other materials with a fractal structure, using the approach of Eqs. (2) and (3). Although it is therefore impossible to obtain a physically meaningful value for the inner surface, the time-course of D_m during age-hardening provides some information about the possible solidification process of the (CSH) cement gel. The enhancement

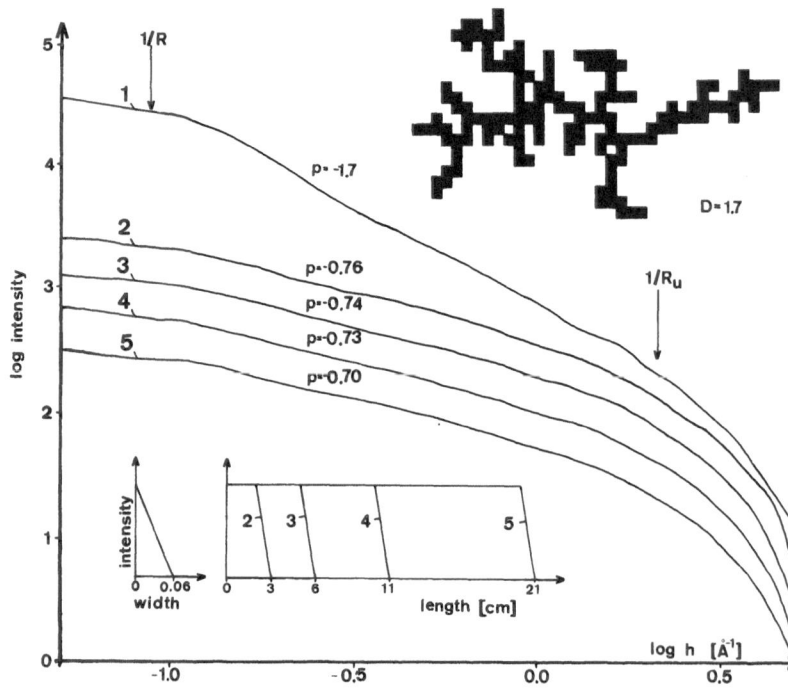

Fig. 3. Theoretical scattering curves of a two-dimensional statistical fractal with $D_m = 1.7$ (curve 1). This theoretical scattering curve has been convoluted with symmetrical beam-profiles 2–5 (insert shows one-half of each profile) and the resulting "smeared" curves (2–5) have slopes p as indicated

of D_m can be interpreted in terms of a development from a very ramified open structure to a more compact, closed one, induced by an increasing crosslinking and decreasing extension of the particles into its surrounding. This process is paralleled by an increase of the compressive strength of Portland cement, shown in quantitative measurements of cement compounds by Williamson [1].

The high value of D_s indicates a very rough and porous surface structure which is largely conserved during age hardening with only a slight trend to become smoother. The crossover point at $h = 0.07$ Å^{-1} in the double-logarithmic $I(h)$ plot, which is at virtually the same position in all measurements corresponds to a size dimension of about 15 Å and can be associated with the average radius of gyration of the structural units [16], building up the mass fractal structure and forming the fractal surface of Portland cement.

The results of our SAXS measurements are comparable to those obtained by Allen et al. [7] and Pearson et al. [6] by neutron scattering. Their neutron scattering curves also exhibit a crossover-point around $h = 0.08$ Å^{-1} and a linear region for $h < 0.08$ Å^{-1} in the log-log plots with the slopes -2.7 [6] (after hardening and 25 days storage underwater) and -2.5 [7] (after 28 days age hardening), respectively.

It should be mentioned that there are two possible sources of error which can influence the value of D_m or D_s: First, polydispersity, which modifies Eq. (1) [17], and second, the finite beam slit-length. The correction to add 1 to the logarithmic decay exponent p, which we made in this study, is sufficient if the slit-length is three-times larger than the linear h-range in the log-log plot, as can be seen in the simulation in Fig. 3, which matches approximately our experimental conditions. We did not consider the problem of polydispersity any further at this stage, since the extended linear regions in the log-log plots are highly persuasive for us to treat the structure in terms of fractal geometries and since polydispersity models would additionally require quite specific and detailed assumptions on size distributions. On the other hand, it appears to be a very attractive idea to treat and to classify the internal growth and diffusion processes in this complex system by two simply definable parameters: the mass and surface fractal dimensions.

The salient conclusion from this study is the important notion that the concept of describing the microporosity of cement by an internal specific surface can not be upheld. It will be important therefore to establish technologically useful relationships between fractal dimensions and material properties. SAXS, by its ease of determining these parameters, is certainly a highly valuable tool in this direction. The inclusion of absolute scale measurements, presently in progress in our laboratory, may further enhance this potential.

Acknowledgement

This work has been supported by the Österreichischer Fonds zur Förderung der Wissenschaftlichen Forschung through grant no. P6287C (P.L.).

References

1. Williamson RB (1972) Prog Mater Sci 15:189-286
2. Powers TC, Brownyard TL (1946) Proc Americ Concr Inst 43:469-504
3. Winslow DN, Diamond S (1973) J Colloid Interface Sci 45:425-426
4. Winslow DN, Diamond S (1974) J Am Ceram Soc 57:193-197
5. Vollet D, Craievich A, Regourd M (1984) J Am Ceram Soc 67:315-318
6. Pearson D, Allen AJ (1985) J Mater Sci 20:303-315
7. Allen AJ, Oberthur RC, Pearson D, Schofiel Wilding CR (1987) Philos Mag B 56:263-288
8. Mandelbrot BB (1982) The Fractal Geometry of Nature. Freeman, San Francisco
9. Avnir D, Farin D (1984) Nature 308:261-263
10. Sander LM (1986) Nature 322:789-793
11. Blinc R, Lahajnar G, Zumer S, Pintar MM (1988) Phys Rev B 38:2873-793
12. Keefer KD, Schaefer DW (1985) Phys Rev Lett 56:2376-2379
13. Porod G (1951) Kolloid-Z 124:83-114
14. Debye P (1915) Ann Physik 46:809-823
15. Glatter O (1977) J Appl Cryst 10:415-421
16. Hurd AJ, Schaefer DW, Martin JE (1987) Phys Rev A 35:2361-2363
17. Martin EJ (1986) J Appl Cryst 19:25-27

Received November, 1988;
accepted January, 1989

Authors' address:

M. Kriechbaum
Institut für Röntgenfeinstrukturforschung
Steyrergasse 17
A-8010 Graz, Austria

Progress in Colloid & Polymer Science

Progr Colloid Polym Sci 79:106–111 (1989)

Asphalt emulsions: experimental study of the cationic surfactant adsorption at the asphalt-water interface

J. E. Poirier[1]), M. Bourrel[2]), P. Castillo[3]), C. Chambu[4]), and M. Kbala[2])

[1]) C.N.R.S. U.A 235 Minéralurgie, Vandoeuvre,
[2]) S.C.P./D.P.A.M. Centre de Recherches de Lacq. Artix,
[3]) L.T.E.M.P.M. Université de Pau et des pays de l'Adour. Pau, and
[4]) CECA S.A. Centre de recherches. Levallois-Perret, France

Abstract: Cationic asphalt-in-water emulsions are widely used for road construction and surfacing. These systems are an interesting model for the basic study of emulsions because the high asphalt viscosity allows the system to afford some perturbations without readily producing coalescence. The adsorption of a cationic surfactant, an octadecyl-propylene-diamine, at asphalt-water interface is investigated. The adsorption isotherm was determined using two different experimental procedures. The reversibility of adsorption was studied. A plateau occured on the isotherm at the CMC of surfactant. Its value was used to calculate the molecular packing area of the surfactant. A value of 110 Å^2 was found, compared to the 160 Å^2 found at the liquid-air interface. This value showed that no more than a monolayer was adsorbed. Due to the adsorbed surfactant cations, the asphalt droplet electrokinetic potential was positive. The zeta potential values and adsorbed amounts were strongly correlated. The zeta maximum value, 105 mV, was reached at the maximum adsorbed amount, 1.5 micromol/m^2. An attempt was made to study the time dependence of adsorption.

Key words: Asphalt, emulsion, adsorption, surfactant.

Introduction

Asphalt is widely used for road construction. Its high viscosity and sticky properties give binding characteristics to this material, but make its handling a tricky and dirty job. This difficulty can be eased by preparing concentrated asphalt-in-water emulsions. This emulsification presents several problems [1]. In particular, the emulsion has to be stable enough to be stored over a period of months without breaking. However, the emulsion must break soon after its application to the mineral aggregates [2–4]. Thus the concentration of surface active agent included must be carefully controlled in order to achieve a quick break when the emulsion is used, while preserving its stability. To obtain such opposite properties several physico-chemical phenomena, among them the surfactant adsorption at the asphalt-water interface, have to be controlled.

The aim of this work was to explain in quantitative terms this adsorption. A number of emulsions were prepared using a laboratory colloidal mill. Their characteristics and the surfactant adsorption are studied first. Attention has been paid to demonstrate the reversibility of the adsorption phenomenon. The adsorption at the asphalt-water interface is then related to the electrokinetics properties of the asphalt droplets.

As already pointed out this work deals with the use of asphalt for road construction. Furthermore, the asphalt in water emulsions is an interesting model for basic studies on emulsions. Because of the high asphalt viscosity, the system can afford some perturbations without readily producing coalescence. Experiments can be carried out in order to get data about the fundamental mechanisms which must be taken into account to modelize the emulsion behavior.

As stated by Lane [3], the bitumen in the emulsified state has been considered to be an inert phase. This assumption is reasonably supported by the following fact: the leaching of inorganic ions from the asphalt occurs at such a slow rate in water that it makes an insignificant contribution to the ionic strength of the aqueous phase compared to the one due to the surface active agents.

1. Experimental

1.1 Materials

The distilled water used was doubly distilled with an all-pyrex apparatus. All inorganic reagents were Prolabo Normapur grade. The surfactant used is the Dinoram S provided by CECA Prochinor S.A. and referred to as *DS*. *DS* is a blend of alkylpropylenediamine (90 %) and alkylpropylenemonoamine (8 %). The length of the alkyl chain is distributed around a value of 18 CH_2 groups according to a Poisson's law. At low pH, the surfactant bears positive charges due to the protonation of the amine functions. Thus, the pH values of solutions used throughout this work were close to 2.0. The *DS* solution surface tension was determined using the platinum plate technique. Its evolution as a function of the logarithm of the solution concentration is plotted in Fig. 1. The shape of the curve is well known. The CMC value of *DS*, corresponding to the break in the curve, was 1.2 mmol/l. For concentration below the CMC a linear variation of the surface tension vs. the logarithm of the concentration is observed. The slope of this variation, 2.7 mN/m, together with the GIBBS adsorption isotherm equation allow calculation of the molecular packing area of *DS*, σ, at air-water interface. σ was found equal to 166 $Å^2$/molecule.

The asphalt used was a 180/220 penetration grade.

1.2 Experimental methods

Emulsion preparation. The emulsions were prepared using a laboratory colloidal mill emulbitume provided by Corlay. The maximum rotation speed, ω_{max}, of the colloidal mill was 6 400 rpm. No further attention will be paid to the hydrodynamical aspects of emulsion preparation. The rotation speed, ω, is expressed as a rotation rate t_R, with $t_R = \omega/\omega_{max}$. The asphalt emulsion content was 60 % weight. The temperature of the initial aqueous surfactant solution was 60 °C, that of the asphalt one 140 °C. According to the volume ratio and to the heat capacity of the two phases, the temperature of emulsion formation was around 90 °C. At the outlet of the emulbitume the emulsions were cooled down to 5 °C. Due to the experimental conditions this temperature was reached within 15 min. This quenching procedure was choosen to obtain emulsions with reproducible characteristics.

Final surfactant concentration determination. In an emulsion the surfactant is not totally concentrated at the interface between the continuous aqueous phase and the asphalt droplets, but is distributed between the bulk aqueous solution and the asphalt surface. The concentration of the continuous phase was determined by analyzing the supernatant obtained after centrifugation. The size distribution of the concentrated emulsion was then checked and compared to that of the original emulsion, before centrifugation. For the emulsions reported here, no change in size distribution was observed.

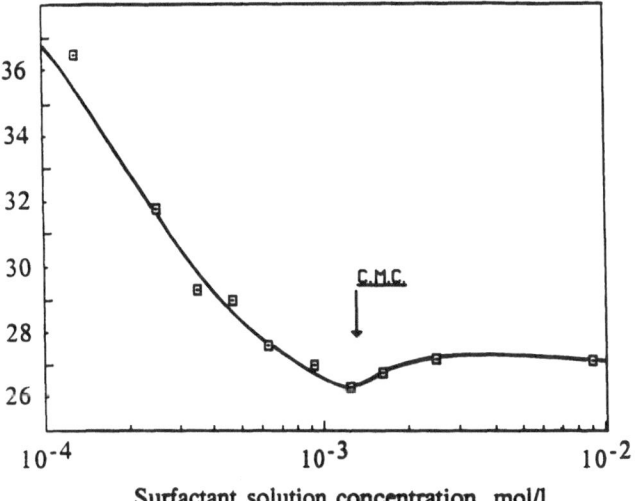

Fig. 1. Surface tension of Dinoram S solution vs. solution concentration

This shows that centrifugation did not result in emulsion breaking. The surfactant concentration was determined through the Epton two-phase procedure [5].

Electrokinetic potential. The electrophoretic mobilities of the asphalt droplets were measured at 25 °C with a Rank Brothers Mark II apparatus, equipped with a cylindrical cell. The measured mobilities were converted into electrokinetic potential, ζ, using the Smoluchowsky equation. The electrophoresis, and also the size distribution measurements required the use of diluted emulsions, typically asphalt content below 1 %. Thus, the initial concentrated emulsions were diluted with an aqueous surfactant solution having the concentration of the concentrated emulsion continuous phase.

Size distribution measurements. The size distribution of the asphalt droplets was determined by light scattering measurements using a CILAS laser granulometer. Typical curves of cumulative weight vs. droplet diameter are shown on Fig. 2. The D_{50} of the size

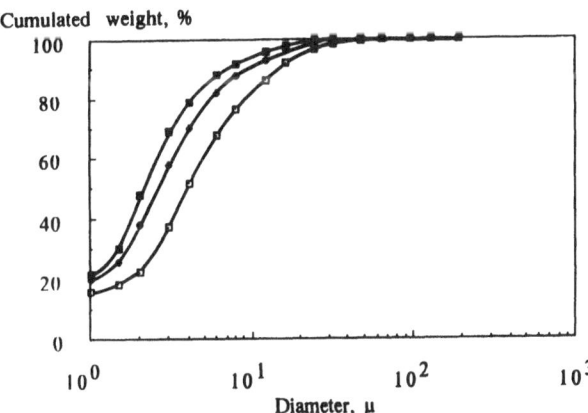

Fig. 2. Size distribution of asphalt droplets. Influence of colloid mill rotation rate. $t_R = 0.6$ □; $t_R = 0.8$ ◆; $t_R = 1.0$ ■

Surface Equivalent Diameter, μ

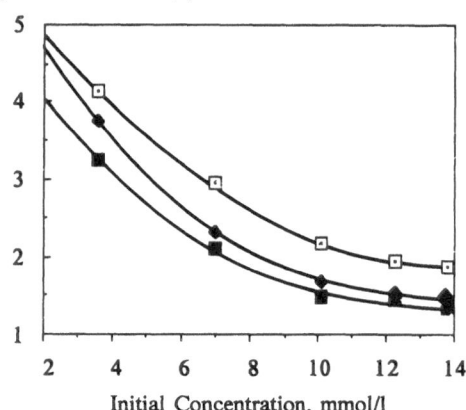

Fig. 3. Surface equivalent diameter of asphalt emulsion as a function of the initial concentration of surfactant solution. $t_R = 0.6$ ▣; $t_R = 0.8$ ◆; $t_R = 1.0$ ■

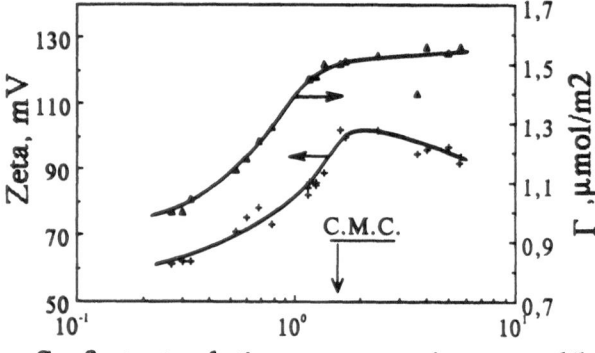

Surfactant solution concentration, mmol/l

Fig. 4. Evolution of the electrokinetic potential of the asphalt droplets (+), and of the adsorbed amount (▲) onto the asphalt/water interface as a function of the surfactant continuous phase concentration

distribution decreases as t_R increases from 0.6 to 1.0. The specific area of the asphalt droplets was calculated from the cumulative weight curve. For that purpose the size distribution was divided into 16 classes. The mass fraction of the class i, f_i, was assumed to have the diameter d_i. As the asphalt droplets are spheres, the specific surface area of the class i, $A_{s,i}$ in m^2/g, is given by $A_{s,i} = 6/(d_i \cdot \varrho)$. ϱ is the asphalt volumic weight, 1000 kg/m³ at 20 °C. The specific surface area of the asphalt droplet distribution, A_s, is given by $A_s = \Sigma m_i A_{s,i}/\Sigma m_i$, where m_i are the mass fraction of the class i.

A_s is the relevant parameter to take into account the adsorption phenomena. Sometimes the emulsions are easier to picture in terms of diameter. Thus, the surface equivalent diameter of an actual emulsion, SED, is defined as the diameter of a hypothetical monodisperse emulsion with the same specific surface area as the actual emulsion.

In order to study the adsorption phenomena a number of emulsions have been prepared using various rotation rates and initial surfactant concentrations, C_i. The size distribution, and thus A_s depends on these preparation conditions. SED is drawn in Fig. 3 as a function of C_i and t_R. SED decreases linearly as C_i increases. A break occurs in this curve for a value of C_i close to the CMC. As already observed, at a given value of C_i, SED decreases as t_R increases. The study of this variations will be developed in a further paper.

Adsorption measurement. For a given emulsion, the amount of adsorbed surfactant, Γ, mol/m², was calculated as

$$\Gamma = ((C_i - C_e) \cdot V)/(m \cdot A_s).$$

C_i is the initial surfactant concentration of the aqueous phase, mol/l; C_e the equilibrium surfactant concentration of the aqueous phase, mol/l; V the aqueous phase volume of one liter of emulsion, 0.4 l; m the asphalt content of one liter of emulsion, 600 g; A_s the specific surface area of the asphalt droplets, m²/g.

When using this equation, it is assumed that no surfactant molecules were dissolved in the asphalt phase. This hypothesis can be supported by some experimental results. A surfactant solution, pH = 2.0, of known volume and concentration was kept in a sealed

bottle together with a known volume of asphalt, at room temperature. No change of solution concentration could be detected within the experimental errors even after 10 days.

Adsorption-desorption procedure. In order to check the adsorption reversibility two concentrated emulsions were prepared, referred to as A and B. The continuous phase of A, C_i^A and B, C_i^B were 0.15 mmol/l and 2 mmol/l, respectively (below and above the CMC). Aliquots (lower than 1 ml) of A or B were immersed in large volumes (0.2 l) of surfactant solutions having concentrations larger, (respectively smaller), than C_i^A, (respectively C_i^B). These new emulsions, now diluted with respect to the asphalt content (less than 1 % weight), were kept at room temperature in sealed bottles for 48 h. No mechanical stirring was used during this step. The size distribution of the asphalt droplets were checked in order to be sure that no emulsion breaking nor specific surface area change occured in the dilution step. The ζ potentials of the asphalt droplets were determined. As the surfactant content of the concentrated emulsion aliquots were always small with regard to the surfactant content of the large volumes, the final continuous phase of the diluted emulsions is assumed to be very close to C_i^A or C_i^B. Some diluted emulsions were used to monitor ζ as a function of time.

2. Results and discussion

2.1 Adsorption isotherm and electrokinetic potential

The curve Γ vs. *DS* equilibrium concentration is plotted in Fig. 4. The lowest concentration studied was 0.4 mmol/l; Γ was then equal to 1 μmol/m². Γ increased smoothly as C_e increased. The Γ maximum value, Γ_{max}, was 1.5 μmol/m², and was obtained for values of C_e close to the CMC. Finally Γ was independent of the concentrations higher than the CMC — this fact is certainly not surprising for an adsorption phenomenon; Γ depends only on the equilibrium concentration of monomers in solution. The micellization thermodynamics (pseudo-phase separation model [6] or multi-step equilibrium [7]) predicts that the monomer concentration must be constant above the CMC; a further increase of surfactant concentration leads only to an increase of the number of micelles. Γ_{max} was used to evaluate the *DS* molecular packing area at the asphalt/water interface, $\sigma_{A/W} \cdot \sigma_{A/W}$ was equal to 110 Å^2. This value is slightly lower than the one obtained for liquid/air interface, but points out clearly that there is not more than a monolayer at the interface.

The following fact should be noted: This value of $\sigma_{A/W}$ is an average one, and the fact that the maximum value of Γ occurs at the CMC raises the question of what is the state of the adsorbed layer at the asphalt droplet surface? In other words, if this $\sigma_{A/W}$-value is governed by steric hindrance, the adsorbed layer should then be condensed and no further molecules could be added to the adsorbed layer. On the contrary, if the $\sigma_{A/W}$-value is governed by energetic factors (no further molecule could adsorb because the surfactant chemical potential could not be raised by the concentration increase), thus the adsorbed layer is not in a condensed state. Roughly speaking, there is still room at the surface for surfactant adsorption.

The evolution of the ζ potential as a function of C_e is drawn in Fig. 4. The asphalt droplets were positively charged. The lowest value of ζ potential was 60 mV. The ζ potential increased then with C_e in the same way as did Γ. The maximum value (105 mV) was reached for C_e equal to the CMC. For concentrations above the CMC, the ζ potential was constant. A slight decrease, as concentration increased, could even be noticed. These observations support the following conclusions: The *DS* adsorption causes the interface between asphalt and water to be positively charged. As long as Γ is constant the charge remains constant as the electrokinetic potential generated by the charge. The slight ζ potential decrease above the CMC is due to the following fact: As the ionic strength was not kept constant (for instance with a large indifferent electrolyte concentration), it increased with the *DS* concentration. Thus, at constant surface charge, the ionic strength increase involves a ζ-potential decrease. It is the *double layer* compression effect described by the *electrical double layer theory* [8].

2.2 Adsorption reversibility

The changes in electrokinetic potential were monitored whether they are due to an increase or to a

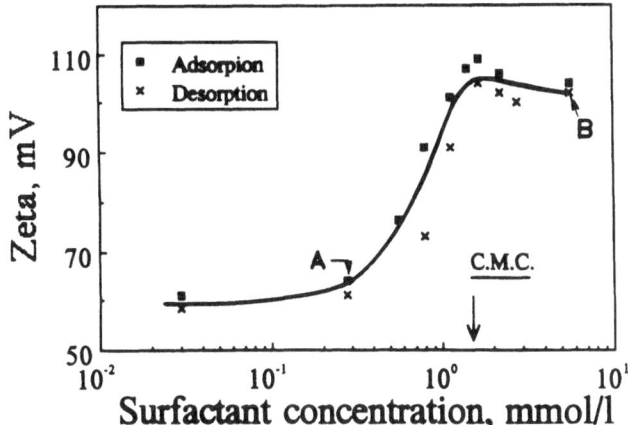

Fig. 5. Electrokinetic potential of the asphalt droplets vs. surfactant solution concentration. ζ variation as the concentration increased: ■; ζ variation as the concentration decreased: ×

decrease of the surfactant solution concentration. These variations are plotted in Fig. 5. The "A" and "B" letters show the electrokinetic potential values of the initial concentrated emulsions *A* and *B*. Within the experimental errors the variations of ζ potential as a function of solution concentration are the same whether the concentration increases or decreases. As ζ potential and Γ are strongly related, this result clearly supports the evidence that the adsorption of surfactant at the asphalt/water interface is reversible.

For concentrations above the CMC the ζ potential was constant or decreased slightly. The average value was 105 mV. Below the CMC the ζ potential decreased as concentration decreased. The lowest ζ potential value was 60 mV at a concentration close to 0.05 mmol/l. These variations were in good agreement with the ζ potential vs. C_e curve shown in Fig. 4, though the experimental procedures were very different. In one case the adsorption occured during the course of emulsion preparation, (i.e., during the formation of the interface between solution and asphalt droplets); in the other case ζ potential and thus Γ, were monitored by solution concentration variations. Again this result clearly supports the conclusion about the reversibility of adsorption. The basic feature of emulsions is to be out of equilibrium, however the surfactant distribution between solution and interface ought to be considered as the result of an adsorption equilibrium.

2.3 Time dependence of adsorption

The ζ potential increase due to a concentration increase was monitored as a function of time. The ini-

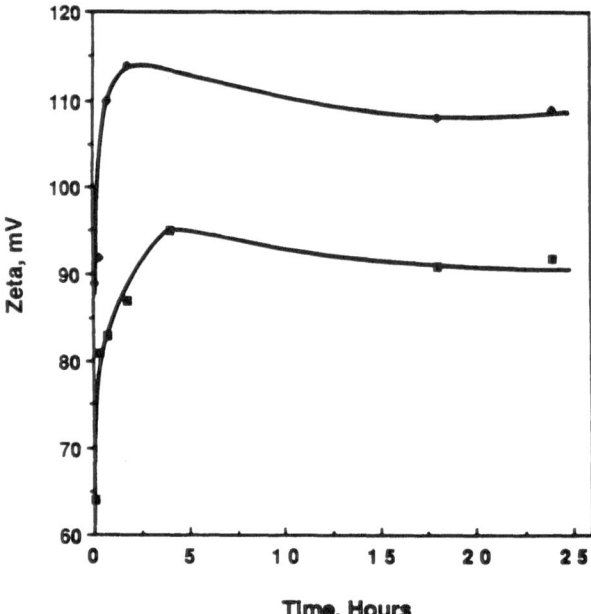

Fig. 6. Time dependence of ζ due to an increase of the bulk surfactant concentration. Initial surfactant concentration: 0.31 mmol/l; final surfactant concentration: 0.83 mmol/l □; 2.2 mmol/l ◆

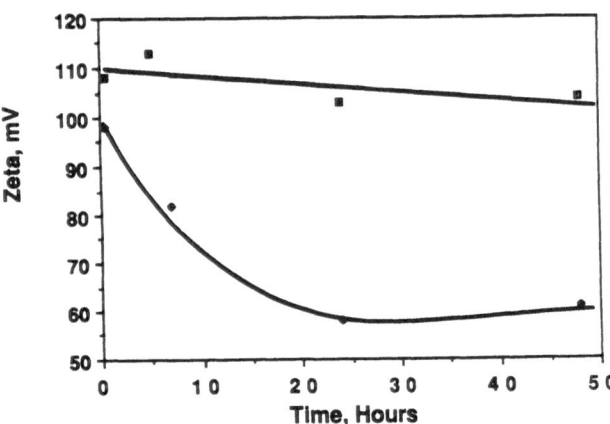

Fig. 7. Time dependence of ζ due to an decrease of the bulk surfactant concentration. Initial surfactant concentration: 1.53 mmol/l; final surfactant concentration: 1.20 mmol/l □; 0.08 mmol/l ◆

tial concentration of the emulsion continuous phase was 0.31 mmol/l. The final concentration of the surfactant solution were 0.83 and 2.2 mmol/l. The ζ potential vs. time curves are plotted in Fig. 6. A sharp variation was noticed during the first minutes. The ζ potential reached its maximum value within 3 h and was then constant up to 48 h. The ζ potential variation is due to the adsorption of the surfactant. In fact the

increase of surfactant concentration causes a shifting of the adsorption equilibrium towards higher adsorbed amounts. The decrease of ζ potential, due to a decrease of the bulk surfactant concentration, is plotted in Fig. 7 as a function of time. As the final concentration was equal or above the CMC, no ζ potential change could be recorded. This corresponds to the fact that Γ was independent of the concentration. For a concentration lower than the CMC, the ζ potential decreased slowly with time, and 25 h were needed to reach equilibrium value. The decrease of surfactant concentration in the bulk shifted the adsorption equilibrium towards lower adsorbed amounts. Thus the surfactant cations desorbed from the interface and diffused into the bulk surfactant solution.

Due to the experimental conditions chosen, the experimental kinetics was the sequence of two elementary mechanisms. In the case of adsorption, first the diffusion of the surfactants from the bulk solution to the interface, and secondly the adsorption itself. A more detailed experimental work should be done in order to determine the two elementary kinetics laws.

3. Conclusions

In order to study the asphalt emulsion stability, the adsorption of a cationic surfactant at the asphalt/water interface was studied. The adsorption reversibility was experimentally yielded. Adsorption is involved in the distribution of surfactant between the continuous aqueous phase of the emulsion and the interface. Thus, after the emulsion preparation the adsorbed amount is monitored by concentration variations in the bulk aqueous phase.

At the CMC of the surfactant, the adsorption isotherm presents a plateau. Its value was used to calculate the molecular packing area of the alkyldiamine used in this study: 110 Å^2. This value is slightly lower than the 166 Å^2 occupied at the liquid-air interface. No multilayer adsorption occurred but the question of the completion of the monolayer is still open to discussion. The adsorbed amount and the electrokinetic potential variations were strongly correlated. At the maximum adsorbed amount, 1.5 µmol/m², the ζ potential was equal to 105 mV.

Acknowledgement

The authors thank CECA S.A. for financial support of the research and for authorization to publish.

References

1. Les emulsions de Bitume (1974) Bull Liaison du Laboratoire des Ponts et Chaussées, special W, juin 1974
2. Bourrel M, Chambu C (1987) On the breaking mechanisms of cationic asphalt emulsions. 2nd International Slurry Seal Association World Congress, Genève, Mars 1987
3. Lane AR, Ottewill RH (1976) In: Smith AL (ed) Proceedings of Symposium: Theory and Practice of Emulsion Technology. Academic Press, pp 157-177
4. Scott ANJ (1976) In: Smith AL (ed) Poceedings of Symposium on, Theory and Practice of Emulsion Technology, Academic Press, pp 179-200
5. Epton SR (1947) Nature, London, 160, p 795 and AFNORNF T73-258 procedure, april 1972
6. Shinoda K (1964) In: Shinoda K, Tamamushi B, Ikeda T (eds) Colloidal Surfactant. Academic Press, New York
7. Tanford C (1979) In: Tanford C (ed) The hydrophobic effect. John Wiley and sons, New York, p 60-77
8. Overbeek JThG (1952) In: Kruyt (ed) Colloid Science. Elsevier 1952

Received November, 1988;
accepted Ferburary, 1989

Authors' address:

Dr. J. E. Poirier
C.N.R.S. U.A. 235 Minéralurgie
B.P. 40
F-54501 Vandoeuvre Cedex, France

Progress in Colloid & Polymer Science   Progr Colloid Polym Sci 79:112–119 (1989)

An overview of recent results on rheology of concentrated colloidal dispersions

D. Quemada

Laboratoire de Biorhéologie et d'Hydrodynamique Physico-chimique, Unité associée au CNRS, Univ. Paris 7, Paris, France

Abstract: From some data for well-characterized systems (as latex or silica sphere suspensions), some recent results on rheological properties of concentrated colloidal dispersions will be discussed. These properties have been related to the changes in some microstructure, which depends on the concentration, size and shape of particles and the type and magnitude of particle interactions, including strong effects due to the presence of polymer molecules in the suspending fluid.

Up to now, exact theories either failed in accounting for all these interactions, excepted in the dilute limit, or took them only partly into account. On the other hand, phenomenological theories for shear viscosity dependences in both the volume fraction ϕ and the reduced shear rate $\dot{\gamma}_r$ have been developed. For $\eta(\phi)$, special mention must be made of the Krieger's equation, as a mean field result, which involves a packing volume fraction of particle, ϕ_m. Alternatively, considering the dispersion as a microscopically inhomogeneous binary mixture, Effective Medium Theories have been recently revisited, but they still remain limited to small ϕ-values, since they give a too small packing fraction ($\phi_m \simeq 0.40$). Such a discrepancy can be seen as resulting from the presence of clusters, the size of which increases as ϕ increases, and for equilibrium conditions, which could reach some macroscopic scale (especially close to the percolation threshold). For steady shear, dynamic equilibrium between rupture and formation of clusters leads to a shear-rate dependent mean cluster size.

Some improvement in rheological description of colloidal dispersions can be expected from considering i) the system as a suspension of clusters, and ii) the internal (micro)-structure of a typical cluster (which may be a fractal) — especially the type of packing, i. e., the solid fraction within a cluster — as mirroring the properties of the system at the microscale, especially those of primary particles (size distribution, shape, nature of interactions, etc.).

Key words: Concentrated colloidal dispersions, viscosity equation, structural models, fractal flocs.

1. Introduction

During the last three decades flow properties of dispersions have been a subject of increasing interest owing to the great variety of such systems found in nature and industry, especially in the case of highly concentrated suspensions of colloidal particles, as polymer colloids.

Different kinds of latices have been prepared in order to obtain well-defined and well-controlled systems, (i. e., monodisperse spheres, with selected surface properties). These systems not only are extensive-ly used in practice but also, as claimed by Krieger [1], constitute a very attractive experimental model for fundamental studies with two aims: i) verifying rigorous fluid mechanical and physico-chemical theories of dilute suspensions [2], and ii) in the case of concentrated dispersions, testing the validity of several approximations introduced in theoretical approaches [3] and that of new concepts (percolation, fractals, etc.) as new tools in this field [4, 5]. Beyond these aims towards theories of simple fluids and more generally of disordered systems, latices offer experimental testing of numerical models used in molecular dynamics [6].

Indeed, most recent progress in knowledge and understanding of concentrated colloidal dispersions has resulted from improvements in experimental models and methods available, theories of interactions between the different components of the system, theoretical approaches based on statistical mechanics, and phenomenological modelling. Moreover, after a number of studies, mainly carried out on (thermodynamically) stable dispersions, an increasing interest in studying flocculated dispersions recently appeared. In this brief survey some examples will be given in order to illustrate this progress, with special emphasis on flocculated systems.

2. Model systems

In stable colloidal dispersions, the rheological behavior appears clearly dominated by three classes of factors: i) particle factors (such as volume fraction, size, shape, etc., and their distributions), ii) solution factors (type, viscosity, temperature, pH, ionic strength, polymer concentration and solvency, etc.), and iii) nature and magnitude of interparticle and particle-solvent potentials. The latter class plays a very crucial role in the dispersion properties.

As well-defined and well-controlled colloidal systems, latices play a dominant role since they offer the possibility to select precisely not only the size but also the shape of the particles, as well as their surface charge, the type of ions and counterions, and the nature of surfactants (e. g., amphoteric ones [7] giving very narrow particle size distributions). Generally, such properties are strongly dependent on the preparation method (see recent reviews [8, 9]).

For a long time, electrostatic stabilization has been used to get stable latices in aqueous media (see, for instance [10]). Unfortunately, as the particle volume fraction increases, the opacity of the dispersion grows, discarding experimental methods using optical measurements (e. g., SALS). Steric stabilization from adsorbed (or grafted) polymer chains onto the particle surface offers the double advantage of being insensitive to the aqueous or non-aqueous character of the solvent and exhibiting a repulsive potential more abrupt than the electrostatic one, i.e., closed to the hard sphere (HS) potential. As for electrostatically stabilized latices, it is possible to obtain a very great diversity of sterically stabilized systems, depending on the preparation methods (see, for example [11]).

3. Hard sphere suspensions: the shear-thinning behavior

A model colloidal dispersion and the hard sphere model

A recent model system consists of a dispersion of monodisperse colloidal silica particles in a good solvent like cyclohexane, the particles being sterically stabilized with octadecyl chains terminally grafted onto the surface silanol groups [12]. Rheo-optical measurements (essentially static and dynamic small-angle light scattering) which can be performed on this system by using an index-matching suspending fluid (allowing measurements at high volume fraction), have clearly established that equilibrium properties of such dispersions agree with the HS model [13].

Recently, rheological properties of such dispersions have been extensively studied by de Kruif et al. [12, 14, 15]. Couette and capillary viscometry showed a Newtonian behavior at low volume fraction, followed by a shear-thinning behavior beyond some "critical" volume fraction ϕ_s (such as $0.10 \leq \phi_s \leq 0.30$, depending on the system studied). Fitting the data on a viscosity vs. shear stress relationship $\eta(\tau)$, (the well-known Krieger-Dougherty equation, with a = sphere radius, K = Boltzman constant, T = temperature, and b = empirical constant) $(\eta_0 - \eta_\infty)/(\eta - \eta_\infty) = 1 + (b\tau a^3/KT)$ led to the limiting viscosities η_0 and η_∞ at zero and infinite shear rate, respectively, as function of ϕ. These results discussed hereafter, have been compared [12] to the existing theories of HS suspensions.

a) In the semi-dilute limit, $\phi < 0.07$, both η_0 and η_∞ compare well with Batchelor's result [2], $\eta_r = 1 + 2.5 \phi + 6.2 \phi^2$, and, as expected for HS suspensions, only differ at $O(\phi^n)$ with $n > 2$.

b) As ϕ increases, difficulties arise in taking into account both hydrodynamic and Brownian many-body interactions. Theories which only accounted for a part of these interactions failed in describing high concentration data. For example, taking into account all hydrodynamic interactions but missing the Brownian ones led [16] to η-values only satisfactory up to $\phi \sim 0.3$, (as shown in Fig. 11 of [12]). On the contrary, a statistical mechanical approach [17] recently extended the Batchelor calculation of the zero shear viscosity η_0, taking into account, through the ϕ dependence of the radial distribution function $g(r, \phi)$, all non-hydrodynamic interactions in terms of a mean field potential $V \sim KT Ln(g)$. However, this theory still remains limited to i) a pairwise additive approximation for hydrodynamic interactions − i.e., taking only $O(\phi^2)$ terms, and ii) weak flows − i.e., at first order in the

Peclet number, $0(Pe)$ –. In the case of HS suspensions this theory leads to a ϕ-dependence of η_0 which shows the increasing contribution of thermodynamic interactions beyond the shear thinning onset. Nevertheless, despite a η-divergence (absent from [16]) at high ϕ-values, this theory still led to too small η-values (see Fig. 2 of [17]).

c) An interesting result found in [12] was the ϕ- dependence of the critical stress $\tau_c = KT/ba^3$ involved in the Krieger-Dougherty equation. Indeed τ_c exhibited an abrupt change close to $\phi = 0.5$, which was believed to mirror the HS liquid-solid transition at $\phi_{HS} = 0.55$. No satisfactory interpretation of this change was obtained [17] from linear-viscoelastic theory, which relates τ_c to the high frequency shear modulus.

d) At high volume fractions, fitting the Krieger equation on zero and infinite shear rate data, led to the corresponding limiting packing values $\phi_0 = 0.63 \pm 0.02$ and $\phi_\infty = 0.71 \pm 0.02$, respectively. Quite similar values were obtained [12] from data fitting of a structural model which writes $\eta_{ro} = F(\phi/\phi_0)$ and $\eta_{r\infty} = F(\phi/\phi_\infty)$ in the zero and infinite shear limits and, more generally at constant shear rate,

$$\eta_r = F(\phi/\phi_m), \qquad F(x) = (1-x)^{-2} \qquad (1)$$

with a "maximum" packing fraction ϕ_m and where the functional form of $F(x)$ was deduced from minimization of the viscous energy dissipation [18]. Moreover these limiting packing values were found independent of the particle size [14].

e) No shear-thickening effects were observed with silica dispersions [12, 14], contrary to many examples in the case of latices (e. g., [10, 36, 38]).

f) Very recent results from torsion resonator method [15] allowed to measure the complex viscosity in the frequency range ~ 4–2400 Hz. The high frequency viscosity shows, as expected, good agreement with the HS theory at $Pe \rightarrow \infty$ [16].

4. Clustering effects in concentrated suspensions: phenomenological approaches

For a number of disperse media, it is well-established now that shear induced changes in their inner structure are responsible for most of the "anomalous" features in their rheological properties. As seen in the previous section, a large amount of work has been devoted to identify the so-called *microstructure* and to study its characteristics through the variations of the radial distribution function $g(r, \phi)$, especially the shear-induced distortions, such as those found in [17] and already observed in numerical studies on either molecular fluids [19] or colloidal suspensions [20]. Since all macroscopic quantities can be calculated once the radial distribution is known, the difficulty seems limited to the determination of this function. Indeed, the microstructure concept reinforced the belief that a system can be considered homogeneous at the macroscale whenever the scale of the system is large enough compared to the particle size, as assumed in Cell Models and Effective Medium Theories.

On the contrary, a number of experimental studies and computer simulations on binary systems have shown that increasing the volume fraction of the dispersed component led the system to exhibit *clustering effects* with clusters increasing in size up to some percolation threshold, $\phi = \phi^*$ at which one observes an equilibrium structure composed of a sample-spanning cluster (the so-called "infinite" cluster) embedded in a distribution of smaller ones. As evidence, the presence of interaction potentials likely should modify the percolation characteristics. Moreover, such clustering effects have been found at any scale, including the macroscale, that renders questionable the results obtained from cell-models and classical effective medium theories (EMT), in which homogeneity at the macroscale is a basic assumption.

It seems obvious that such a clustered system cannot keep its equilibrium structure unchanged under an applied shear, which certainly should induce the rupture of the infinite cluster and cause changes in the size distribution of small clusters. Therefore, a (non-equilibrium) steady structure, characterized by a (more or less narrow) size distribution of (more or less spherical) clusters is expected under steady shear. (Note that clustering likely overshadows size and shape effects of primary particles.) Such a structure can be seen as resulting from a dynamical equilibrium between rupture and formation of clusters. In HS suspensions, the former is mainly due to hydrodynamic shear stresses and the latter to Brownian motion that results in a Pe-dependent steady structure [21]. Indeed, in absence of cohesive forces, one can define a transient cluster as a group of spatially correlated particles during a characteristic time τ_c (naturally, correlation length ξ and time τ_c will constitute the mean "size" and "lifetime" of the cluster). In more realistic systems in which some (generally short range) cohesive force between particles in a cluster exists, a shear-dependent critical cluster size R_c should result from the balance between these cohesive forces f_i and the shear force on the cluster particles

$(f_H \sim \eta a^2 \dot{\gamma})$ leading to the relation (in dimensionless form)

$$R_c = a\, \Psi(\dot{\gamma}_r), \quad \dot{\gamma}_r = f_H / f_i, \tag{2}$$

where $\dot{\gamma}_r$ plays the role of a dimensionless shear rate. If Brownian effects dominate all other interactions then $f_i \sim KT/a$ and the Peclet number dependence is recovered ($\dot{\gamma}_r = Pe$). In the presence of various interactions (such as van der Waals, electrostatic or steric ones) f_H should be compared to the dominant (effective) force f_i resulting from superimposition of these interactions.

A very important feature of clustering lies *in fluid immobilization in the voids between* the particles forming a cluster, leading to an effective volume of a cluster larger than the solid volume of particles within the cluster. Given the shear rate and the type of interactions some relation should obviously exist between the compactness φ_c — as the solid volume fraction within the cluster — and the size of a cluster R_c, the form of the relation $\varphi_c(R_c)$ depending of the inner structure, as, for instance, in the case of a fractal cluster

$$\varphi_c = (R_c/a)^{D-3} \tag{3}$$

with a fractal dimension D. Since, as discussed above, (and Eq. (2)), R_c is shear-dependent, so is the compactness, and owing to such relationships one can expect to obtain, at given $\dot{\gamma}$, a cluster size distribution narrow enough to lead to a shear-dependent mean cluster compactness $\varphi_c(\dot{\gamma})$. Therefore, considered as a suspension of clusters, the dispersion should exhibit a shear rate-dependent effective volume fraction

$$\phi_{eff} = (\phi/\varphi_c) = \phi_{eff}(\dot{\gamma}). \tag{4}$$

As η is an increasing function of the volume fraction, a qualitative interpretation of rheological behavior of dispersions results from $\phi_{eff}(\dot{\gamma})$. Indeed, since large structures generally are formed at low shear, the shear thinning behavior can be thought of as resulting from shear-induced changes, with either cluster rupture or rearrangement of particles leading to a closer packing. Such structural changes lower the amount of trapped fluid: the higher the shear rate, the lower the trapped fluid volume. As a consequence, ϕ_{eff}, so η decreases. On the contrary, shear thickening can be seen as a shear-induced lowering of the particle packing within the cluster (i.e., giving looser clusters, in agreement with the naïve concept of dilatancy), therefore increasing ϕ_{eff} and hence, η.

A more quantitative approach [21] can start from Eq. (4), considering the dispersion as a suspension of clusters. At given shear rate, a viscosity equation, Eq. (1) for instance, can be retained; however, with ϕ_{eff} as the volume fraction of the dispersed phase, it is required to put $x = \phi_{eff}/\phi_m$ in Eq. (1). As a function of the volume fraction of primary particles the viscosity is finally written

$$\eta_r = (1 - x)^{-2}, \quad x = \phi/\phi_p, \quad \phi_p = \varphi_c \phi_m \tag{5}$$

where ϕ_p is an *effective packing fraction*, product of the cluster maximum packing ϕ_m by the mean cluster compactness. This effective packing fraction appears as a *basic structural variable*, which is both ϕ- and $\dot{\gamma}$- dependent [21]. The viscosity equation, Eq. (5) succeeded in interpretation of a viscosity-salinity scan in several microemulsion (Winsor) systems [22]. Salinity dependence of effective packing fraction, $\phi_p(S)$, was deduced from viscosity data through Eq. (5). Such dependence was interpreted in terms of structural changes from single micelles (at low and high S-values) to bicontinuous structure (in the middle of the scan), with intermediate structural states resulting from micelle clustering, progressive partial coalescence of micelles on the inside of cluster, and finally cluster-cluster coalescence. These changes were associated with quantitative variations of the effective packing (see Fig. 1).

An explicit form of the shear rate dependence in Eq. (5) has been deduced from a kinetic equation which describes the above-mentioned dynamic equilibrium at given shear rate, leading [21] to a non-Newtonian shear viscosity equation

$$\eta = \eta_\infty \left[\frac{1 + \dot{\gamma}_r^{1/2}}{\chi + \dot{\gamma}_r^{1/2}} \right]^2$$

where

$$\chi = \frac{1 - \phi/\phi_0}{1 - \phi/\phi_\infty} = \pm (\eta_\infty/\eta_0)^{1/2} \tag{6}$$

valid for shear-thinning, shear-thickening or plastic behavior, according to χ-values. Good fittings were obtained on HS data both for shear-thinning and shear-thickening systems, with $\dot{\gamma}_r \equiv Pe$ [21], and for plastic (Casson) behavior of deionized latices [23], taking in the latter case $\dot{\gamma}_r$ as the ratio of hydrodynamic energy to effective electrostatic energy in a highly concentrated system of strongly repulsive Debye

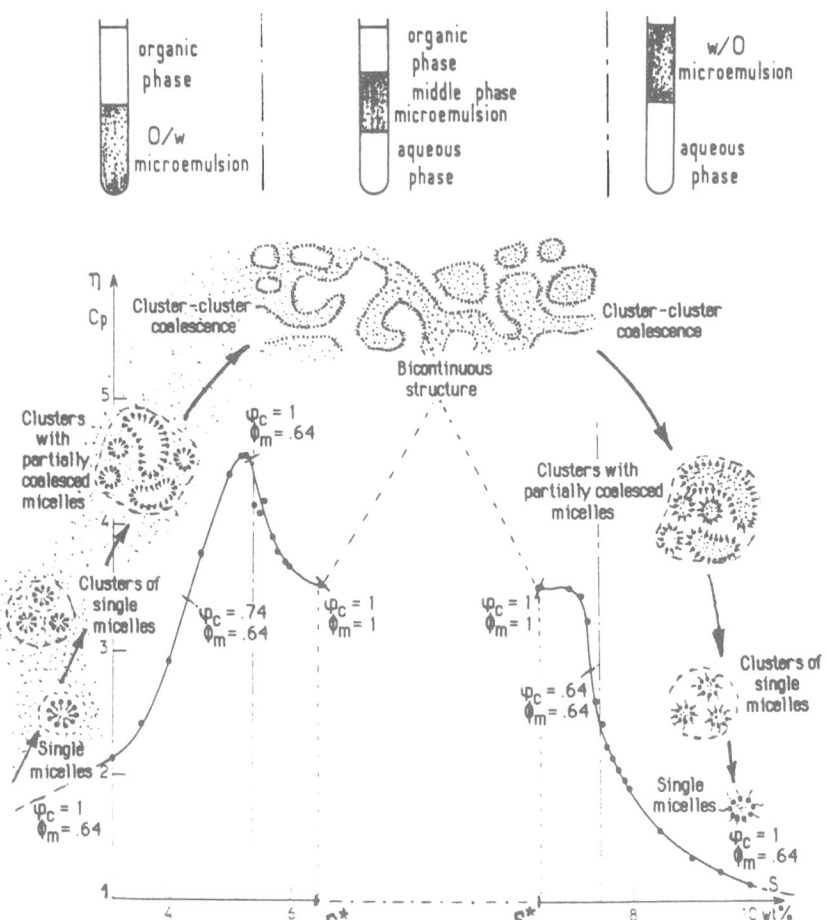

Fig. 1. Shear viscosities of the microemulsion phases vs. water salinity for the mixtures brine + toluene + SDS + butanol

spheres. Further improvements of such a modelling could be expected from refining the kinetic equation which governs the structural variable, as well as the shear dependence of the characteristic times involved. (Note that, through this kinetic equation, time-dependence effects, such as thixotropy, can be described and extended to non-linear viscoelasticity [42].)

5. Flocculated suspensions and suspensions of flocs

Adding a flocculant to a concentrated suspension, at rest or stirred very gently, generally leads to a sample spanning network, i. e., a (gel-like) flocculated suspension, which exhibits viscoelastic properties, most often linear ones owing to the very smallness of the maximum strain beyond which one observes the network rupture. On the contrary, a suspension of flocs is expected from either applying the shear during flocculation or shearing the filling-volume network formed

after flocculation is achieved. Therefore, rheological behavior must be studied at two levels: i) the rheology of flocs and networks, which are, in general, plastic solids, and ii) the rheology of a suspension of such flocs. In both cases, floc rheology studies are relevant. However, in the second case the problem seems insurmountable since the (simplest) case of concentrated suspensions of monodisperse HS still remains unsolved. There again some improvement could be expected from phenomenological approaches.

5.1. Floc rheology

A number of recent studies have been devoted to the problem of floc rheology, either directly, as shear induced formation (especially in the presence of Brownian motion [24]), and rupture of flocs [25–27], or indirectly through the effect of different types of interactions, as aggregation by depletion in sterically stabilized dispersions [28] or surfactant capillary bridging [29]. Some examples will be discussed hereafter, with special emphasis on the floc structure [25–27].

a) Floc formation: Many computer simulations of the inner structure of isolated flocs formed by different growth rules (essentially diffusion-limited aggregation or cluster-cluster aggregation) have established their fractal structure, with a fractal dimension D which depends on the rule and the dimension of the system, d (see a recent review [5]). However, effects of particle interactions and particle concentration ϕ have to be taken into account in practice [30]. Such effects have been confirmed, in $(d = 2)$-systems, by Brownian dynamics simulation of colloidal dispersions with different DLVO potentials [31]. Moreover, it seems obvious that, at very high ϕ, aggregates must be more compact, which increases D up to values close to d.

A number of (new) processes such as depletion-flocculation have been observed (especially in non-aqueous media) in the presence of polymer molecules. Extensions of the DLVO theory including this new attractive potential allowed a coherent interpretation of the observed phase transitions, and floc morphology deduced from photometric and sedimentation measurements [32]. A recent experimental and theoretical study [33] confirmed a previous theory from statistical mechanics [34] and was found to be in good agreement with observations [32]. A more recent study on latice phases separated by dextran [35] supported all these experimental and theoretical findings.

In the case of sterically stabilized particles a mean field theory for depletion-flocculation followed by restabilization was recently given [36]. However, up to now neither the inner structure of aggregates which would result from this process nor the effects of shearing have been studied.

b) Shear induced floc breakup: On the other hand, recent experimental studies have shown the importance of shear on the floc size, determined by SALS experiments carried out on PS-latices [25]. This investigation demonstrated that isolated flocs, broken down by steady shear, exhibit an average number of particles per floc, N, and a mean radius of gyration R_g, both shear-rate-dependent, yielding a relationship $N \sim R_g^D$, i. e., a fractal floc structure with a fractal dimension $D = 2.48$. This (rather large) value certainly mirrors the fact that shear stress should rearrange the particles within the floc, leading to a more compact structure (indeed, lower D-values were found at very low shear rates).

A companion paper [26] presented a theory for breakup of fractal aggregates with compactness $\varphi_c \sim (R/a)^{D-3}$. A mechanical model of the floc assumed a linearly elastic structure, with elastic moduli varying as φ_c^n, this porous solid being percolated by the viscous fluid. A R vs. $\dot{\gamma}$ relationship was then obtained from assuming that floc rupture occurs if the local distortion energy exceeds a critical value U_c. In the limit of non-draining flocs, this theory led to $R \sim \dot{\gamma}^z$, where $z = (D - 1)/2n(D - 3)$. Fitting this result on floc breakdown data with the observed D-value led to $n = 4.45$ in good agreement with other n-values, $n = 4.03$ (from direct (SALS) estimations of $R_g(\dot{\gamma})$ [25]), and $n = 4.4$ (from direct measurement of the shear modulus G, varying as φ_c^n, [27]). A critical floc radius R_c at given shear stress τ then results. Writing $e_{\max} = \tau/G$ for the maximum strain at floc rupture and taking $\nu = 0$ for the Poisson ratio, the τ vs. R_c relation becomes

$$\tau = S_0 \varphi_c^n = S_0 (R_c/a)^{n(D-3)} \tag{7}$$

where S_0 is closely related to U_c. Finally, in order to justify the dependence $G \sim \varphi_c^n$, a model based on percolation theory applied to a disordered elastic network (a partially filled *fcc* lattice) was recently proposed [37] and gave only qualitative agreement with several data.

c) Plastic behavior: All flocculated suspensions exhibit plastic behavior and associated linear viscoelasticity under strain at small amplitude. Although quite different in the type of interactions involved, "gelled suspensions" were obtained with PMMA particles dispersed in silicone fluids [38]. Casson behavior and thixotropy were observed "when the molecules of the dispersing medium were comparable in length to average spacings between particle surfaces". Hence, the authors attributed this gelation-like process to some "elastic-steric" repulsive interaction, which to some extent, resembles the electrostatic repulsion responsible for colloid crystal formation in deionized latices.

5.2. Rheology of concentrated suspensions of flocs

The phenomenological model, Eqs. (5) and (6), based on the existence of clusters, should apply directly to concentrated suspensions of flocs whenever these flocs are supposed to be approximately uniform in size with a (shear-dependent) mean compactness $\varphi_c(\dot{\gamma}_r)$, and a packing fraction of flocs ϕ_m that leads to the viscosity equation, Eq. (5). Up to now, although φ_c appears as the main structural variable in describing suspension, rheology, some arbitrariness remained in choosing on one hand the kinetic equation (which governs φ_c) as a relaxation equation, and on the other hand the $\dot{\gamma}$-dependence of the characteristic times involved in this equation [21].

In any event, once the floc radius vs. shear and the compactness vs. size are known, a rheological equation will be obtained from inserting the resulting expression $\phi_{eff}(\dot{\gamma})$ or $\phi_{eff}(\tau)$ in the viscosity equation $\eta_r(\phi_{eff})$. As evidence, this rheological equation should depend only on the selected relations $\eta(\phi_{eff})$, $\varphi_c(R_c)$, and $R_c(\dot{\gamma}$ or $\tau)$. Progress could be likely expected in assuming fractal structure of flocs as done in two recent papers [39, 40].

The first one started from a relative viscosity equation $\eta_r = (1 - \phi_m x)(1 - x)^{-2}$, where $x = \phi_{eff}/\phi_m$, deduced from a cell-model and assuming a fixed packing fraction $\phi_m = 4/7$ (for HS) [39]. A $R_c(\tau)$ relationship resulted from assuming elastic flocs having and ad hoc "adhesive energy" depending on the coordination number between particles within a floc. However, such a relationship failed in the limit $\tau \to \infty$, and in order to get $R_c \to a$ in this limit the authors arbitrarily changed the $R_c(\tau)$ relation that was lastly written

$$(R_c/a) = 1 + (\tau_c/\tau)^s, \quad \text{with } s = 1/(4 - D). \quad (8)$$

From Eqs. (3) and (4) the expression $\phi_{eff}(\tau)$ was obtained hence a shear stress dependent viscosity. An interesting feature lies in the recovering of a Casson-like equation in the case of a fractal dimension $D = 2$, and in the low stress limit $\tau < \tau_c$. Note that such a feature appeared in the viscosity Eq. (5) if $\phi_0 \to \phi$ (i.e., $\eta_0 \to \infty$), leading to very similar ϕ-dependences in yield stress and plastic viscosity [21a].

The second approach [40] is a self-consistent field model in which a radially dependent internal volume fraction $\varphi_{fl} \sim (r/a)^{D-3}$ within the floc is postulated. Equation (7) was taken as the $R_c(\tau)$ relation. However, as in the previous work, this relation fails at $\tau \to \infty$ (giving $R_c \to 0$); the author added an ad hoc minimum floc size R_0, changing Eq. (7) to

$$\tau \, AS_0[(R_c/a)^p - (R_0/a)^p]^{-1} \quad (9)$$

where A is a (known) numerical constant and $p = n(3 - D)$. Equations (9), (3), and (4) lead to $\phi_{eff}(\tau)$. This expression, introduced in a viscosity relation, gave a rheological equation which, in the weak flow

limit, indicated a "creep at constant τ while the flocs break down" and a Bingham plastic behavior at high stress. Nevertheless, the EMT-expression choosen for the viscosity ($\eta_r = (1 - 2.5 \, \phi_{eff})^{-1}$) certainly should limit the validity of the rheological equation obtained in the high ϕ-range.

Finally, the assumption of simple fractal floc structure wass recently discussed [41] and found inadequate to describe experimental data giving φ_c as a function of the mean particle number N in a floc. Indeed a multifractal structure was found, with D increasing from 2.4 to 3 as N grows up to ∞. Moreover, a $\varphi_c(\dot{\gamma})$ relation was obtained if the N-value, resulting from balance between floc aggregation and shear-induced rupture, is inserted in the empirical φ_c vs. N relationship. Equation (6) is then recovered in the form of Eq. (5), with slight differences in notation.

Coming back to phenomenological approaches, it is worth noting that, whichever viscosity equation is chosen, with explicit compactness dependence, further progress can be expected from more realistic expressions for $\varphi_c(R_c)$ and $R_c(\dot{\gamma}$ or $\tau)$, the choices of which are consolidated in accounting for some direct information about the cluster structure (e. g., from microscopy, sedimentation, SALS or SANS, ...) allowing φ_c to be quantified, *whether the cluster is a fractal or not*. Beyond that, accounting for time-dependent effects through the kinetics of the structure, i. e., a time-dependent compactness, appears promising [42].

6. Conclusive remarks

This brief (and inevitably incomplete) review of recent work on rheological properties of polymer colloids has been mainly devoted to flocculated dispersions. Special emphasis on fundamental aspects of the increasing interest for such media have been given, although they are very important for practical uses. Most of works concerned the relationship between rheological properties and inner structural characteristics, with the help of percolation and fractal concepts and took into account new types of interactions, especially those due to the presence of bounded or free polymer molecules. Perhaps the more promising aspect for future progresses lies in i) introduction of these new tools in phenomenological approaches, and ii) disposal of new experimental dispersion model and new techniques to allow very precise determination and/or control of the system variables.

References

1. Krieger IM (1972) Adv Colloid Interface Sci 3:111–136
2. Batchelor GK (1977) J Fluid Mech 83:97–113
3. Russel WB (1980) J Rheol 24:287–317
4. Kirkpatrick S (1979) In Ballian R et al (eds) Ill Condensed Matter. North Holland Pub Cy, pp 324–403
5. Meakin P (1988) Adv Colloid Interface Sci 28:249–326
6. Hayter JB (1983) Faraday Discuss Chem Soc 76:9–18
7. Essadam H, Pichot C, Guyot A (1988) Colloid Polym Sci 266:462–469
8. Sugimoto T (1987) Adv Colloid Interface Sci 28:65–108
9. Bernhardt C (1988) Adv Colloid Interface Sci 29:79–139
10. Laun HM (1984) Die Angew Makromol Chem 123/124:335–339
11. Daniel JC (1985) Makromol Chem Suppl 10/11:359–368
12. De Kruif CG, van Iersel EMF, Vrij A, Russel WB (1985) J Chem Phys 83:4717–4725
13. Vrij A, Jansen JW, Dhont JKG, Pathmamanoharan C, Kops-Werkhoven MM, Fijnaut HM (1983) Faraday Discuss Chem Soc 76:19–36
14. Van der Werff JC, de Kruif CG (1989) J Rheol 33:421–454
15. Van der Werff JC, de Kruif CG, Blom C, Mellema J (1988) In: Cazabat AM et al (eds) Hydrodynamics of Dispersed Media. Euro Physics Conf Abstracts, EPS, Geneva, pp 18–19; (1989) Phys Rev A 39:795–807
16. Beenakker CWJ (1984) Physica 128A:48–81
17. Russel WB, Gast AP (1986) J Chem Phys 84:1815–1826
18. Quemada D (1977) Rheol Acta 16:82–94
19. Hess S (1983) Physica 118A:79–104
20. Bossis G, Brady JF (1984) J Chem Phys 80:5141–5154
21. Quemada D (1982) In: Casas-Vasquez J, Lebon J (eds) Lecture Notes in Physics 164: Stability of Thermodynamic Systems. Springer, Berlin, pp 210–247; (1984) In: Mena B et al (eds) Advances in Rheology: vol 2. UNAM, Mexico City, pp 571–582
22. Quemada D, Langevin D (1986) J Theor Appl Mech, No Spec 1985:201–237 (see also in Mittal et al (eds) Surfactant in Solutions. Proc Intern Conf, New Delhi, 1986, Plenum Press, New York, in press)
23. Quemada D (1986) J Theor Appl Mech, No Spec 1985:289–301
24. Feke DL, Schowalter WR (1985) J Colloid Interface Sci 106:203–214
25. Sonntag RC, Russel WB (1986) J Colloid Interface Sci 113:399–413
26. Sonntag RC, Russel WB (1987) J Colloid Interface Sci 115:378–389
27. Sonntag RC, Russel WB (1987) J Colloid Interface Sci 116:485–489
28. Gast AP, Leibler L (1986) Macromolecules 19:686–691
29. Koh PTL, Uhlherr PHT, Andrews JRG (1985) J Colloid Interface Sci 108:95–101
30. Ball RC, Warren PB, Thomson BR, Weitz DA (1988) In: Cazabat AM et al (eds) Hydrodynamics of Dispersed Media. Euro Physics Conf Abstracts, EPS, Geneva, pp 48–49
31. Ansell GC, Dickinson E (1985) Chem Phys Let 122:594–598
32. Sperry PR (1984) J Colloid Interface Sci 99:97–108
33. Gast AP, Russel WB, Hall CK (1986) J Colloid Interface Sci 109:161–171
34. Gast AP, Hall CK, Russel WB (1983) J Colloid Interface Sci 90:251–267
35. Patel PD, Russel WB (1987) J Rheol 31:599–618
36. Laun HM (1988) In: Giesekus H et al (eds) Progress and trends in Rheology II. Steinkopff Verlag, Darmstadt, pp 287–290
37. Mall S, Russel WB (1987) J Rheol 31:651–681
38. Choi DM, Krieger IM (196) J Colloid Interface Sci 113:101–113
39. Mills P, Snabre P (1988) In: Giesekus H et al (eds) Progress and trends in Rheology II. Steinkopff Verlag, Darmstadt, pp 105–108
40. Russel WB (1986) In: El-Aasser MS, Fitch RM (eds) Future directions in polymer colloid. NATO ARW, Racine, Wi (USA), pp 1–39
41. Lapasin R (1987) Chem Biochem Eng Q 1:143–149
42. Quemada D (1984) Biorheol 21:423–436

Received November 1988;
accepted February 1989

Author's address:

Prof. D. Quemada
LBHP – Tour 33/34
Université Paris 7
2 Place Jussieu
F-75251 Paris Cedex 05, France

Progress in Colloid & Polymer Science   Progr Colloid Polym Sci 79:120–127 (1989)

Correlation of viscoelastic properties of concentrated dispersions with their interparticle interaction

Th. F. Tadros

ICI Agrochemicals, Jealott's Hill Research Station, Bracknell, Berks., England

Abstract: In this overview the correlation between viscoelastic properties of concentrated dispersions and their interparticle interaction has been demonstrated. Three model polystyrene latex dispersions were studied. The first was an electrostatically stabilised dispersion whereby the viscoelastic properties were measured as a function of volume fraction, ϕ, at two NaCl concentrations namely 10^{-5} and 10^{-3} mol dm^{-3}. The viscoelastic properties could be related to double layer repulsion which at a given ϕ was stronger in 10^{-5} compared to 10^{-3} mol dm^{-3} NaCl. The second system was a sterically stabilised dispersion produced either by grafting or physically adsorbing a polymer layer. A predominantly elastic response was produced when the surface-surface separation became less than twice the adsorbed layer thickness, under which condition strong steric repulsion occurs. The third system was flocculated latex dispersions. Weakly flocculated dispersions were obtained by addition of free (non-adsorbing polymer) to a sterically stabilised latex. At a critical free polymer volume fraction, ϕ_p^+, the dispersion showed non-Newtonian flow. It was possible to relate the extrapolated yield stress to the energy of interaction, E_{Sep}, between the particles. Comparison of E_{Sep} with theoretical caculation of G_{dep} (the free energy of interaction due to depletion) showed reasonable agreement. Strongly flocculated dispersions were obtained either by reduction of solvency for a polymer stabilised dispersion or addition of electrolyte to an electrostatically stabilised dispersion. Again good correlation between the viscoelastic properties and the energy of interaction was obtained in this case.

Key words: Concentrated dispersions, interparticle interaction, polystyrene latex, viscoelastic properties, sterically stabilised dispersions, steric repulsion, flocculated latex dispersions, energy of interaction.

Introduction

Concentrated dispersions of the solid/liquid (suspension) or liquid/liquid (emulsion) types usually show non-Newtonian behavior as a result of interparticle interaction. This is particularly the case when the volume fraction ϕ reaches a value whereby strong interaction occurs as a result of significant double layer overlap (for electrostatically stabilised systems) or polymer layer overlap (for sterically stabilised dispersions). Under these conditions the flow behavior of the dispersion shows a predominantly elastic response provided the experimental time scale of measurement is comparable to the relaxation time of the system. Such experimental time scale can be controlled using dynamic (oscillatory) measurements whereby one is able to change the frequency of the applied (sinusoidal)

deformation. By fixing the frequency range the system may show a predominantly viscous or predominantly elastic response depending on the magnitude of the interaction.

The above rheological technique provides a powerful tool for studying the properties of concentrated dispersions. The aim of this paper is to show how the viscoelastic properties can be correlated with the interaction forces between the particles in a concentrated dispersion. For this purpose three different systems have been chosen: an electrostatically stabilised dispersion, a sterically stabilised dispersion and dispersions where the net interaction is attractive (i. e. a flocculated system). In the latter case weakly and strongly flocculated dispersions have been investigated. As model dispersion polystyrene latex was chosen since it can be prepared monodisperse with well characterised surface properties. The dynamic (oscillatory) measurements were obtained using a computer controlled instrument that enables one to obtain measurements as a function of applied strain (at constant frequency) and as a function of applied frequency (at constant strain). These measurements were supplemented in some cases with steady state shear stress-shear rate measurements and shear modulus measurements.

Outline of experimental techniques

Latex dispersions

Two main polystyrene latex dispersions were prepared. The first was an electrostatically stabilised latex that was prepared using the surfactant free emulsion polymerisation technique described by Goodwin et al. [1]. The latex was extensively dialysed and then concentrated. The particles were fairly monodisperse with an average radius of 700 nm, as measured using photon correlation spectroscopy and Coulter Counter. Rheological measurements were carried out as a function of the volume fraction of the latex at two NaCl concentrations, namely 10^{-5} and 10^{-3} mol dm^{-3}.

The above latex was used to prepare a sterically stabilised dispersion containing physically adsorbed poly(vinyl alcohol) (PVA) with a weight average molecular weight of 45 000. This polymer was previously characterised [2] and its adsorption measured on polystyrene latex. The saturation adsorption was found to be 3.0 mg m^{-2}. The adsorbed layer thickness was obtained using photon correlation spectroscopy [3, 4] and found to be 46 nm.

A second latex with grafted poly(ethylene oxide) (PEO) with an average molecular weight of 2 000 was prepared using dispersion polymerisation [5, 6]. The adsorbed layer thickness of the PEO is in the region of 20 nm, as recently estimated from viscosity measurements [7]. Flocculated or coagulated latex dispersions were obtained as follows. For the PVA stabilised latex flocculation was induced by addition of electrolyte (Na$_2$SO$_4$ or KCl) at constant temperature or by increasing the temperature at constant electrolyte concentration. Coagulated polystyrene latex dispersions were produced by addition of 0.2 mol dm^{-3} NaCl, i. e. well above the critical flocculation concentration.

Rheological measurements

Three main rheological techniques were applied. The first was dynamic (oscillatory) measurements carried out using a Bohlin VOR (Bohlin Rheologie, Lund, Sweden) interfaced with a Facit or an IBM personal computer. In this instrument a sinusoidal strain is applied to the outside cup and the stress simultaneously measured on the bob that is connected to interchangeable torsion bars. Details of the experimental procedure was given elsewhere [7]. From the amplitude of stress and strain (τ_o and γ_o respectively) and the phase angle shift δ, the complex modulus G^*, the storage modulus G' and the loss modulus G'' are obtained, i. e.

$$G' = G^* \cos \delta \qquad (1)$$

$$G'' = G^* \sin \delta \qquad (2)$$

$$G^* = G' + i\, G'' \qquad (3)$$

and

$$G^* = (G'^2 + G'^2)^{1/2} \qquad (4)$$

where i is equal to $(-1)^{1/2}$.

The second rheological method was steady state shear stress τ, shear rate $\dot{\gamma}$ measurements which were carried out using a Haake-Rotovisco (Haake, FRG) and concentric cylinder platens.

The $\tau/\dot{\gamma}$-curves were analysed using the Bingham [8] and Casson [9] models, i. e.,

$$\tau = \tau_\beta + \eta_{\text{app}}\, \dot{\gamma} \qquad (5)$$

$$t^{1/2} = \tau_c^{1/2} + \eta_c^{1/2}\, \dot{\gamma}^{1/2} \qquad (6)$$

where τ_β is the extrapolated Bingham yield stress, whereas τ_c is the Casson's yield value.

The third rheological technique was shear modulus measurement obtained using the pulse shearometer (Rank Brothers, Bottisham, Cambridge, UK). The dispersion is placed between two plates connected to piezoelectric crystals. A small amplitude ($< 10^{-4}$ rad) high frequency (~ 200 Hz) pulse is applied to one of the plates and the velocity v of the generated shear wave is measured in the dispersion [10]. From v and the density of the dispersion, the shear modulus G_∞ is obtained using the simple formation,

$$G_\infty = v^2 \varrho. \qquad (7)$$

All measurements were carried out at $25 \pm 0.1\,°C$, except for the shear modulus which was measured in a constant temperature room at $20 \pm 0.5\,°C$.

Results and discussion

Details of the results obtained using the various systems described in the experimental section were either published elsewhere [7, 11] or will be published in the future. In the present paper few examples were selected with the objective of correlating the viscoelastic properties to the interaction forces between the particles. These are described below.

1. Electrostatically stabilised polystyrene latex dispersions

As mentioned in the experimental section results were obtained as a function of ϕ at two NaCl concentrations, namely 10^{-5} and 10^{-3} mol dm^{-3} [12]. These concentrations were chosen to give an order of magnitude difference in the double layer extension. For 10^{-5} mol dm^{-3} this extension (or thickness $1/\varkappa$, where $\varkappa$ is the Debye-Hückel parameter) is 100 nm, whereas for 10^{-3} mol dm^{-3} it is only 10 nm. For particles with radius 700 nm, a double layer thickness of 100 nm produces a significant effect on particle-particle interaction when ϕ is increased. In other words the dispersion should show "soft" (long range) interaction. In contrast, in 10^{-3} mol dm^{-3}, a double layer thickness of 10 nm will have a small effect on particle interaction till a near random of hexagonal close packing is reached (ϕ = 0.64 or 0.74 respectively). The dispersion in this case will therefore behave as a near hard-sphere dispersion.

The above predictions are realised from the viscoelastic properties of the polystyrene dispersions in 10^{-5} and 10^{-3} mol dm^{-3} NaCl shown in Fig. 1. Several features may be identified from the plots shown in Fig. 1. Firstly, all moduli show a rapid increase above a critical ϕ value which is smaller for dispersions in 10^{-5} mol dm^{-3} compared to those in 10^{-3} mol dm^{-3}. Secondly, at any given ϕ, the moduli are orders of magnitude higher in 10^{-5} mol dm^{-3} when compared with the values obtained in 10^{-3} mol dm^{-3}. Thirdly, within the volume fraction range studied, all dispersions in 10^{-5} mol dm^{-3} NaCl are more elastic than viscous. In contrast, in 10^{-3} mol dm^{-3} the dispersions are either slightly more elastic than viscous (in the range 0.35 $\leq \phi \leq$ 0.5) or more elastic than viscous (at ϕ > 0.5).

The correlation between the viscoelastic properties of the dispersions and their double layer interaction can be clearly seen if one calculates the effective volume fraction ϕ_{eff} of each dispersion. ϕ_{eff} is given by the expression,

$$\phi_{eff} = \phi \left[1 + \frac{(l/\varkappa)}{R} \right]^3 . \tag{8}$$

In 10^{-5} mol dm^{-3} NaCl, ϕ_{eff} = 1.48 ϕ, whereas in 10^{-3} mol dm^{-3} NaCl ϕ_{eff} = 1.04 ϕ. Thus, in 10^{-5} mol dm^{-3}, the ϕ range investigated is 0.463–0.525 which corresponds to a ϕ_{eff} range of 0.7–0.77. In other words within the volume fraction range studied all dispersions have an effective ϕ that is above the maximum random packing fraction (ϕ = 0.64). In all these dispersions there is therefore significant double layer overlap and hence the dispersions are predominantly more elastic than viscous. Clearly the higher the ϕ value, the more elastic the dispersion. Indeed at the highest value studied, namely 0.525, ϕ_{eff} = 0.77 and $G' \sim G^*$, i.e. the dispersion behaves as a near elastic body.

In contrast the ϕ range studied in 10^{-3} mol dm^{-3} NaCl, namely 0.273–0.566 corresponds to a ϕ_{eff} range of 0.284–0.588 which is well below the maximum random packing fraction. Under these conditions, double layer overlap is insignificant and the dispersions are predominantly viscous. With such thin double layer and large particles one needs to go to higher ϕ values before an elastic response is produced.

Thus, the viscoelastic properties of charge stabilised systems can be accounted for in terms of double layer interaction. The main parameters are the dimensionless number $[(1/\varkappa)/R]$ and the volume fraction ϕ. Therefore viscoelasticity is determined by the magnitude of electrolyte concentration particle radius and volume fraction.

2. Sterically stabilised dispersions

In this case the interaction also depends on the dimensionless quantity (δ/R) (δ is the adsorbed layer thickness) and the volume fraction ϕ. As with electrostatically stabilised dispersions, an effective volume fraction ϕ_{eff} can be defined, i.e.

$$\phi_{eff} = \phi[1 + (\delta/R)]^3 . \tag{9}$$

It is clear that the larger (δ/R), the larger ϕ_{eff} at any given ϕ. This means that small particles with thick adsorbed layers should show strong interaction at relatively small ϕ values. Under these conditions the transition from predominantly viscous to predominantly

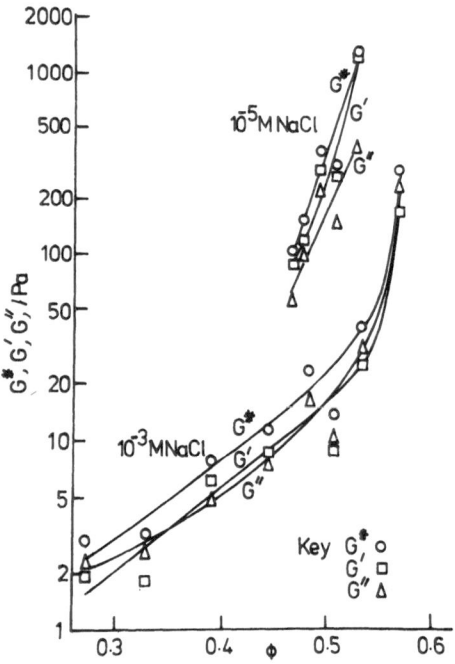

Fig. 1. Variation of G^*, G', G'' with ϕ for polystyrene latex dispersions at two NaCl concentrations (10^{-5} and 10^{-3} mol dm^{-3})

elastic response will occur at low values. By way of illustration, let us take the case of $(\delta/R) = 0.5$. In this case $\phi_{eff} = 3.3\,\phi$ and hence interaction becomes strong at $\phi > 0.2$. This has been recently illustrated using PVA stabilised latex with a small radius of 92 nm, whereby δ was 46 nm [13]. In contrast where (δ/R) is small such transition occurs at higher ϕ value. This is illustrated in Fig. 2 which shows plots of G^*, G', G'' versus ϕ for polystyrene latex ($R = 175$ nm) containing grafted PEO with δ in the region of 20 nm. The results in Fig. 2 show a transition from predominantly viscous to predominantly elastic response at ϕ of 0.49 which corresponds to ϕ_{eff} of 0.669. This is higher than the maximum random packing fraction of 0.64. However, in the calculation of ϕ_{eff} we have assumed that δ does not change with increase in ϕ. Recent viscosity results [7] showed that δ decreases gradually with increase in ϕ as a result of compression of PEO tails. Indeed at $\phi = 0.49$, δ was found to be reduced to 15 nm. Using δ value, ϕ_{eff} at the cross-over point of Fig. 2 is equal to 0.63 which is very near to the maximum random packing fraction.

Another example of sterically stabilised dispersion is that of PVA stabilised polystyrene latex dispersions. The results of G^*, G', G'' vs. ϕ are shown in Fig. 3 [14]. In this case the adsorbed layer thickness is relatively large (46 nm) due to the presence of long dangling PVA tails. However, since R in this case is 700 nm, interaction becomes strong at relatively high ϕ values of ~ 0.53. This corresponds to ϕ_{eff} of 0.64 which is the maximum random packing fraction.

Thus, the above results for viscoelastic properties of sterically stabilised dispersions can be correlated with

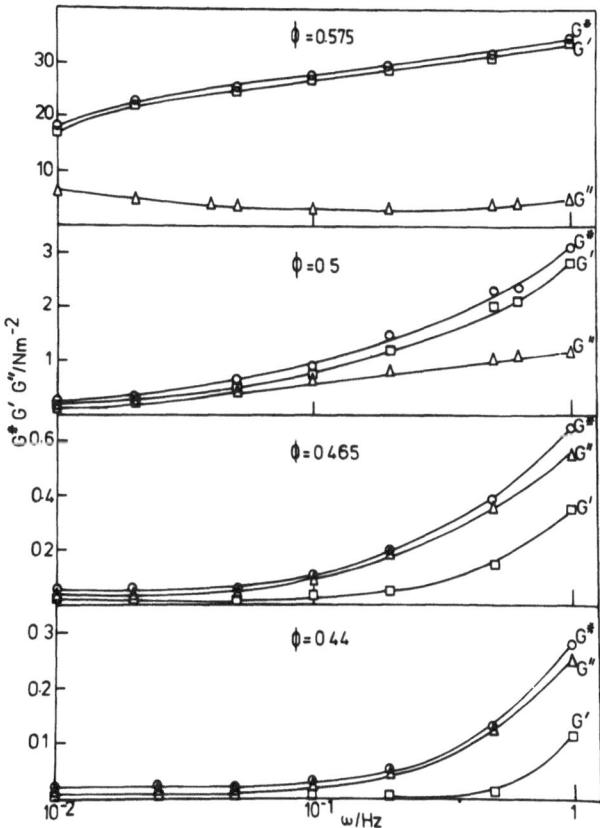

Fig. 2. Variation of G^*, G', G'' with ϕ for polystyrene latex dispersions containing grafted PEO chains with $M_w = 2\,000$

steric interaction provided information is available on the adsorbed layer thickness. The latter may change with volume fraction as a result of chain compression, particularly when the adsorbed layer is not dense.

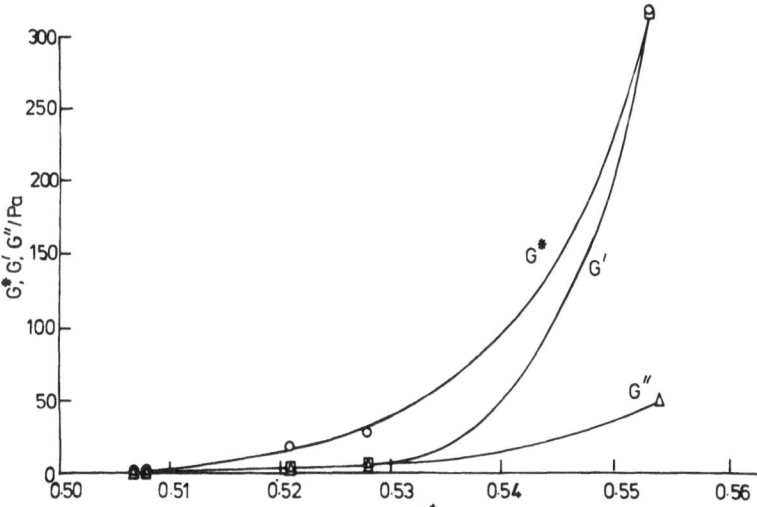

Fig. 3. Variation of G^*, G', G'' with ϕ for polystyrene latex dispersions containing physically adsorbed PVA chains with $M_w = 45\,000$

Thus for quantitative interpretation of rheological properties one should have accurate information on the structure of the adsorbed or grafted polymer layer.

3. Flocculated systems

3.1. Weakly flocculated dispersions

These are exemplified by sterically stabilised dispersions to which a free (non-adsorbing) polymer is added to the continuous phase. An attractive model system studied recently in our laboratory [6] was that of polystyrene latex with grafted PEO chains (with molecular weight of 2 000) to which free PEO (with $M_w = 20\,000$, $35\,000$ and $90\,000$) was added at various concentrations. Grafting of the PEO chains to the particles ensures non displacement of the polymer layer by the higher molecular weight added free polymer. Under these conditions depletion flocculation could be studied using various rheological techniques. As an illustration, Fig. 4 shows the variation of G_∞, τ_β and τ_c with ϕ_p (free polymer volume fraction) for PEO with $M_w = 20\,000$. All rheological parameters show a rapid

increase above a critical ϕ_p value (to be denoted ϕ_p^+) which decreases with increase in M_w of PEO as expected. A summary of ϕ_p^+ values for the three polymers studied is given in Table 1. This table also give the values of ϕ_p^*, the semidilute volume fraction at which the polymer coils begin to overlap. The latter was calculated using the expression,

$$\phi_p^* = \frac{M}{b\,\langle s^2 \rangle^{3/2}\,N_{av}\,\varrho} \tag{10}$$

where $\langle s^2 \rangle^{1/2}$ is the radius of gyration of the polymer coil, N_{av} is the Avogadro's number and ϱ is the density of the polymer. It can be seen that ϕ_p^+ is lower than ϕ_p^* and this may be accounted for by the polydispersity of the polymer.

It is also possible to use the rheological results to calculate the energy of separation of particles, E_{Sep}, in the floc structure. For example, the Bingham yield stress τ_β may be related to E_{Sep} by the expression [15, 16]

$$\tau_\beta = N\,E_{Sep} \tag{11}$$

where N is the number of contact points of particles in a floc, i. e.

$$N = \frac{1}{2}\left(\frac{3\,\phi\,n}{4\,\pi\,R^3}\right) \tag{12}$$

where n is the average contact points, i. e. the coordination number. Combining Eqs. (11) and (12) one obtains

$$\tau_\beta = \frac{3\,\phi\,n\,E_{Sep}}{8\,\pi\,R^3}. \tag{13}$$

Thus, E_{Sep} may be calculated from τ_β provided an assumption is made for n. The maximum value of n is probably 8 which corresponds to a compact floc structure. Most likely n is in the region of 4 since the floc structure is more open. Values of E_{Sep} calculated using an assumed value of $n = 4$ are given in Table 2.

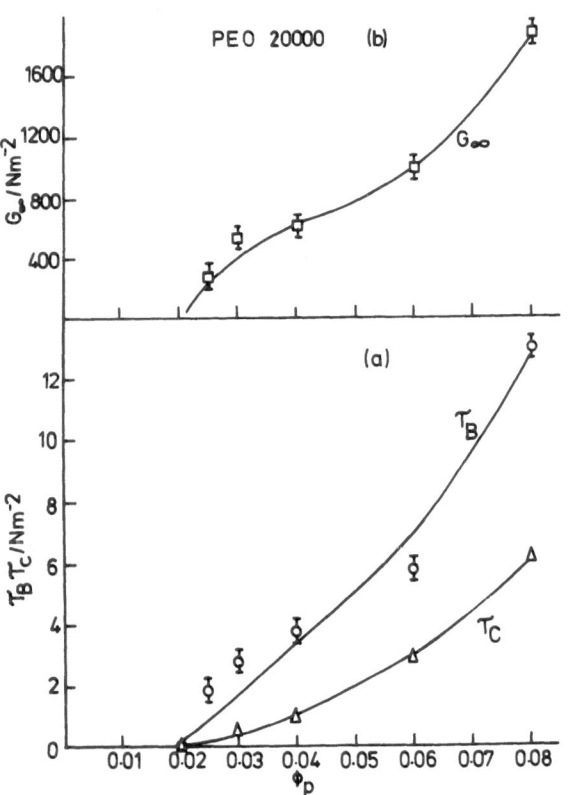

Fig. 4. Variation of G_∞, τ_β and τ_c with ϕ_p (PEO, $M_w = 20\,000$) for a sterically stabilised latex dispersion ($\phi = 0.3$)

Table 1. Summary of the results of ϕ_p^+, ϕ_p^* for PEO with various molecular weights

M_w	$[\eta]$ Rg/nm	ϕ_p^+	ϕ_p^*
20 000	5.52	0.02	0.029
35 000	7.59	0.01	0.020
90 000	12.90	0.005	0.010

Table 2. Comparison of volumes of E_{Sep} calculated from τ_β (assuming $n = 4$) and G_{dep} calculated according to [17]

ϕ_p	τ_β / Nm^{-2}	E_{Sep}/kT	G_{dep}/kT
a) PEO; $M_w = 20\,000$			
0.025	2.0	18.2	25.3
0.03	2.8	25.4	30.3
0.04	3.8	34.6	40.5
0.06	5.8	52.8	60.7
0.08	13.1	111.2	80.9
b) PEO; $M_w = 35\,000$			
0.015	2.3	21.0	28.6
0.02	4.4	40.0	38.1
0.03	7.0	63.8	57.1
0.04	11.7	115.0	76.2
c) PEO; $M_w = 90\,000$			
0.01	1.2	11.0	55.0
0.015	2.8	24.5	82.6
0.02	4.4	40.0	110.1
0.025	5.9	53.8	137.6

It is interesting to compare E_{Sep} with the free energy of interaction due to depletion flocculation, G_{dep}. The latter can be calculated using simple models of hard-sphere dispersions assuming that the depletion free energy results from an osmotic force exerted by the polymer chains that are "squeezed out" from between the particles. Using this simple model, Asakura and Oosawa [17] derived the following expression of G_{dep},

$$G_{dep} = -(3/2)\,\phi_2\,\beta\,x^2; \quad 0 < x < 1 \qquad (14)$$

where ϕ_2 is the volume concentration of the free polymer chains that is equal to $(4/3)\,\pi\Delta^3\,N_2/v$, where Δ is the depletion thickness (that is approximately equal to the radius of gyration Rg), N_2 is the total number of polymer chain and v is the total volume of the solution. β is equal to R/Δ, whereas x is given by the expression

$$x = [\Delta - (h/2)]/\Delta. \qquad (15)$$

Clearly when $h = 0$, i.e. at the point where the polymer chains are "squeezed out", $x = 1$.

Values of G_{dep} calculated using Eq. (14) are given in Table 2. It can be seen that inspite of all assumptions and approximations it is of the same order as E_{Sep}, particularly for the lower M_w PEO polymers (20 000 and 35 000). However, the deviation between E_{Sep} and G_{dep} is significant for the highest molecular weight

PEO studied. This is not surprising since the calculation of G_{dep} is based on a hard-sphere model for the particles and polymer coils, clearly an unrealistic model. Moreover, E_{Sep} is calculated based on the assumption that $n = 4$ and that at a shear rate above a critical value, all flocs are broken down to individual particles. Both assumptions are obviously questionable and hence quantitative agreement between E_{Sep} and G_{dep} is not to be expected.

3.2. Sterically stabilised dispersions brought to conditions of incipient flocculation

When the solvency of the medium for the stabilising chains is brought to worse than θ-conditions, incipient flocculation occurs [18]. This phenomenon can also be investigated using viscoelastic measurements. This is illustrated in Fig. 5 which shows the variation of G^*, G', G'' with $C_{Na_2SO_4}$ for a PVA stabilised latex suspension ($\phi = 0.5$) [19]. In can be seen that the moduli initially decrease with increase in Na_2SO_4 concentration, reach a minimum and then increase sharply above a critical electrolyte concentration. The initial reduction must be correlated with reduction in the adsorbed layer thickness as a result of reduction in solvency. This has been demonstrated before for PVA [4]. Reduction in δ means a reduction in ϕ_{eff} and hence a reduction in moduli values. The sudden increase above a critical Na_2SO_4 concentration of 0.15 correlates with the onset of flocculation of the dispersion. Thus, 0.15 mol dm^{-3} Na_2SO_4 may be considered as

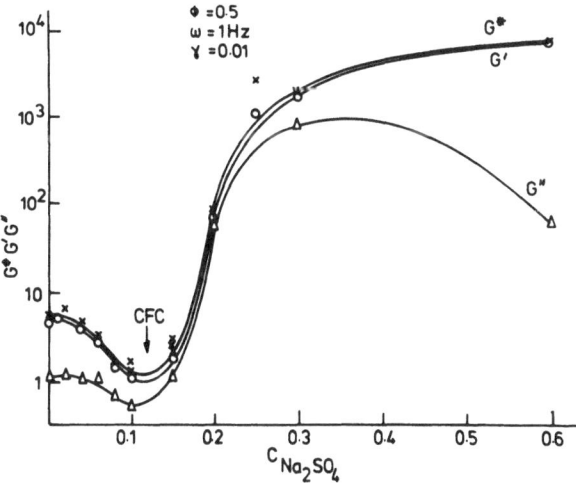

Fig. 5. Variation of G^*, G', G'' with $C_{Na_2SO_4}$ for PVA stabilised latex dispersion ($\phi = 0.5$)

the critical flocculation concentration (CFC) for the PVA stabilised latex at 25 °C.

Incipient flocculation of PVA stabilised latex may also be produced by raising the temperature of the dispersion at constant electrolyte concentration. This is illustrated in Fig. 6 at 0.15 mol dm^{-3} for a PVA stabilised latex with $\phi = 0.5$. The results in Fig. 6 show also the variation of moduli with temperature on cooling the dispersion after passing the flocculation point. Two main important conclusions may be drawn from the results of Fig. 6. Firstly, viscoelastic measurements can be applied to obtain the critical flocculation temperature (CFT) of the concentrated dispersion without any need to dilute it. Secondly, flocculation in such concentrated dispersions is not completely reversible. This is to be contrasted with the case of dilute dispersion whose flocculation is normally reversible. The lack of reversibility with concentrated dispersions is to be expected since other phenomena may occur when the particles are very close to each other e. g. bridging flocculation or polymer entanglements. Both would play a role in maintaining some flocculation after cooling the dispersion.

Thus, the above results of viscoelastic measurements can be correlated with particle interaction that occur under conditions of worse than θ-conditions. The attraction of rheological measurements is that one is able to obtain the information without the need to

dilute the dispersion, which may disrupt the flocculated structure. It would be of interest to relate the measured moduli value to the energy of attraction under conditions of incipient flocculation. This will be attempted in the near future.

3.3. Strongly flocculated (coagulated) systems

These are exemplified by polystyrene latex dispersions in the presence of an electrolyte concentration well above the CFC e. g. 0.2 mol dm^{-3} NaCl. Results were recently obtained at various ϕ values [20]. Two main types of investigations were made, namely strain and oscillatory sweep measurements. By changing the strain amplitude, at constant frequency, one is able to measure the critical strain γ_{cr} at which the system changes from linear viscoelastic (where G^*, G' and G'' are independent on the applied strain) to non-linear viscoelastic responses. γ_{cr} is a measure of the deformation above which the flocculated structure starts to become partially disrupted. Results for such strain sweep measurements at various ϕ values are given in Fig. 7.

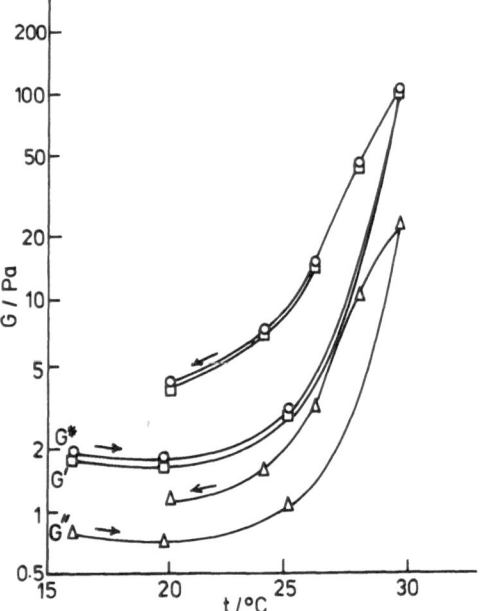

Fig. 6. Variation of G^*, G', G'' with temperature for PVA stabilised latex dispersion ($\phi = 0.5$) at $C_{Na_2SO_4} = 0.15$ mol dm^{-3}

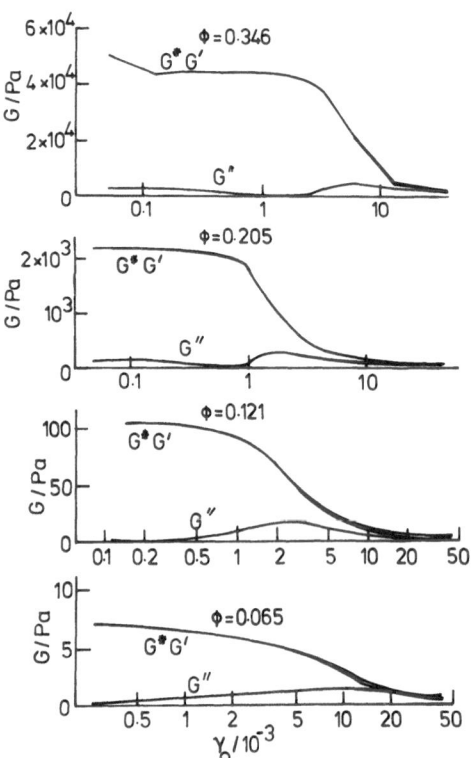

Fig. 7. Strain sweep results for latex dispersions coagulated with 0.2 mol dm^{-3} NaCl at various volume fractions

Log-log plot of G' (in the linear region) vs. ϕ is as shown in Fig. 8. The linear plot could be represented (using least square methods) by the following empirical relationship

$$G' = 1.98 \times 10^7 \, \phi^{6.0} . \tag{16}$$

The high power in ϕ is indicative of the strong interaction in such coagulated suspensions.

From γ_{cr} and G' (in the linear region) it is possible to obtain the cohesive energy E_c in the flocculated structure [21], i.e.

$$E_c = \int_0^{\gamma_c} \sigma \, d\gamma . \tag{17}$$

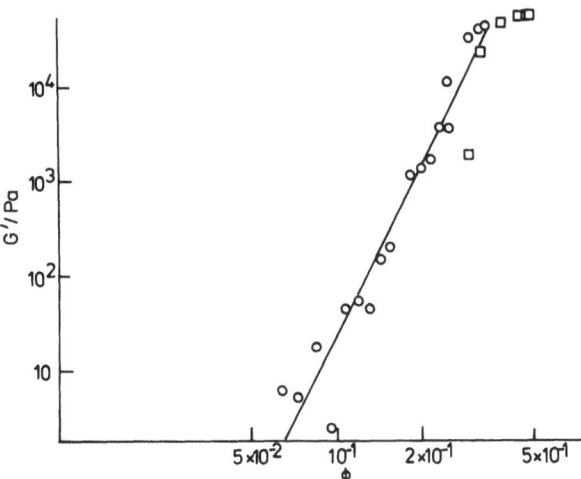

Fig. 8. Log-log plot of G' vs. ϕ for coagulated latex dispersions

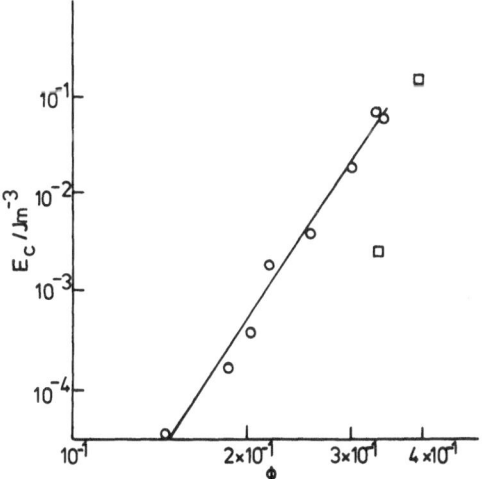

Fig. 9. Log-log plot of E_c vs. ϕ for coagulated latex dispersions

Since $\sigma = G'\gamma$, then

$$E_c = \int_0^{\gamma_c} G'\gamma \, d\gamma = \frac{1}{2} \, G'\gamma_c^2 . \tag{18}$$

A log-log plot of E_c vs. ϕ is shown in Fig. 9, which can also be represented by a power law, i.e.

$$E_c = 1.02 \times 10^3 \, \phi^{9.1} . \tag{19}$$

Again the high power in ϕ is indicative of the strongly flocculated structure. Thus, the correlation between viscoelastic properties and particle-particle interaction is also possible for coagulated systems.

References

1. Goodwin JW, Hearn J, Ho CC, Ottewill RH (1974) Colloid Polym Sci 252:464
2. Garvey MJ, Tadros ThF, Vincent B (1974) J Colloid Interface Sci 49:66
3. Garvey MJ, Tadros ThF, Vincent B (1976) J Colloid Interface Sci 55:440
4. Van den Boomgaard Th, King TA, Tadros ThF, Tang H, Vincent B (1978) J Colloid Interface Sci 66:68
5. Bromley C (1985) Colloids and Surfaces 17:1
6. Prestidge C, Tadros ThF (1988) Colloids and Surfaces 31:325
7. Prestidge C, Tadros ThF (1988) J Colloid Interface Sci 124:660
8. Whorlow RW (1980) Rheological Techniques. Ellis Horwood Ltd, Chichester
9. Casson N (1959) In: Hill CC (ed) Rheology of Disperse Systems. Pergamon Press, Oxford
10. Buscall R, Goodwin JW, Hawkins MW, Ottewill RH (1982) J Chem Soc Faraday Trans I 78:2873, 2889
11. Tadros ThF (1988) Rheology of Concentrated Stable and Flocculated Suspensions. Proceedings of Engineering Foundation Symposium, Florida, in press
12. Hopkinson A, Tadros ThF (submitted for publication)
13. Tadros ThF (1984) ACS Symposium Series 246:411
14. Hopkinson A, Tadros ThF (submitted for publication)
15. Gillespie T (1960) J Colloid Sci 15:219; Hunter RJ (1982) Adv Colloid Interface Sci 17:197
16. Luckham P, Vincent B, Tadros ThF (1983) Colloids and Surfaces 6:101
17. Asakura S, Oosawa F (1958) J Polym Sci 33:183
18. Napper DH (1983) Polymeric Stabilisation of Colloidal Dispersions. Academic Press, London
19. Hopkinson A, Tadros ThF (submitted for publication)
20. Hopkinson A, Tadros ThF (submitted for publication)
21. Ramsay JDF (1986) J Colloid Interface Sci 109:441

Received November 1988;
accepted February 1989

Author's address:

Dr. Th. F. Tadros
ICI Agrochemicals
Jealotts Hill Research Station
Bracknell, Berks. RG12 6EY, England

Progress in Colloid & Polymer Science
Progr Colloid Polym Sci 79:128–134 (1989)

Synthesis and magnetic properties of manganese and cobalt ferrite ferrofluids

F. Tourinho, R. Franck, R. Massart[1]), and R. Perzynski[2])

[1]) Laboratoire de Physicochimie inorganique*), Université Pierre et Marie Curie, Paris, France
[2]) Laboratoire d'Ultrasons*), Université Pierre et Marie Curie, Paris, France

Abstract: We present here new magnetic liquids: aqueous sols of $MnFe_2O_4$ and $CoFe_2O_4$ particles. The chemical synthesis of these ionic ferrofluids is performed in two steps:
— coprecipitation of chemically stable fine grains;
— ionic stabilization of these grains to get stable colloidal aqueous solutions.

These ferrofluids contain anionic or cationic particles, of typical size ranging from 5 nm to 15 nm, with a volume fraction which can be as large as 25%. Superparamagnetic behavior of liquid solutions allows to characterize particle internal structure. Magnetization measurements on magnetic grains trapped in a gelatin array clearly show the large difference in anisotropy constant of the two ferrites. Strong remanence of $CoFe_2O_4$ particles is turned to account, realizing a crystal growth of microscopic needles under magnetic field.

Key words: Magnetic colloid, $MnFe_2O_4$, $CoFe_2O_4$, fine grain synthesis, magnetization.

1. Introduction

Ferrofluids are colloidal suspensions of subdomain magnetic grains dispersed in a liquid carrier. They remain colloidally stable and flowable if immersed in a magnetic field. Union of both fluid and magnetic properties leads to numerous industrial applications such as rotating seals, loudspeakers, and so on [1, 2].

However, ferrofluid elaboration remains a limiting factor. An easy and convenient chemical synthesis of ionic ferrofluids in aqueous solutions was proposed by R. Massart [3] in 1980 and improved by V. Cabuil and R. Massart [4] in 1987. Manganese and cobalt ferrites are chosen here, according to their magnetic properties: large specific magnetization and, for cobalt, large anisotropy constant. We present suitable parameters to prepare stable magnetic particles of convenient size to obtain colloidal solutions in acidic as well as in alkaline media. Then we investigate magnetic properties of these ferrofluids in order to characterize these new magnetic liquids. Indeed, these sols are of particular interest:

— highly concentrated solutions may be directly synthesized with volume fractions of the order of 25%. This allows some new surface instabilities, specific to ferrofluids, to be observed, as their magnetic threshold is then substantially lowered (e. g. magnetic wetting transition [5]);

— on the other hand, because of the large anisotropy constant of $CoFe_2O_4$, monocrystalline rod particles may be obtained by crystal growth under magnetic field, bringing out many industrial applications (e. g. data storage on magnetic tapes).

2. Chemical synthesis

A *stable magnetic colloid* needs:
— *small particles* (~ 10 nm), chemically stable, in order to prevent particle sedimentation; this is realized in the first part of the chemical synthesis;
— *repulsive particle-particle interaction* (here electrostatic) to prevent particle agglomeration; this is realized in the second part of the synthesis by coating particles with a suitable charge surface density and properly choosing the counterions in solution.

2.1 Elaboration of particles

$MnFe_2O_4$ or $CoFe_2O_4$ magnetic particles are coprecipitated in alkaline medium from aqueous mixtures of Fe(III) and Mn(II) chlo-

*) Associated with the Centre National de la Recherche Scientifique.

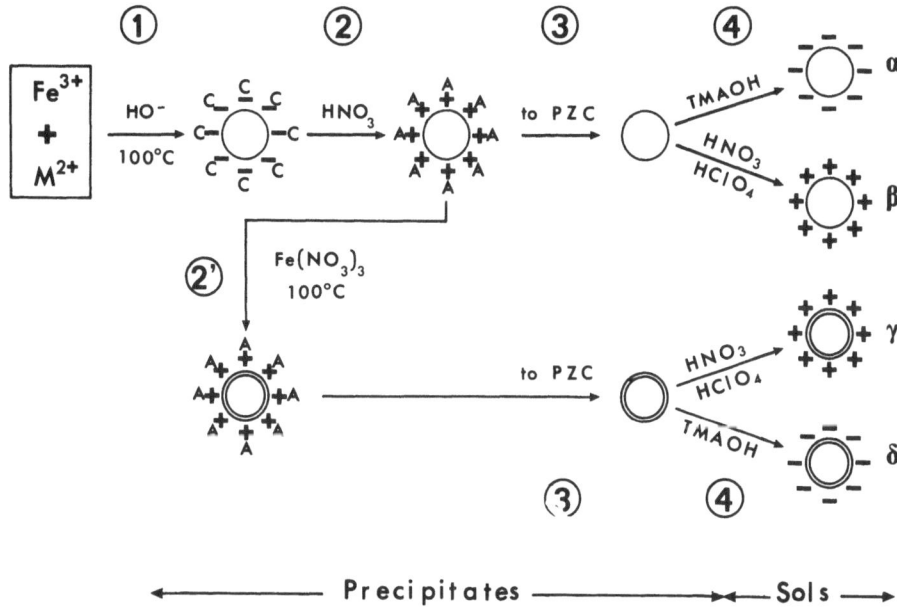

Fig. 1. Synthesis diagram (see text). $A = NO_3^-$; $C = Na^+$, NH_4^+ or $CH_3NH_3^+$; α and δ = anionic stable sols; β = cationic unstable sols; and γ = cationic stable sols

rides or of Fe(III) chloride and Co(II) nitrate (cf., Fig. 1, step 1). Different relevant parameters act on magnetic yield and size distribution of particles [6,7], namely:

— the starting M(II) molar fraction; the best value corresponds to the concerned ferrite stoichiometry.

— the coprecipitation temperature: synthesis must be performed at 100 °C.

— the way of mixing the different reagents: the base must be poured as quickly as possible into the Fe(III)–M(II) mixture, under vigorous stirring.

— the nature of the base (i. e., NaOH, NH₃, CH₃NH₂).

— the initial Fe(III)–M(II) mixture acidity.

At the end of this step, the molar ratio X is equal to 0.33, with:

$$X = \frac{[M(II)]}{[M(II) + Fe(III)]}. \tag{1}$$

2.2 Preparation of colloidal solutions

Figure 1 shows the different steps required to get stable anionic and cationic ferrofluids (particles are elaborated during *step 1*):

— *step 2*, after coprecipitation, particle surface must be cleaned of adsorbed counterions by HNO₃ (2 mol/l);

— *step 3*, a stay at the point of zero charge (PZC) of the precipitate allows a better, further charge control;

— *step 4*, alkaline sols are prepared in TMAOH (tetramethylammonium hydroxide) and acidic ones in HClO₄ or HNO₃. Anionic ferrofluids in TMAOH are stable in time. On the contrary, cationic ferrofluids in HClO₄ or in HNO₃ are degraded in a few hours.

— *step 2'*, fortunately, a ferric nitrate treatment [6] allows to get stable particles for acidic sols preparation. However, particles then exhibit a different ratio $X = 0.25$.

Thus, stable aqueous solutions of these ferrofluids are available in acidic or in alkaline medium, in a large range of concentration (cf part 3).

3. Physico-chemical aspects

Two kinds of ferrites were investigated: MnFe₂O₄ and CoFe₂O₄. They were chosen because of the large saturation magnetization $\overline{ms}$ of these bulk materials and because of their largely different anisotropy constants K [8]:

— for bulk MnFe₂O₄ $\overline{ms} = 386$ kA/m and $K = 4.10^3$ J/m³;

— for bulk CoFc₂O₄ $\overline{ms} - 422$ kA/m and $K = 2.10^5$ J/m³.

Synthesized particles have a typical mean-size ranging from 5 nm to 15 nm. They are roughly spherical, as observed through electron micrographs obtained on a transmission electronic microscope JEOL 100 CX2. Samples are never monodisperse. A log-normal distribution P, for the diameter D of the particles is usually assumed [9]:

$$P(D) = \frac{1}{\sigma D \sqrt{2\pi}} \exp - \left(\frac{\ln^2 (D/Do)}{2\sigma^2} \right) \tag{2}$$

where σ is the standard deviation and $\ln Do$ the mean value of $\ln D$. The maximum of $P(D)$ corresponds to a

Table 1. Physico-chemical sample characteristics

Sample	Nature	d_X (nm)	d_E (nm)	$\overline{ms}$ (kA/m)	ms (kA/m)	e_1 (nm)	d_M (nm)	e_2 (nm)
A	MnFe$_2$O$_4$	6.0	5	386	63	1.2	3.1	1.0
B	MnFe$_2$O$_4$	10.7	9.4	386	261	0.6	5.9	1.7
C (C')	CoFe$_2$O$_4$	8.1	5.8	422	166 (/)	0.8 (/)	3.6 (2.7)	1.1 (1.6)
D (D')	CoFe$_2$O$_4$	13.8	10.4	422	305 (235)	0.6 (1.0)	5.2 (4.6)	2.6 (2.9)
			$\sigma_E \sim 0.2$				$\sigma_M \sim 0.45$	

Samples A, B, C, D are anionic samples of molar ratio $X = 0.33$ obtained through path α of Fig. 1; samples C' and D' are cationic samples of molar ratio $X = 0.25$ obtained through path y of Fig. 1; d_X is the diameter deduced from Debye-Scherrer x-ray measurements; d_E and σ_E are the most probable diameter and the standard deviation deduced from electron micrographs; $\overline{ms}$ is bulk saturation magnetization and ms the experimental value for particles; d_M and σ_M are the most probable diameter and the standard deviation deduced from magnetization measurements. σ_M is as usual [11] greater than σ_E: all the small aggregates in solution contribute to the magnetic measurement; e_1 and e_2 are thicknesses of the non-magnetic shell deduced from two different measurements (see text).

diameter $Dmp = Do \exp(-\sigma^2)$. Histograms from electron micrographs are fitted to $P(D)$, leading to d_E, value of the most probable diameter Dmp, and σ_E, value of the standard deviation σ [$d_E = do_E \exp(-\sigma_E^2)$].

Complementary Debye-Scherrer x-ray measurements were performed. They led to a diameter d_X, usually related to d_E through $d_X = d_E \exp(3.5 \sigma_E^2)$, in agreement with the present results (cf. Table 1).

To prevent aggregation due to Van der Waals and magnetic interactions, ferrofluid particles bear superficial charges. The surface charge density is of the order of 0.2 C/m^2 [6, 7], negative in alkaline media and positive in acidic ones. The colloidal stability is ensured by low polarizing counterions (cf. part 2 and Fig. 1). As for previously studied ionic ferrofluids made of γ-Fe$_2$O$_3$ particles [10], a separation into two liquid phases of different particle concentrations occurs if the ionic strength exceeds a given threshold depending on the ferrofluid characteristics. This phase separation may also be induced by a magnetic field or lowering the temperature. Typically, sample B (MnFe$_2$O$_4$) of Table 1 remains monophasic, at room temperature, if ionic strength is less than 10^{-1} mol/l even if immersed in a 8.10^5 A/m magnetic field, and sample D (CoFe$_2$O$_4$), if ionic strength is less than 10^{-2} mol/l. Under these conditions, monophasic samples may be obtained with a volume fraction ϕ of magnetic particles of the order of 25%, which is two-times larger than what is usually obtained with monophasic γ-Fe$_2$O$_3$ ferrofluids.

4. Magnetic properties

Various kinds of magnetic properties of the ferrofluid solutions are here explored. Behavior of liquid solutions as a function of magnetic field provides determination of saturation magnetization and size distribution of magnetic particles. Influence of anisotropy constant K is detected through magnetization curves of ferrofluid solutions frozen in a gelatin array. K values are correlated to experimental results of crystal growth under magnetic field. Magnetic measurements are performed using a classical Foner device [11]. The accuracies ΔH on magnetic field and ΔM on magnetization are, respectively, 80 A/m and 3 A/m.

4.1 Liquid solutions

If interparticle interactions are negligible, that is if the ferrofluid solution is dilute enough, (namely $\phi < 1\%$ [12]), magnetization M of the monophasic ferrofluid may be described by Langevin formalism: the solution is superparamagnetic. Each particle is a magnetic monodomain of permanent moment $\vec{\mu}$ free to align along the magnetic field $\vec{H}$. For an ideal monodisperse solution, M is given by

$$M(D,H) = Ms \; \alpha(\mu H/kT) \tag{3}$$

where k is Boltzmann constant, T temperature, α Langevin function, and

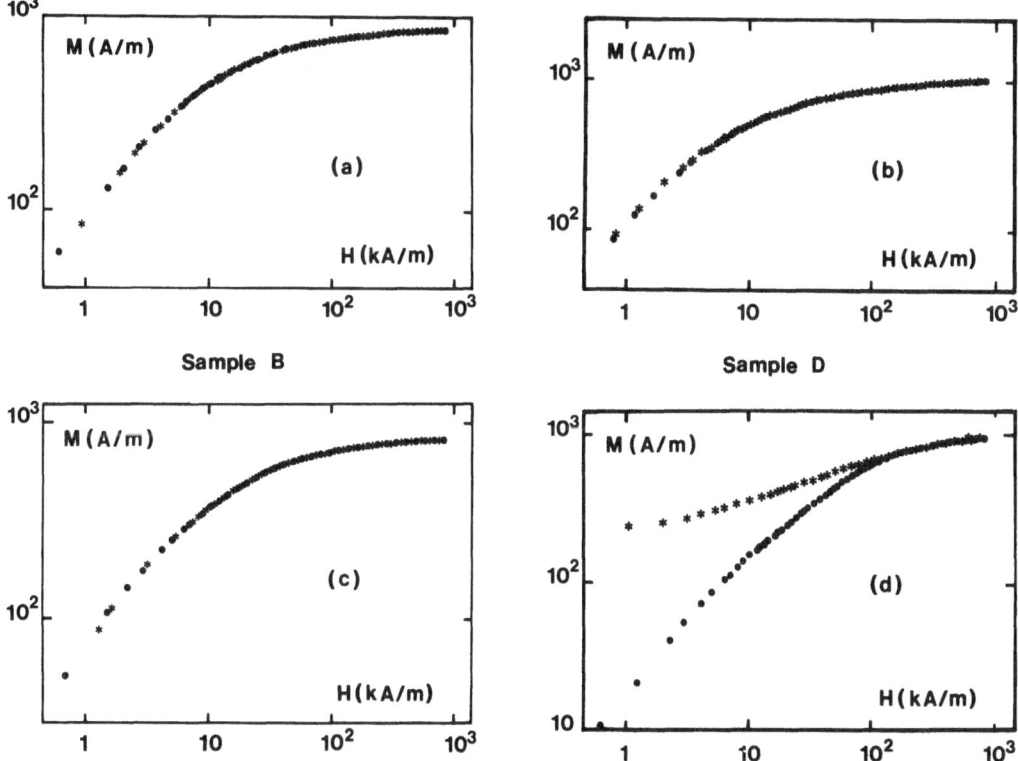

Fig. 2. Magnetization curves of sample B (MnFe$_2$O$_4$; $\phi = 0.3\%$) and sample D (CoFe$_2$O$_4$; $\phi = 0.3\%$). (a), (b) liquid solutions; (c), (d) frozen solutions. ● (resp. ✳) are measurements with increasing (resp. decreasing) fields

$$Ms = ms \cdot \phi \qquad (4)$$

with ms saturation magnetization of particles, different of bulk value $\overline{ms}$.

Experimental magnetization curves (cf. Figs. 2a and 2b) lead to Ms saturation magnetization of solutions. ms is then deduced from typical plots of Fig. 3: Ms vs. ϕ. For particles made of a given material, ms largely depends on two factors:

— the *particle size* (cf. in Table 1, samples A and B, or samples C and D). This is in agreement with [13].

— the *chemical treatment* (cf. in Table 1 and in Fig. 3, homologous samples D and D′ obtained from paths α and y of Fig. 1). ms is proportional to the molar ratio X of the sample (cf. expression 1).

We may propose an internal structure to explain these points: a bulk-like magnetic core surrounded by a non-magnetic shell, the shell thickness e being function of the chemical treatment undergone by the particles. An evaluation of this thickness $e_1 = do_E [1 - (ms/\overline{ms})^{0.33}]/2$, is presented in Table 1; e_1 is roughly constant, whatever the anionic sample A, B, C or D.

Moreover, a measurement of the diameter of the bulk-like magnetic core is possible through an analysis

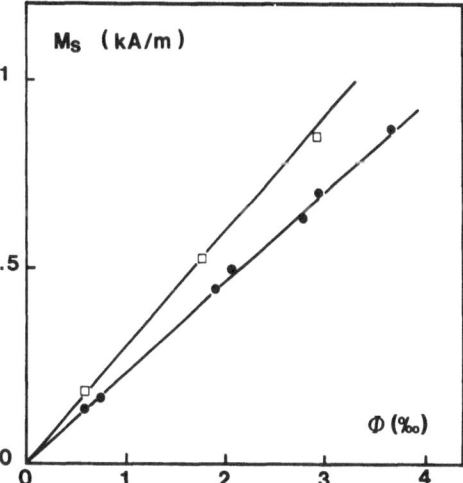

Fig. 3. Saturation magnetization of ferrofluid solutions vs volumic fraction for CoFe$_2$O$_4$ samples. □: anionic sample D (path α in Fig. 1, $X = 0.33$, $ms = 305$ kA/m); ●: cationic sample D′ (path y in Fig. 1, $X = 0.25$, $ms = 232$ kA/m)

of the shape of the magnetization curve [11] leading to a most probable diameter d_M and a standard deviation σ_M for the magnetic core. These measurements are also collected in Table 1: d_M is systematically smaller than d_E. A second evaluation of the shell thickness from $e_2 = (d_E - d_M)/2$ is in complete agreement with the first one, within the experimental accuracy. For anionic samples (path α in Fig. 1) mean value of e is $\sim$ 1.2 nm. This non-magnetic shell seems due to a superficial alteration during HNO_3 treatment (step 2 of Fig. 1). For cationic samples (path γ in Fig. 1), the shell thickness is always $\sim$ 0.4 nm larger, whatever the calculation method. This is very probably due to an additional superficial alteration of particles during the $Fe(NO_3)_3$ treatment of step 2' of Fig. 1.

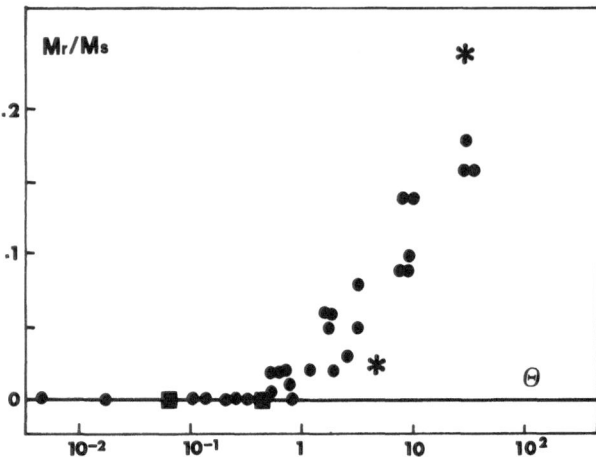

Fig. 4. Reduced remanence vs θ parameter. ●: from [15]; ■: $MnFe_2O_4$ samples; $*$: $CoFe_2O_4$ samples

4.2 Frozen solutions

If ferrofluid particles are frozen in a polymeric matrix, they are mechanically trapped in the matrix and particle Brownian rotation is inhibited. In liquid ferrofluid solutions, Brownian rotation is usually the dominant process which tends to align the magnetic moment of particles along the magnetic field direction [11]. Thus, in frozen solutions, the mechanism of moment rotation is different and is due to thermal fluctuation of magnetic moment within the crystalline particle. This process [14] is hindered by an anisotropy energy barrier KV, K and V being, respectively, the anisotropy constant and the volume of particles. In dilute frozen solutions ($\phi \leq 1\%$), three different energies are involved for each particle: anisotropy energy KV, magnetic energy due to magnetic field μH and energy of thermal motion kT. Several magnetic regimes correspond to different orderings of KV, μH and kT; in particular, for samples of anisotropy axis direction at random, magnetization curve may exhibit an hysteresis with a remanence Mr, only function of one reduced parameter $\theta = KV/kT$ [15]. Experimen-

tal variations (from [15]) of Mr/Ms vs. $\theta = (K \cdot \pi d_E^3/6)/kT$ are presented in Fig. 4. Two different regimes may be distinguished: Mr/Ms = 0 and Mr/Ms > 0. The θ value limiting these two regimes is a function of the time scale of the experiment [14]; our measurement scale (as in [15]) is of the order of 100 s and then: Mr/Ms = 0 for $\theta \leq 0.4$ and Mr/Ms > 0 for $\theta \geq 0.4$.

Samples of Table 1 are trapped in zero magnetic field in a gelatin array (5 % gelatin) of mesh size smaller than the magnetic particle size [6]. Magnetization measurements are performed at room temperature. In agreement with the calculated θ values (cf. Table 2), Mn-based samples do not exhibit any remanence, according to their low K value (cf. Fig. 2c). On the contrary (cf. Fig. 2d), Co-based samples exhibit a large Mr (cf. Table 2), the large K value of $CoFe_2O_4$ leading to $\theta >$ 0.4. In agreement with measurements of [15] realized on γ- Fe_2O_3 samples at low temperature, present measurements on $MnFe_2O_4$ and $CoFe_2O_4$ particles lie on the same master curve of Fig. 4. Bulk anisotropy constants of these two materials allow to explain the magnetic behavior of these frozen solutions.

4.3 Crystal growth under magnetic field

Data storage on magnetic tapes requires fine particles of large remanence and large anisotropy; usually acicular particles of large shape anisotropy are employed. We show here a simple way to obtain such monodomain acicular particles of $CoFe_2O_4$. The

Table 2. θ Parameter and remanence of the samples

Sample	$K(J/m^3)$	θ	Mr/Ms
A	4.10^3	0.065	0
B	4.10^3	0.43	0
C	2.10^5	5.1	0.02
D	2.10^5	29.5	0.24

K is the bulk anisotropy constant of the magnetic material; θ parameter is equal to $K(\pi d_{O_E}/6)/kT$; Mr and Ms are, respectively, remanence and saturation magnetization of frozen solutions.

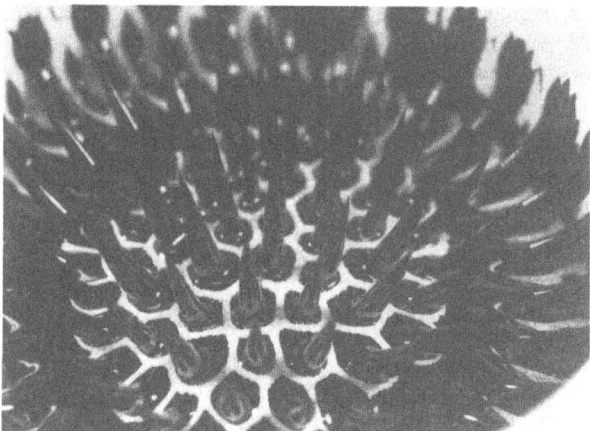

Fig. 5. Spike instability of a flat ferrofluid interface

large crystalline anisotropy constant of this material allows a crystal growth under magnetic field.

The method is based on a specific and very spectacular property of magnetic liquids. If a flat surface of ferrofluid is submitted to a large enough magnetic field perpendicular to the surface, this flat interface bristles with a whole series of spikes, a few centimeters high,

(cf. Fig. 5). If such liquid spikes made of $CoFe_2O_4$ or $MnFe_2O_4$ highly concentrated ferrofluids ($\phi = 25\%$) are dried off, allowing the liquid carrier to evaporate under a magnetic field of the order of 200 kA/m, one obtains solid spikes of different magnetic behaviors, depending on the material of the magnetic particles [6].

Solid spikes made of $MnFe_2O_4$ do not exhibit any remanence, nor do the corresponding frozen solutions. On the contrary, and as is expected from results of section 4.2, $CoFe_2O_4$ solid spikes behave like permanent magnetic needles. Moreover, if the spikes are ground to powder, electron micrographs of such $MnFe_2O_4$ powders lead to roughly spherical particles very similar to the initial ones. On the contrary, electron micrographs of such $CoFe_2O_4$ powders exhibit microscopic needles (cf. Fig. 6a and 6b) well crystallized (cf. Fig. 6c): monodomain crystals of typical dimensions 13 nm × 130 nm are thus obtained.

5. Conclusion

New magnetic liquids of very large concentration ($\phi \sim 25\%$), namely ionic ferrofluids of colloidal $MnFe_2O_4$ and $CoFe_2O_4$ particles, are thus available.

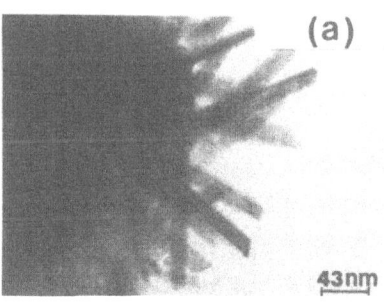

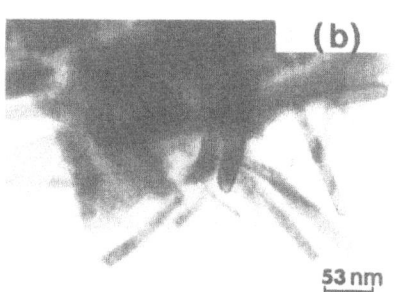

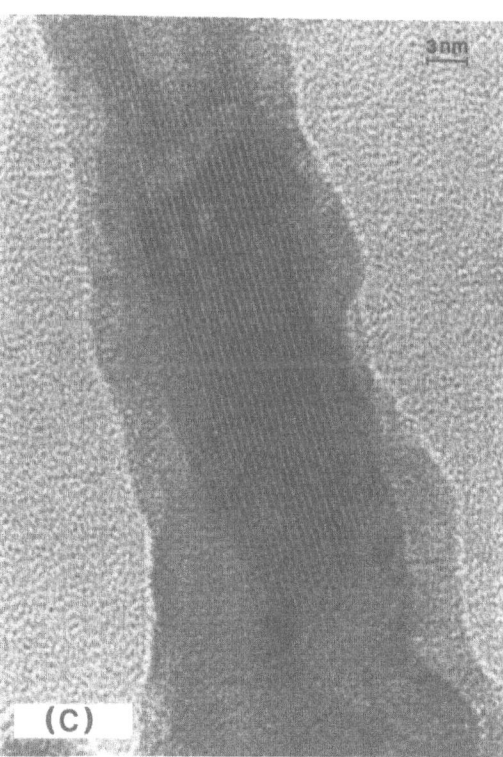

Fig. 6. Electron micrographs of $CoFe_2O_4$ spikes dried under magnetic field and ground to powder. (a) and (b): general views of particles; (c): enlargement of a microscopic needle.

Present chemical synthesis allows to prepare stable anionic and cationic solutions. Magnetic properties of solutions are consistent with an internal structure of particles: a bulk magnetic heart surrounded by a non magnetic shell of thickness ranging from ~ 1 nm to 2 nm depending on the chemical treatment undergone by the particles. In close correlation with bulk anisotropy constant K of $MnFe_2O_4$ and $CoFe_2O_4$, frozen solutions exhibit or not a strong remanence. Thus, large K value of $CoFe_2O_4$ allows to very easily realize a crystal growth under magnetic field of $CoFe_2O_4$ acicular particles.

Acknowledgement

We are greatly indebted to Mrs. M. Carpentier and M. R. Rajaonarison for technical assistance. Electron micrographs have been performed in the Groupement Régional de Mesures Physiques. We thank Professors J. C. Bacri, G. Djega-Mariadassou and D. Salin for helpful discussions.

References

1. Rosensweig RE (1985) In: Ferrohydrodynamics. Cambridge University Press
2. Bacri JC, Perzynski R, Salin D (1988) Endeavour 12:76–83
3. Massart R (1979) Brevet français 7918842; (1982) US Patent 4329241; Massart R (1980) C R Acad Sc Paris 291 C:1–3; (1981) IEEE Trans Magn MAG-17 n°2:1247–1248
4. Massart R, Cabuil V (1987) J Chim Phys 84 n° 7–8:967–973
5. Bacri JC, Perzynski R, Salin D, Tourinho FA (1988) Europhys Lett 5(6):547–552
6. Tourinho FA (1988) Ferrofluides à base de ferrite de cobalt et de ferrite de manganèse. Elaboration, comportement physicochimique et magnétique. Thesis, Université Pierre et Marie Curie, France
7. Tourinho FA, Franck R, Massart R, submitted to publication
8. Kupricka S, Novak P (1982) In: Wohlfarth EP (ed) Ferromagnetic Materials, vol 3. North Holland publishing Company, pp 296
9. Grandqvist CG, Buhrman RA (1976) J Appl Phys 47:2200–2219
10. a) Cabuil V, Massart R, Bacri JC, Perzynski R, Salin D (1987) J Chem Res (S):130–131
    b) Bacri JC, Perzynski R, Salin D, Cabuil V, Massart R (1989) J Colloid Interface Sci
11. a) Bacri JC, Perzynski R, Salin D, Cabuil V, Massart R (1986) J Magn Magn Mat 62:36–46
    b) Bacri JC, Perzynski R, Salin D, Servais J (1987) J Physique 48:1385–1391
12. Pynn R, Hayter JB, Charles SW (1983) Phys Rev Lett. 51:710–713
13. Sato T, Iijima T, Seti M, Inagaki M (1987) J Magn Magn Mat 65:252–256
14. Néel L (1949) C R Acad Sci, Paris 228:664–666, (1949) Ann Geophys 5:99
15. Bacri JC, Perzynski R, Salin D (1988) J Magn Magn Mat 71:246–254

Received October, 1988;
accepted March, 1989

Authors' address:

Dr. R. Perzynski
Laboratoire d'Ultrasons
Tour 13
Université Pierre et Marie Curie
4, place Jussieu
F-75252 Paris, Cedex 05, France

Progress in Colloid & Polymer Science   Progr Colloid Polym Sci 79:135–141 (1989)

C. Wetting, adsorption and interfaces

Adsorption of cetyltrimetyl ammonium at the free surface and at the mercury electrode

K. Bennis[1]), P. Martinet[1]), D. Schuhmann[2]), and P. Vanel[2])

[1]) Laboratoire d'Electrochimie Organique, Université de Clermont, Aubière, France
[2]) Laboratoire de Physicochimie des Systèmes Polyphasés, C.N.R.S., Montpellier, France

Abstract: Variations of adsorption energy and CMC with salt concentration and nature are explained by salting out of monomers in the bulk. A monolayer is formed at the free surface, its molecular area is consistent with the presence of counter-ions. A bilayer is formed at mercury from direct adsorption of micelles. A transition in the surface curves at about 50 CMC is tentatively explained by a phase separation due to structuration of water molecules around micelles under the influence of dipolar effects. The presence of very high effective dipole moments is evident from analysis of charging curves.

Key words: Salting out and micelles, cationic surfactant, adsorption, micellar transition, micelles and dipolar effects.

Introduction

This work was carried out in relationship with the influence of surfactant on the kinetics of electrochemical reactions [1–3] and in view of understanding the influence of electrostatic interactions upon micellization equilibria, using the mercury electrode as a model system [4–7]. Similarities between the behaviors of zwitterionic surfactants [7] and symmetrical tetraalkyl ammonia [8] led us to study the adsorption of the cetyltrimethyl ammonium ion (CTA$^+$) at a polarized mercury electrode and at the air/solution interface.

It is well known that inorganic salts have the CMC of ionic surfactants decreased [9–17]. With pure CTAB (B = bromide), the accepted value is 9×10^{-4} M [18–22] while with 0.5 M KBr, the value 2.5×10^{-5} M has been found [23] using a fluorescence technique [24]. Having shown by thermodynamic [25] and kinetic [26] considerations that the CMC should depend on the activity coefficient of dissolved monomers, we thought that salting out effects could be involved.

The mean aggregation number n and the degree of ionization α of cationic micelles depend on the surfactant concentration c [27–31] and for n, on the salt concentration C_s [13]. From comparison between anionic micelles and adsorption layers at mercury [6] it was concluded that with correlated changes of n and α the surface charge density of micelles remain constant. Cationic micelles become rodlike in the presence of salts containing Br$^-$ [32–34], even for concentrations near the CMC, while with CTAB (C = chloride), micelles remain spherical up to 38% of surfactant [31]. This difference has been related to that between the sizes and/or hydrophobicity of Cl$^-$ and Br$^-$ ions [35].

Experimental

The break in surface or interfacial tension curves (γ vs. log c) allows to determine the CMC. It has been found with anionic [4, 36], nonionic [37], and zwitterionic [7] surfactants forming adsorbed monolayers that the same value of CMC is obtained from measurements at free surfaces and Hg electrode. Moreover, the same isotherms were found at both interfaces if the latter is not charged.

The surface tension was measured with a Prolabo tensiometer. The purity of the surfactant was verified from Elworthy and Mysels' test [38].

Measurements with KBr were carried out with a buffer at pH 2.5 [41] as in previous kinetic studies and at an ionic strength I 0.5 M. The interfacial tension was measured using a dropping mercury electrode, the dropping time being proportional to γ. The drop-time depends on the capillary radius and height of the mercury column (which can fluctuate). A relative procedure is thus used: the capillary is calibrated very often, measuring the drop-time with a reference solution (0.1 M KCl) whose electrocapillary curve (y vs. E) is known with great accuracy. Further details on the technique and procedure are reported in another paper [39]. The mean error was about 0.5 mN/m, save for too low concentrations (say 10^{-5} M) because of diffusion effects (40). CTAB, KCl, and KBr were Merk products used without any further purification. Given the KCl concentrations used (0.5 M or more) and the low values of c (less than 10^{-3} M), data obtained with KCl solutions are identical in the limits of accuracy to those resulting from experiments with CTAC.

Results

Adsorption at the free surface

The surface excess Γ is given by Gibb's equation:

$$RT\Gamma = -\delta y/\delta \ln c_1 \tag{1}$$

c_1 being the concentration of the species actually adsorbed. The temperature 29 °C allowed a good solubilization in most cases and was chosen in concordance with the previous kinetic studies.

CTAB + KCl solutions

y vs. log c curves are plotted in Fig. 1. The values of CMC $\times 10^5$ (M) are 2.2, 1.1, and 0.25, the surface tensions y_0 for $c = 0$ are 73.5, 73.7, and 75.8 mN/m, the limiting values y_∞ for $c >$ CMC are 35.5, 34.0, and 33.5 mN/m, for $c_S = 0.5$, 1 and 2 M, respectively. The maximum excess Γ_m deduced from the constant slopes

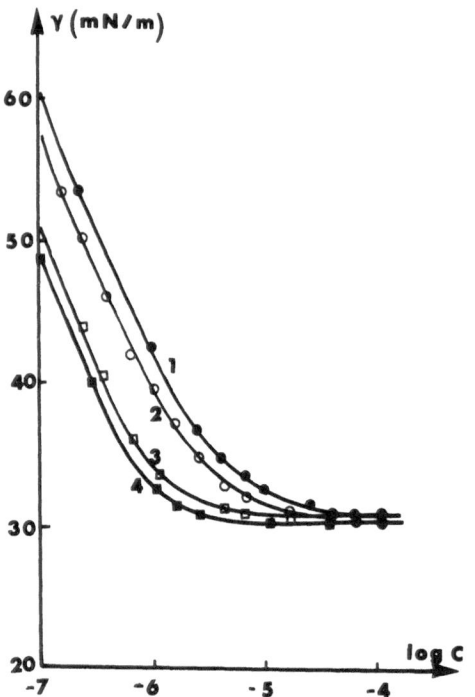

Fig. 2. Variations of the surface tension with the logarithm of the surfactant concentration. CTAB + KBr solutions. 29 °C; C (KBr) (M); 1:0.5, 2:1, 3:2, 4:35 °C, 1 M

for $c <$ CMC is 3×10^{-10} mole/cm^2 and the corresponding molecular area A is 0.55 nm^2. The horizontal shift Δ log c_i allowing to have the curves coincided for $c <$ CMC, linearly depends on c_S: $\delta \Delta \log c_i/\delta c_S = -0.7$ l/mole $\approx \delta$ CMC/δc_S.

CTAB + KBr solutions

y vs. log c curves are plotted in Fig. 2. y_0 is equal to 75 ± 1 mN/m for the same three values of c_S studied. The surfactant solubilization was still slow at 29 °C and $c_S = 0.5$ M: the points reported for this value of c_S correspond to the smallest values obtained in a series of experiments. One finds $\delta \Delta \log c_i/\delta c_S = -0.35$ l/mole and $\delta y_\infty/\delta c_S \approx 0$ in the limits of accuracy. From the curve at 35 °C, as expected, the adsorption energy is found temperature dependent. A is still equal to 0.55 nm^2. The transition at CMC is less sharp than with KCl. The end of the transition zone corresponds to the value of CMC determined by a fluorescence technique [23] and it corresponds to the value found with KCl.

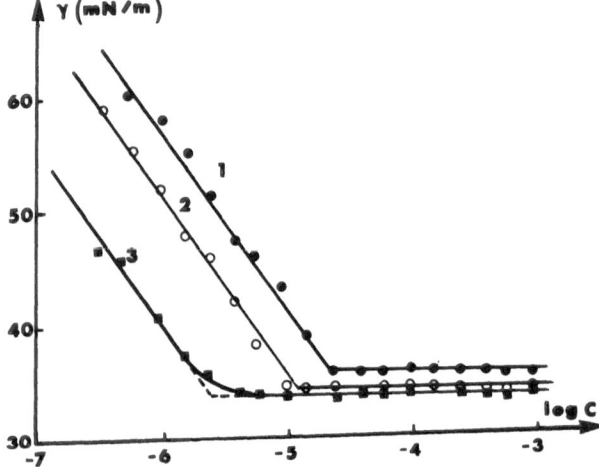

Fig. 1. Variations of the surface tension with the logarithm of the surfactant concentration. CTAB + KCl solutions. C (KCl) (M); 1:0.5, 2:1, 3:2

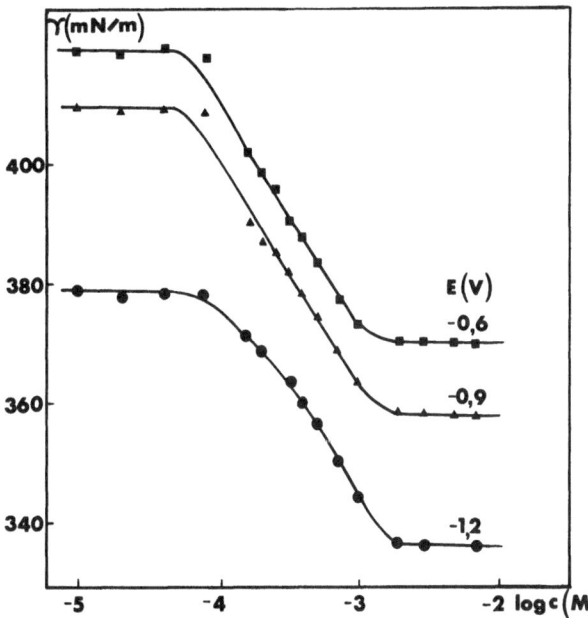

Fig. 3. Variations of the interfacial tension with the logarithm of the surfactant concentration at various potentials applied to the mercury electrode. CTAB + KBr solutions

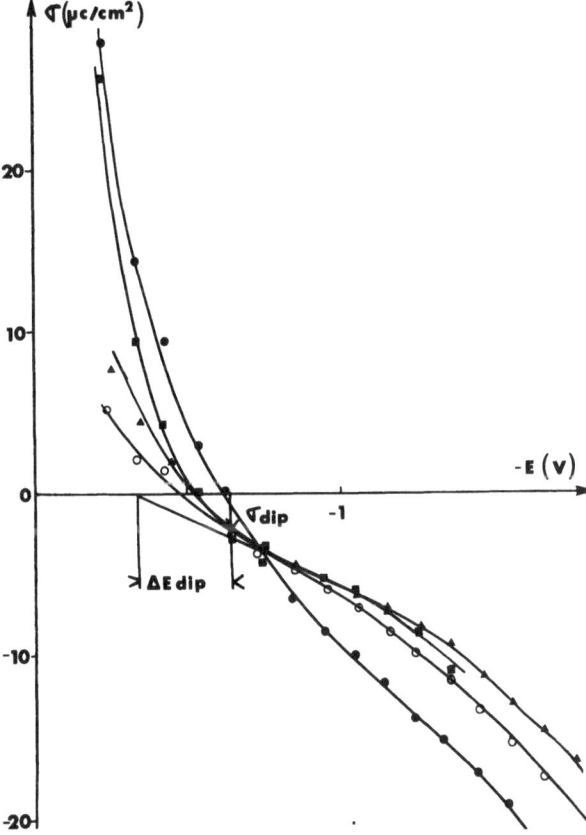

Fig. 4. Variations of the interfacial tension with the logarithm of the surfactant concentration at various potentials applied to the mercury electrode. CTAB + KCl solutions

Adsorption at the mercury electrode

Interfacial tension curves

These are reported in Figs. 3 and 4 for KBr and buffer ($I = 0.5$ M) and 1 M KCl solutions, respectively. Gibb's equation can still be applied at constant potential E; E was measured with respect to a $N/10$ calomel reference electrode. For $c < 10^{-4}$ M, $y(E)$ remains practically equal to $y_0(E)$, $y_0(E)$ being equal to 424.1, 418.4, and 411 mN/m for the three values of E and KCl solutions. Relevant data having been obtained with other longchained surfactants at $c = 10^{-4}$ M, it may be concluded that adsorption of CTAB or CTAC is very low for $c <$ CMC. For c between roughly 10^{-4} and 10^{-3} M, y strongly decreases. After a transition, sharp with KBr and smooth with KCl, y remains constant whatever E. Meakins [42, 43] found similar results and attributed the transition to CMC. This is not consistent with our data for the free surface reported above. In the range where the slope of the curves is constant, application of Gibb's law (discussed below) leads to $A = 0.26$ and 0.22 nm^2 for KBr and KCl solutions, respectively.

Charging curves

The surface charge density (at the metal side) is given by Lippmann's equation, $\sigma = - \delta y/\delta E$. Information on the electrostatic properties can be deduced from the charging curves σ vs. E as shown by Frumkin et al. [44]. The slope $\delta\sigma/\delta E$ is the differential capacity C. One has $\delta C/\delta \Gamma < 0$. Charging curves without and with surfactant are plotted in Figs. 5 and 6. The minimum value of C for $\Gamma = \Gamma_m$ is ≈ 10 µF/cm^2 and the corresponding shift of the potential of zero charge denoted ΔE_{dip} (see Fig. 5) is ≈ 0.3 V for both salts.

Discussion

Adsorption at the free surface

Effects of the inorganic salt

For almost all classes of organic species studied so far, adsorption isotherms may be fitted with the empiric Frumkin's expression:

$$\theta \exp (- 2a\theta)/(1 - \theta) = Bc_1 \qquad (2)$$

with $\theta = \Gamma/\Gamma_m$; a is a constant, B the adsorption equilibrium constant, and c_1 the concentration of the species which actually adsorbs (monomers). From Eqs.

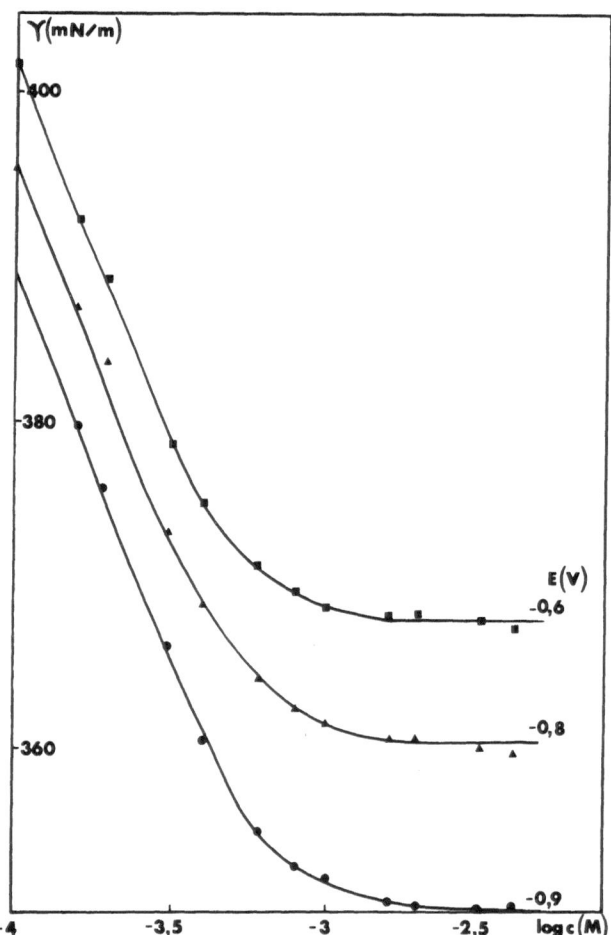

Fig. 5. Some charging curves for x M CTAB + 0.5 M KBr solutions.
●: $x = 0$; ■: $x = 2 \times 10^{-4}$; O: $x = 10^{-3}$; ▲: 2×10^{-3}

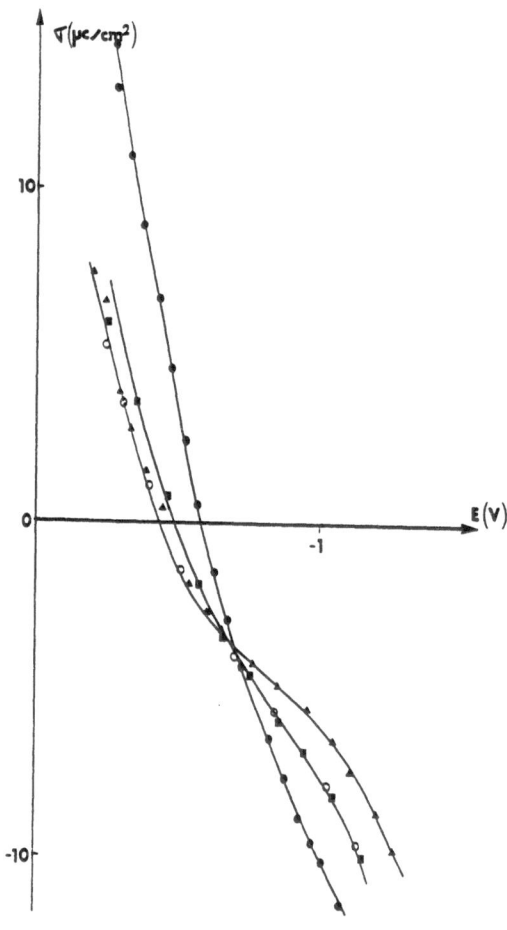

Fig. 6. Some charging curves for x M CTAB + A M KCl solutions.
(Same symbols as in Fig. 5)

(1) and (2), and by integrating, one obtains the „equation of state" $y_0 - y$ vs. θ. In the region where $\Gamma \approx \Gamma_m$, $c_1 \approx c$, one finds:

$$y_0 - y = RT\Gamma_m \left[a + \ln(BC) \right]. \tag{3}$$

The abscissa of the intercept with the horizontal line $y = y_0$ of the linear part in the surface pressure curve is thus $- \log c_i$: $0.434\ a + \log B$. If $\delta a / \delta c_s = 0$, given the constancy of y_0, variations of $\log c_i$ are directly related to those of $\log B$. For $c > $ CMC, if the limiting monomer concentration is constant and noted c_1^∞, one obtains:

$$y_0 - y^\infty = PT\Gamma_m \left[a + \ln (Bc_1^\infty) \right] \tag{4}$$

and from Eqs. 3 and 4, c_1^∞ is equal to the CMC.

For KBr, Γ_m, y_0 and y do not depend on c_s. Equation (4) suggests that the same is true for a: the shape of the isotherm would not depend on c_s. For KCl, variations of y_0 suggest a small negative salt adsorption without any surfactant, while the slight decrease of y with c_s could be due to a small change in α.

It has been shown that the equilibrium constants B depend on the activity coefficient f of the surfactant in the bulk [46, 25]. They are thus affected by salting out or in [47] described by Setchenov's coefficient k:

$$\log f/f_0 = kc_s \tag{5}$$

where k depends on the salt and solute natures. From the above data for CTA$^+$, one has $k = 0.7$ and 0.35 l/M for KCl and KBr, respectively. Reported values of k are indeed lower for KBr than for KCl, while K$^+$ and

Na^+ have similar effects: for benzene, one has $k = 0.156$, 0.150, and 0.063 for NaCl, KCl and KBr, respectively [48]. For longchained alcohols, k linearly depends on the number of carbon atoms in the chain n_c. Extrapolating to $n_c = 16$ the values found for NaCl at $25\,°C$ and n_c between 3 and 6 [49], one would find $k \approx 0.5\,l/mole$. The dipole moment of neutral molecules has no effect on k [49]. The values found in the present work for CTA^+ are thus quite reasonable and variations of B and CMC with c_s can be attributed to salting out of monomers in the bulk. The effect on CMC is explained by the fact that micellization is a stepwise process [26] defined by equilibria such as $S_s + S_1 \leftrightarrows S_{s+1}$, where S is the surfactant, S_1 its monomer form, and s is the transient degree of aggregation. The equilibrium constants K_s, and thus also CMC depend on the activity coefficient of the monomer. Variation of CMC with n_c can be explained in the same way.

The molecular area

The value $A = 0.55\,nm^2$ shows that the maximum excess is not limited by the aliphatic chains ($A = 0.2\,nm^2$) but by the surfactant heads. A value of the same order ($0.5\,nm^2$) for the head group area was recently found for CTAB by measuring the forces between two layers adsorbed on mica [50].

The ionic radii for Me_4N^+, Cl^-, and Br^- are, respectively [51]: $R = 0.347$, $r_C = 0.18$, and $r_B = 0.194\,nm$. The presence of counter-ions favors the quadratic array. Computation for the more compact array gives for $\alpha = 1$, $A = 0.48\,nm^2$ and for $\alpha = 0$, $A = 0.56\,nm^2$ and $0.58\,nm^2$ with Cl^- and Br^-, respectively, the computation being made under the assumption that the centers of the head groups anions are in the same plane. The value found for A suggests that α is low, as was recently found in the case of adsorption on mica [50]. This value is thus quite reasonable and shows that the chains are adsorbed in a nearly normal orientation, as in the case of other surfactants [12].

Adsorption at the mercury electrode

Structure of the adsorbed layer

Symmetrical tetraalkylammonium ions [8, 52, 53] are adsorbed even at positive values of σ and alcaloids, compounds bearing a nitrogen site are chemisorbed at mercury [54]. The low value of Γ found with CTAB and $c <$ CMC may thus be explained by interactions of the heads with mercury and hydrophobic interac-

tions leading to a parallel orientation. The value of A found for $c >$ CMC, being about one-half that found at the free surface, suggests that a bilayer is formed by direct adsorption of micelles. The same occurs at mica [50] where the layer thickness δ could be determined: $\delta = 3.2 \pm 0.2\,nm$ while the length of the extended molecules is $2.5\,nm$. This value of δ shows that the chains of both monolayers are tangled, which is possible given the different areas of polar heads and aliphatic chains, the polar heads of the first monolayer being in contact with mercury and those of the second one turned towards the solution. If micelles with a constant aggregation number n are directly adsorbed, for $c \gg$ CMC, $c_m \approx c/n$ and $\Delta\log c_m = \Delta\log c$, c_m being the micelle concentration. Gibb's law thus gave the correct values of Γ_m and A. Meakins' data [43] also suggested the presence of a bilayer, but the use of a false value of CMC led to the belief that the bilayer was formed by adsorption of monomers.

Behavior at high concentrations

The transition was also observed by Meakins et al. [42, 43]. The data recalled in the introduction and others [55] prevent attributing it to a change in the shape of micelles, even if the γ vs. $\log c$ curves behave as in the case of phase transition. Formation of hexagonal or lamellar phases requires concentrations much higher than $10^{-3}\,M$. The results obtained with mica without any inorganic salt [50] show that bilayers have the same value of α (0.22). A lower value is expected in the presence of salt concentration of the order of $1\,M$ and is suggested by the value found for Γ_m. As shown below, this leads to high dipole moments in adsorbed layers and thus, on the surface of micelles. These dipole moments could influence the structure of neighboring water molecules to an extent different with Cl^- and Br^- ions. The structure of water around micelles would prevent their „floculation" but their relative closeness could enhance the structuration of water between them. „Association" of further micelles could then proceed with the same phenomenology as that describing stepwise micellization. The assumption that the transition is due to some ordering in parts of the solution [56] is quite tentative but may be put forward as a working hypothesis for further investigations.

Analysis of the charging curves

The theory for neutral surfactant reported by Frumkin et al. [44] was reexamined by one of us [57,

58]. The surface potential due to adsorbed dipoles is given by $\Delta E_{dip} = \mu / (A\varepsilon_0\varepsilon_r)$ where μ is the normal component of the dipole moment, ε_0 the permittivity of vacuum, and ε_r the relative dielectric constant of the medium where dipoles are embedded (the adsorbed layer). ε_r is forgotten in the Helmholtz formula. The theory generalized for partly polar and partly charged layers [59] shows that the shift in the potential of zero charge is not only due to dipoles and the apparent surface potential is $F\Gamma(z\alpha/C_0 + h\beta/C)$ with $z = +1$ for CTA$^+$, C_0 is the double layer capacity measured at high salt concentrations without any surfactant [60, 61], C the adsorbed layer capacity, $\beta = 1 - \alpha$ and $h = \mu/\delta e$, e being the electronic charge. In the potential range where C is minimum (Γ is maximum), Figs. 5 and 6 show that C_0 is approximately equal to 16 and 35 μF/cm^2 for KBr and KCl solutions. However, C and the apparent ΔE_{dip} have identical value while $F\Gamma/C_0$ is of the order of 0.5 to 1 V. This suggests $\alpha \approx 0$. In fact, with 1 M solution the Debye length is equal to 0.3 nm. As far as effective moments are concerned, it is therefore difficult to distinguish between actually bound and free counter-ions. Taking into account the contributions of both monolayers, ΔE_{dip} may be taken equal to $(\mu_2/A_1 + \mu_1/A_2)/(\varepsilon_0\varepsilon_r)$. Assuming approximately $A_1 = A_2 = 2A \approx 0.5$ nm^2 and taking $\delta = 3.2$ nm as in the case of bilayers adsorbed on mica with $C = \varepsilon_0\varepsilon_r/\delta \approx 10$ μF/cm^2, one finds $\mu_2 + \mu_1 \approx 50 \times 10^{-30}$ Cm, which is a very high value; for a monolayer of alcohol, one finds $\mu \approx 3 \times 10^{-30}$ Cm [57].

If a_i is the algebraic distance between the centers of anions and cationic groups, positive when the former are farther from the surface than the latter, with $\mu_i = a_i e$, one finds $a_1 + a_2 \approx 0.3$ nm. If the anions in monolayer 1 were in contact with mercury and those in monolayer 2 at the top of the cationic groups, one would have $a_1 - r = R$ and $a_2 = r + R$ and thus, $a_1 + a_2 = 2r \approx 0.4$ nm. Recent statistical treatments show that less hydrated halides tend to displace the water molecules in contact with a hard charged wall without assuming a specific adsorption with the wall [62]. The above analysis is thus quite consistent and the assumption of a strong dipolar effect on the surroundings of adsorbed layers and micelles does not seem unreasonable. It can be shown that the dipoles have a strong effect on the adsorbed layer itself. With the values of C and δ used above, $\varepsilon_r \approx 35$, while for alcohols the value extrapolated to compact layer is 2, corresponding to paraffin [57]. With a monolayer of dodecylbetaine, a zwitterionic surfactant, ε_r was of the order of 15 to 20 [63]. Polar solutes increase the

dielectric constant of solutions in agreement with Buckingham's theory [64]. A higher dipolar effect for bilayers and micelles than for monolayers and solutions is not surprising.

From the consistency of the above analysis, dipolar effects on the surroundings of adsorbed layers and micelles may indeed be expected.

Acknowledgements

R. Zana is thanked for his suggestion of an ordering explaining the transition observed at mercury which led us to go deeper in the analysis of dipolar effects.

References

1. Pouillen P, Martre AM, Martinet P (1981) Electrochim Acta 26:1035-1040
2. Honnorat A, Martinet P (1983) Electrochim Acta 28:1703-1712
3. Bennis K, Martinet P, to be published
4. Naficy G, Vanel P, Schuhmann D, Bennes R, Tronel-Peyroz E (1981) J Phys Chem 85:1037-1042
5. Schuhmann D, Vanel P, Tronel-Peyroz E, Raous H (1984) In: Mittal KL, Lindman B (eds) Proc Intern Symp Surfactants in Solution, Plenum Press, New York, pp 1233-1247
6. Schuhmann D, Vanel P (1988) Surface Sci Technol 4:139-167
7. Belambri NO, Vanel P, Schuhmann D (1987) J Colloid Interface Sci 120:224-228
8. Verdier E, Naficy G, Vanel P (1973) J Chim Phys 70:160-166
9. Lindman B, Wennerström H (19??) Physics Reports 52:1-86
10. Funasaki N, Hada S (1979) J Phys Chem 83:2471-2481
11. Hayano S, Shinozuka S (1970) Bull Chem Soc Japan 43:2083-2089
12. Sengh HN, Swarup S, Saleem SA (1979) J Colloid Interface Sci 68:128-134
13. Anacker EW (1979) In: Mittal KL (ed) Solution Chemistry of Surfactants. Plenum, New York, 1:247-260
14. Miyashita Y, Hayano S (1982) J Colloid Interface Sci 86:344-349
15. Pareds S, Tribout M, Sepulveda L (1984) J Phys Chem 88:1871-1874
16. Ikeda S, Ozeki S, Tsunodo MA (1980) J Colloid Interface Sci 73:27-37
17. Venable RL, Nauman RV (1964) J Phys Chem 68:3498-3503
18. Wennerström H, Lindman B (1979) Physics Report 2:1-86
19. Shinoda K, Nakagama I, Tamamushi B, Isurema T (1963) Colloidal Surfactants Some Physicochemical Properties. Chap 1, Academic Press, New York
20. Mukerjee P, Mysels KJ (1971) Critical Micelle Concentrations of Aqueous Surfactant Systems. NSRDS-NBS 36, Washington DC US Government printing office
21. Cha-Dara M (1973) Thesis, Montpellier
22. Zana R, Yiv S, Strazielle C, Lianos P (1981) J Colloid Interface Sci 80:208-223
23. Mousty C, Mousset G (1988) private communication
24. Matsuo T, Yudate K, Nagamura T (1981) J Colloid Interface Sci 83:354-360
25. Hamdi M, Schuhmann D, Vanel P, Tronel-Peyroz E (1986) Langmuir 2:342-349

26. Schuhmann D (1988) Comm to the Second European Colloid and Interface Society Conference, September 19-22, Arcachon, France
27. Croonen Y, Gelade E, Van der Zegel M, Van der Auwerae M, Van Derndriessche H, De Schryver FC, Almgren M (1983) J Phys Chem 87:1426-1431
28. Hayter JB, Penfold J (1983) Colloid Polym Sci 261:1022-1030
29. Ekwall P, Mandell L, Solyom P (1971) J Colloid Interface Sci 35:513-526
30. Lindblom G, Lindman B, Mandell L (1973) J Colloid Interface Sci 42:400-408
31. Reiss-Husson F, Luzzati V (1964) J Phys Chem 68:3504-3511
32. Porte G, Appel J (1983) In: Mittal KL, Lindman B (eds) Surfactants in Solution. Plenum, New York, vol 2, pp 805-823
33. Candau SJ, Hirsh E, Zana R (1984) J Physique 45:1263-1270
34. Porte G, Appel J, Poggi Y (1980) J Phys Chem 84:3105-3110
35. Schuhmann D (1983) Annal NY Acad Sci 404:463-470
36. Vanel P, Raous H (1982) CR Acad Sci Paris, Ser II, 295:857-862
37. Tronel-Peyroz E, Schuhmann D, Raous H, Bertrand C (1984) J Colloid Interface Sci 97:541-551
38. Elworthy PH, Mysels KJ (1966) J Colloid Interface Sci 21:331-347
39. Hamdi M, Vanel P, Schuhmann D, Bennes R (1982) J Electroanal Chem 136:229-250
40. Guidelli R, Moncelli MR (1978) J Electroanal Chem 89:261-270
41. Elving PJ, Markowitz JM, Rosenthal I (1956) Anal Chem 28:1179-1186
42. Meakins RJ (1967) J Appl Chem 17:157-161
43. Meakins RJ, Stevens MG, Hunter RJ (1969) J Phys Chem 73:112-117
44. Damaskin BB, Petrii OA, Batrakov VV (1971) Adsorption of Organic Compounds on Electrodes. Plenum, New York
45. Schuhmann D (1987) Electrochim Acta 32:1331-1336
46. Butler JAV (1937) Trans Faraday Soc 33:229-241
47. Mairanovsky SG, Klochko NP, Orechova VV (1984) Elektrokhim 20:690-692
48. Treiner C, Chattopadhyay AK (1983) J Chem Soc, Faraday Trans 79:2915-2927
49. Treiner C (1981) Can J Chem 59:2518-2526
50. Pashley RM, McGuiggan PM, Horn RG, Ninham BW (1988) J Colloid Interface Sci 126:569-578
51. Robinson RA, Stokes RH (1959) Electrolyte Solutions. Butterworths, London
52. Piro J, Bennes R, Bou Karam E (1974) J Electroanal Chem 57:399-412
53. Bou Karam E, Bennes R, Bellostas D (1977) J Electroanal Chem 84:21-32
54. Tronel-Peyroz E, Schuhmann D, Jubault M (1981) J Electroanal Chem 129:265-284
55. Porte G (1988) Thesis, Montpellier
56. Zana R (1988) Private Communication
57. Schuhmann D (1986) J Electroanal Chem 201:247-261
58. Schuhmann D (1988) J Electroanal Chem 239:447-451
59. Schuhmann D (1988) J Electroanal Chem 252:1-10
60. Henderson D, Blum L (1982) J Electroanal Chem 132:1-13
61. Bhuyian LB, Blum L, Henderson D (1983) J Chem Phys 78:443-445
62. Russier V, Badiali JP, Rosinberg ML (1987) J Electroanal Chem 220:213-224
63. Kirchnerova J, Farrel PG, Edward JT (1976) J Phys Chem 80:1974-1980
64. Buckingham AD (1953) Austral J Chem 6:93-103

Received October, 1988;
accepted March, 1989

Authors' address:

D. Schuhmann
LPCSP, CNRS
BP 50 51
F-34033 Montpellier, France

Progress in Colloid & Polymer Science Progr Colloid Polym Sci 79:142–149 (1989)

The kinetics of wetting: the dynamic contact angle

M. Bracke, F. De Voeght, and P. Joos

Universitaire Instelling Antwerpen, Department Biochemistry, Wilrijk, Belgium

Abstract: From measurements of dynamic contact angles, whereby a continuous solid strip is drawn into a large liquid pool, we obtained an operational equation:

$$\cos \theta_d = \cos \theta_o - 2(1 + \cos \theta_o)\, Ca^{1/2}$$

which predicts that under air entrainment conditions, the velocity is independent from the nature of the strip.

This empirical equation is also applicated to two other methods, namely the wetting of a vertical plate and the spreading of small drops.

Key words: Moving contact line, dynamic contact angle, air entrainment, wetting of a plate, spreading of a drop.

Introduction

The kinetics of wetting are very important for the technical process of coating. But it also is a challenge for a theorist to understand the mechanism of wetting: what about the boundary condition at the moving contact line and the fact that just there the pressure becomes infinite without a slip condition [1]. Until now the theoretical aspect was unresolved. Nevertheless, a large number of papers report measurements on dynamic contact angles [2–8], drawing a solid strip into a liquid [9–18], on spreading of drops on solid [19–21]. Different authors claim that the dynamic contact angle is ruled by macroscopic hydrodynamics [13–18, 22, 23], e. g., the capillary number, while others find it necessary to introduce molecular considerations [9, 10, 24].

In this paper we first present our own experiments, i. e., dynamic contact angle measurements when a continuous solid strip is drawn into a large pool of liquid. Out of these data we distil an operational equation and compare this with reanalyzed published results. In order to show its universality, we apply this equation to two other practical problems: the wetting process of a vertical solid plate for which we use our own results, and the spreading of a liquid drop on a solid, for which we use published data.

Experimental

We used two different methods to determine the dynamic contact angle:

1. A continuous solid strip is drawn into a large pool of liquid. The dynamic contact angle θ_d, i. e., the angle between the solid and liquid at the moving contact line, changes with strip velocity: θ_d is increasing with increasing speed. The apparatus for measuring these dynamic contact angles is similar to the one used by Burley and Kennedy [13, 14]. The liquid profile is illuminated by a Helium Neon laser and dynamic contact angles are photographed in function of the strip velocity. For each liquid this velocity is varied from $v = 0$ till $v = v_{max}$, which is the velocity where air entrainment sets in. Air entrainment can easily be observed [10–13, 15] because the liquid-solid contact line changes from a straight line into a broken zigzag curve.

The materials used for the continuous strip are: smooth and rougher polyethylene (PE), polyethylenetherephthalate (PET), and for the liquids: aqueous glycerine solutions, aqueous ethyleneglycol solutions and ordinary corn oil.

Surface tensions are measured using the Wilhelmy plate technique and the viscosities using a capillary viscometer.

2. The second series of experiments concern the dynamic contact angle during the wetting of a vertical, initially dry platina Wilhelmy plate. This plate is coupled with a transducer (Gould Statham Universal Transducing cell UC 3 with a level arm, a UL 5 accessory), which is connected with a Gould Digital Storage Oscilloscope (OS 4000 with output unit 4001). The outgoing signal is finally recorded on a strip chart recorder. Since for high viscosity liquids the use of the storage oscilloscope is not necessary, the transducer is directly connected with the pen recorder.

At time $t = 0$, when we make contact between the dry plate and the liquid, a trigger signal starts the registration of the change in weight, caused by the climbing of the liquid on the plate. This force profile $F(t)$ is stored by the oscilloscope, after which it is recorded. After a sufficient time $t = \infty$, i.e., when the weight remains constant, we also note $F(\infty)$. It is the ratio between $F(t)$ and $F(\infty)$ which gives information about the dynamic contact angle θ_d,

$$\frac{F(t)}{F(\infty)} = \cos \theta_d. \tag{1}$$

As liquid we use silicon oils with a viscosity in the range $\mu = 3.5$ P(oise) to 588 P, and a surface tension $\sigma = 20$ dyne/cm.

Results and discussion

1. The dynamic contact angle for a strip drawn into a large pool of liquid

For a smooth PE strip and the aqueous glycerol solutions we plotted $\cos \theta_d$ as a function of the square root of the strip velocity $v^{1/2}$; an example is shown in Fig. 1. It is seen that a linear relationship is obtained. At $v = 0$, the curve extrapolates to the static advancing contact angle; $\theta_d = \theta_o$ [25]. At $\cos \theta_d = -1$, i.e., $\theta_d = \pi$, the curve extrapolates to a velocity v_π, which corresponds to the velocity at which air entrainment sets in, or in other words $v_\pi \approx v_{max}$. This is in agreement with the findings of other authors [10, 13, 18]. For 10 other viscosities lying in the range from $\mu = 0.104$ P to 1.84 P, and with almost the same surface tension $\sigma \approx 63$ dyne/cm, we obtain analog graphs. The slopes, $k = -d \cos \theta_d / v^{1/2}$, from each of these curves

are calculated and plotted as a function of the square root of the viscosity $\mu^{1/2}$ (Fig. 2). Again a linear relation between k and $\mu^{1/2}$ is obtained, suggesting that the dynamic contact angle depends on the square root of the capillary number, $Ca^{1/2} = (\mu v / \sigma)^{1/2}$. All the results with the smooth PE strip and the aqueous glycerol solutions are taken together in a $\cos \theta_d$ vs. $Ca^{1/2}$ graph (Fig. 3.). This linear relationship is expressed in the next equation

$$\cos \theta_d = \cos \theta_o - \gamma Ca^{1/2} \tag{2}$$

with $\gamma = -d \cos \theta_d / d Ca^{1/2}$, being the corresponding slope.

An analog analyzing method is used for other combinations of the different strips with the different liquids. The final results are given in Table 1.

We can conclude from this that:

i) $\cos \theta_d$ decreases linearly with the square root of the capillary number as shown in Eq. (2).

ii) At $v = 0$, the dynamic contact anlge extrapolates to the static advancing one obtained through the method of Wolfram [25, 26]. In Table 2, where we compare this extrapolated value θ_o to the static ones θ_A, we find good agreement.

iii) At $\cos \theta_d = -1$, i.e., $\theta_d = \pi$, the extrapolated velocity v_π corresponds to the directly observed velocity v_{max}.

iv) The slopes γ, given by the linear relationship of Eq. (2), are not constant but are a function of the extrapolated contact angle θ_o. In Fig. 4, where, beneath our own data, the results of other investigators are shown,

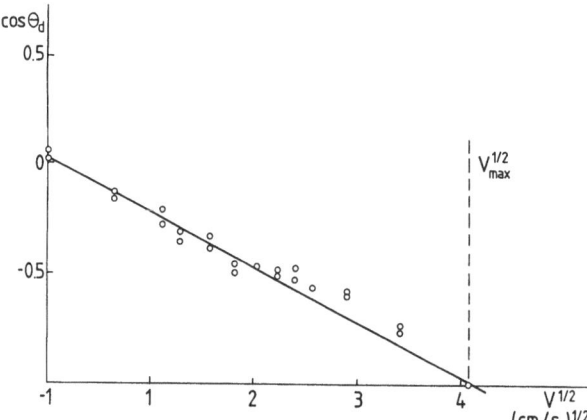

Fig. 1. Smooth polyethylene strip drawn into an aqueous glycerol solution: $\mu = 1.506$ P, $\sigma = 63$ dyne/cm. $\cos \theta_d - (0.02 \pm 0.04) = (-0.25 \pm 0.01) v^{1/2}$; $\theta_o = 89° \pm 2°$; regression coefficient = 0.96

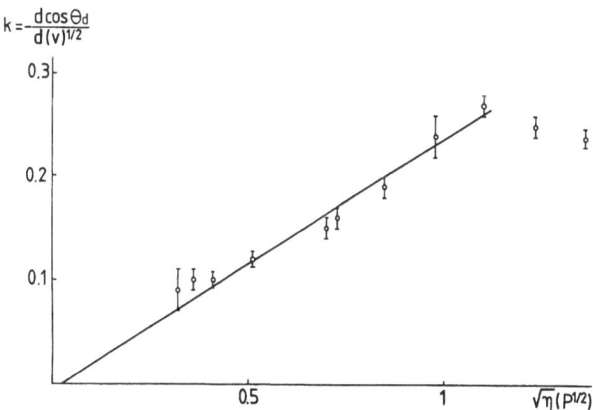

Fig. 2. Smooth polyethylene strip drawn into aqueous glycerol solutions. $\cos \theta_d = \cos \theta_o - 0.23 \pm 0.01 (\mu v^{1/2})$; regression coefficient = 0.98

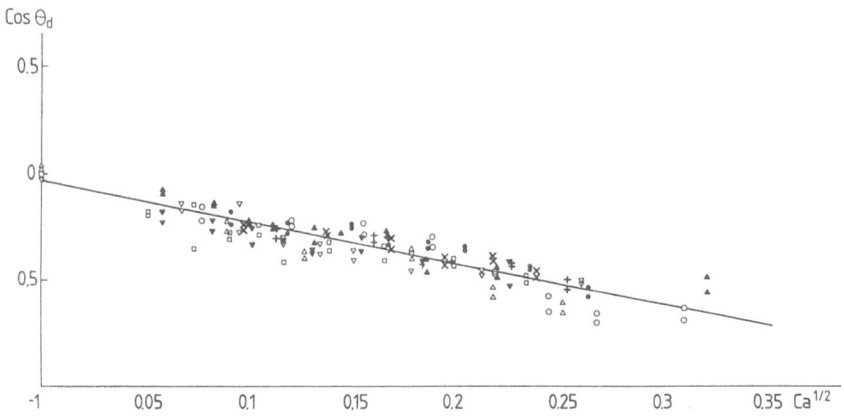

Fig. 3. Smooth polyethylene strip drawn into aqueous glycerol solutions. $\mu = 1.214$ P:$\triangle$; $\mu = 0.910$ P:$\bigcirc$; $\mu = 0.723$ P:$\times$; $\mu = 0.533$ P:$\bullet$; $\mu = 0.492$ P:$+$; $\mu = 0.265$ P:$\blacktriangle$; $\mu = 0.173$ P:$\triangledown$; $\mu = 0.130$ P:$\blacktriangledown$; $\mu = 0.104$ P: $\cos\theta_d - (-0.03 \pm 0.06) = -1.95 \pm 0.06$ $Ca^{1/2}$; $v_{max} = 15.5 \pm 0.9$ cm·s^{-1}/μ tot $\mu = 1.2$ P; $\theta_o = 92° \pm 3°$; regression coefficient = 0.87

Table 1. Results from $\cos\theta_d - Ca^{1/2}$ analysis method: $\cos\theta_d = \cos\theta_o - \gamma Ca^{1/2}$

	$\theta_o \pm$ SD	$\gamma \pm$ SD
aqueous glycerol solutions:		
pe, smooth	$92° \pm 3°$	1.95 ± 0.06
pe, rough	$97° \pm 2°$	1.61 ± 0.05
pet	$77° \pm 5°$ [a]	2.62 ± 0.05
aqueous ethyleneglycol solutions:		
pe, smooth	$80° \pm 4°$	2.17 ± 0.05
pe, rough	$84° \pm 4°$	2.03 ± 0.06
pet	$65° \pm 5°$	2.69 ± 0.07
corn oil:		
pe, smooth	$37° \pm 10°$	3.7 ± 0.2
pe, rough	$45° \pm 7°$	3.9 ± 0.2
pet	$66° \pm 4°$	2.3 ± 0.2

[a] Experimental results from Blake [10] for five viscosities are also considered in this case.

Table 2. Comparison between the static advancing angle θ_A and the extrapolated value θ_o. θ_Y is the Young contact angle and α is a measure of the roughness of the substratum, both obtained using the tilted plate method [25, 26]

	θ_Y	α	$\theta_A = \theta_Y + \alpha$	$\theta_o \pm$ SD
aqueous glycerol solutions:				
pe, smooth	$82°$	$13°$	$95°$	$92° \pm 3°$
pe, rough	$82°$	$20°$	$102°$	$97° \pm 2°$
pet	$65°$	$8°$	$72°$	$77° \pm 5°$
aqueous ethyleneglycol solutions:				
pe, smooth	$63°$	$13°$	$76°$	$80° \pm 4°$
pe, rough	$63°$	$20°$	$83°$	$84° \pm 4°$
pet	$48°$	$8°$	$56°$	$65° \pm 5°$

corn oil: no experiments with the tilted plate method are available

we have plotted γ vs. $\cos\theta_o$, and again we get a linearity which can be written as

$$\gamma = 2(1 + \cos\theta_o). \qquad (3)$$

So, Eq. (2) becomes

$$\cos\theta_d = \cos\theta_o - 2(1 + \cos\theta_o)\, Ca^{1/2}. \qquad (4)$$

Here refer here to Tanners equation [27], which gives a relation between small dynamic contact angles θ_d and the capillary number Ca:

$$\theta_d = (3\Phi)^{1/3}\, Ca^{1/3} \qquad (5)$$

with Φ nearly constant, being a logaritmic function of Ca. Tanner [27] also gives some theoretical evidence for spreading on a dry as well as on a wet surface, making use of the lubrication method for thin films.

Developing Eq. (4) for small angles, i. e., $\theta_d \approx 1 - \theta_d^2/2$, we get

$$\theta_d = \Omega\, Ca^{1/4} \qquad (6)$$

with $\Omega = \Omega(\theta_o)$ a function of the static contact angle θ_o and $\Omega(\theta_o = 0) = \sqrt{6}$.

We can say that Eqs. (5) and (6) are comparable for small capillary numbers Ca.

The first very important practical consequence of the operational Eq. (4) is that under air entrainment conditions, i. e., $\cos\theta_d = -1$ and $v = v_{max}$,

$$v_{max} = \frac{\sigma}{4\mu}. \qquad (7)$$

v_{max}, the air entrainment velocity is independent of the nature of the solid strip. This important conclusion is

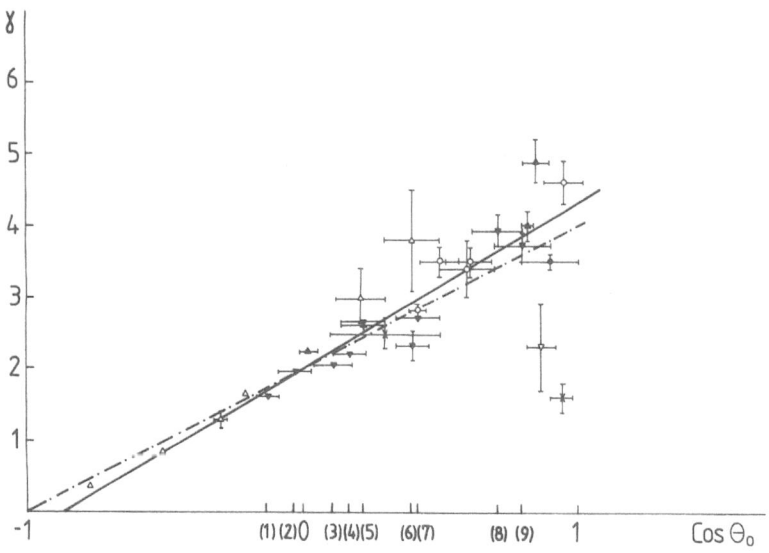

Fig. 1. ——: $\cos\theta_d$ $\cos\theta_o = -2(\cos\theta_o + 1)$ $Ca^{1/2}$ (4); $v_{max} = \sigma/4\mu = 15.8$ dyne $\cdot$ cm$^{-1}/\mu$. pe, smooth: □, pet: ×, pet 1: ○, pet 2: ●, pet 3: ■, triacetaat: +; pet 1, pet 2, and pet 3 are polyethyleneterephthalate with different gelatin layers

also stated by Burley et al. [14, 17] and Gutoff et al. [18]. The intrinsic value of Eq. (7) is also pointed out in Fig. 5 where we have plotted v_{max} vs. μ for different

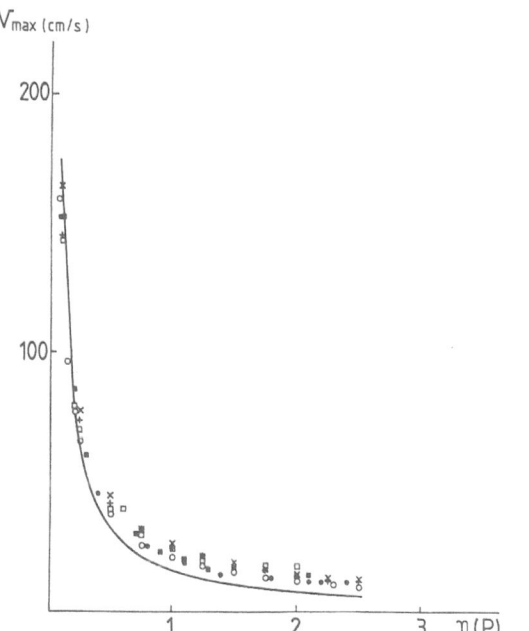

Fig. 5. ——: Linear regression, $y = (2.3 \pm 0.1) \cos\theta_o + (2.0 \pm 0.2)$; regression coefficient = 0.94; $-\cdot-\cdot-$: $y = 2(\cos\theta_o + 1)$. [18]: ○, [5,6]: ●, [6]: ×, [10, this work]: +, [4]: ✳, [14, 16, 17]: △, [8]: ▲, [3]: ▽, [this work]: ▼. Our experimental results: 1) rough pe/aq. glycerol, 2) smooth pe/aq. glycerol, 3) rough pe/aq. ethyleneglycol, 4) smooth pe/aq. ethyleneglycol, 5) pet/aq. glycerol, 6) pet/corn oil, 7) pet/aq. ethyleneglycol, 8) rough pe/corn oil, 9) smooth pe/corn oil

strip materials drawn into an aqueous glycerol solution. For aqueous ethyleneglycol solutions we get an analog result.

Without going into details it should be noted that these empirical findings cannot be applied on the rougher PE strip for viscosities larger than about 1 P. Since Eq. (7) does not explain either the minimum in the $v_{max} - \mu$ curve or the increasing evolution for $\mu > 1$ P. Burley and Jolly [14] also reported anomalous behavior for surfaces of a certain roughness.

2. The wetting of a vertical solid plate

For a strip drawn into a large liquid pool, Burley and Kennedy [16] showed that for dynamic contact angles less than $\pi/2$ the correct liquid profile is obtained by only taking gravitational and surface forces into account. This means that viscous forces in the bulk may be neglected. De Gennes [28] comes to the same conclusion; he stated that all the free energy S, i.e., $S = \sigma_s - \sigma_{s1} - \sigma_1$, is "burned up" in the precursorfilm. This means that the height of the liquid on a vertical plate is given by

$$h = h_L(1 - \sin\theta_d)^{1/2} \qquad (8)$$

with $h_L = (2\sigma/\varrho g)^{1/2}$, the Laplace length, and ϱ is the liquid density, and g is the gravitational acceleration.

The viscosity is implicitly included in the dynamic contact angle θ_d given by Eq. (4), which can be regarded as the boundary condition of the wetting problem. Furthermore, we consider the dynamic contact angle

θ_d as a macroscopic reflection of all molecular considerations.

The rate of wetting is given by Eq. (4)

$$v = \frac{dh}{dt} = \left[\frac{\cos\theta_d - \cos\theta_o}{2(1 + \cos\theta_o)}\right]^2 \frac{\sigma}{\mu} \qquad (9)$$

and also be Eq. (8)

$$\frac{dh}{dt} = -h_L \frac{\cos\theta_d}{2(1 - \sin\theta_d)^{1/2}} \frac{d\theta_d}{dt}. \qquad (10)$$

After equalizing Eq. (9) to Eq. (10) and taking into account that at $t = 0$, $\theta_d = \pi/2$, we find

$$T = \int_{\theta_o}^{\pi/2} \frac{(1 + \sin\theta_d)^{1/2}}{2(\cos\theta_d - \cos\theta_o)^2} d\theta_d \qquad (11)$$

with $T = t/\tau_R$ the dimensionless time, τ_R being the relaxation time given by

$$\tau_R = 4\mu (1 + \cos\theta_d)^2 \left\{\frac{2}{\varrho g \sigma}\right\}^{1/2}. \qquad (12)$$

Equation (11) can be solved by numerical integration.

In our case where the platina Wilhelmy plate is completely wetted by the silicon oils, i. e., $\theta_o = 0$ and thus, $\cos\theta_o = 1$, we have to rewrite Eqs. (11) and (12) as

$$T = \int_0^{\pi/2} \frac{(1 + \sin\theta_d)^{1/2}}{2(\cos\theta_d - 1)^2} d\theta_d \qquad (13)$$

$$\tau_R = 16\mu \left\{\frac{2}{\varrho g \sigma}\right\}^{1/2}. \qquad (14)$$

The relaxation time for water is extremely small, i. e., $\tau_R = 8.5 \, 10^{-4}$ s. Thus, in order to prove Eq. (13) experi-

mentally, we have to use rather high viscosity fluids. Therefore, we have chosen silicon liquids with their wide range of viscosities and their low surface tension. The experimental results, presented in Fig. 6, match very well with the theoretical Eq. (13). The fact that at low viscosities, i. e., $\mu < 3$ P, the experimental points lie under the theoretical ones is due to the slowness of our measuring device (especially to the Statham transducer).

We remark that our present theory and experiments apply to the wetting of a clean but dry, vertical plate. The rate of wetting is, as expected, much faster for a prewetted plate. This explains the deviation from the previous results of Rillaerts and Joos [4], who performed their experiments in a prewetted capillary (Fig. 5).

3. Spreading of a small drop on a solid

Making use of Eq. (4), we can perfectly describe the spreading of a drop on a solid substrate. For small drops the effect of gravity is negligible and we can use the spherical approximation. The use of large drops does not make the situation different but complicates only the calculations.

We express here the same point of view as in the previous paragraph, that is that the viscosity effects are implicitly included in the dynamic contact angle and can be neglected in the bulk liquid.

A small spherical drop with radius r_o and volume V is placed on a solid surface. This drop spreads from an initial contact angle $\theta_d = \pi$ until an angle $\theta_d = \theta_o$. Because the volume V remains constant, and from geometric arguments, the radius of the drop becomes after a time t

$$r = 4^{1/3} r_o \sin\theta_d \left[(1 - \cos\theta_d)^2 (2 + \cos\theta_d)\right]^{-1/3}. \qquad (15)$$

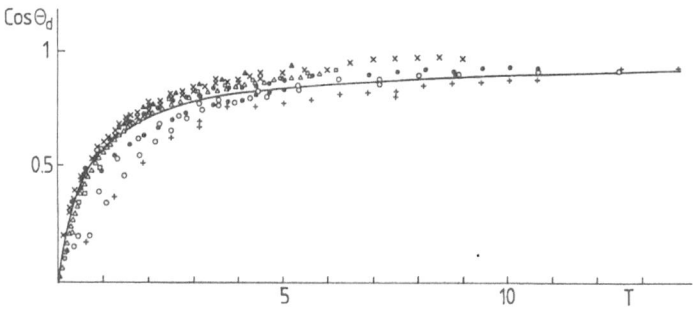

Fig. 6. Climbing of silicon liquids on a platina Wilhelmy plate: $\mu = 3.43$ P: $\bigcirc$, $\mu = 4.85$ P: $+$, $\mu = 9.7$ P: $\bullet$, $\mu = 10$ P: $\square$, $\mu = 122.5$ P: $\times$, $\mu = 294$ P: $\triangle$, $\mu = 588$ P: $\blacktriangle$; ——: Eq. (13)

The rate of spreading is given by

$$\frac{dr}{dt} = -4^{1/3} r_o \frac{(1-\cos\theta_d)^2}{[(1-\cos\theta_d)^2 (2+\cos\theta_d)]^{4/3}} \frac{d\theta_d}{dt} .$$

(16)

Equalizing Eq. (16) to Eq. (9), we get

$$T = \int_{\theta_o}^{\pi} \frac{d\theta_d}{[(1-\cos\theta_d)^2 (2+\cos\theta_d)]^{1/3} (2+\cos\theta_d)(\cos\theta_o - \cos\theta_d)^2}$$

(17)

with the relaxation time τ_R given by

$$\tau_R = 4^{1/3} \frac{4\mu}{\sigma} r_o (1+\cos\theta_o)^2 .$$

(18)

Under complete wetting conditions, i. e., $\theta_o = 0$ and thus $\cos\theta_o = 1$ Eq. (17) and Eq. (18) become

$$T = \int_{\theta}^{\pi} \frac{d\theta_d}{[(1-\cos\theta_d)^2 (2+\cos\theta_d)]^{4/3}}$$

(19)

$$\tau_R = 4^{1/3} \frac{16\mu}{\sigma} r_o .$$

(20)

We did not perform our own experiments because there are enough data on the spreading of small drops available in literature. We remark that all authors started their experiments by $\theta_d = \pi/2$.

Radigan et al. [19] studied at a temperature of 1000 °C the spreading of glass drops of different masses, $m = 0.7$ mg; 4.5 mg and 11.3 mg on Fermico metal. Those authors claim, in contrast to our Eq. (20), that the spreading rate does not depend on the drop size. This conclusion can be blamed on the large scatter of their experimental values. Despite this wide scat-

ter, we see that in Fig. 7, where we have plotted $\cos\theta_d$ vs. $T - T_{90}$°, some agreement between theory and experiments is achieved. It is also important to mention that they observed, by use of scanning electron microscopy, a precursor film with a height of about 1 μm.

Schonhorn et al. [20] placed small drops of polyethylene and ethylene vinyl acetate copolymers on high-energy (alumina and mica) and low-energy (teflon) surfaces. The high-energy surfaces wet completely, i. e, $\theta_o = 0$; for teflon they found a contact angle of $\theta_o = 75$ °C. Schonhorn et al. defined a sort of relaxation time, namely $1/a_T = L_w \mu/\sigma$, but they claimed as well that L_w is independent of r_o, m, or ϱg. Nevertheless, in order to analyze their experiments we put forward that $L_w = 2r_o$. We have plotted $\cos\theta_d$ vs. $T - T_{90}$° for the high

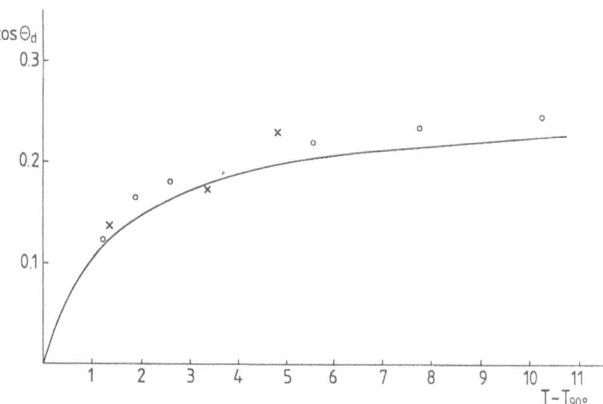

Fig. 8. Spreading of a drop, (from [20]): $\bigcirc$: Elvax 220 op Al: 170 °C; $\bullet$: Elvax 220 op Al/ 151 °C; $\triangle$: Elvax 220 op Al: 123 °C; $\blacktriangle$: Elvax 220 op Al: 118 °C; $\square$: Elvax 220 op Mica: 151 °C; $\wedge$: Dylt PE op Al: 151 °C; ——: Eq. (19)

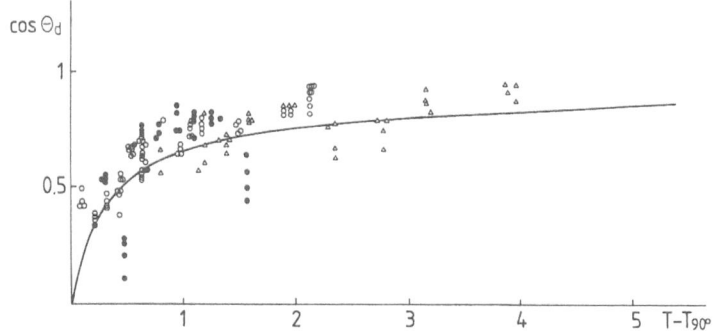

Fig. 7. Spreading of a drop, glass on Fermico metal: 1000 °C, (from [19]): $\blacktriangle$: $m = 0.7$ mg; $\bigcirc$: $m = 4.5$ mg; $\bullet$: $m = 11.3$ mg; ——: Eq. (19)

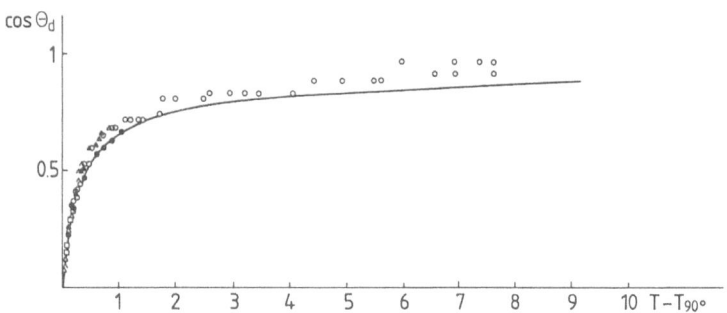

Fig. 9. Spreading of a drop, (from [20]): ○: Elvax 220 op teflon: 170°C; ×: Elvax 220 op Telfon: 151°C; $\theta_o = 75°$; ——: Eq. (17)

energy substrata in Fig. 8 and we can say that our theory, i. e., Eq. (19), fits their results well.

In Fig. 9 we made the analog graph for the spreading of drops on teflon. Our theory, i. e., Eq. (17), predicts the right progress of the dynamic contact angle in function of time.

Van Oene et al. [21] also performed spreading experiments with Dylt drops on alumina. In contrast with the two other authors his results lead to the conclusion that spreading is depending on the drop radius. We see in Fig. 10 that his results lie on the same master curve if plotted as a function of the dimensionless time. However, the spreading rate is about three times faster as that predicted by Eq. (19). Van Oene also stated that his rate is faster as compared with the similar results of Schonhorn. The reason for this deviation is, in our opinion, that Van Oene modified the surface by his intensive cleaning process so that the spreading coefficient $S = \sigma_s - \sigma_{s1} - \sigma_1$ is larger than zero, $S > 0$. This means that we are in a situation of non-equilibrium and the Young equation is no larger valid.

In previous research [29] we found that there is a limiting value above which the small drops in these systems cannot be refered to as spherical, or in other words, gravitational effects have to be taken into account. This limiting value is given by $\log V_L^* \approx 0$, with $V_L^* = V_L / a^3$ the dimensionless volume and $a = (\sigma/\varrho g)^{1/2}$.

For glass drops, Radigan [19] gives the values: $\varrho = 2.21$ g/cm³, $\sigma = 278$ dyne/cm, which means that $a = 0.358$ cm and $V_L = 0.046$ cm³. The drops with respective masses $m = 0.7$ mg, 4.5 mg, and 11.3 mg have respective volumes of $V = 0.347$ cm³, 2.036 cm³, and 5.113 cm³, and thus, we cannot neglect the gravitational effect anywhere. This partially explains the deviation between his experiments and our theory (Fig. 7).

For Elvax the limiting volume (V_L) is 0.01 cm³, with $a = 0.212$ cm. Since Schonhorn [20] used volumes of $V = 0.0067$ cm³, 0.0246 cm³, and 0.0282 cm³, the spherical approximation can only be applied to the first volume.

Van Oene [21] used for Dylt drops ($\varrho = 0.91$ g/cm³, $\sigma = 30$ dyne/cm) masses of $m = 0.0015$ g and 0.0167 g, which lead to volumes of $V = 0.0016$ cm³ and 0.0184 cm³. With $a = 0.183$ cm, the limiting volume becomes $V_L = 0.0061$ cm³ and only the first drop volume is small enough.

Conclusions

The experiments with a strip drawn into a liquid lead to the empirical equation

$$\cos \theta_d = \cos \theta_o - 2(1 + \cos \theta_o)\, Ca^{1/2}. \qquad (4)$$

This formula is tested by comparision with published results, and by application to two other systems, those being the wetting of a vertical solid plate and the spreading of a drop. It is obvious that our empirical approach allows a successful description of different dynamic wetting phenomena. A theoretical base is not available, but we have strong arguments that the viscosity effect can be neglected in the bulk liquid and is

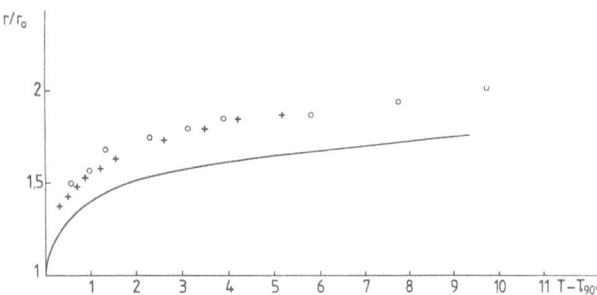

Fig. 10. Spreading of a drop, (from [21]): Dylt drops on alumina: 120°C; small drop: $r_o = 0.073$ cm; large drop: $r_o = 0.164$ cm; ——: Eq. (19)

only intrinsicly included as a boundary condition in the dynamic contact angle, given by Eq. (14).

Acknowledgement

We thank Rik Geerts for assistance with the strip experiments.

References

1. Huh C, Scriven LE (1971) J Colloid Interface Sci 35:85
2. Elliot GEP, Riddiford AC (1967) J Colloid Interface Sci 23:389
3. Hansen RJ, Toong TY (1971) J Colloid Interface Sci 36:410
4. Rillaerts E, Joos P (1980) Chem Eng Sci 35:883
5. Rose W, Heins RW (1962) J Colloid Interface Sci 17:39
6. Hoffman R (1975) J Colloid Interface Sci 50:228
7. Newman S (1968) J Colloid Interface Sci 26:209
8. Schwartz AM, Tejada SB (1972) J Colloid Interface Sci 38:359
9. Blake TD, Haynes JM (1969) J Colloid Interface Sci 30:421
10. Blake TD (1988) In: AIChE International Symposium on the Mechanics of thin film coating: Wetting Kinetics — How do Wetting lines Move?
11. Blake TD (1973) Ver Deut Ing Berichte 182:117
12. Blake TD, Ruschak KJ (1979) Nature 282:489
13. Burley R, Kennedy BS (1976) Chem Eng Sci 31:901
14. Burley R, Jolly RPS (1984) Chem Eng Sci 39:1357
15. Burley R, Brady PR (1972) J Colloid Interface Sci 42:131
16. Burley R, Kennedy BS (1977) J Colloid Interface Sci 62:48
17. Burley R, Kennedy BS (1976) Polymer Journal 8:140
18. Gutoff EB, Kendrick CE (1982) AIChE Journal 28:459
19. Radigan W, Ghiradella H, Frisch HL, Schonhorn H, Kwei TK (1972) J Colloid Interface Sci 49:241
20. Schonhorn H, Frisch HL, Kwei TK (1966) J Appl Physics 37:4967
21. Van Oene H, Chang YF, Newman S (1969) J Adhesion 1:54
22. Hansen RJ, Toong TY (1971) J Colloid Interface Sci 37:196
23. Wilson SDR (1975) J Colloid Interface Sci 51:532
24. Hoffman RL (1983) J Colloid Interface Sci 94:470
25. Bracke M, Joos P (1988) Progr Colloid Polym Sci 76:251
26. Wolfram E, Faust R (1977) In: Paddy JF (ed) Wetting, spreading and adhesion. Academic Press Inc, London
27. Tanner LH (1979) J Phys D, Appl Phys 12:1473
28. de Gennes PG (1985) Rev Mod Phys 57:827
29. Bracke M, Joos P, unpublished results

Received November, 1988;
accepted February, 1989

Authors' address:

M. Bracke
Universitaire Instelling Antwerpen
Department Biochemistry
Universiteitsplein 1
2610 Wilrijk, Belgium

Progress in Colloid & Polymer Science   Progr Colloid Polym Sci 79:150–154 (1989)

Interaction of metallic cations with the hydrous goethite (α-FeOOH) surface

M. Djafer, I. Lamy, and M. Terce

I.N.R.A., Station de Science du Sol, Versailles, France

Abstract: Adsorption of Cu, Pb, and Cd on goethite was studied as a function of pH and metal ion concentration in KNO_3 medium with complementary methods. Measures of the interfacial electrical properties of goethite in the presence and absence of metallic cations were obtained from microelectrophoretic mobilities and compared with those derived from potentiometric titrations. A quantification of the adsorption was made by the more classical solution depletion method which was used to calculate the amounts of adsorbed heavy metals at the adsorption equilibrium as a function of pH. The main results showed that i) the isoelectric point (iep) of the goethite (microelectrophoresis) was comparable to the point of zero charge (pzc) (potentiometry) in the absence of specific ion adsorption; ii) the shift of the iep towards higher pH values in presence of each metallic cation was confirmed by the shift of the pzc towards lower pH values, indicating a specific adsorption; iii) the adsorption of metallic cation increased with the increase in pH of the suspension and took place even when the surface charge was positive.

Key words: Adsorption, metallic cation, goethite, potentiometric titration, electrophoresis.

Introduction

The adsorption of metal ions by oxides in aqueous media has been the subject of continued attention in relation to various applied problems, such as in soil science where iron oxides are common minerals which are known to be important factors affecting trace metal transport.

In the case of oxides, the adsorption of H^+ (or OH^-) from electrolyte solutions is measured either by employing a potentiometric titration technique or by studying the effect of pH on the electrophoretic mobilities of mineral particles. Relatively few investigations of the sorption of metallic cations are concerned with these two methods [1, 2]; the solution depletion method is more generally employed [3–9].

The aim of this work is to compare the results of adsorption studies obtained by three complementary methods:

— pH titrations allow the determination of the point of zero charge (pzc) in the presence and absence of metallic cation;

— electrophoretic mobilities at different pH values give the isoelectric point (iep) and its shifts in presence of metallic cation; and

— the more classical solution depletion method is used to determine the amount of adsorbed cation in function of pH.

The oxide used is a well-known goethite synthesized in our laboratory; copper, cadmium, and lead are taken as the adsorbing metallic cations.

No attempt has been made in this paper to use one or more of the published models to describe the adsorption phenomena.

Materials and methods

Preparation of goethite

Goethite was synthesized following the technique outlined by Schwertmann et al. [10]. KOH (900 ml of 7/9M) was added to $Fe(NO_3)_3$ (100 ml of 1M). The precipitate was aged at 60 °C for 7 days. The oxide was washed repeatedly by centrifugation to remove the excess of salt until the conductivity of the supernatant remained constant. It was stored as aqueous suspension at about

5 °C in PTFE bottles. The oxide was verified through powdered x-ray and IR spectrophotometry. The sample had a BET nitrogen surface area of $34 \, m^2 \cdot g^{-1}$ and electron micrographs show that goethite particles exist mainly as needles with a mean length of 1 500 to 2 000 nm and a thickness of about 200 nm.

Experimental

Merck pro analysi reagents were used without further purification. Solutions of cadmium, copper, and lead nitrate, and potassium nitrate as supporting electrolyte were prepared just before use. Milli-Q water of high purity was used for all experiments.

The potentiometric titrations were made with a constant quantity of suspension of $4.24 \, g \cdot dm^{-3}$ in 0.01, 0.1, or 1 M KNO_3 medium. Initially, HNO_3 was added to give a pH of ca 3 and the system was then titrated in the presence and absence of metallic cations with 0.1 M KOH under nitrogen atmosphere at 25 °C. The pH was monitored with a Metrohm glass electrode and a Tacussel Ag/AgCl reference electrode with a bridge filled with 1 M KNO_3, connected to a "Tacussel TT Processor" automatic titrator. Two freshly prepared buffers and blank titrations allowed the determination of electrode parameters and in presence of solid, the surface excess of proton, Q, in $mol \cdot g^{-1}$ is calculated by

$$Q = [C_A - C_B - [H^+] + [OH^-]]/a \qquad (1)$$

where C_A and C_B are concentrations in $mol \cdot dm^{-3}$ of strong base and strong acid added, and a is the quantity of goethite used in $g \cdot dm^{-3}$.

Electrophoretic measurements were carried out using a modified laser Zee Meter model 501 (PEM KEM apparatus) with a rotating prism, a TV monitor, and a flat cell entirely made of quartz with two platinum electrodes. $1 \, mg \cdot dm^{-3}$ of goethite was suspended in solution containing 10^{-4} M or 10^{-3} M KNO_3, with or without metallic cation added, at different pH values adjusted with either HNO_3 or KOH.

In the adsorption studies by the solution depletion method, 2.5 $g \cdot dm^{-3}$ of goethite were mixed with initial fixed concentrations of metallic cations of 10^{-6}, 10^{-5} or 10^{-4} M in 10^{-3} M KNO_3 medium, at different pH values. The experimental conditions were chosen to avoid hydrolysis phenomena. The suspensions, contained in plastic tubes, were centrifuged after stirring for 24 h at a constant temperature of 20 °C. The metallic cation concentration of the supernatant for each pH value was determined by atomic absorption spectroscopy. The amount adsorbed was calculated by difference.

Results and discussion

Determination of goethite pzc and iep

Figure 1 shows the potentiometric titration behavior of goethite for several concentrations of KNO_3. A common intersection point is observed which indicates that the system pzc is at 8.5 ± 0.1. Electrokinetic mobilities were also measured as a function of pH and KNO_3 concentration (Fig. 1). The iep appears clearly at the same pH as the pzc. This suggests that the surface is free of any impurity and of specific adsorption of the ions of the electrolyte [11].

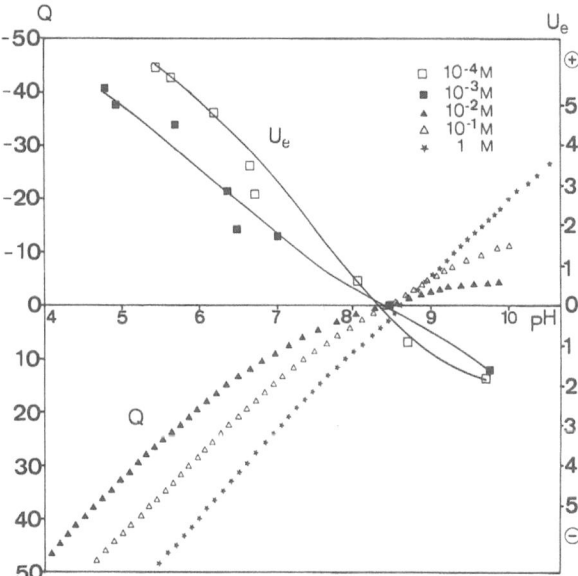

Fig. 1. The surface excess of proton (Q, in $\mu mol \cdot g^{-1}$) and the electrophoretic mobility (U_e in $10^{-4} \cdot cm^2 \cdot V^{-1} \cdot s^{-1}$) of goethite as a function of pH in aqueous solutions of potassium nitrate

Electrophoretic results are presented as mobilities instead of zeta potential because the particles of goethite are non-spherical which presents a particularly difficult matter when converting mobility into electrokinetic or zeta potential. As expected, the adsorption of the so-called "potential determining ions" (H^+ and OH^-) is enhanced, and the electrophoretic mobility is decreased with the increase in ionic strength.

Adsorption of metallic cations

Figure 2 shows the results obtained from potentiometric titrations of goethite suspension in the presence of three initial concentrations of each metallic cation, nearly identical for cadmium, copper, and lead. It can be seen that when a metal ion is added to the system, the pzc shifts towards acid pH. This corresponds to an increase of OH^- consumption by oxide surface and/or by bulk solution, apart of hydrolysis metallic cation phenomena. As proposed by several authors, this can be due to a competition between protons and metallic cations for the surface sites [12], the co-adsorption of OH^- and metallic cations [13], or the hydrolysis of metallic cations followed by adsorption [14].

For a given metallic cation, the shift of pzc is higher when the initial concentration of the metallic cation increases. This is compatible with anyone of the three

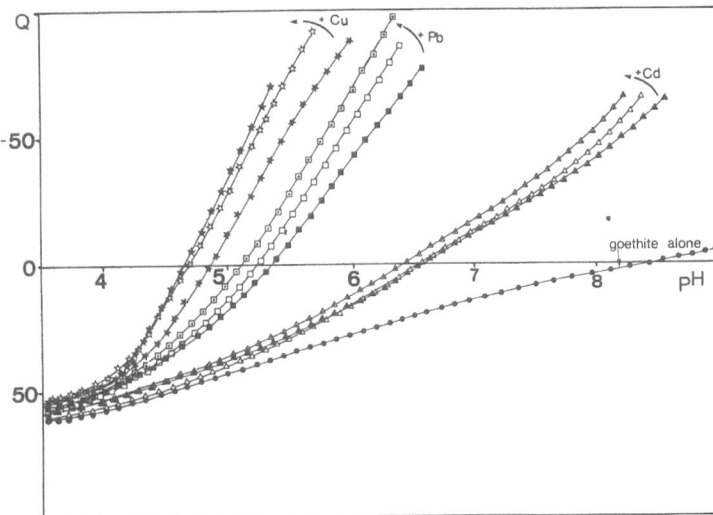

Fig. 2. Potentiometric titration behavior of goethite in absence and in presence of metallic cation (KNO$_3$ 0.1 M): copper (★) $3.2 \cdot 10^{-4}$ M; (☆) $6.4 \cdot 10^{-4}$ M; (✩) $9.7 \cdot 10^{-4}$ M; lead (■) $3.1 \cdot 10^{-4}$ M; (□) $6.3 \cdot 10^{-4}$ M; (▢) $9.4 \cdot 10^{-4}$ M; cadmium (▲) $3.1 \cdot 10^{-4}$ M; (△) $6.2 \cdot 10^{-4}$ M; (△) $9.3 \cdot 10^{-4}$ M

mechanisms suggested in the literature. For equal concentrations of added metallic cation, the shift of pzc follows the order Cu > Pb > Cd (when total metal

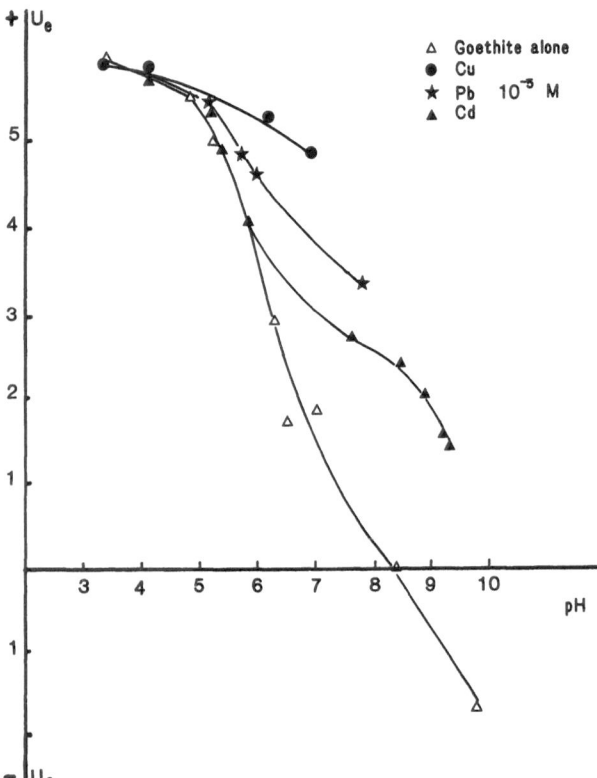

Fig. 3. Electrophoretic mobilities of goethite in function of pH in absence and in presence of metallic cation (KNO$_3$ 10^{-3} M)

concentration is superior to 10^{-4} M). This could suggest that copper is more strongly specifically adsorbed than lead and than cadmium [15].

Figure 3 shows the electrophoretic mobility measurements of goethite particles in the presence of an initial concentration of 10^{-5} M for each of the three metallic cations. The experiments were carried out at a range of pH chosen to avoid the complications of hydrolysis of the cations and precipitation. Consequently, the quantification of the shift of the iep in presence of metallic cation could not be made. But it is clear that the same order: Cu > Pb > Cd is observed for the increase in electrophoretic mobilities (Fig. 3) than for the decrease in surface charge (proportional to the surface excess of proton, Fig. 2). The same assumptions as above are made to interpret the increase in electrokinetic mobilities in basic medium. Opposite shifts of pzc and iep in the presence of metallic cation are a characteristic criterion for specific adsorption [16].

The results obtained by solution depletion method confirm those presented above. Figure 4 shows the percentage of metallic cations adsorbed on goethite at constant ionic strength and temperature. As expected [14], the adsorption is strongly pH-dependent, and the curves shift to the acid side upon lowering the concentration of the adsorbate. The comparison between Fig. 4a, b, and c shows that for a same pH and a same initial concentration of metallic cation, the amount of adsorbed cation follows the order Cu > Pb > Cd. The same results with goethite have been obtained by Forbes et al. [17]. The value of the pH of 50 % adsorp-

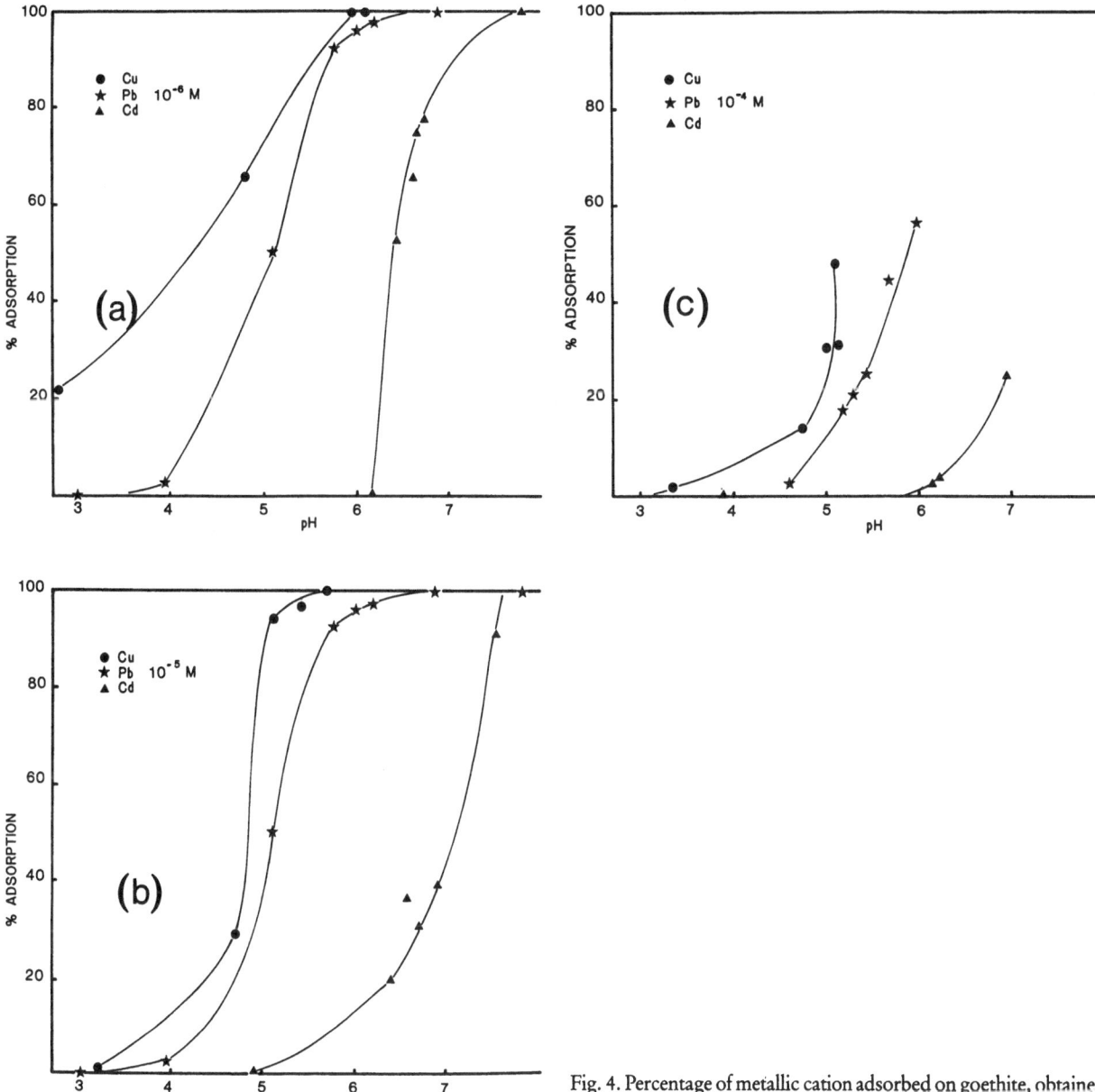

Fig. 4. Percentage of metallic cation adsorbed on goethite, obtained by solution depletion method (KNO_3 10^{-3} M), in function of pH

tion increases as the tendency of the corresponding metal ion to hydrolyse decreases [18]. The adsorption of metallic cations takes place even when the net surface charge of goethite is positive: this could mean that chemical forces in addition to simple electrostatic ones occur between metallic cation and surface sites. But the solution depletion method cannot provide direct information on the nature of the involved interactions.

Conclusion

The three experimental methods show that there is little reaction at low pH with metallic cations, but interaction occurs even when the surface of goethite is positively charged.

The solution depletion method cannot be used to characterize the nature of the interaction, contrary to the two other methods. Indeed, from the results of the potentiometric and electrokinetic titrations, the specificity or non-specificity of the adsorption can be deduced. The adsorption of the metallic cation with the goethite surface expressed in terms of proton exchange (potentiometric method) can suppose strong interaction and may be chimisorption. Furthermore, the changes in electrokinetic charges (electrophoresis) confirm the short-range interaction with the surface.

The oxide used in this study is a basic one with high value of pH for the pzc and iep (in absence of specific adsorption). Therefore, quantification of the shift of pzc in presence of metallic cation can be made by potentiometric method because of the shift in acidic direction. On the other hand, the shift of iep in basic medium observed with electrokinetic methods is an experimental difficulty, but this is the reverse of the study of anion adsorption which emphasizes the complementarity of the two methods.

Acknowledgement

This work was supported in part by Environmental Ministry Project No. 86191.

References

1. Blesa MA, Larotonda RM, Maroto AJG, Regazzoni AE (1982) Colloids Surfaces 5:197-208
2. Jang HM, Fuerstenau DW (1986) Colloids Surfaces 21:235-257
3. Forbes EA, Posner AM, Quirk JP (1974) J Colloid Interface Sci 49:403-409
4. James RO, Stiglich PJ, Healy TW (1976) Faraday Discuss Chem Soc 59:142-156
5. Kinniburgh DG, Sridhar K, Jackson ML (1977) ERDA Symp Ser No Biol Implic Met Environ, Proc Annu Hanford Life Sci Symp 15th vol 42:231-239
6. Davis JA, Leckie JO (1978) J Colloid Interface Sci 67:90-107
7. Balistrieri LS, Murray JW (1981) Am J Sci 281:788-806
8. Benjamin MM, Leckie JO (1981) J Colloid Interface Sci 83:410-419
9. Tiller KG, Gerth J, Brümmer G (1984) Geoderma 34:17-35
10. Schwertmann U, Cambier P, Murad E (1985) Clays and Clay Minerals 33:369-378
11. Kallay N, Babic D, Matijević E (1986) Colloids Surfaces 19:375-386
12. Hohl H, Stumm W (1976) J Colloid Interface Sci 55:281-288
13. Fokkink LGH, de Keizer A, Lyklema J (1987) J Colloid Interface Sci 118:454-462
14. Tamura H, Matijević E, Meites L (1983) J Colloid Interface Sci 92:303-314
15. Lyklema J (1984) J Colloid Interface Sci 99:109-117
16. Lyklema J (1983) In: Parfitt GD, Rochester CH (eds) Adsorption from solution at the solid/liquid interface. Academic Press INC, London, pp 223-246
17. Forbes EA, Posner AM, Quirk JP (1976) J Soil Sci 27:154-166
18. Schindler PW, Fürst B, Dick R, Wolf PU (1976) J Colloid Interface Sci 55:469-475

Received November, 1988;
accepted April, 1989

Authors' address:

Dr. M. Terce
I.N.R.A.
Station de Science du Sol
Route de Saint-Cyr
F-78026 Versailles Cedex, France

Progress in Colloid & Polymer Science   Progr Colloid Polym Sci 79:155–161 (1989)

The surface viscoelasticity of surfactant solutions and high frequency capillary waves

J. C. Earnshaw and A. C. McLaughlin

The Department of Pure and Applied Physics, The Queen's University of Belfast, Belfast, Northern Ireland

Abstract: The high frequency ($1 \times 10^4 < \omega < 4 \times 10^5 \, s^{-1}$) surface viscoelasticity of aqueous solutions of some primary alcohols has been studied by light scattering from capillary waves. The principal conclusion is that the surface dilational viscosity appears to have negative values. From the effects of such negative surface viscosity it is shown that the experiments could reliably distinguish between positive and negative values. Certain features of the results can be accounted for by a straightforward extension of the theory of diffusion controlled surfactant exchange between bulk and surface. However, quantitative agreement is not found and further theoretical work is required.

Key words: Surfactant solutions, surface viscoelasticity, surface dilational viscosity, negative, light scattering.

1. Introduction

The surface viscoelasticity of surfactant solutions has been related to dynamic processes involved in the exchange of molecules between bulk and surface, and in the adsorption process [1]. These dynamic aspects are particularly reflected in the frequency dependence of the dilational modulus. There has thus been considerable interest in the surface viscoelasticity of surfactant solutions, and its theoretical interpretation in terms of dynamic processes. Whilst a wide frequency range would be advantageous, *direct* studies of dilational waves are limited to $f < 10$ Hz [1]. However, at a solution-air interface the dilational waves couple to transverse (capillary) waves [1] which *can* be observed over a wide frequency range, permitting the dilational properties to be inferred, albeit indirectly. A few studies have been reported [2].

This paper concerns a study of the high-frequency dilational properties of the surfaces of dilute solutions of the primary alcohols. The results display rather unexpected features and do not easily fit into the existing theoretical framework. A possible explanation for the major novel feature of our data is advanced, but the quantitative details remain unexplained.

2. Experimental and data analysis

A fluid interface supports transverse (capillary) and longitudinal (compressional or dilational) waves of thermal origin; the former can be detected by light scattering, which should carry information upon the dilational properties of present interest.

Such capillary fluctuations were studied by laser light scattering using a heterodyne spectrometer which has been described in detail elsewhere [3]. Photon correlation was used to measure the temporal evolution of waves of well-defined wavenumber (q); experimental correlation functions were typically of low noise ($\sim 1\%$).

A correlation function is the Fourier transform of the spectrum of thermally excited waves of wavenumber q. This spectrum, $P(\omega, q \mid X)$ where X is the set of relevant material properties of the system, is theoretically well established [4]. The fluid density and viscosity are assumed known, and X comprises the moduli governing the surface response to transverse shear stress (γ, surface tension) and in-plane dilational stress (ε, surface elastic modulus). Both of these moduli can be expanded as linear response functions to incorporate dissipative effects [5]:

$$\gamma = \gamma_0 + i\omega\gamma' \text{ and } \varepsilon = \varepsilon_0 + i\omega\varepsilon' \qquad (1)$$

where γ' and ε' have the dimensions of surface viscosities. The exact status of γ' and ε' is not yet clear; they are perhaps best regarded as surface excess quantities [5].

In general $P(\omega \mid X)$ is a complicated function of $X(= \gamma_0, \gamma', \varepsilon_0, \varepsilon')$, which approximates to a Lorentzian shape; it has been shown to be correct for various systems [6–8]. The Lorentzian approximation

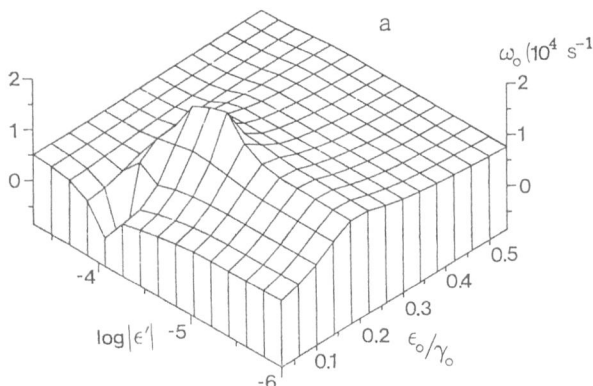

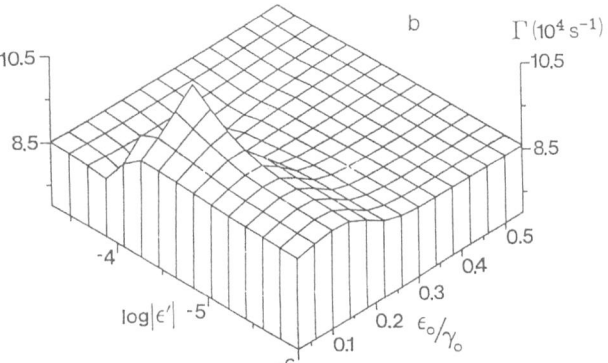

Fig. 1. The variations of ω_0 (a) and Γ (b) with ε_0 and $\log|\varepsilon'|$, for negative values of ε'. The other surface properties were $\gamma_0 = 60$ mN/m and $\gamma' = 0$ ($q = 500$ cm^{-1})

permits estimation of the complex frequency ($\omega = \omega_0 + i\Gamma$) of the capillary waves from the spectrum (or correlation function) of light scattered by given q [9]. However, such analysis only yields *two* observables, from which up to *four* surface properties (X) must, in general, be inferred. These difficulties have led to a novel method of analysing observed correlation functions, by fitting with the functional form [8]

$$g(\tau) = B + A\{FT[P(\omega, q|X)]\}\, e^{-\beta^2 \tau^2/4}, \qquad (2)$$

where A and B are amplitude and background terms and the final Gaussian accounts for an instrumental function. As we are interested in the frequency dependence of ε, the observed correlation data were also analysed in terms of ω.

Several studies of insoluble monolayers [8, 10] have demonstrated the reliability of this 'direct' analysis; the four surface properties were usually $\gtrsim 0$, as expected (in the analysis these quantities are initially constrained to be positive, although this constraint can be relaxed). In some few situations where one or other surface viscosity was of very small magnitude, relaxation of the positivity requirement led to that viscosity being scattered about zero within

errors [10]. This accords with the common (and usually implicit) assumption that the surface viscoelastic properties must be positive (although Lucassen has described circumstances in which the ε_0 may be negative). However, the present studies of aqueous solutions of alcohols raise the possibility that the dilational viscosity ε' can be negative.

Long-chain alcohols (Aldrich, > 99.8 % purity) were dissolved at known concentrations in ultra-pure water (Millipore Milli-Q); the solutions were degassed. The samples for light scattering were placed in a very shallow cell which was sealed by a cover, provided with optical windows, to minimize evaporation over the experimental duration (several hours). Light scattering observations at the first q-value studied were repeated at the end of the experiment to check that there were no changes in surface properties due to evaporation.

3. Background

The effects which negative values of ε' would have upon capillary waves, and upon light scattered from them will be considered briefly. The coupling between transverse and longitudinal waves at a liquid surface causes a resonance [1, 4] which gives rise to a maximum in the variation of the capillary wave frequency (ω_0) and in that of the damping (Γ) with ε_0 (for fixed γ_0). Well above the resonance, at large ε_0, both ω_0 and Γ vary only slowly with ε_0. Positive ε' reduces these variations, ultimately causing the resonance to disappear as the dilational waves become overdamped.

The effects of negative ε' upon capillary waves are summarised in Fig. 1: as ε' becomes more negative the resonance peaks in both ω_0 and Γ sharpen, until overdamping again sets in, above about 10^{-4} mN · s/m, leaving ω essentially independent of both ε_0 and ε'. More detailed calculations indicate that when $\varepsilon_0 > 30$ mN/m negative ε' has very small effects upon ω_0 and Γ for experimentally relevant values of tension.

However, even in these circumstances the scattered light still shows evidence of the negative ε'. Figure 2 shows the differences between theoretical correlation

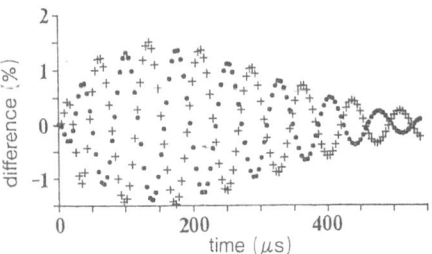

Fig. 2. The differences between theoretical correlation functions computed for $\varepsilon' = \pm 5 \times 10^{-5}$ mN · s/m and one for $\varepsilon' = 0$ (+: $\varepsilon' > 0$, ●: $\varepsilon' < 0$). The other properties used were $\gamma_0 = 60$ mN/m, $\varepsilon_0 = 30$ mN/m and $\gamma' = 0$ ($q = 500$ cm^{-1})

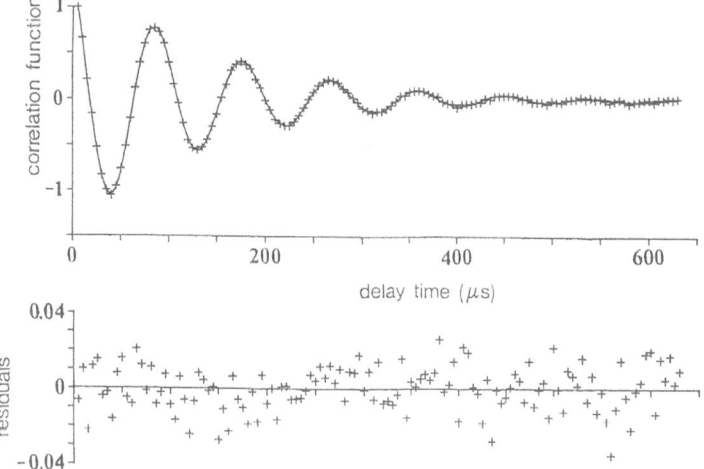

Fig. 3. A typical correlation function observed for a 5 × 10^{-5} M solution of decanol, $q = 381$ cm^{-1}. The fit (ε' not constrained) is discussed in the text

functions, computed for positive and negative ε', and one for $\varepsilon' = 0$. The two sets of differences are clearly quite distinct, showing different phases and magnitudes. These differences reach 1.5 % of the amplitude, indicating that measured correlation functions having noise levels typical of our experimental data should easily be able to distinguish the two cases.

4. Results

Figure 3 illustrates how the data seem to demand that ε' take negative values. An experimental correlation function is shown, together with the best-fit function of the form of Eq. (2). Other fits, discussed below, would not appear different to the eye, although their residuals do differ.

Three separate fits to the data will be discussed.

1) In the first, all four surface properties were constrained $\gtrsim 0$: both surface viscosities were determined as zero, whilst the surface tension (fitted value 61.7 mN/m) agreed within errors with the value measured using a Wilhelmy plate. Two different avenues were then pursued, involving separately relaxing the positivity constraint on γ', and then on ε'.

2) In the first case the fitted value of γ_0 fell by 2.6 mN/m, while ε_0 dropped from 35 to 16 mN/m. This fit is a secondary minimum in the sum-of-squares hyperspace, and is quite unphysical. We thus discard the possibility of negative γ'.

3) In the second case the fitted values of both γ_0 and ε_0 were essentially unchanged from the values of the original fit. The negative ε' (-3.5×10^{-5} mN · s/m) had a large uncertainty, reflecting the difficulty in

determining this quantity from light scattering from transverse waves. In this case $\gamma' = 0$, and γ_0, as noted, agreed with its equilibrium value.

Several statistical measures of the goodness of fit were improved in fit 3, as compared to fit 1, including the sum of squared residuals (S), the correlation coefficient of the residuals and the norm of the gradient of S at the fitted point. These improvements were small, but were observed for all the present data.

We do not associate this with an experimental problem. The same apparatus and method of data analysis have been used to study insoluble monolayers at the air-water interface [8, 10], for which surface tension and dilational elasticity were always reliably recovered. Negative values of ε' were very occasionally returned by the analysis routine, but only in monolayer states where this viscosity seemed to be negligible, in which case the fitted values tended to be scattered about zero if the positivity constraint was relaxed. For the present aqueous solutions of alcohols, this was not evident, ε' systematically falling below zero. Extensive simulations show that in most circumstances ε' is not correlated with the rest of X [10].

The frequency dependence of the dilational modulus ε is much more informative than an individual value. We show two sets of data. Figure 4 shows the light scattering values of ε_0 and ε' for a decanol solution as a function of the capillary wave frequency (ω_0) — i.e. for different wavenumbers. Figure 5 shows data, for a limited number of wavenumbers, for an octanol solution. In both cases the errors upon the data are quite large, due to the difficulty of determining dilational properties from observations of capillary waves. In

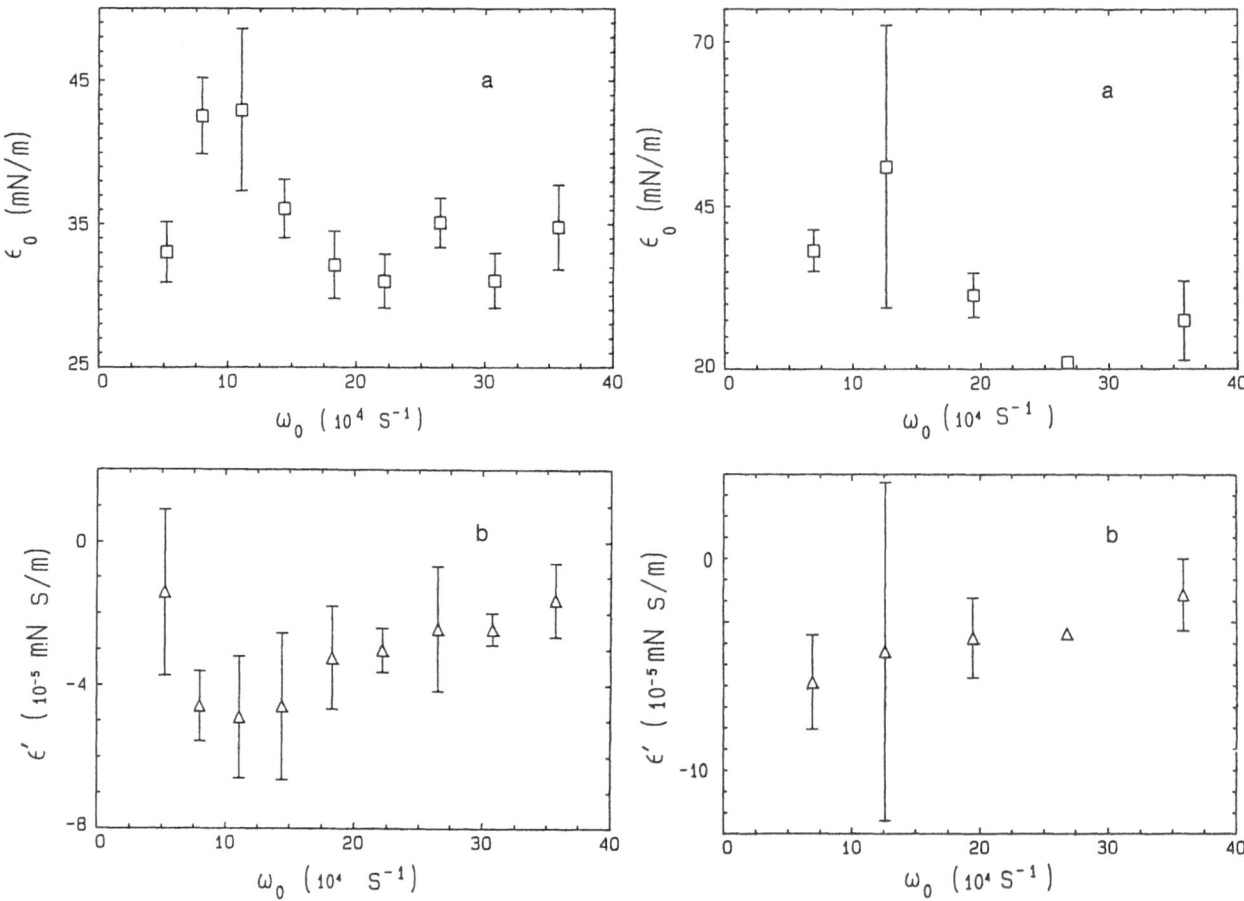

Fig. 4. The frequency dependence of ε_0 (a) and ε' (b) measured for a 5×10^{-5} M decanol solution

Fig. 5. The frequency dependence of ε_0 (a) and ε' (b) measured for a 2.5×10^{-5} M octanol solution

particular the effects of ε' upon the capillary waves are so indirect that it can often only be determined to within an order of magnitude unless the measured correlation functions are of quite extraordinarily low noise (precision dependent upon the magnitudes of ε_0 and ε'). The errors shown derive from the Hessian matrix of S; the minimization routine used sometimes estimated this matrix incorrectly [11], contributing somewhat to the size of the errors shown.

5. Discussion

For soluble monolayers diffusion limited exchange of surfactant molecules between solution and surface must be significant, as considered by Lucassen and Van den Tempel [12]. They have shown that, with certain assumptions, such diffusion causes the surface

dilational modulus to be freqency dependent, the predicted variations being

$$\varepsilon_0 = \varepsilon_\infty \frac{1 + \phi}{1 + 2\phi + 2\phi^2} \tag{3}$$

$$\varepsilon' = \frac{\varepsilon_\infty}{\omega} \frac{\phi}{1 + 2\phi + 2\phi^2}, \tag{4}$$

where ε_∞ is that value of the dilational modulus found for $\omega \to \infty$, ω being here the frequency of the fluctuation studied. The other parameter of these equations is a dimensionless diffusion parameter,

$$\phi = \frac{dc}{d\Gamma} \sqrt{\frac{D}{2\omega}}, \tag{5}$$

where D is the diffusion constant of the surfactant molecules, and c and Γ are, respectively, the surfactant

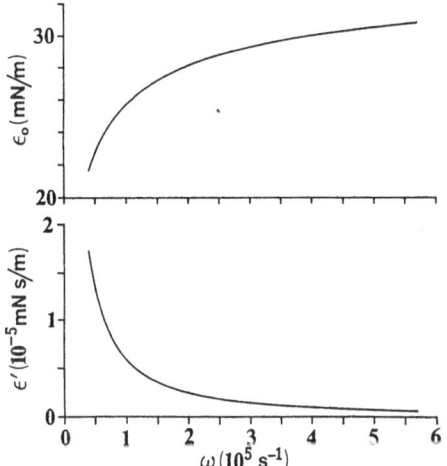

Fig. 6. Theoretically predicted variations of ε_0 and ε' for diffusion-controlled exchange of surfactant molecules between the surface and the sub-surface liquid layer

concentrations in the liquid layer immediately below the surface and in the surface film. It is clear that one would normally assume that $\phi > 0$. The variations of Eq. (3) and (4) with frequency (assuming constant D and c) are shown in Fig. 6: ε_0 increases asymptotically to ε_∞, while ε' decreases continuously.

If the negative values of ε' found in the present study are correct, what modifications would be necessitated in the above formalism? Firstly, if Lucassen's equations are to be retained, then it is straightforward to show that if $\varepsilon' < 0$ whilst $\varepsilon_0 > 0$ then

$$-1 < \phi < 0. \qquad (6)$$

The consequent effect upon ε_0 and ε' are shown in Fig. 7. Rather than reaching an asymptote as $\omega \rightarrow \infty$, ε_0 passes through a maximum and then falls. The variation of ε' shows a minimum. Translating these ϕ dependences into variations with ω (taking $dc/d\Gamma$ such as to yield $\varepsilon_0 \sim$ experimental values, and D equal to the bulk diffusion constant of a C-8 alkyl chain molecule [13]) yields the forms sketched. The frequency range covered was chosen to reflect the experimental range of ω_0; note that the minimum in ε' is not apparent, being at very low ω.

Certain features of the variations of Fig. 7b are qualitatively similar to aspects of the observed data of Figs. 4 and 5. In particular in both data sets ε_0 appears to pass through a maximum at relatively low ω_0, before declining steadily at higher frequencies. However, the magnitudes of these variations considerably exceed that predicted by our simple adaptation of Lucassen's theory. A second shortcoming is best illustrated by the ε' data of Fig. 4. Despite the rather large errors on these data, it does appear that there is an initial drop in ε' before the increase (which *is* predicted) starts above $\sim 10^5$ s^{-1}. While Fig. 7a suggests that ε' may pass through a minimum, the frequency at which this occurs is far from that of the maximum in ε_0. The variations observed for the octanol solution (Fig. 5) are much less marked than for decanol, despite the fact that the solution concentrations were chosen to give similar surface tensions, and that the magnitudes of ε_0 were similar. The increase in ε' for octanol is not statistically significant but the trend of the data does agree with expectation.

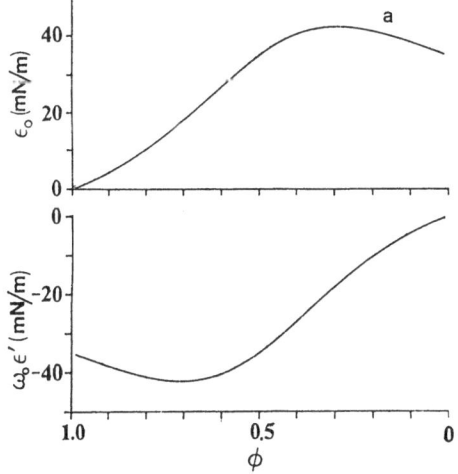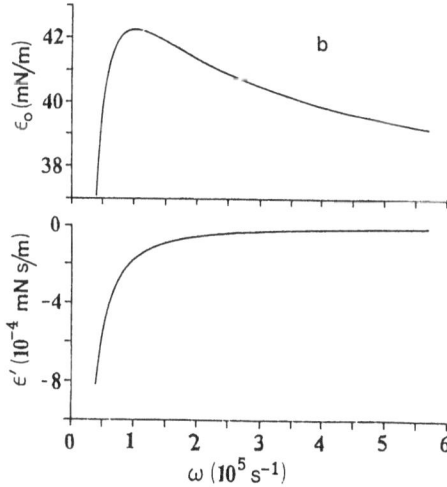

Fig. 7. The variations for $\varepsilon' < 0$, plotted (a) as functions of the diffusion parameter ϕ and (b) in terms of wave frequency

Is there a plausible physical explanation of the negative values of ε'? Its incorporation into Lucassen's theory would require $-1 < \phi < 0$. From Eq. (5) this implies

$$dc/d\Gamma < 0. \tag{7}$$

However, Lucassen assumed $dc/d\Gamma$ to be the *equilibrium* rate of change of the surfactant concentration in the immediate sub-surface layer with surface concentration. In the present non-equilibrium situation this equilibrium differential may be inappropriate. Let us suppose that there exist some (as yet undefined) process which introduces a phase difference between a change in c and the corresponding change in Γ. In particular let c and Γ vary due to the surface waves as

$$c = c_m e^{i\omega t} \text{ and } \Gamma = \Gamma_m e^{i\omega(t - \tau)} \tag{8}$$

where τ is the relaxation time of the hypothetical process. Then, taking $dc/d\Gamma = (dc/dt)/(d\Gamma/dt)$, we find

$$\frac{dc}{d\Gamma} = \frac{c_m}{\Gamma_m} e^{i\omega\tau} \equiv \frac{c_m}{\Gamma_m} \cos(\omega\tau). \tag{9}$$

Thus $dc/d\Gamma$, and hence ϕ, may be negative if $\pi/2 < \omega\tau < 3\pi/2$. For the experimental range of ω_0 (approximately $4 \times 10^4 - 4 \times 10^5 \text{ s}^{-1}$) this implies that the relaxation time of the hypothetical process must be of the order of microseconds. For fixed τ an increase of ω by an order of magnitude would cause $dc/d\Gamma$ to oscillate in sign, which we do not observe. This again suggests that the phenomena reported are not consistent with our simple modification of the Lucassen model, allowing $dc/d\Gamma$ to be negative.

Discrepancies from the Lucassen model have been observed before for long-chain surfactants [14, 15]. These workers used the equilibrium $dc/d\Gamma$ and interpreted their data in terms of an apparent diffusion constant D. For long-chain molecules this D differed from the accepted value [14, 15] and appeared to be a function of surfactant concentration [15]. It may be concluded that either the diffusional theory breaks down for the long-chain case or the equation of state used to evaluate $dc/d\Gamma$ is inappropriate (or both). Our analysis involves no equation of state and so tend to indicate that the model is incomplete. The shortcomings may include possible contributions to ε from intra-monolayer relaxation processes such as have been observed for insoluble monolayers [8, 10]. Such monolayer processes would make a (positive) contribution to ε'

which would fall at high frequencies as ω^{-2}. Detailed studies over a wide range of frequencies will be required to elucidate the potentially complex variations of ε' due to the several processes which may be present, including surface-bulk interchange and monolayer relaxation.

6. Conclusions

To summarize, our data raises, for the first time, the possibility that the surface dilational viscosity ε' may assume negative values. It has been shown that light scattering from thermally excited transverse surface waves is indeed capable of distinguishing between positive and negative values of ε'. In an attempt to understand these negative dilational viscosities we have explored the consequences of incorporating them into Lucassen's theory of diffusion limited bulk-surface exchange. This analysis yields qualitative agreement with our data, but differences remain. It is clear that this approach does not fully describe the dynamic processes involved, but supports previous indications that the diffusional exchange theory [12] is an incomplete description in certain circumstances.

Acknowledgements

One of us (A.C. McL.) wishes to thank the Department of Education for Northern Ireland and Unilever Research for financial support.

References

1. Lucassen-Reynders EH, Lucassen J (1969) Adv Colloid Interface Sci 2:347-395
2. Thominet V, Stenvot C, Langevin D (1988) J Colloid Interface Sci 126:54-62; Ting L, Wasan DT, Miyano K, Xu S-Q (1984) J Colloid Interface Sci 102:248-259
3. Earnshaw JC, McGivern RC (1987) J Phys D 20:82-92
4. Langevin D, Meunier J, Chatenay D (1984) In: Mittal KL, Lindman B (eds) Surfactants in solution. Plenum, New York, vol 3, pp 1991-2014
5. Goodrich FC (1981) Proc Roy Soc Lond A374:341-370
6. Bouchiat M-A, Langevin D (1971) C R Acad Sci Paris B272:1357-1359
7. Wu ES, Webb WW (1973) Phys Rev A 8:2077-2084
8. Earnshaw JC, McGivern RC, Winch PJ (1988) J Phys France 49:1271-1293
9. Earnshaw JC, McGivern RC (1988) J Colloid Interface Sci 123:36-42
10. Winch PJ (1988) PhD thesis (unpublished), Queen's University of Belfast

11. NAG E04KBF, Numerical Algorithm Group, Oxford, error note IER 447
12. Lucassen J, Van den Tempel M (1972) Chem Eng Sci 27:1283-1291
13. Lindman B, Puyal MC, Kamenka N, Rymden R, Stilbs P (1984) J Phys Chem 88:5048-5057
14. Lucassen J, Hansen RS (1967) J Colloid Interface Sci 23:319-328
15. Stenvot C, Langevin D (1988) Langmuir 4:1179-1183

Received October, 1988;
accepted March, 1989

Authors' address:

Dr. J. C. Earnshaw
The Department of Pure and Applied Phyics
The Queen's University of Belfast
Belfast BT7 1NN, Northern Ireland

Progress in Colloid & Polymer Science   Progr Colloid Polym Sci 79:162–166 (1989)

Aggregation in interfacial colloidal systems

J. C. Earnshaw and D. J. Robinson

The Department of Pure and Applied Physics, The Queen's University of Belfast, Belfast BT7 1NN, Northern Ireland

Abstract: The structure and growth dynamics of two dimensional aggregation of 1 μm polystyrene latex particles trapped at the air-water interface have been observed. For some time after induction of growth rather compact clusters grew slowly; thereafter the clusters became much more mobile and rapidly aggregated into more ramified structures. The clusters were fractal, the $R_g^2 - N$ plot yielding a Hausdorf dimension of 1.46 ± 0.02, typical of diffusion limited cluster aggregation. The growth dynamics displayed scaling appropriate to this model. The dynamic scaling exponents were rather imprecisely determined, but did satisfy the scaling relation proposed for cluster-cluster aggregation. The observed static scaling exponent of the size distribution function and the crossover of cluster statistics were in excellent agreement with expectation.

Key words: Interfacial colloids, aggregation, cluster-cluster, fractal aggregation, dynamic scaling in aggregation, diffusion limited cluster aggregation.

1. Introduction

The formation of random clusters from small basic subunits by non-equilibrium processes such as aggregation, coagulation, and flocculation, is the main feature of many important phenomena in science and technology [1]. These processes have been widely investigated, because of their practical significance and their fundamental interest. In recent years colloidal particles have been seen as providing ideal model systems to investigate such processes, to test current theories. This paper concerns the use of such particles for experiments on two-dimensional aggregation. Studies in spaces of different dimensionality provide wider tests of theoretical predictions than are possible in three dimensions alone.

Colloidal particles can be trapped very effectively at the surface of a liquid (here aqueous solutions), forming a monodisperse monolayer of charged particles [2]. This provides a rather well-defined model system in two dimensions (it should be noted, however, that the electrostatic fields of the particles will extend into the adjoining media, so that some effects of the third dimension may be retained). Such colloidal particles have interactions which are, in principle, exactly

known so that well-controlled experiments should be possible. The different interactions can be selectively modified to induce the macroscopic behavior of interest, here diffusion limited cluster-cluster aggregation (DLCA). In our experiments aggregation of the particles can be triggered by the addition of electrolyte to the liquid subphase, screening out the electrostatic repulsive forces.

After a brief review of certain recent theoretical predictions concerning the dynamic aspects of the aggregation, and an outline of our experimental set-up, we present preliminary results concerning the structure of the DLCA clusters and of the dynamics of the growth.

2. Theoretical background

Random aggregation has been the subject of considerable theoretical interest, largely involving computer simulations rather than analytical calculations. Like much of the experimental work, this activity has tended to concentrate upon the morphology of the aggregates, expressed, for example, via their fractal dimension (=1.44 for two-dimensional DLCA [1]); this conveys only a limited amount of information about the aggregation process. Growth is, however, an intrinsically dynamic process and several recent simulations have concentrated upon the dynamics of the cluster-cluster aggregation process [3, 4]. Scaling forms reminiscent of those arising in the description of critical phenomena have been found for the cluster size distribution and its time dependence.

For DLCA the normalised distribution function of cluster sizes (s) is found to follow a scaling behavior with s and time (t) as [3]

$$n_s(t) = s^{-\tau} t^{-w} f(s/t^z), \qquad (1)$$

where τ, w and z are, respectively, a static and two dynamic scaling exponents which describe the growth process. The scaling function $f(x)$ acts as a cut-off in s and t, behaving as $f(x) \sim 1$ for $x \ll 1$ and $f(x) \ll 1$ for $x \gg 1$. The first term in Eq. (1) describes how the number of clusters varies as a function of size for all times. At sufficiently large times (i.e. $x \ll 1$) a scaling relationship of $n_s(t)$ as $s^{-\tau}$ is obtained, up to s values such that $f(x)$ cuts off. Further, at the cut-off point the distributions, $n_s(t)$, for different times display a common tangent which can be shown to have an exponent -2 due to the scaling behavior; this result arises from the dynamic aspects [4]. Again, the t dependence of $n_s(t)$ exhibits scaling (as t^{-w}) at sufficiently small s and long t. It is easy to show that Eq. (1), with the form of $f(x)$ quoted above, implies that the mean cluster size scales with time as t^z. These scaling laws have been recovered in simulations in the limit of large sizes and times: in two dimensions $\tau = 0.75 \pm 0.15$, $w = 1.70 \pm 0.2$ and $z = 1.4 \pm 0.2$ for the case where the diffusion coefficient of the clusters is assumed to be independent of s [3, 4].

All three exponents are non-classical and differ from those found in other models. The most significant feature of the DLCA model is that the exponents satisfy the relationship

$$w = (2 - \tau) z \qquad (2)$$

and thus satisfy the initial scaling assumption of Vicsek and Family [3] (Eq. (1)).

3. Experimental

In our experiments monodisperse polystyrene particles (diameter 1.088 µm, s.d. = 0.079 µm) were spread on the surface of a liquid (thermostatted at 25.0 ± 0.2 °C) contained in a PTFE Langmuir trough with an optical window in its base which was mounted on the stage of an inverted optical microscope. The apparatus rests upon a vibration isolated optical table.

Colloidal particle monolayers were formed by carefully syringing a suspension of microspheres in methanol onto the liquid surface. This suspension spread quickly and evenly across the whole trough area. The most stable monolayers appeared to be formed by spreading onto a slightly salty subphase (10^{-3} M CaCl₂) rather than pure water. Care was taken to remove the electrostatic charge on the PTFE trough, as this would drastically affect the spreading process. The monolayers formed were slightly inhomogeneous, but displayed large diffuse regions (covering about one-third of the trough area) which were very homogeneous. The particle number density only varied by about 10 % between these regions. The field of view ($\sim$ mm) was much smaller than these regions, so that an essentially uniform monolayer was observed. The field of view contained a large number of particles, providing good statistics. Over the entire monolayer the thermodynamic limit must be approached.

The particles effectively formed a two-dimensional system, being trapped at the air-water interface by a surface tension [2] (and electrostatic [5]) potential several orders of magnitude greater than kT. The polystyrene particles have an inherent negative charge due to bound surface groups, so that the main interparticle forces acting are repulsive (dipole) electrostatic [6] and attractive short-range Van der Waals forces. There may also, in principle, be an attractive capillary force due to the deformation of the surface by the particles [7], but for individual 1 µm particles this is negligible.

Particle aggregation was induced by carefully adding CaCl₂ solution to the subphase using an infusion pump. Smaller quantities of the doubly charged Ca⁺⁺ ion had to be added to initiate aggregation, and the CaCl₂ seemed to crystallize out of the subphase less than did NaCl. During the aggregation process micrographs of the sample were stored using a video recorder. The images were then digitised using a frame-grabber of 768 × 512 pixel resolution. One pixel covered 1.5 µm² on the image, corresponding to approximately one particle. Selected frames were then analysed using a VAX 8650 computer.

4. Results and discussion

After the latex monolayer had been spread and allowed to equilibrate, aggregation was induced by the addition of CaCl₂ solution. The growth process seemed relatively insensitive to the salt concentration and to the particle surface number density. It always appeared to proceed via two stages: in the first, which lasted some 6 h, relatively compact clusters were formed by slow aggregation; thereafter the clusters suddenly became much more mobile and cluster-cluster aggregation started, leading ultimately to the typical ramified structure. What caused the initial growth to be slow, or what precipitated the change to faster aggregation is not yet clear. In our analysis all times were measured from the moment of addition of the CaCl₂ to the subphase. During the later rapid growth stages the clusters tended to move about on the surface, so that successive micrographs (taken at 30 min intervals) were of different populations. Given the sta-

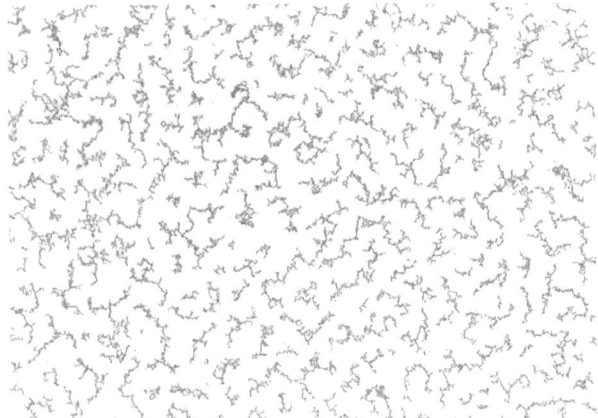

Fig. 1. A typical micrograph taken during the later stages (cluster-cluster) of aggregation of 1 μm polystyrene latex particles at the water-air surface

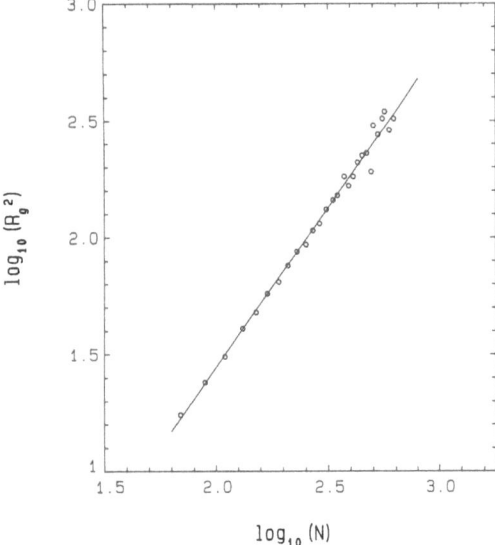

Fig. 2. A double logarithmic plot of radius of gyration squared vs. number of pixels per cluster. The line is a least squares linear fit to the data

tistical self-similarity involved, this is of no consequence.

Here we present a preliminary analysis of one set of data for a surface particle density ~ 0.1 on a subphase of aqueous 0.7 M CaCl$_2$. A typical micrograph taken in the later (cluster-cluster) stages of growth is shown in Fig. 1. We concentrate upon two aspects of the image analysis, the structure of the aggregates and the growth dynamics.

Considering first the structure of the aggregates, the number of pixels (N) contained in each cluster and the

principal radius of gyration (R_g) of the cluster were evaluated for all clusters in the digitised images observed throughout the experimental duration. These quantities were found to scale in the usual manner: Fig. 2 shows a log-log plot of R_g^2 vs. N based upon about 17 000 clusters. A lower cut-off in cluster size was imposed, as small clusters tended to deviate from the straight line behavior. This is probably due to digitizing effects, and does not affect the main result. To reduce the statistical fluctuations involved, each point represents N and R_g^2 averaged over a range of 20 pixels in N. The line shown is the best fit straight line scaling law, from which we infer that the fractal dimension D

$$D = 1.46 \pm 0.02 \tag{3}$$

in good agreement with the accepted value for DLCA.

Turning now to the dynamics of the growth, we examine the scaling behavior (Eq. (1)) predicted by Vicsek and Family [3]. First we present data on the unnormalised cluster size distribution function, $n_s(t)$, as a function of size s. A log-log plot of this (Fig. 3) shows the expected behavior: an initial power law decline (of exponent which appears independent of t)

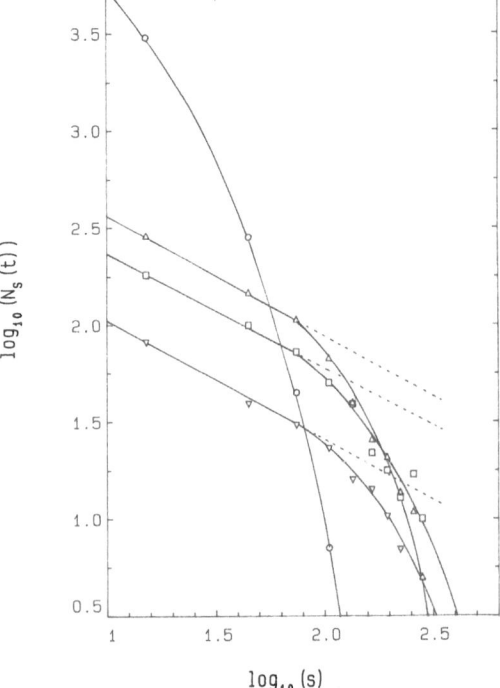

Fig. 3. The cluster size distribution (s in pixels) determined for four different times: $t = 180$ (O), 450 (△), 480 (□) and 510 min (▽)

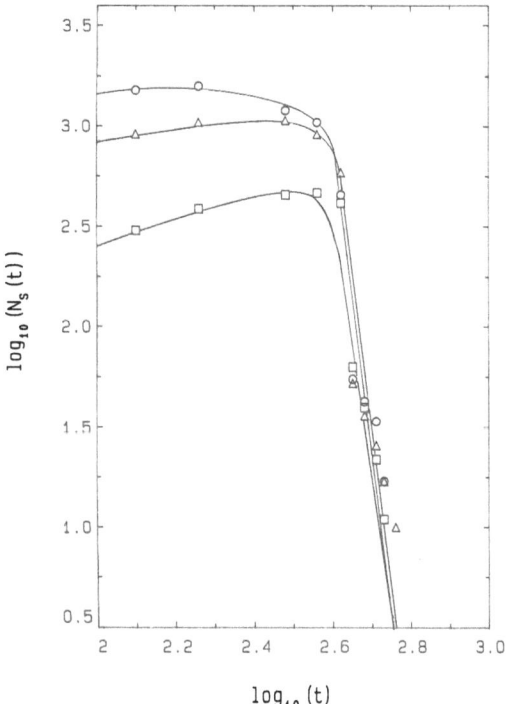

Fig. 4. The evolution of the cluster size distribution function with time (minutes) for three different sizes: $s = 1$–10 ($\bigcirc$), 11–20 ($\triangle$) and 21–30 pixels ($\square$)

Considering the dynamic aspects of the aggregation process, there is some difficulty in comparing the behavior of $n_s(t)$ as a function of t with the simulations, in which cluster-cluster aggregation was evident from the shortest times, whereas such growth apparently only commenced after some time in our experiments. The experimental values of scaling exponents will thus not be comparable with those from simulations, due to the difference in time scales. Further problems may arise in comparing the present data, for rather high surface particle densities, with theoretical results expected to be strictly applicable only as the density $\rightarrow 0$ [3]. However, in this preliminary analysis we seek asymptotic scaling behavior; whilst the exponents will not be strictly correct their relative values may still have some merit.

The scaling behavior is not so evident in the case of the variation of $n_s(t)$ with t (shown in Fig. 4) as for the dependence on s, but by averaging the gradients of the apparently linear portions at large t, the value of the dynamic scaling exponent can be estimated:

$$w = 10.3 \pm 1.8, \tag{5}$$

which, as expected, is far from the theoretical value.

We finally consider the variation of the mean cluster size $(S_0(t))$ with time. In Fig. 5 the number average cluster size,

$$S_0(t) = \frac{\Sigma_s \, sn_s(t)}{\Sigma_s \, n_s(t)}, \tag{6}$$

is plotted rather than the weight average treated by Vicsek and Family. However, the expected scaling behavior (as t^z) is independent of the weighting used, being determined by the universal function $f(s/t^z)$ in Eq. (1). The log-log plot shows a linear variation at large times, from which a scaling exponent of

$$z = 6.6 \pm 0.2 \tag{7}$$

is obtained. Again, this is far from theoretical expectation for two-dimensional DLCA.

5. Conclusions

Morphologically the clusters observed at long times in these experiments conform to expectation for diffusion limited cluster-cluster aggregation in two dimensions. Our observations of the dynamic aspects of the growth afford some measure of confirmation of

leading to a cut-off and very rapid fall. The static scaling exponent τ was estimated from the average of the slopes of the 3 linear fits shown:

$$\tau = 0.61 \pm 0.05. \tag{4}$$

Further, the envelope of the size distribution functions (the common tangent) has a gradient equal, within errors, to the expected value of -2.

Taken together, these two results provide the strongest evidence that the present data obey the scaling hypothesis, the static exponent τ being in good agreement with expectation, while the common tangent includes dynamic aspects of the scaling. The difference between the τ values from simulations and experiment is less than the combined error. However, Meakin et al. [4] have shown that if the diffusion coefficient of the clusters falls as s increases (as is physically realistic) then τ should fall in magnitude. The present experimental value of τ would be consistent with a rather slow decrease of the diffusion constant with s. More precise data are required to confirm this point.

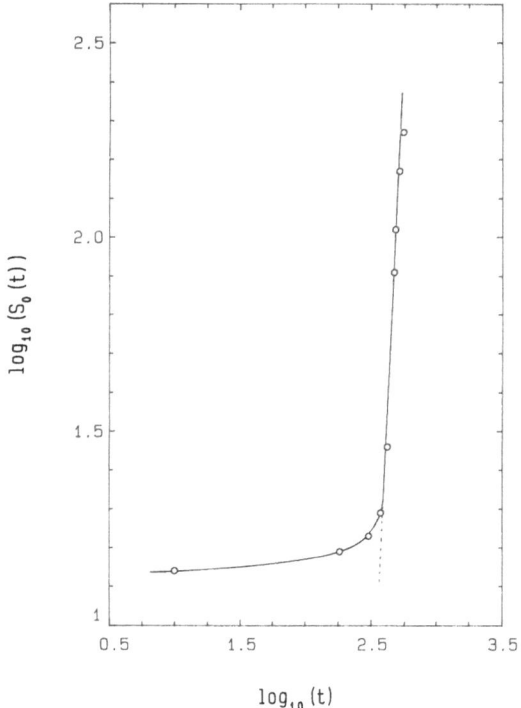

Fig. 5. The number averaged mean cluster size (Eq. (6)) as a function of time (minutes)

Vicsek and Family's scaling hypothesis for this process [3]. The values found for τ and for the common tangent to the variations of Fig. 3 accord well with expectation. Despite the uncertainties concerning the absolute values of the dynamic scaling exponents w and z, the three exponents determined do satisfy the relationship (Eq. (2)) expected from simulations of cluster-cluster aggregation. A more sophisticated analysis might provide more reliable values of w and z. The present results provide the first experimental evidence for the scaling hypothesis concerning DLCA in two dimensions.

The major question concerning these results concerns the change of character of the aggregation from compact to ramified clusters after about 6 h. As stated, the cause of this change is still uncertain. It may be that the initial, small clusters form a meta-stable superstructure on the surface [8], somewhat resistant to free motion of the clusters. Alternatively, leaching of surfactants from the latex particles could lead to increas-

ing capillary forces which might ultimately overcome the electrostatic repulsive forces on the particles. Neither hypothesis is entirely satisfactory, and further experiments are required.

Appendix

Some progress has been made in resolving the origin of the change in character of the aggregation which was observed after about 6 h. In the experiments described in the paper $CaCl_2$ solution was slowly and carefully added close to the bottom of the subphase ($\sim$ 5 mm deep). Calculation suggests that several hours would be required for the Ca^{++} ions to diffuse to the subphase surface where they could effectively screen the charges on the latex particles. In subsequent experiments gentle stirring of the subphase after addition of the $CaCl_2$ solution caused the cluster-cluster aggregation to commence much earlier than before, although some time-lag was still apparent. This remnant effect may arise from one or more of the causes discussed in Section 5 of the paper.

Acknowledgements

This work has been supported by the SERC. One of us (D.J.R.) gratefully acknowledges financial support from the Department of Education for Northern Ireland and from Unilever Research.

References

1. Jullien R, Botet R (1987) Aggregation and Fractal Dynamics. World Scientific, Singapore
2. Pieranski P (1980) Phys Rev Lett 45:569-572
3. Vicsek T, Family F (1984) Phys Rev Lett 52:1669-1672
4. Meakin P, Vicsek T, Family F (1985) Phys Rev B 31:564-569
5. Earnshaw JC (1986) J Phys D 19:1863-1868
6. Hurd AJ (1985) J Phys A 18:L1055-1060
7. Chan DYC, Henry JDJr, White LR (1981) J Colloid Interface Sci 79:410-418
8. Andelman D, Brochard F, de Gennes PG, Jouanny J-F (1985) C R Acad Sci Paris 301:675-678

Received October, 1988;
accepted March, 1989

Authors' address:

Dr. J. C. Earnshaw
Department of Pure and Applied Physics
The Queen's University of Belfast
Belfast BT7 1NN, Northern Ireland

Progress in Colloid & Polymer Science   Progr Colloid Polym Sci 79:167–171 (1989)

Stability of silica colloids and wetting transitions

V. Gurfein, F. Perrot, and D. Beysens

Service de Physique du Solide et de Résonance Magnétique CEA-CEN Saclay, Gif-sur-Yvette, France

Abstract: When silica particles (diameter 500–2 000 Å) are immersed in a binary fluid (water + 2–6 lutidine) the stability of the colloids is affected by capillary interactions. For particular values of the mixture composition and of the fluid temperature, a flocculation of the colloids can be observed. This process is reversible and is accompanied by a large increase of an absorbed layer of either lutidine or water, according to the origin of silica. This phenomenon, which is expected to be universal, can be related to the wetting and adsorption properties of solids in contact with fluid mixtures.

Key words: Silica colloids, adsorption, wetting, wetting transitions, flocculation.

1. Introduction

The properties of wetting and adsorption on a substrate in contact with a gas-liquid system or a two-component fluid have recently been the subject of a great deal of attention [1]. Among the numerous theoretical predictions, the first-order wetting transition is the most frequently encountered in experiments [2]. This transition corresponds to a discontinous jump in film thickness from a microscopic to a macroscopic value along a path at coexistence. However, the associated prewetting line between the weak and the strong adsorption regimes in the one-phase region has not been observed up to now (Fig. 1). Because of its large specific area, the use of a colloidal substrate seems to be a convenient way to study these phenomena. This choice has several consequences. The curved geometry of the particles makes it impossible to observe the divergence of the wetting layers at coexistence and limits the thickness of the adsorption layers. Moreover, the transitions expected in the one phase-region can modify the interactions between the colloidal particles [3]. It is the aim of this contribution to report an experimental study concerning the role of wetting and adsorption phenomena on the stability of silica colloids when immersed in a binary fluid of 2–6 lutidine and water (L–W). Two different kinds of silica colloids (powder from fused quartz and spheres of Stöber silica) have been used. Our results show that whatever the origin of the colloids, a reversible flocculation can be observed for particular values of the composition of the binary mixture. This flocculation can be related to general wetting properties and is closely connected to the adsorption properties of the silica surface.

2. Experimental

The phase diagram of the 2–6 lutidine and water mixture is reported in Fig. 2. The coexistence curve is inverted, with a lower critical point at $Tc = 34\,C$ and a lutidine concentration $c_L = c_C \simeq 0.29$ wt % [3]. The colloid concentration was always low, typically in the range of $10^{11} - 5 \times 10^{12}$ cm^{-3} in number density. Flocculation and adsorption have been studied mainly through optical techniques, that is, turbidity and light scattering measurements [4]. We recall that turbidity is the integral over all solid angles (Ω) of the intensity of the scattered light (I) through

$$\tau = \int I \langle \Omega \rangle \, d\Omega. \tag{1}$$

It is also the rate of decrease of light propagating in the sample. With L being the sample thickness, and I_T and I_0 the transmitted and incident light intensity, one gets

$$\tau = (-1/L) \ln (I_T/I_0). \tag{2}$$

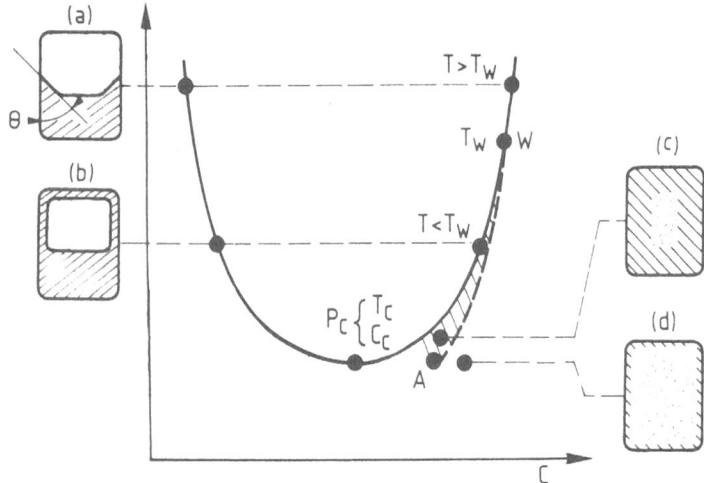

Fig. 1. Phase diagramm of a binary fluid with a lower consolute point (e.g., 2–6 lutidine and water). Solid curve: coexistence curve with P_c (c_c, T_c), the liquid-liquid critical point. At the wetting temperature T_w a transition may occur between partial and complete wetting of the wall by one of the coexisting phases. a) $T > T_w$, partial wetting, contact angle $\theta \neq 0$; b) $T < T_w$, complete wetting, $\theta = 0$. In the case of a first-order wetting transition there should exist a prewetting line (dotted curve) which starts in W and ends in a critical point A. This line corresponds to a transition between strong adsorption (c) and weak adsorption (d) regimes

In order to study the flocculated phase we have also used a new interferometric technique that is described below.

The experimental procedure is as follows: lutidine is added at room temperature to a sample of colloids immersed in water until a concentration c_L is reached. Here the binary mixture is homogeneous and the colloids are stable and dispersed. Then the temperature T is increased by steps until the mixture phase separates, thereby denoting the coexistence temperature T_{cx}. It turns out that the adsorption phenomena at the surface of the colloids and their stability are quite different according to the sign of the difference $(c_L - c_c)$.

3. Results obtained with fused quartz particles

A dry powder of fused quartz is first immersed in deionized water, and the largest particles are removed

by sedimentation. In these conditions a stable sol is obtained, and a electron microscopy study shows that it is composed of particles with an average size of order 100 nm. Polydispersity is however rather large and the particles are not perfectly spherical in shape. The stability of these colloids was found to be a function of the sign of the difference $c_L - c_c$. For L-concentrations lower than the critical one $(c_L - c_c < 0)$, the colloids remain stable until the binary fluid phase separates. On the other hand, when $c_L - c_c > 0$, a flocculation phenomenon occurs at a temperature T_A which can be defined to within a few mK. This transition is detected by a sudden drop of the light transmittivity in the sample, which is accompanied by a very rapid sedi-

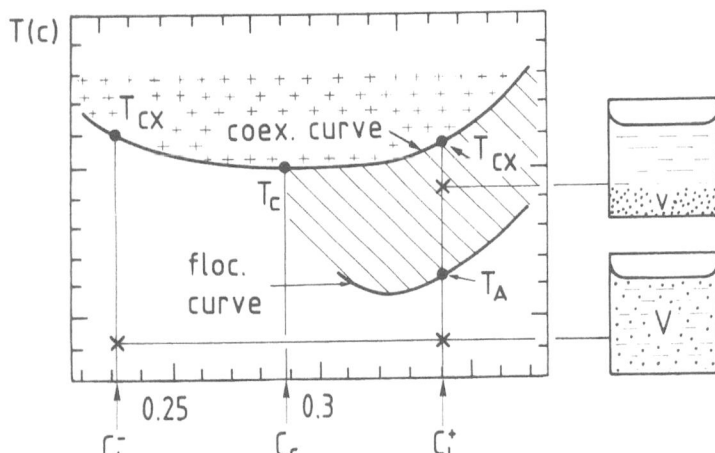

Fig. 2. Schematic of the observations with fused-quartz particles in the 2–6 lutidine and water mixture. T is the temperature and c_L the mass fraction of lutidine

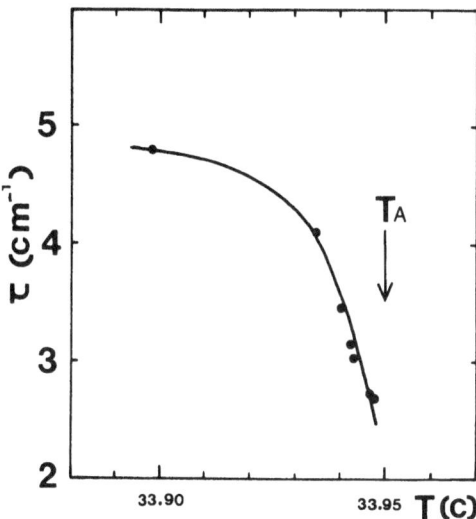

Fig. 3. Temperature variations of the light turbidity measured with fused-quartz particles in 2–6 lutidine and water. The decrease of τ shows that water is preferentially adsorbed

mentation of the particles. These particles form a condensed phase at the bottom of the cell. The corresponding characteristic time is much shorter (typically by a factor of 100) than the sedimentation time corresponding to a single particle. This flocculation is reversible: once the condensed phase is formed, if the temperature is lowered below T_A and the sample is gently shaken, the colloidal system becomes dispersed and stable again, which is not the case if the same manipulation is performed when the temperature is higher than T_A. Here a condensed phase forms again as soon as the shaking has been stopped.

The reversibility of this flocculation and the fact that it occurs only on one side of the coexistence curve of the binary fluid suggests that it is related to the formation of a large adsorbed layer at the surface of the particles (Fig. 1). The presence and the nature of this layer can be determined by turbidity measurements in the stable colloid. In fact, we observe that slightly before reaching T_A the turbidity of the whole sample decreases when T increases, although the bulk turbidity of the fluid increases in the same temperature region (Fig. 3). This variation can be explained only by the formation of a layer with low refractive index at the surface of the particles. According to the respective values of the refractive indices (n_S, n_L, n_W) of silica, lutidine and water respectively, that is [3]

$$n_W = 1.33 < n_S = 1.46 < n_L = 1.48.\qquad(3)$$

The fact that τ decreases can only be due to the formation of a large W-rich layer on the silica particles before flocculation.

This indicates that this flocculation phenomenon is closely related to the preferential adsorption of water at the surface of the colloids. It is worth noticing that this observation is quite consistent with the detection of a W-rich wetting layer on a planar fused quartz substrate [5] immersed in the same mixture but in its two-phase region where a wetting transition has been observed.

With this close relation between the stability of the colloids and their wetting properties, one wonders if inverting these properties by changing the nature of the substrate will also invert the location of the flocculation in the phase diagram. It turns out that this inversion can be performed by using another kind of silica particles.

4. Results obtained with Stöber colloids

These colloids are prepared directly in an aqueous medium by a nucleation process (Stöber method [6]). This method provides spherical particles with a weak polydispersity. We have used particles of mean radius $\bar{R}$ in the range of 65–200 nm.

With this new kind of silica colloid the same flocculation phenomenon can be observed, but on the opposite side of the coexistence curve with respect to that of fused quartz particles, that is for $c_L < c_C$. This flocculation is also reversible.

The flocculation reversibility indicates that in the flocculated phase the spheres are not in contact. This can be confirmed by direct measurements in the floc, made possible here because of the well-defined size of the Stöber particles. Given the number of particles and from a measurement of the volume of the condensed phase, the mean distance d between the centers of particles can be estimated as

$$d = (3.0 \pm 0.3)\,\bar{R},\qquad(4)$$

which shows that the spheres are indeed surrounded by a liquid phase in the floc. We have been able to estimate the composition of this liquid phase by an interferometric technique, which uses the experimental cell itself as an interferometer [7]. When the temperature was decreased from $T > T_A$ to $T < T_A$, a desorption front from the condensed phase was detected. The corresponding variation of the refractive index shows that the condensed phase was richer in lutidine than the liquid phase above.

The inversion of the flocculation location in the phase diagram is thus associated now with the preferential adsorption of lutidine. This can be also verified before flocculation through the turbidity behavior. In contrast to what happens with the quartz particles, the turbidity was seen to increase with temperature in the region near T_A. Following the same reasoning as before, this can be interpreted as the appearance of a L-rich layer on the spheres.

All these results show clearly that the flocculation of Stöber colloids is associated with the preferential adsorption of lutidine. This can be made more quantitative by light scattering measurements which can be performed in the sol, i.e., before flocculation. From these measurements one can determine the adsorbed layer thickness, assuming a very simple slab model of pure lutidine layer (Fig. 4). As the refractive indices of

lutidine and silica are very close [see Eq. (3)], the results can be interpreted in terms of an increasing sphere radius. This equivalent layer thickness l is drawn in Fig. 4 as a function of the temperature difference $(T - T_{cx})$. In order to compare with a typical lengthscale of the mixture, we have measured the correlation length by the same light scattering technique (Fig. 4). It can be seen that the layer thickness l increases significantly before reaching T_A and that it is always much larger on the side of the phase diagram where flocculation occurs.

5. Flocculation mechanism

It seems clear that flocculation is closely related to the adsorption properties of the silica colloids. What is the mechanism responsible of this phenomenon? For the moment we are unable to give a definitive answer but we can suggest a number of mechanisms.

It is well known that the stability of charged silica colloids (in this mixture pH remains around 9 and the colloids are negatively charged) is the result of the balance between electrostatic repulsion and van der Waals attraction [8]. Attractive interactions due to the adsorbed layer which forms and grows on the particles can modify this balance and cause flocculation. A possible mechanism corresponds to a continuous increase of l: for a given "critical" value l_o, the secondary minimum of the potential energy curve could become deep enough to cause flocculation. However, the observed flocculation is a very sharp transition (it occurs within a few mK) and this sharpness rather suggests a gap in adsorption.

This gap could correspond to two different transitions: either a pure surface transition, i.e., prewetting, or capillary condensation [9]. Prewetting corresponds to a jump in adsorption at the surface of each sphere. This jump involves a large increase of the surface energy associated with each particle, which can be minimized by the gathering of the spheres and then by flocculation. Capillary condensation is a finite-size effect which corresponds to a bulk phase transition occuring off-coexistence when a fluid is confined between two closely spaced walls. It could also be relevant here because, by Brownian motion, the particles can approach each other to a small distance. This local phase separation can induce flocculation. However, we must stress that, though very likely, a jump in adsorption cannot be directly detected by the techniques that we used.

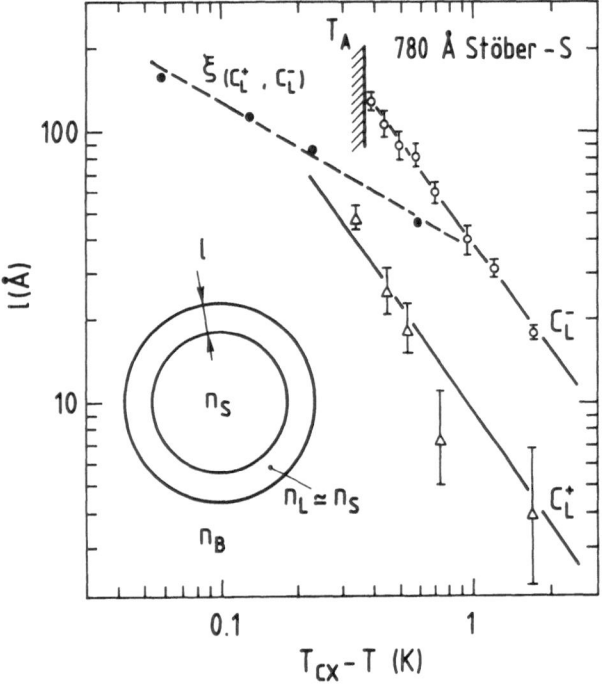

Fig. 4. Thickness 1 of the adsorption layer as a function of the temperature difference $(T_{cx} - T_A)$ measured on Stöber silica spheres in 2–6 lutidine and water mixture. Lutidine is preferentially adsorbed on these particles. A slab model assuming a pure lutidine layer is used. For comparison, the correlation lengths of the bulk concentration fluctuations are reported. The lines are guides for the eye

6. Conclusion

The flocculation phenomenon that has been observed with silica colloids from different origins when immersed in the partially miscible mixture of water and 2–6 lutidine is always reversible and occurs only on the "wetting" side of the coexistence curve. This phenomenon is due to the relevance of a new type of colloid interactions, closely related to the surface properties of the silica colloids and universal aspects of wetting phenomena. In this sense the significant increase of adsorption and the observable flocculation process should be a general phenomenon that should be detected in other colloids and in other mixtures.

References

1. Dietrich S (1988) In: Domb C, Lebowitz JL (eds) Phase Transitions and Critical Phenomena. Vol 12, Academic, London
2. Beysens D, In: Charvolin J, De Gennes PG, Helfrich W, Joanny JF (eds) Liquids at Interfaces. Les Houches, Plenum, to appear
3. Beysens D, Esteve D (1988) Phys Rev Lett 54:2123 and Refs therein
4. Gurfein V, Beysens D, Perrot F, Phys Rev A (submitted for publication)
5. Pohl DW, Goldburg WI (1982) Phys Rev Lett 48:185
6. Stöber W, Fink A, Bohn E (1968) J Colloid Interface Sci 26:62
7. Beysens D (1979) Rev Sci Instrum 50:509
8. Shaw DJ (1986) In: Introduction to Colloid and Surface Chemistry. Butterworths, London
9. Evans R, Marconi UMB, Tarazona P (1986) J Chem Soc Faraday Trans II82:1763 and Refs. therein

Received November, 1988;
accepted January, 1989

Authors' address:

Dr. D. Beysens
SPSRM CEN-Saclay
F-91191 Gif-sur-Yvette Cedex, France

Progress in Colloid & Polymer Science
                          Progr Colloid Polym Sci 79:172–177 (1989)

Quasi-elastic light scattering study of poly(acrylic acid) networks swollen with water

F. Ilmain and S. J. Candau

Laboratoire de Spectrométrie et d'Imagerie Ultrasonores, Université Louis Pasteur, Strasbourg, France

Abstract: Poly(acrylic acid) networks swollen with water and having different ionization degrees were investigated using quasi-elastic light scattering. The polymer concentration was 10^{-1} g $\cdot$ cm^{-3} and the ionization was varied between 0 and 0.8. Three different ionization regimes can be detected from the variation of the cooperative diffusion constant with the ionization degree.

Key words: Poly(acrylic acid), polyelectrolytes, gels, quasi-elastic light scattering.

Introduction

Quasi-elastic light scattering (QELS) studies on semi-dilute and concentrated solutions of flexible polyelectrolytes without an added low molecular weight electrolyte have revealed a complex behavior [1–6]. Two relaxation processes were observed in the concentrated $c \gg c^{**}$ regime, c^{**} being defined as the concentration at which the lattice melts and the system is in the isotropic phase of wormlike chains. The fast mode corresponds to the cooperative diffusion associated with concentration fluctuations of polyions and counterions [7]. The slow mode characterized by a relaxation time decreasing upon increasing the polyion concentration and/or the ionization degree is not yet understood but it is thought that it is correlated with a chain reptation process or a diffusion of clustered chains [1–6]. The collective fluctuations in polyelectrolytes can be conveniently studied in swollen crosslinked networks. The presence of crosslinks prevents the occurence of slow modes, thus allowing a good characterization of the fast mode.

In this paper we report a QELS study of poly(acrylic acid) networks swollen with water and ionized at extents varying from 0 to 0.8, the polymer concentration being fixed at 10^{-1} g cm^{-3}.

Both the scattered intensity and the cooperative diffusion coefficient obtained from the autocorrelation function of the scattered light are studied as a function of the ionization degree α.

1. Experimental section

1.1. Samples characteristics

Gels were prepared by radical copolymerization in aqueous solution of acrylic acid and methylene bisacrylamide as a crosslinking agent. The gelation is initiated by ammonium peroxydisulfate in an oven at 70 °C.

We define the degree of ionization α of the poly(acrylic acid) as the ratio of carboxylate groups to the total number of monomers.

Poly(acrylic acid) being a weak acid, the ionization degree can be varied over a very wide range by changing the pH of the medium. The contour distance b between two charges for vinylic polyelectrolytes is $b = 2.5/\alpha$ Å [8].

Poly(acrylic acid) in aqueous solution has a non-zero ionization degree due to the acido-basic equilibrium

$$+CH_2-CH(OOH)+ \; + \; H_2O$$
$$\overset{K_a}{\rightleftarrows} +CH_2-CH(COO^-)+ \; + \; H_3O^+ \qquad (1)$$

At the concentration used in this study (10^{-1} g $\cdot$ cm^{-3}), the dissociation of poly(acrylic acid) is very low and we have approximated the dissociation constant to that of the monomeric acrylic acid: $K_a = 5.6 \cdot 10^{-5}$. We obtain $\alpha \simeq 0.65 \cdot 10^{-2}$.

High ionization degrees ($\alpha > 10^{-2}$) are obtained by addition of NaOH to the solution in order to partially neutralize the polyacid to a given stoechiometric ionization degree according to

$$+CH_2-CH(COOH)+ \; + \; Na^+ \; + \; OH^-$$
$$\rightarrow +CH_2-CH(COO^-)+ \; + \; H_2O \; + \; Na^+. \qquad (2)$$

This leads to an ionic strength $I \equiv \alpha$, due to the counterions Na$^+$. For very low ionization rates ($\alpha < 5 \cdot 10^{-3}$), we add HCl to the solution to shift the dissociation equilibrium of the weak acid toward the acidic form

$$+CH_2-CH(COO^-)+ + H_3O^+ + Cl^-$$
$$\overset{K_a}{\rightleftharpoons} +CH_2- CH(COOH)+ + H_2O. \qquad (3)$$

In this case, the ionic strength is due to the added HCl and to the counterions: $I \equiv \alpha + 2\,[\text{HCl}]$. At low pH, α becomes negligeable compared to [HCl]. α is calculated using the mass action law

$$\frac{[A^-]\,[H^+]}{[AH]} = K_a$$

and by computing the pH for aqueous solutions of mixtures of strong acid HCl and weak acid poly(acrylic acid).

1.2. Sample preparation

A standard procedure was used to prepare the gels. To 10 g of acrylic acid we add 0.1 g of methylene bisacrylamide and 33 mg of ammonium peroxydisulfate and complete with 90 g of a solution of water containing the amount of sodium hydroxide or hydrochloric acid to get a given ionization degree. The solution is filtered with 0.2 µm filters to get rid of dust particles. A few cm³ are then poured into a light-scattering cell. Nitrogen is bubbled in the solution to remove the dissolved oxygen which would inhibit the radical reaction. The gelification is carried out at 70 °C during 12 h.

1.3. Correlation time measurements

Hydrodynamic theories of gels [9–11] predict that the time-dependent term of the auto-correlation function of light scattered from a swollen polymer network is a single exponential with a time constant τ_c given by

$$\tau_c = (2k^2 D_c)^{-1} \qquad (4)$$

where D_c is the cooperative diffusion coefficient. The magnitude of the scattering vector $\bar{k}$ is given by

$$k = (4\pi n \sin{(\theta/2)})/\lambda \qquad (5)$$

where θ is the scattering angle, λ is the wavelength of the incident light in a vacuum, and n is the index of refraction of the scattering medium.

The optical source on the light scattering apparatus is a Spectra-Physics argon ion laser operating at 4 880 Å. The time-dependent correlation function of the scattered intensity is derived by using a 64-channel digital correlator (Brookhaven BI 2030). All the measurements were performed at 25 °C.

In previous papers from this laboratory [11] it was shown that the intensity scattered from longitudinal fluctuations of swollen networks is generally heterodyned to some extent by the static component due to non-random long-range spatial fluctuations. In order to check whether the scattered signal is fully heterodyned, we measured the autocorrelation function obtained by mixing the scattered signal with an external oscillator using a Michelson-type interferometer. The comparison between the results obtained with and without reference beam is discussed in the next section.

Intensity correlation data were routinely processed by using the method of cumulants [12] to provide the average decay rate $\langle\Gamma\rangle$ and the variance $\bar{v}$. The latter is a measure of the width of the distribution of the decay rates and is given by

$$\bar{v} = (\langle\Gamma^2\rangle - \langle\Gamma\rangle^2)/\langle\Gamma\rangle^2 \qquad (6)$$

$\langle\Gamma^2\rangle$ is the second moment of the distribution. The scattering from microscopic swelling inhomogeneities of the gel generally exhibits long-time fluctuations, which may produce an error in the determination of the flat background B of the autocorrelation function. Therefore, we have considered B as an adjustable parameter in the fitting procedure. For the investigated gels the ratio between the fitted and computed values of B ranges from 1.002 to 1.015.

2. Experimental results

2.1. Intensity correlation

The autocorrelation function of light scattered from the investigated gels contains, in addition to the exponential term, a quasi-static component which has been previously attributed to the effect of microscopic swelling inhomogeneities. As a general rule the static scattering increases as k decreases, whereas the time dependent contribution is k-independent. This can be inferred from the observed increase of the total scattered intensity and the correlative decrease of the normalized amplitude of the autocorrelation function as k decreases. The time-dependent term of the autocorrelation function is heterodyned by the static scattering to an extent which varies with k, and depends on the structure of the gel.

In Figs. 1a and 2a, we have represented the k dependence of $\langle\Gamma\rangle/2k^2$ for gels with ionization degree of, respectively, 0.1 and 0.2. At low k, $\langle\Gamma\rangle/2k^2$ tends to half the value it has at high k, while the ratio I_d/B of the dynamic intensity to the background decreases (Figs. 1b and 2b). Experiments performed with a two beam interferometer (cf. experimental section) show that, in the low k range, the time-dependent term is fully heterodyned by the static scattring. When a reference beam is superimposed on the scattered signal at large k value, we obtain a value of $\langle\Gamma\rangle/2k^2$ equal to that at low k without reference beam.

In the heterodyne regime, the average cooperative diffusion constant D_c is related to the time constant τ_c of the exponential term through

$$\tau_c = (k^2 D_c)^{-1}. \qquad (7)$$

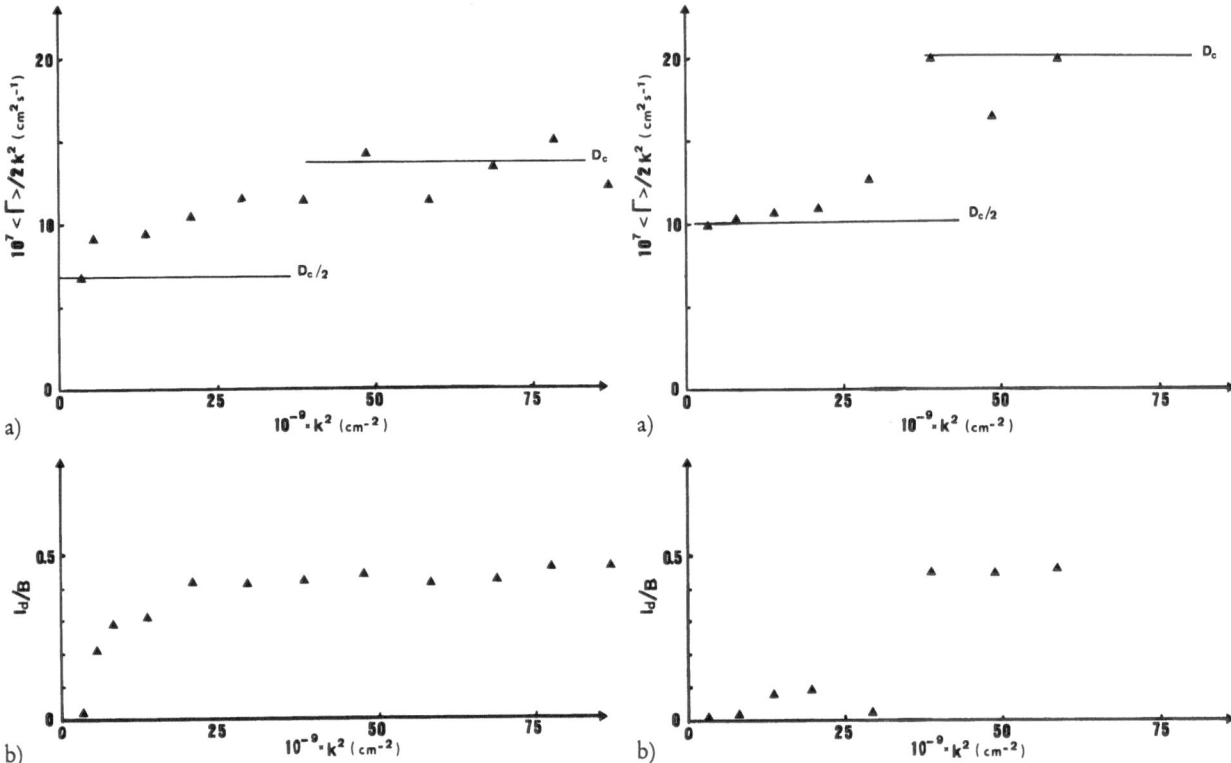

Fig. 1. a) Variation of $\langle \Gamma \rangle/2k^2$ vs. k^2 for poly(acrylic acid) gel at $c = 10^{-1}$ g·cm^{-3} and ionized at $\alpha = 0.1$. b) Variation of the ratio I_D/B of the dynamic intensity to the background vs. k^2

Fig. 2. a) Variation of $\langle \Gamma \rangle/2k^2$ vs. k^2 for poly(acrylic acid) gel at $c = 10^{-1}$ g·cm^{-3} and ionized at $\alpha = 0.2$. b) Variation of the ratio I_D/B of the dynamic intensity to the background vs. k^2

The quality of the exponential fit is found to be very good, as it can be ascertained from the very low values of the variance $v \sim 0.05$ and of the mean square deviation per channel between the experimental curve and the exponential, which is typically of the order of 5×10^{-3}.

It must be emphasized that this behavior is very different from that of semidilute polyelectrolytes solutions, which generally exhibit non-exponential autocorrelation function [1–6]. Nevertheless, the cooperative diffusion obtained in crosslinked networks is likely to be the same as that determined from the fast mode in semi-dilute solution, as D_c was found to be insensitive to the crosslinking degree for a given ionization degree [13]. Figure 3 shows a log-log plot of D_c vs. the ionization degree α of the investigated poly-(acrylic acid) networks.

2.2. Scattered intensity

Measurements of the scattered intensity form the dynamic concentration fluctuations of gels are difficult to perform because of the presence of static scattering.

Methods based on the measurement of the amplitude of the correlation function or frequency filtering of the scattered signal allow the determination of the time-dependent contribution with a rather poor accuracy [11]. Comparison of the decay times obtained with or without the reference beam shows that at large k the scattering regime is very close to a pure homodyne one. Therefore we have performed the intensity measurements at a scattering angle of 90° ($k^2 = 5.86 \ 10^{10}$ cm^{-2}) without correction for the static scattering. The results are reported in Fig. 4 which shows the variation of the scattered intensity excess with respect to that of water as a function of $1/\alpha$.

3. Discussion

3.1. Scattered intensity

The photocurrent i_s due to the polarized scattered light from the collective excitations of a network is given by

$$i_s = A_o \frac{I_o k_B T}{M_{os}} \Phi^2 \left(\frac{\partial \varepsilon}{\partial \Phi} \right)^2 \tag{8}$$

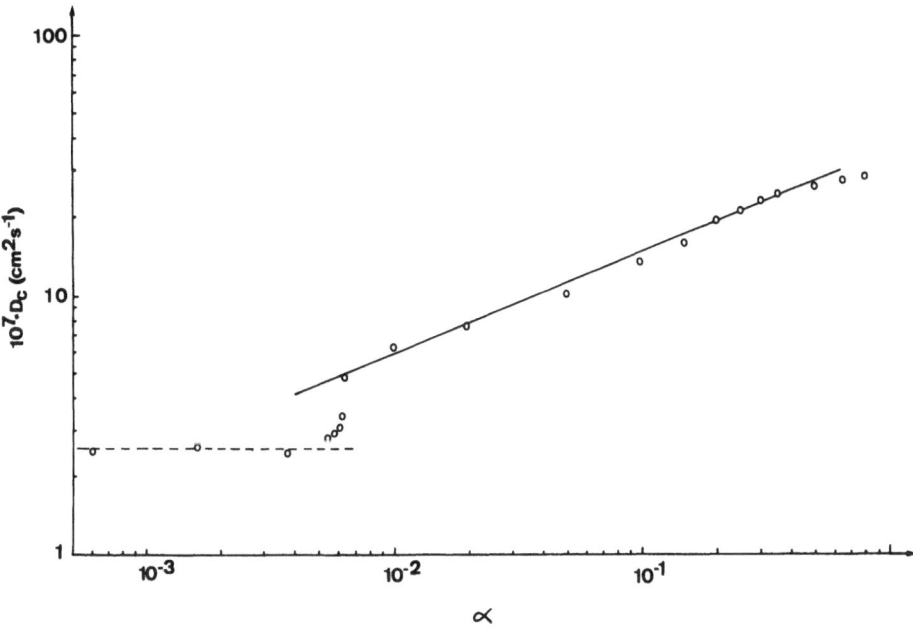

Fig. 3. Log-log plot of D_c vs. the ionization degree α for poly(acrylic acid) gels at $c = 10^{-1}$ g $\cdot$ cm^{-3}; α varies between $6 \cdot 10^{-4}$ and 0.8

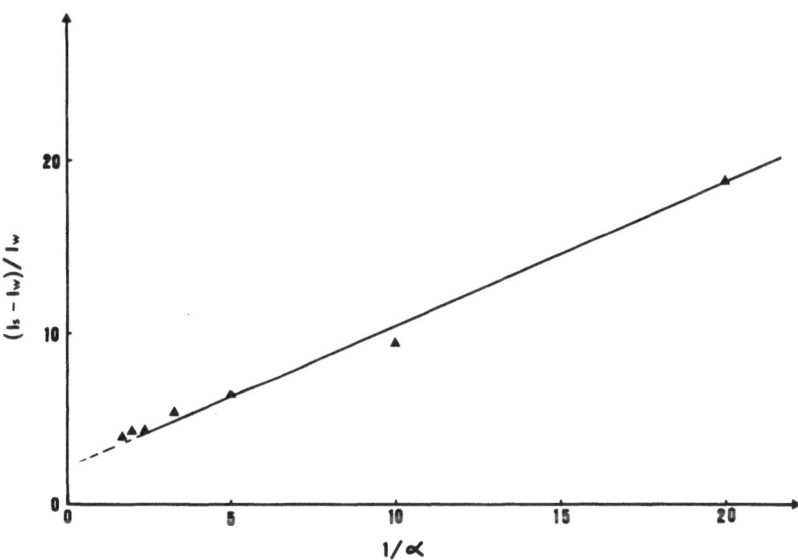

Fig. 4. Normalized excess of scattered intensity $(I_s - I_w)/I_w$ with respect to that of water I_w for a poly(acrylic acid) gel at $c = 10^{-1}$ g $\cdot$ cm^{-3} vs. $1/\alpha$. The straight line is the best least square fit to the data

where A_o is a constant which depends on the wavelength of the incident light, the geometry used and the quantum efficiency of the photomultiplier; ε is the dielectric constant, I_o the incident light intensity, k_B the Boltzmann constant, T the temperature, Φ the polymer volume fraction and M_{os} the longitudinal osmotic elastic modulus given by

$$M_{os} = K_{os} + \frac{4}{3}\mu \tag{9}$$

where μ is the shear modulus and K_{os} the osmotic compressional modulus. Whereas for neutral gels in good solvent K_{os} and μ are of the same order of magnitude, in ionic gels $K_{os} \gg \mu$ because of the dominant contribution of the ionic osmotic pressure [14]. The latter is given by [15]

$$\pi_1 = \nu k_B T \frac{\Phi}{\Phi_o} \alpha \qquad (10)$$

where Φ is the volume fraction of the network, Φ_o is the volume fraction of the network at the condition the constituent polymer chains have random-walk configuration, and ν is the number of constituent chains per unit volume at $\Phi = \Phi_o$.

The osmotic modulus is related to the ionic osmotic pressure by

$$(K_{os})_1 = \Phi \left(\frac{\partial \pi_1}{\partial \Phi} \right). \qquad (11)$$

Combining Eqs. (8)–(11) gives

$$i_s = \nu \frac{\Phi}{\Phi_o} \frac{k_B T}{\alpha}. \qquad (12)$$

The results of Fig. 4 are in agreement with the above expression. The best least square fit of a straight line to the data has a positive intercept. A possible explanation of this result is that the intensity scattered from density fluctuations of the gels could be considerably larger than that of the pure water. Also one cannot discard a possible contribution from the inhomogeneities of the gel. In this respect one must note that the acceptance angle in the scattered intensity measurements is much larger than in correlation measurements. Therefore the inhomogeneities play a more important role in the former experiments.

We also remark that for $\alpha > 0.4$ the Manning condensation occurs (cf. § 2.2.c) so that the effective ionization degree is overestimated and the scattered intensity does not decrease anymore.

Studies performed on different sets of samples reproduced the results of Fig. 4 with good accuracy.

3.2. Cooperative diffusion coefficient

The results of Fig. 3 allow to distinguish roughly three regimes, depending on the α range.

a) $0 \leq \alpha \leq 5 \cdot 10^{-3}$

In that range the electric charges are very scarce and their effect is screened by the ions introduced to decrease the pH (Cl^- and H^+).

For the value $\alpha = 5 \cdot 10^{-3}$ obtained at pH = 2, below which D_c has a constant value, the average spatial distance between 2 charges in the gel taken as the cubic root of the charge density is ~ 70 Å, and is therefore longer than the Debye length at pH = 2 ($\varkappa^{-1} = 35$ Å). The average contour distance A between two charges on the polymer chain and that between two neighboring crosslinks are respectively ~ 500 Å and ~ 250 Å. The latter has been calculated by assuming a random distribution of the crosslinks, total conversion, and a length of the monomer equal to 2.5 Å. In that regime the gel can be considered as neutral, the electrostatic interactions being totally screened. The cooperative diffusion coefficient is nearly constant as α varies. The correlation length ξ which gives an estimate of the mesh size of the network [10, 11] can be obtained from the following relationship

$$\xi = \frac{k_B T}{6\pi \eta_o D_c} \qquad (13)$$

where η_o is the viscosity of the diluent. One obtains $\xi = 85 \pm 2$ Å.

b) $\sim 10^{-2} \leq \alpha \leq 0.4$

In that range the effect of electrical charges becomes predominant. However the contour length A between two successive charges on the chain is larger than both the Bjerrum length L_B and the Debye-Hückel length $\varkappa^{-1}$. A scaling approach to this weak coupling limit has been proposed by Pfeuty [16] who predicted that the correlation length ξ should vary as $\alpha^{-1/3}$. If the correlation length ξ is the length scale that determines the values of D_c then D_c should scale as $\alpha^{1/3}$. The best fit of a straight line to the results of Fig. 3 is:

$$D_c = 36.5 \ 10^{-7} \ \alpha^{0.43} \ (\text{cm}^2 \ \text{s}^{-1})$$

which leads to the following variation of D_c vs. α by using Eq. (13)

$$\xi = 6.7 \ \alpha^{-0.43} \ (\text{Å}).$$

The exponent of the power law of ξ as a function of α is found to be significantly larger than the above prediction. In fact, it is found that the variation of ξ with α is correlated instead to that of $\varkappa^{-1}$, which is

$$\varkappa^{-1} = 3.7 \; \alpha^{-0.5} \; (\text{Å}).$$

c) $\alpha \geq 0.4$

For neutralization degrees exceeding 0.4, the effective ionization degree remains approximately constant, due to the Manning condensation. The ionization corresponding to a distance between charges equal to the Bjerrum length $L_B = 7$ Å is 0.35 and therefore beyond this value D_c tends to level off.

Conclusion

The results reported in this paper show that the fluctuations in a swollen polyelectrolyte network are dominated by the effect of ionic pressure. The variation of the cooperative diffusion coefficient with the effective ionization degree for relatively high ionization degrees seems to be correlated with that of $\varkappa^{-1}$. Experiments where $\varkappa^{-1}$ is varied by addition of salt are in progress.

Acknowledgements

This work was supported by ORKEM-NORSOLOR. The authors are indebted to Dr. Robinet and Dr. Crétenot for valuable discussions on the sample preparation. They acknowledge the helpful assistance of Dr. Hirsch for light-scattering experiments. They are very grateful to Dr. P. Pincus for interesting suggestions relative to the interpretation of the data.

References

1. Grüner F, Lehman WP, Fahlbuch H, Weber R (1981) J Phys A14:L307
2. Koene RS, Mandel M (1983) Macromolecules 16:973
3. Drifford M, Dalbiez JP (1985) J Phys (Fr) 46:L311
4. Drifford M, Dalbiez JP, Tabti K, Tirant P (1985) J Chim Phys 82:571
5. Drifford M, Dalbiez JP (1985) Biopolymers 24:1501
6. Stepanek P, Konak C (1984) Adv Colloid Interface Sci 21:194
7. De Gennes PG, Pincus P, Velasco RM, Brochard F (1976) J Phys (Fr) 37:1461
8. Manning GS (1969) J Chem Phys 51:934
9. Tanaka T, Hocker L, Benedek GB (1973) J Chem Phys 59:5151
10. De Gennes PG (1976) Macromolecules 9:587
11. Candau SJ, Bastide J, Delsanti S (1982) Adv Polym Sci 44:27 and references therein
12. Koppel DE (1972) J Chem Phys 57:4814
13. Ilmain F, Candau SJ (1988) Proceedings of the 9th International Meeting of the Polymer Networks Group, Freiburg
14. Tanaka T (1981) Scientific American 244:110
15. Hirotsu S, Hirokawa Y, Tanaka T (1987) J Chem Phys 87:1392
16. Pfeuty P (1978) J Phys (Paris) Colloq 39:C2 149

Received October, 1988;
accepted April 1989

Authors' address:

F. Ilmain
LSIU Université Louis Pasteur
4, rue Blaise Pascal
F-67070 Strasbourg Cedex, France

Progress in Colloid & Polymer Science   Progr Colloid Polym Sci 79:178–183 (1989)

Measurement of the bending elasticity of a monolayer: ellipsometry and reflectivity

J. Meunier and B. P. Binks

Laboratoire de Physique Statistique de l'ENS, Paris, France

Abstract: The bending elasticity of a monolayer at a liquid interface can be deduced from the analysis of the optical properties of the interface which are sensitive to interfacial thermal fluctuations of short wavelengths. There are two techniques; the first is ellipsometry which is very sensitive to short wavelengths and therefore is an accurate method to measure the bending elasticity but needs large thermal fluctuations. It is limited to liquid interfaces of low surface tension ($y \sim 10^{-2}$ mN/m) and low bending elasticity ($K \sim k_B T$) and our analysis takes into account the coupling between thermal modes through a renormalization of the surface tension and the bending elasticity with the scale. The second method consists of the measurement of x-ray reflectivity which is sensitive to the whole spectrum of the thermal fluctuations and, consequently, is less accurate than ellipsometry; however, it allows study of monolayers of large surface tension at the free surface of a liquid. It has recently given interesting information about the jumps of the bending elasticity of a monolayer at some phase transitions.

Key words: Bending elasticity, thermal fluctuations, liquid surfaces, monolayers, ellipsometry, x-ray reflectivity.

1. Introduction

Optical techniques such as ellipsometry and reflectivity measurements are well adapted to the study of liquid interfaces: They are non-pertubative and many liquids are reasonably optically transparent. X-rays or neutrons can take the place of light, giving more information at low scale because of their low wavelength. As x-rays are strongly absorbed by liquids, up to now their use has been limited to the free surface of liquids. Neutron experiments are not sensitive because of the weak flux and because of the low contrast of materials (lower than with x-rays) but they add two advantages: neutron beams are weakly absorbed by most of the materials and the contrast can be varied by isotopic substitution.

In a reflectivity measurement (light, x-rays, neutron beam reflectivity or ellipsometry), the scattering vector Q is normal to the interface, allowing one to probe the refractive index along the normal. The origins of the index variation along the normal to the interface are the structure of the interface (diffuse interface layer) and the roughness of the interface.

In the case of an interface between two structureless fluid phases, with or without a monolayer, the roughness originates in the thermal fluctuations (or spontaneous fluctuations). The average shape of the interface is a plane, but it is constantly distorted. The distance of a surface point to the average plane $z = 0$ is $\zeta (r, t)$, where ζ, $r(x, y)$ are the coordinates of the point. One can write ζ as a sum of Fourier components

$$\zeta(r, t) = \sum_q \exp (iq \cdot r) \, \zeta_q(t). \tag{1}$$

ζ_q are the thermal modes. This procedure is correct as far as the continuum hypothesis holds, i. e., as far as $q < q_{max}$, where q_{max} is a cut-off determined by the typical molecular size.

1.1 Ellipsometry and reflectivity

E_j is the incident electric field and E_{ij}^r the reflected field

$$E_{ij}^r = r_{ij}(\theta) \, E_j \tag{2}$$

where θ is the incidence angle and i, j are equal to s or p according to whether the electric field is perpendicular or parallel to the incidence plane.

Ellipsometry consists of measuring the ratio $(r_{pp}^r + r_{ps}^r)/(r_{ss}^r + r_{sp}^r)$. This ratio is positive for $\theta = 0$ and negative for $\theta = \pi/2$ because of a change of phase of π between the two polarizations of the reflected field. For a Fresnel interface (i. e., an interface without thickness and roughness: the refractive index changes at $z = 0$ from n_1, the index of one phase, to n_2, the index of the other phase). This ratio vanishes at the Brewster angle θ_B: $\tan \theta_B = n_2/n_1$. For a real interface this ratio is complex and its real part vanishes but its imaginary part keeps a finite value. This imaginary part is called the ellipticity $\bar{\varrho}_B$ at the Brewster angle and is very sensitive to the interfacial structure and roughness because it is a deviation from zero. In the case of thin interfaces (much thinner than the wavelength λ of the light); no further information can be obtained by varying the incidence angle and $\bar{\varrho}_B$ is the sum of two terms: 1) a structural term $\bar{\varrho}_B^L$ whose expression depends upon $n(z)$, the refractive index in the interface [1], and upon a possible anisotropy of the refractive index [2]; and 2) a roughness term [3–4]: $\bar{\varrho}_B^R \sim \sum_q q \langle \zeta_q^2 \rangle$.

An ellipsometric measurement gives only a global value $\bar{\varrho}_B$. In many cases one of the two terms ($\bar{\varrho}_B^L$ or $\bar{\varrho}_B^R$) is smaller than the other and can be neglected or roughly estimated with a model. In other cases it is possible to vary one parameter of the system which affects only one of the two terms — for instance, the surface tension which affect only the roughness term — and therefore to obtain information about the roughness and the structure of the interface.

In a reflectivity measurement the signal is the ratio of the reflected intensity to the incident intensity of the light: the phase behavior is lost. This ratio R_j depends upon the polarization of the incident light j and takes a value $R_j^F(\theta)$ for a Fresnel interface. It is the ratio $R(\theta) = R_j/R_j^F$ which contains the information about the interface. It is equal to 1 for a Fresnel interface and is lower for a real interface. This technique is sensitive only when the wavelength of the incident beam is not very much larger than the thickness and the roughness of the interface. For most of the liquid interfaces this needs short wavelength radiation such as x-ray and neutron beams. The ratio R has then a very simple expression for θ, smaller than the angle of total reflection, because the index variations along the vertical axis are very small for these radiations: it is the product of a structural term R_0 and a roughness term $\exp -$

$Q^2 \langle \zeta^2 \rangle$ ($\langle \zeta^2 \rangle$ is the average of ζ^2). By varying the incidence angle it is possible to separate the structural term and the roughness term and get, in addition, accurate information about the structure because of the small wavelength of the x-rays ($\lambda \sim 1.5$ Å): minima (or peaks) in the curve $R(\theta)$ are observed for destructive (or constructive) interferences between the radiation reflected at different levels in the thickness of the interface.

1.2-The thermal roughness term in ellipsometry and reflectivity

The reflectivity technique is sensitive to: $S_1 = \langle \zeta^2 \rangle = \sum_q \langle \zeta_q^2 \rangle$, while ellipsometry is sensitive to $S_1 = \sum_q q \langle \zeta_q^2 \rangle$. The reflectivity is equally sensitive to each thermal mode while the ellipsometry is more sensitive to the thermal modes of high q. The calculation of S_1 and S_2 is a statistical mechanical problem. The probability of a configuration ζ is proportional to:

$$\exp - E(\zeta)/k_B T \tag{3}$$

where $E(\zeta)$ is the energy of the configuration ζ and k_B the Boltzman constant. E is the sum of three terms: 1) the gravitational energy:

$$E_G(\zeta) = \int_A dA \int_0^z \Delta\varrho\, g\, z\, dz \tag{4}$$

where $\Delta\varrho$ is the density difference between the two phases, g the acceleration due to gravity, and A is the projection of the area S of the interface on the horizontal equilibrium plane; 2) the capillary energy:

$$E_c(\zeta) = \int_S \gamma\, (dS - dA) \tag{5}$$

where γ is the interfacial tension; and 3) the curvature energy [5]:

$$E_k(\zeta) = \int_S \frac{K}{2}\, C^2\, dS \tag{6}$$

where K is the bending elasticity and C the mean curvature of the interface.

The calculation of $\langle \zeta_q^2 \rangle$ needs approximation. We first examine the approximation of independent modes and the limits for its validity, then we examine a coupled modes theory.

a) Approximation of independent modes

The roughness of liquid interfaces and its slope are generally small: $\nabla \zeta^2 \ll 1$. In most cases a development of the interfacial energy to first order in ζ^2 is a sufficient approximation. The energy of a configuration ζ is then the sum of the energy of each mode and the energy equipartition theorem gives:

$$\langle \zeta_q^2 \rangle = \frac{k_B T}{A(\Delta \varrho \, g + \gamma \, q^2 + K \, q^4)}, \quad \text{for } q < q_{max} \quad (7)$$

This formula leads to:

$$\langle \zeta^2 \rangle = \frac{k_B T}{2\pi\gamma} Ln\left(\frac{q_R}{q_0}\right) \quad (8)$$

$$\sum_q q \, \langle \zeta_q^2 \rangle = \frac{k_B T}{2\pi\gamma} \, q_e \quad (9)$$

where $q_R = \sqrt{\frac{\gamma}{K}}$ and $q_e = \frac{\pi}{2}\sqrt{\frac{\gamma}{K}}$ if $\sqrt{\frac{\gamma}{K}} < q_{max}$, and $q_e = q_R = q_{max}$ if $\sqrt{\frac{\gamma}{K}} > q_{max}$; q_o is a cut-off at large scale introduced by the aperture of the beams in the experiment or by the gravity energy: $q_o = (\gamma/\Delta \varrho \gamma)^{-1/2}$.

The experimental determination of the roughness terms which are functions of the bending elasticity in the case $\sqrt{\frac{\gamma}{K}} < q_{max}$, is a way to measure this parameter on interfacial monolayers. The roughness term in a reflectivity measurement is a logarithmic function of the bending elasticity. The accuracy on the K value deduced from such a measurement is weak. This is not the case in ellipsometry where the roughness term is proportional to $(\gamma K)^{-1/2}$. Therefore, x-ray reflectivity measurements are only sensitive to large values of K ($K \sim 100 \, k_B T$) of a monolayer at the free surface of a liquid while ellipsometry is sensitive to small K values ($K \sim k_B T$) of a monolayer at an interface of small surface tension ($\gamma \sim 10^{-2}$ dyne/cm) where the roughness ellipticity $\bar{\varrho}_B^R$ is large. But in this last case, where the bending elasticity is small, the roughness is large and the independent modes approximation fails: the coupling between modes introduces a scale dependence of K [6]. The bending elasticity K is measured at the scale q_e the value of which depends on the surface tension γ. In order to compare different monolayers, the K values must be deduced at the same scale; the coupling between modes must be taken into account.

b) A coupled modes theory

The next terms in the development of the energy must be taken into account. For instance,

$$E_c = \gamma \int_A \left[\frac{1}{2} \nabla \zeta^2 - \frac{1}{8} (\nabla \zeta^2)^2\right] dA. \quad (10)$$

The second term in the integral is a coupling term. One can write:

$$E_c = \frac{\gamma}{2} A \sum_q q^2 |\zeta_q|^2 \left(1 - 3 \sum_{q' > q} q'^2 |\zeta_{q'}|^2\right). \quad (11)$$

In a surface tension measurement we do not measure γ. We observe a thermal fluctuation of wavevector q_1 and we assume that the average of the potential energy of such a fluctuation is $k_B T/2$. The measured surface tension is the coefficient in the equation giving the surface energy as a function of the increase of area due to the mode of wavelength q_1:

$$E(\gamma_1, q_1) = \int \gamma_1 (dS - dA) = \frac{1}{2} \gamma_1 \zeta_{q_1}^2 \, A. \quad (12)$$

Observing at a different scale q_2, we have the same equation with a coefficient γ_2. Of course, in the approximation of independent modes $\gamma_1 = \gamma_2$, because the increase of the area due to a mode q_1 is only q_1-dependent. If we take into account the coupling between modes, a mode q_1 is coupled to all the modes $q' > q_1$ and a mode q_2 is coupled to all the modes $q' > q_2$. The couplings are not the same for the two observed modes and there is no reason for the measured surface tensions to be identical at the scales q_1 and q_2. The surface tension is scale-dependent, as is the bending elasticity; they are renormalized. The calculation gives [7]:

$$\gamma(q) = \gamma_\infty + a \, k_B T \, q^2 \quad \text{with } a = \frac{3}{8\pi} \quad (13)$$

where γ_∞ is the surface tension measured at a macroscopic scale. $K(q)$ has no analytical expression and some solutions are given in Fig. 1 in reduced parameters:

$$Q = \sqrt{\frac{k_B T}{\gamma_\infty}} \, q \quad \text{and} \quad H(Q) = \frac{K(q)}{k_B T}. \quad (14)$$

At large scale, when the capillary energy dominates ($E_c \gg E_K$), the bending elasticity is a constant K_∞. At

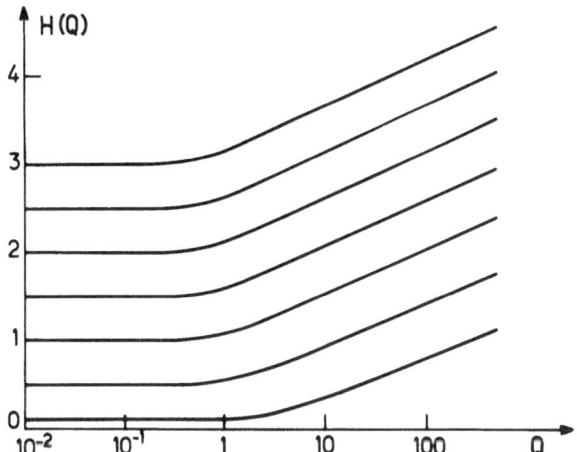

Fig. 1. Solutions for the reduced bending elasticity constant H vs. the reduced wave vector Q for interfaces with microscopic surface tension $\gamma_\infty \neq 0$

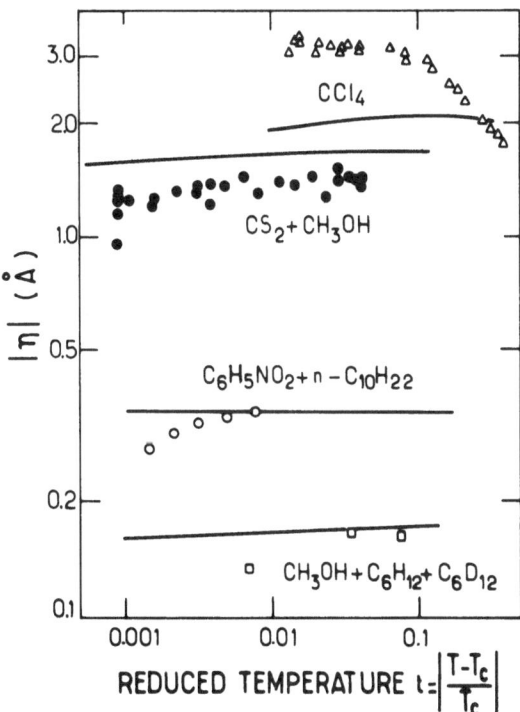

Fig. 2. The ellipsometric parameter of four different critical interfaces: Experimental points from [11] and [12] and theoretical fit (solid line) deduced from the coupled-mode theory. An unexplained discrepancy appears for CCl_4; further experiments are needed for this critical system where ellipsometric measurements are very difficult because of the high pressures involved

small scale, the curvature energy dominates ($E_c \ll E_K$) and one gets:

$$K(q) = K_1 + a\, k_B T\, Ln[q^2 K_1 / q_1^2 K(q)] \qquad (15)$$

where K_1 is the bending elasticity observed at the scale q_1. This reduction of K with the increasing observation scale agrees with the results obtained by Peliti and Leibler [8] if we neglect the K variation in the logarithm, i. e., if we neglect the surface tension term.

2. Experiments

2.1 Critical interfaces

It must be remarked that the second term in $\gamma(q)$ (Eq. (13)) varies as q^2. The corresponding surface energy varies as q^4, i. e., as a curvature term. This means that the curvature energy never vanishes, even for $K = 0$. In this last case, one obtains:

$$q_e = \frac{\pi}{2} \sqrt{\frac{\gamma}{a\, k_B T}} \qquad (16)$$

which is smaller than q_{max} for low γ.

This formula allows one to calculate the ellipticity $\bar{\varrho}$ of a critical interface (low γ and $K = 0$) with macroscopic parameters: γ_∞, $n_1 - n_2$ and $n_1 + n_2$. A good agreement is obtained between these calculated values and the measured ones at the same interfaces (Fig. 2) without adding any intrinsic thickness, possibly due to

the fact that the profile obtained by the summation on the high q modes is equivalent to the intrinsic profile of Fisk and Widom. In our opinion this agreement is important; it means that the theory remains valid at very low K values and these K values can be deduced from ellipsometric measurements.

2.2 Monolayers at the oil-water interface

The interfacial tension can be very small with soluble surfactants, sufficiently small so that the roughness term is large and the bending elasticity can be measured by ellipsometry. The main difficulty is the separation of the structural term $\bar{\varrho}_B^L$ and the roughness term $\bar{\varrho}_B^R$. This is possible with a soluble surfactant because it is possible to vary the surface tension of the monolayer in a large range by varying the ionic force without changing its molecular density and its rigidity.

The simplest case is that of a pure surfactant such as AOT at the heptane-water interface [9]. The ionic

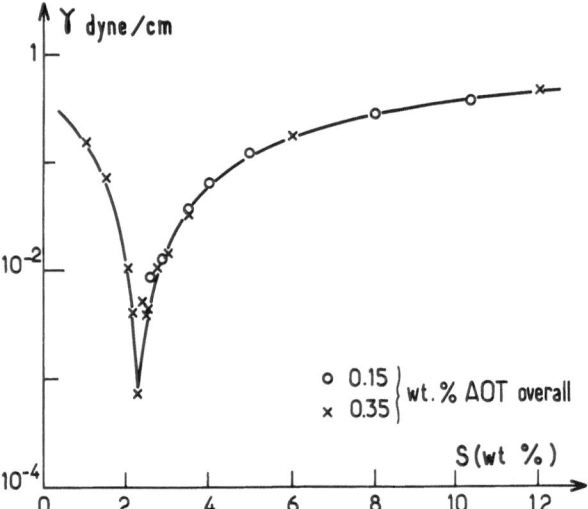

Fig. 3. The interfacial tension γ vs. the brine salinity S in a mixture of heptane, brine, and AOT at 20 °C

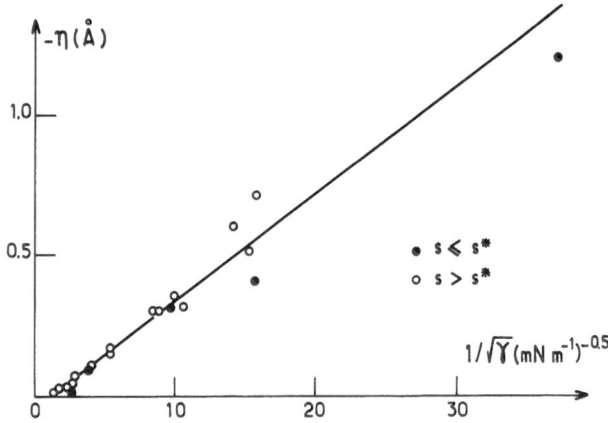

Fig. 5. The ellipsometric parameter vs. $1/\sqrt{\gamma}$ for the oil-water interface in a mixture of heptane, brine, and AOT at 20 °C

force is varied by adding sodium chloride to the water; the quantity of AOT added to the brine-heptan system allows one to have an AOT concentration in brine at just above the CMC. With this concentration the interface is saturated in AOT and the bulk concentration in AOT is low and does not introduce any perturbation in the measurements (the refractive indices of the two phases are those of pure heptane and brine). Figure 3 shows the experimental results of the interfacial tension γ vs. the brine salinity S: a large variation of γ is observed with a minimum for $S^* = 0.225$ %. Figure 4

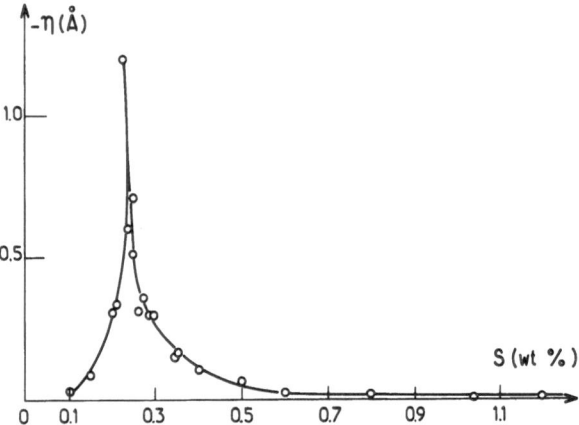

Fig. 4. The ellipsometric parameter η vs. the brine salinity S measured at the oil-water interface in a mixture of heptane, brine, and AOT at 20 °C

shows the experimental results of the ellipsometric parameter η vs. S:

$$\eta = \frac{\lambda}{\pi} \frac{(n_1^2 - n_2^2)}{\sqrt{n_1^2 - n_2^2}} \bar{\varrho}_B . \qquad (17)$$

A large increase of η is obtained at S^* and η varies as $1/\sqrt{\gamma}$, indicating that $\eta^R \gg \eta^1 \cdot \eta$ vs. $1/\sqrt{\gamma}$ is shown in Fig. 5. The experimental points lie on a straight line as is specified by the independent thermal mode approximation. The domain of the scale of measurement q_e, which is proportional to the experimental domain of $1/\sqrt{\gamma}$, is small and the logarithmic deviation from this straight line due to the coupling between modes is not observed: it is less than the uncertainty on each measurement. In the independent mode approximation, we found $K = 1.10\ k_B T$, while in the coupling mode approximation we found $K = 1.65\ k_B T$ at the molecular scale $q = 2.10^7$ cm^{-1}.

2.3 Monolayers at the free surface of water

The ellipsometric measurement of K fails for monolayers of large surface tension because the roughness is too small and the structural term dominates. A technique using radiation with a smaller wavelength than that of the light must be used. The x-ray reflectivity measurements are suitable at the free surface of a liquid. The refractive indices of the liquids for x-rays are close to one and the reflected intensity is very weak as soon as the incident angle θ is smaller than the total

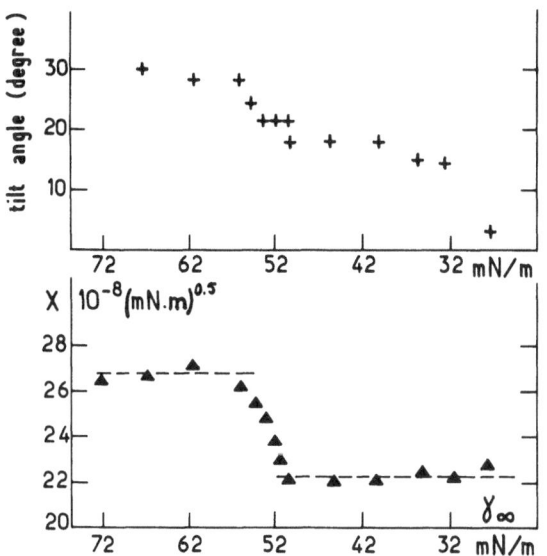

Fig. 6. The tilt-angle of molecules in a behenic acid monolayer and the parameter $X = \sqrt{\gamma \langle \zeta^2 \rangle}$ deduced from x-ray reflectivity measurements vs. the surface tension ($t = 20\,°C$, pH $= 5.5$)

reflection angle θ_L. For this reason the incident beam is always grazing: $0.001 < (\pi/2 - \theta) < 0.1$ radian.

A monolayer of behenic acid on water has been studied in detail with this method [10]. The molecule tilt angle variation, and the quantity $X = \sqrt{\gamma \langle \zeta^2 \rangle}$ vs. the surface tension have been measured (Fig. 6). A phase transition is observed for $\gamma \sim 52$ mN/m with a jump in the tilt angle and in the X value. It is the liquid condensed-solid phase transition which can be identified on the surface pressure-molecular area isotherm of the monolayer. The bending elasticity increases from a low value (that of a liquid monolayer: $K \sim k_B T$) to a large value (that of a solid monolayer: $120\ k_B T < K < 300\ k_B T$). The accuracy on K is low because it appears in a logarithm ($X \sim \ln(K)$), but the jump in K at the phase transition is so large that it is easily observed in X. A second jump is observed on the last experimental point for $\gamma \sim 30$ mN/m, which might be a tilted nontilted phase transition or an artefact due to the vicinity of the collapse pressure.

Conclusions

The bending elasticity of a monolayer at a liquid interface can be deduced from the thermal fluctuations

of the interface studied by optical techniques. Ellipsometry is the more accurate method because it is only sensitive to the small wavelength thermal fluctuations whose energy is dominated by the curvature energy. It is well adapted to the study of monolayers of ultra low surface tension and low bending elasticity whose roughness is large, requiring one to take into account the coupling between modes. Thermal fluctuations on large surface tension monolayers are small and their study needs short wavelength radiation such as x-rays. The sensitivity of the x-ray reflectivity to the thermal fluctuations is independent of their wavelength and, consequently, the accuracy of the bending elasticity measurements is poor by this method, but provides interesting information on two-dimensional phase transitions.

References

1. Drude P (1891) Ann Phys 43:126
2. Azzam RMA, Bashara NM (1977) Ellipsometry and Polarized Light, North Holland, pp 349–358
3. Beaglehole D (1980) Physica 100B:163–174
4. Zielinska BJA, Bedeaux D, Vlieger J (1981) Physica 107A:91–108
5. Helfrich W (1973) Z Naturforsch 28c:693
6. Helfrich W (1985) J Phys 46:1263
7. Meunier J (1987) J Physique 48:1819–1831
8. Peliti L, Leibler S (1985) Phys Rev Lett 54:1690–1693
9. Meunier J, Jerôme B (1986) In: Mittal KL (eds) Surfactants in solution: modern aspects. 6th International Symposium on surfactants in solution, New Delhi, Plenum Press; Binks BP, Meunier J, Abellen O, Langevin D (1989) Langmuir 5:415–421
10. Daillant J, Bosio L, Benattar JJ, Meunier J (1989) Europhysics Letters 8:453–458
11. Schmidt JW, Structure of a fluid interface near the critical point. To appear
12. Beaglehole D (1987) Phys Rev Lett 58:1434–1436

Received November, 1988;
accepted February, 1989

Authors' address:

J. Meunier
Laboratoire de Physique Statistique de l'ENS
24 rue Lhomond
75231 Paris Cedex 05, France

Progress in Colloid & Polymer Science

Progr Colloid Polym Sci 79:184–193 (1989)

Pattern growth during the liquid expanded-liquid condensed phase transition in Langmuir monolayers of myristic acid

K. A. Suresh[1]), J. Nittmann[2]), and F. Rondelez

Dowell Schlumberger, St.-Etienne, France
Université Pierre et Marie Curie Laboratoire Structure et Réactivité aux Interfaces, Paris, France
[1]) Raman Research Institute, Bangalore, 560080, India
[2]) OMV Aktiengesellschaft Technical and Scientific Systems, Taborstraße 1-3, 1020 Vienna, Austria

Abstract: The nucleation and growth of liquid condensed domains into a continuous liquid expanded phase has been monitored by epifluorescence microscopy in monolayers of myristic acid mixed with a fluorescent dye. Various patterns, ranging from circular droplets to highly ramified structures, have been observed over a temperature range of a few degrees C around room temperature. This reflects the complexity of the phase diagram of myristic acid monolayers. For instance the existence of a critical temperature T_c for the two-phase coexistence region around 31 °C has probably a strong influence on the patterns aspects since the line tension between the liquid expanded and the liquid condensed phases vanishes at T_c. We believe that the morphological unstability of the interfacial front is diffusion-controlled, as first predicted by Mullins and Sekerka in binary alloys for the growth of a solid sphere in a uniformly supersaturated melt. On the other hand, thermal diffusion only a small plays role in the present experiments. In the limit of very small line tension, the patterns are self-similar and are characterized by a Hausdorff exponent of 1.8 ± 0.1.

Key words: Liquid condensed domain, liquid expanded phase, Langmuir monolayer, myristic acid, Hausdorff exponent.

Introduction

The study of phase transitions in monolayers of amphiphilic molecules spread at the air-water interface goes back to 1917 when Langmuir [1] described an apparatus allowing measurement of the two-dimensional monolayer pressure as the surface density was continuously varied with a moving barrier. Over the years, a wide variety of compounds have been shown to form the so-called Langmuir monolayers: fatty acids or their salts, phospholipids, long-chain polymers or proteins [2]. They all possess the common feature of hydrophilic heads, e.g., a carboxylic group for fatty acids, attached to long hydrocarbon tails which are hydrophobic and insure insolubility in the aqueous substrate. Although details vary, the associated pressure isotherms show abrupt changes of slope and plateau regions, suggesting phase transitions. The main phases which have been identified are the gaseous phase at very low density (typically above 1000 Å²/molecule), the liquid expanded phase between 40 and 60 Å²/molecule, the liquid condensed phase between 22 and 25 Å²/molecule and finally the solid phase around 20 Å²/molecule. We use here the nomenclature described in Gaines' book [2] and the numerical values are typical for single-chain fatty acids with 13 to 18 carbon atoms in the hydrocarbon tail. Depending on materials, some of the above phases may or may not exist. The phase change between the gaseous and the liquid expanded phases and that between the liquid expanded phase and the liquid condensed phase are generally thought of as being first-order [3]. Indeed, these phase transformations show up in the isotherms as plateau regions, the surface pressure being practically independent of molecular density until one phase has been completely transformed into the other one.

The edges of the plateau region define the limits of the coexistence region. By performing experiments at various temperatures, it is possible, at least in principle, to draw the complete coexistence curve and to identify the upper critical consolute temperature where the two transforming phases become undistinguishable. In practice, it turns out as expected that surface pressure measurements do not provide clear-cut answers. At the gas — liquid expanded transition, several groups working on the same system, i. e., pentadecanoïc acid spread on acidified water, have published boundary values differing by more than a factor of 2 and critical temperatures differing by at least 40 °C [4]. At the liquid expanded — liquid condensed transition, there has been a long-lasting dispute over if the plateau region was strictly horizontal and if the transition was not actually second-order. To explain all these discrepancies, the usual arsenal of sample purity, spreading conditions, etc., was frequently invoked [5]. In our opinion, it just shows that surface pressure measurements are difficult to perform accurately at the gas-liquid expanded transition, where pressures are extremely low.

Also, these measurements are too macroscopic to clearly identify unexpected, albeit important, phenomena such as electrostatic interactions which can occur within the monolayer during the phase changes.

Fortunately, these studies have now been revitalized by the advent of three novel techniques which are sensitive enough to probe even incomplete monolayers of organic molecules at the air-water interface. In 1985, Shen et al. [6] used optical second-harmonic generation to study the reorientation of the C-OH bond upon compression of pentadecanoic acid monolayers through the liquid expanded — liquid condensed transition region. In 1986, Ketterson et al. [7] obtained x-ray diffraction patterns of lead stearate monolayers, using a synchrotron source, and they observed evidence for a solid phase at high molecular densities. Similar experiments have simultaneously been performed by Möhwald et al. [8] on monolayers of phospholipids. Perhaps the most important experimental breakthrough took place in 1981–1983 when von Tscharner and McConnell [9], and Lösche, Möhwald, Sackmann [10] demonstrated that phase transitions in monolayers could be visually observed by epifluorescence microscopy with a spatial resolution of a few μm. The crux of the method is to disperse a small amount of fluorescent dye in the monolayer. If the dye partitions preferentially in one of the phases, this provides a strong optical contrast between, say, the liquid expanded and the liquid condensed phases. It is also possible to use the variations of the dye fluorescence quantum yield due to the changes in molecular orientations in the various phases [11].

In this paper we present preliminary, and therefore mainly qualitative, results on epifluorescence observations during phase transitions in myristic acid monolayers. On compression the following sequence of transitions is observed: gaseous to liquid expanded to liquid condensed. At low temperatures both the gaseous to liquid expanded and the liquid expanded to liquid condensed phase change takes place through a first order transition, characterized by a broad coexistence region. The critical temperature for the liquid expanded to liquid condensed phase transition is found to be 31 ± 0.5 °C. At temperatures below 19 °C, the liquid condensed domains nucleate as dense circles and grow to a size of 100–200 μm. Even close to the maximum surface density of 20 Å^2 per molecule, the circles do not merge together and there is clearly long-range repulsion between them. At temperatures closer to the critical temperature, the liquid condensed domains form highly branched structure with tips and fjords. The growth proceeds by successive irregular splitting of the leading finger tips. If the compression is stopped, the fingered clusters relax to circular shapes over a time scale of a few minutes. We believe that diffusion-limited aggregation, as proposed by Mullins and Sekerka many years ago [12], is the mechanism responsible for the unstable growth. Our branched clusters are very similar to the fractal viscous fingers formed in a Hele-Shaw cell with radial symmetry [13] and also to the computer-generated aggregates of Witten and Sander[14].

Experimental

The experimental set-up consisted of a rectangular (4 × 10 cm^2) Langmuir trough made of teflon and equipped with a movable barrier, also of teflon. The barrier position was precisely controlled by a motorized micrometer screw (NRC, USA). Its shape also allowed to make optical observations right at the barrier edge where nucleation seems to be favored. The temperature of the water subphase on which the monolayer was spread was controlled by circulating fluid from a thermal bath into a heat exchanger immersed in the water. This allowed cooling or heating with a minimum of dead time. A tin oxyde-coated glass plate was used to cover the Langmuir trough. The cover was especially useful to prevent air currents, water evaporation, and to minimize thermal gradients within the monolayer and convection in the subphase. Joule heating of the conductive glass surface was effective to prevent water condensation. All these factors could affect the quality of the optical observations and blur the images, especially for long time exposures. The monolayer was formed by preparing a chloroform solution of myristic acid at high dilution (0.1 mg/cc) and by depositing about 50 μl of this solution at the air-water interface. The pH of the aqueous subphase was adjusted to pH 2 with dilute hydrochloric acid in order to avoid ionic dissociation of the myristic acid carboxylic head groups. The chloroform spreaded rapidly and evaporated, leaving the myristic acid on the surface. The surface pressure was measured by the Wilhelmy hanging plate technique, using a sand-blasted platinum plate which was assumed to be completely wetted (zero contact angle) by the water subphase. The small amounts of fluorescent dye (which were necessary for the epifluorescence microscopy observations) of the order of 0.01 to 0.1 mole %, , barely modified the surface pressure diagrams. NBD – HDA (4-hexade-cylamino-7-nitrobenz-2-oxa-1,3-diazole) was selected because it has a high quantum yield and possesses absorption and emission bands conveniently located in the blue-green part of the visible spectrum. The optical observations were performed using a metal-lurgical microscope (Polyvar-Met, Reichert Jüng, Austria) in the reflective mode. A high pressure 200 W Hg lamp was used as an intense light source. A sensitive Vidicon camera with a silicon-intensified target (LH 4036, Thomson-CSF, France) was used as the detector. The images were viewed on a TV monitor and photo-graphed with a still camera for recording and later analysis. Total magnification was about 200 X. When necessary, the prints were further enlarged and digitized with a hand-held pencil digitizing system. The density-density correlations were then calculated on a Vax 11/785 computer, using a fast Fourier transform algorithm.

molecular orientation is almost certainly parallel to the air-water interface. The quantum yield of the fluorescent probe is known to be very low in polar media and this, coupled with the low overall density of the gas phase, explains the observation of a dark background.

It is very important for the rest of our experiments to realize that the phase boundaries are dependent on temperature. Figure 1 summarizes our results between 5° and 35°C for myristic acid doped with 0.1 % of NBD-HDA. The width of the liquid expanded — liquid condensed region narrows down as temperature increases. Above 31°C, coexisting phases are no longer observed and this temperature has therefore been identified as the critical temperature T_{c2} for the first order liquid expanded — liquid condensed phase transition. Above T_{c2}, there exists only one phase for areas per molecule between 47 and 20 Å² per molecule. The dotted line on the left-hand side of the coexistence region should just be used as a guide to the eye. It indicates the minimum area occupied by one myristic acid molecule standing upright in the "all-trans" configuration. The liquid expanded — gas co-

Results

We started the experiments by drawing the phase diagram of myristic acid as the monolayer was gradually compressed or expanded at fixed temperature. The liquid expanded phase was easily identified as a uniformly bright, homogeneous, phase. The liquid expanded — liquid condensed coexistence region was characterized by the appearance of isolated black domains, corresponding to islands of the liquid condensed phase. Indeed, the dye solubility is lower in the liquid condensed phase than in the liquid expanded phase, which provides a strong optical contrast for the nucleating droplets of the denser phase. The fully condensed monolayer was observed to be almost uniformly dark with occasionally tiny bright spots corresponding to dye crystals expelled from the interface. The liquid expanded — gas coexistence region was identified by the appearance of isolated black droplets indicating formation of gas bubbles. Upon further expansion of the monolayer, the liquid expanded phase reduced to a network of thin bright lines, with a topology characteristic of computer-simulated two-dimensional foams [11]. Eventually these lines vanished, leaving a dark field of view, as expected for a fully gaseous phase. In this highly diluted phase, the

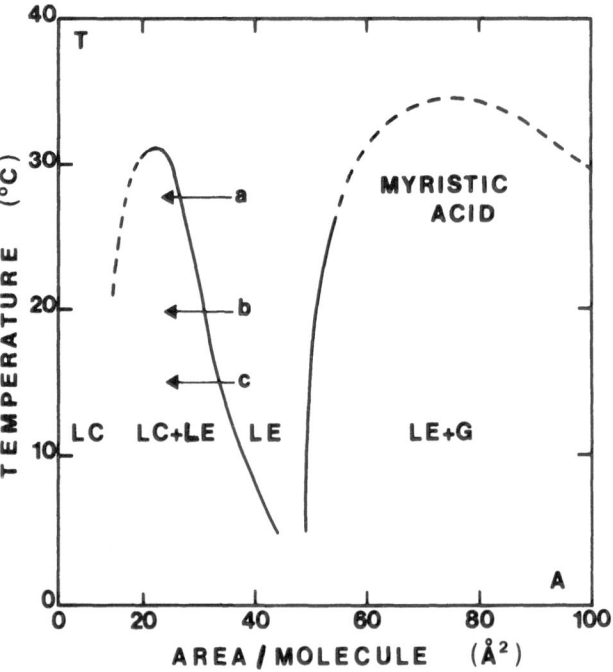

Fig. 1. Phase diagram for myristic acid monolayers at the air-water interface. Subphase is at pH 2. LC = liquid condensed, LE = liquid expanded, G = gas phase. The solid lines correspond to actual measurements, the dashed lines are a guide for the eye

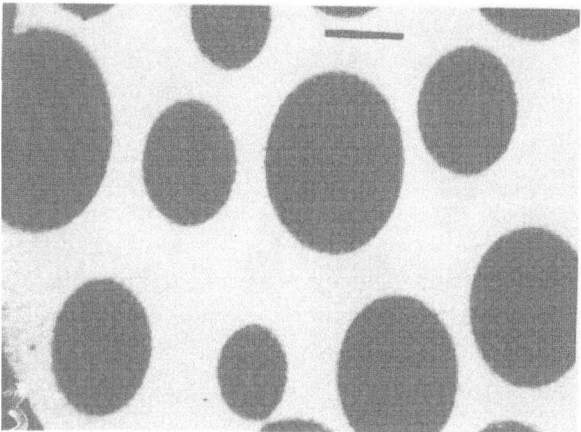

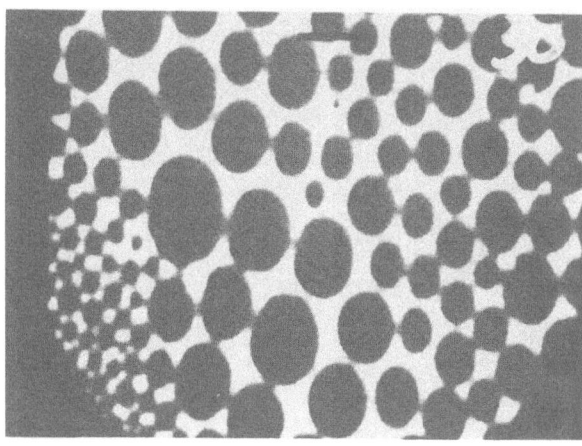

Fig. 2. Optical aspect of the liquid condensed domains in a monolayer of myristic acid labelled with 4-hexadecylamino-7-nitrobenz-2-oxa-1,3-diazole (NBD-HDA). $T = 15\,°C$. Concentration of dye is 0.3 %. Bar length is 75 μm. The domains should be circular but there is a slight surface flow which makes them elliptic. Mean surface density is 30 $Å^2$ per molecule

Fig. 3. Same as Fig. 2 except surface density which is 25 $Å^2$ per molecule. The wide variation in the domain size is due to inhomogeneous compression when decreasing the surface density from 34 to 25 $Å^2$ per molecule. Note the dark bridges ("whiskers") which connect the neighboring liquid condensed domains

existence region is observed at large areas per molecule. On expanding the monolayer, this two-phase region is entered at approximately 45–47 $Å^2$ per molecule, almost independent of temperature. There should exist a critical temperature, T_{c1}, above which the liquid expanded and the gaseous phases become indistinguishable. However, T_{c1} certainly exceeds 40 °C, the maximum temperature accessible with the present set-up. Attempts to reach higher temperatures were not successful due to rapid evaporation of the subphase which, in turn, induces erratic motions within the myristic acid monolayer. Our lower bound estimate for T_{c1} is consistent with the observations of Pethica and Pallas [4] who have estimated T_{c1} to be in the region of 50 °–60 °C for pentadecanoic acid, a fatty acid with one extra methylene unit compared to myristic acid. The liquid expanded — gas co-existence region extends up to very large areas per molecule, of the order of 1 500 $Å^2$ per molecule at 20 °C. This part of the phase diagram has not been reported in Fig. 1 since we are mainly interested here in the liquid expanded — liquid condensed co-existence region.

We shall now describe our microscope observations of the shape of the liquid condensed domains formed as the myristic acid monolayer is compressed at 0.1 $Å^2$ s^{-1}/molecule from the one-phase, liquid expanded, region into the two-phase, liquid expanded — liquid condensed, region. The experiments have been performed at three temperatures, corresponding to the

paths marked a, b, c in Fig. 1 and at increasing distances from the critical point T_{c2}.

1. Path c: Temperature much below T_{c2}

If the compression is performed at a fixed temperature of 15 °C, the liquid condensed domains appear as almost perfet circular disks of typical size 20 μm. Upon further compression, they grow in size (Fig. 2) up to a few hundred μm, but their number stays fixed, which is typical for a nucleation and growth process. The density of nuclei per unit area depends on the conditions of supersaturation [15], namely the local surface pressure excess generated during the compression. Since this transient pressure is not uniform throughout the entire surface of the monolayer [16] the density of nuclei is also not uniform, as shown in Fig. 3. There is one additional feature which is worth noticing at this stage since we will not elaborate further on it. As the mean distance between the circular liquid condensed domains becomes much smaller than their diameter, dark bridges form, which connect neighboring domains at their closest distances of approach. The origin of these "whiskers" is still unknown. It can be a manifestation of the electrostatic repulsion between the liquid condensed domains which had been pointed out earlier by McConnell et al. [17] and Möhwald et al. [18]. The dark domains do not want to grow above a certain size. They minimize the total free energy of the

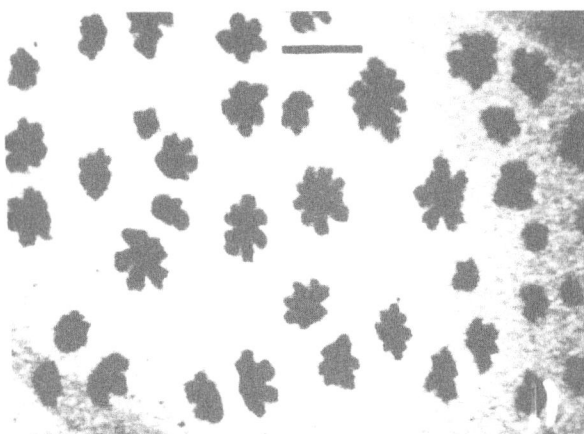

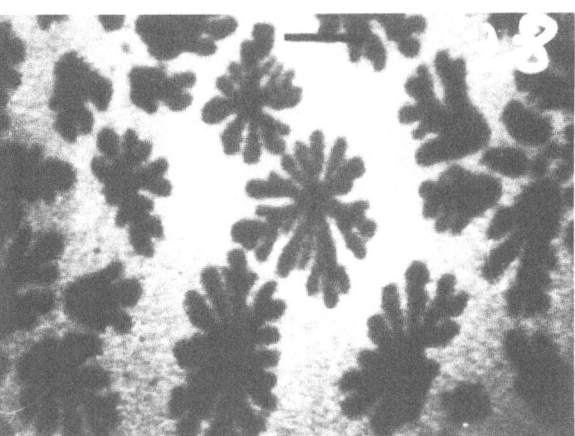

Fig. 4. Same as Fig. 2 except temperature which is 28 °C. The liquid condensed domains have highly carved, irregular, structures. Mean surface density is 25 Å² per molecule

Fig. 5. Same as Fig. 4 except mean surface density which is 21 Å² per molecule. Compression rate was 0.1 Å² s⁻¹ per molecule. Note that the fingered domains of Fig. 4 have grown by tip-slipping

system by interconnecting themselves with narrow bridges. This behavior is not particular of myristic acid monolayers and we have also observed it with pentadecanoic acid [19]. Detailed studies of this unexpected phenomenon are currently under way.

2. Path a: Temperature close to T_{c2}

Pattern formation is radically different if the compression experiments are performed at a temperature of 28 °C i.e., much closer to the critical temperature $T_{c2} = 31$ °C. The liquid condensed domains are now highly carved structures with marked protuberances (Fig. 4). As the growth proceeds, the domains exhibit a characteristic arborescent, tree-like, pattern, with radial symmetry around the initial nucleus. The number of branches increases with the domain size. As can be seen from a comparison of Fig. 4 and Fig. 5, new branches are generated by successive irregular splitting of the leading finger tips. This is different from crystal growth where dendrites show frequently stable-tips with regular side branches [20]. It has been pointed out by Goldenfeld and coworkers that large anisotropy is of central importance for the occurrence of dendritic growth [21]. Our observation of tip-splitting can therefore be taken as an indication that the liquid condensed phase is not ordered, at least over macroscopic distances. The liquid condensed domains will grow until they start to impede each other. It is observable in Fig. 5 that if two initial nuclei were too close

initially, the arborescent structures only develop on the opposite sides. The nearest sides, on the contrary, do not form unstable fronts. Figure 6 shows a monolayer which consists of a dense array of arborescent patterns. There is no interpenetration of the fingers belonging to two different nuclei. The few white dots observed in the image are dye crystals which have phase-separated from the monolayer. As the dye molecules get gradually expelled from the liquid condensed domains, the dye concentration in the bright liquid expanded continuous phase becomes so large than it exceeds the limit of solubility and crystallization occurs.

In Figs. 4 and 5, the finger width is about 10–20 μm. This width will stay constant as long as a finite compression rate is applied and the domains continue to divide and branch out. If the compression is stopped, the growth ceases immediately and the fingers thicken. The arborescent structures relax to a circular shape over a time period of several minutes. The actual time depends on the temperature of the experiment: it is shorter the further away the temperature is from T_{c2}.

It is possible to find experimentally situations where a cluster can grow to a large size while staying fairly isolated from its neighbors. Such a case, shown in Fig. 7, permits a quantitative analysis of the structure. It was chosen to digitize the pattern and to then measure the density-density correlation function $C(R)$. In this method, each point inside the cluster is taken as the center of a circle with radius R. Then the number of cluster sites which lay on the circle perimeter are

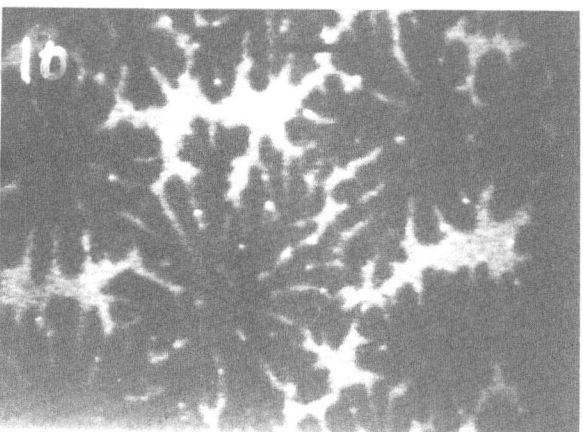

Fig. 6. Aspect of the monolayer following continuous compression across the entire liquid expanded – liquid condensed coexistence region $T = 22\,°C$. The fingered domains do not fuse but stay separated by "white alleys". The bright dots are small dye crystallites which have phase-separated from the monolayer

Fig. 7. Typical fingered pattern for liquid condensed domains at 25° − 30 °C. Dye concentration is 0.1 %

counted. If the structure is self-similar, this number of cluster sites, divided by the total number of lattice points in the cluster for normalization purposes, should scale as $R^{d_f - d}$ where d_f is the so-called Hausdorff dimension and d the Euclidean dimension [14]. The result of the calculation is displayed in Fig. 8. On log-log scales, one observes a linear variation of $C(R)$ over slightly more than one decade in R. The slope is found to be -0.19, yielding $d_f = 1.81$. We estimate the accuracy to be ± 0.1. At larger R, the precipitous drop in $C(R)$ expresses the fact that regions outside the cluster start to be probed and the calculation becomes of course meaningless. On the whole, it seems that the patterns observed close to the critical temperature T_{c2} are self-similar and can be characterized by a fractal geometrical exponent over a reasonable spatial range.

3. Path b: Intermediate temperature range

When the compression experiment is performed at a temperature of 20 °C, the liquid condensed domains have a pattern which is neither totally spherical nor markedly tree-like. The largest domains exhibit a few fat branches while the smallest ones display irregular, noncharacteristic, shapes, as can be seen in Fig. 9. If the commpression is stopped, the structures relax quickly to circular shape. It is obvious that line tension is an important restoring force and drives the system towards compact structures. Path b is clearly an inter-

mediate case between path a and path c. Absolute temperature values however have a physical meaning only if the compression rate is specified. Situations corresponding to path a are observed down to lower and

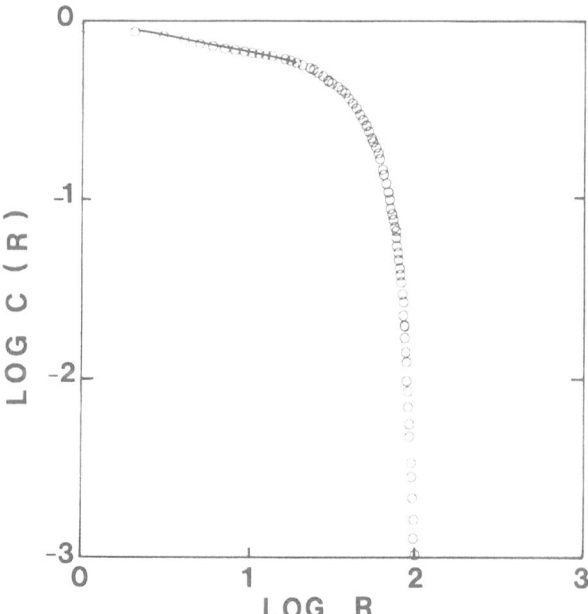

Fig. 8. Plot of the density-density correlation function $C(R)$ as a function of the spatial distance R for the central fingered pattern shown in Fig. 7. The linear variation at low R values indicates that $C(R) \propto R^{d_f - d}$ with $d_f = 1.81 \pm 0.10$. d_f is the Hausdorff dimension and d the Euclidean dimension

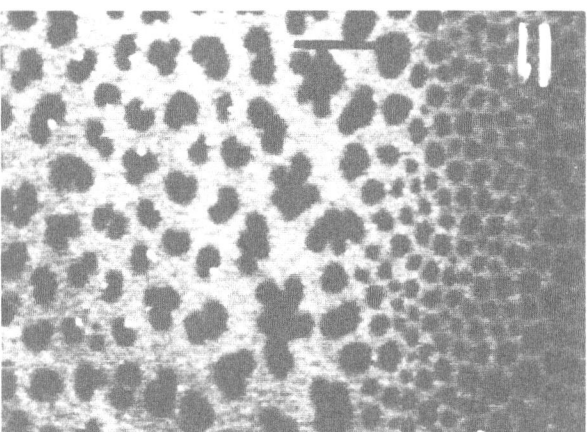

Fig. 9. Liquid condensed domains at $T = 20\,°C$. Note the irregular, non-characteristic shape contrary to the previous figures

lower temperatures as the compression rate is increased. Indeed it is known from previous studies on viscous fingering [13] and solidification [22] that high capillary numbers (ratio of driving forces over surface tension) are essential to obtain highly branched interfaces.

Discussion

The observed patterns are strikingly similar to the viscous fingers observed when a low viscosity fluid displaces a high viscosity fluid in a Hele-Shaw cell [13]. In both cases the growth occurs by successive random splitting of the leading finger tips. The end result is a highly branched cluster whose density decreases with time. These clusters are fractals and can be characterized by their Hausdorff dimension. The published value is 1.70 ± 0.05 for viscous fingering in a radial cell (in order to avoid boundary effects), to be compared to 1.80 ± 0.10 in the present case. We can recall here that the theoretical value of the Witten-Sander computer simulation of diffusion-limited aggregation is 1.68 [14]. In both viscous fingering and growth of liquid condensed domains the basic reason why highly branched clusters can be observed is that interfacial tension has been minimized. In the viscous fingering experiments of Daccord et al. [13], miscible fluids were selected, namely aqueous polymer solutions of schleroglucan, hydroethylcellulose, guar derivatives, were displaced by point injection of pure water under large hydrostatic pressure. This was underlined as the

necessary condition for enabling any small fluctuation of the interface to grow. In the growth of liquid condensed domains by pressure quenching, we have worked in the vicinity of the critical point T_{c2} for the liquid expanded – liquid condensed coexistence curve. It is well known that the line tension σ at T_{c2} must vanish since the two phases become undistinguishable at T_{c2}. Changing the temperature of the experiment allows one continuously to tune the value of σ since $\sigma \sim |T - T_{c2}|^{\mu}$ with μ of the order of unity in two dimensions [23].

Perhaps it should be made clear that a non-zero surface tension might not necessarily prevent fractal growth. It will merely introduce a length scale below which the growth is certainly not fractal. Chuoke et al. [24] have shown that during the displacement of immiscible fluids only perturbations with a wave-length λ greater than the critical wave-length $\lambda_c \sim b \left(\dfrac{\sigma}{\eta v} \right)^{1/2}$ will develop where η is the fluid viscosity, v the flow velocity and b the plate spacing in the Hele-Shaw cell. An analogous result has been recently proposed by Paterson for miscible fluids [25]. The only difference is that λ_c should not depend on flow rate. Daccord et al. [13] have indeed shown that the finger width, which should be proportional to λ_c, varies linearly with b. In our experiments of liquid condensed domains growth, we observed qualitatively that, for a given compression rate, the finger width gets larger as the temperature is decreased (and therefore σ increases). Detailed measurements are currently under way. This notion of a critical wave-length may also explain why small clusters, as obtained during the first stage of the nucleation and growth mechanism, are not fractal. Experimentally, we observe that isolated clusters do not show protuberance below a critical radius of the order of 20–30 μm at 28\,°C. (We insist on the word "isolated" because we have already discussed the fact that the presence of nearby clusters impedes fingering growth).

So far, we have not discussed the mechanism by which the interfacial front between the liquid expanded continuous phase and the growing liquid condensed domains become unstable to tip-splitting. In the literature, two separate mechanisms have been discussed, one in the case of pure compounds and one in the case of alloys. The now conventional thermodynamic model of the solidification of a pure substance from its melt is due to Mullins and Sekerka [12]. They were the first to quantitatively show that the fundamental rate-controlling mechanism is the diffusion of latent heat away from the interface. Indeed, the latent heat released in the solidification process must be removed before further solidification can take place. This removal is more effective around spikes (or protuberances) of the front. This is a morphologically unstable process and a tip and fjord structure will grow preferentially. In three-dimensional systems, the necessity to evacuate the heat generated by the condensation is the dominant effect. However in Langmuir monolayers, the liquid subphase provides a very efficient thermal sink and we can consider the monolayer to be always at thermal equilibrium, with no temperature gradients in the plane of the monolayer. As a consequence, the diffusion of latent heat is not a likely mechanism in the present experiments.

In their seminal paper, Mullins and Sekerka also pointed out that a similar unstability would develop for binary systems during the diffusion-controlled growth of nearly spherical precipitate particle from a supersaturated matrix initially at a spatially uniform concentration. This mechanism is now commonly referred to as constitutional supercooling [26]. It relies on the fact that the solubility of component A into component B is lower in the solid phase than in the liquid phase. As soon as a solid nucleus forms, molecules of the A-type are expelled from the precipitate and tend to diffuse away into the bulk liquid phase. However, since diffusion is not instantaneous, there will be a local concentration increase. This concentration increase forces further solidification to occur at lower temperatures or equivalently at higher external pressure, by a classical supercooling effect. It will eventually relax by mass diffusion at a rate constant which is proportional to the local concentration gradient. If noise-generated protuberances randomly appear on the solidification front, the isoconcentrate lines will bunch together whereas they will be rarefied above the depressions (or fjords). Since the gradient is steeper at protuberances, they will eventually grow unstable because supercooling will be minimized locally. The conditions for the onset of the unstability have been worked out analytically by Mullins and Sekerka, but it was not possible for them to calculate the detailed shape of the front. This was done many years later for dendritic growth by Langer and Müller-Krumbhaar [22], [27]. Even more recently, a discrete counterpart of the dendritic growth model was proposed by Witten and Sander [14]. Using computer simulations, they were able to calculate the geometrical properties of aggregates formed by diffusion of random walkers toward a central seed particle. They were in particular the first to show that the clusters formed by diffusion-limited aggregation were self-similar objects, with scale-independent correlations over a large range of distances.

Since our experiments have been performed on a binary system, i. e., myristic acid molecules mixed with a small amount of fluorescent dye, it is natural to attribute the observation of unstable growth fronts to constitutional supercooling effects. This mechanism has first been invoked by Miller et al. [28] to explain similar patterns in monolayers of a two-chain phospholipid (L-α-dimyristoyl phosphatidyl ethanolamine; DMPE for short) labelled with a porphyrin dye (tetra-3-eicosyl-pyridinium porphyrin bromide; TPyP for short). The prerequisite for constitutional supercooling is the existence of a large excess of dye molecules at the interface between the liquid condensed domains and the liquid expanded continuous phase. In Miller's experiments, this local excess showed up as a bright halo surrounding the liquid condensed domains. Moreover separate surface pressure isotherm measurements showed that the pressure inside the liqid expanded — liquid condensed coexistence region was strongly dependent on the dye concentration. For instance the transition pressure π_c was observed to increase by as much as 1 dyne $\cdot$ cm^{-1} per mole % of fluorescent impurity. Despite an apparent similarity, the present experiments however suggest than it is not necessary to invoke the presence of impurities to explain the unstable growth. A first indication is given by the fact that we do not observe bright halos in any of our images (see Figs. 4–7). Second, the transition pressure dependence on dye content is much smaller, of the order of 0.17 dyne $\cdot$ cm^{-1} per mole % in our case. This is quite consistent with the fact that the molecular structures of NBD-HDA and myristic acid are very similar, contrary to the DMPE-TPyP system. Third, tip-splitting phenomena have been detected down to extremely low dye concentrations, of the order of 0.01 mole %. There seems actually to be no lower limit to the dye concentration for the observation of the patterns, except that imposed by the sensitivity level of our optical detection. Fourth, there are reports in the literature in which unstable growth patterns have been detected with observation techniques such as phase contrast electron microscopy [29] and surface plasmon imaging [30] which do not require the addition of marker molecules.

In view of all these facts, we have tried to find another diffusion mechanism which could be applicable to monolayers of surface-active molecules spread at the air-water interface. The key of the understanding is that we are not dealing with a one component system – the myristic acid molecules – but with a two-component system – the myristic acid molecules plus the water molecules. To transform a liquid expanded region of the monolayer into a liquid condensed domain, it is necessary to displace the bulk water molecules initially located in the interfacial region. We are thus dealing with a mass diffusion process for binary mixtures very similar to the one described by Mullins and Sekerka [12] and Langer and Müller-Krumbhaar [27]. For Langmuir monolayers there are however differences in the quantitative growth description which have been pointed out by De Gennes [31]. In particular, molecular diffusion in the plane of the monolayer induces a viscous dissipation in the fluid subphase and the diffusion coefficient depends on the diffusion length. This changes drastically the dynamics of the system. For instance, the radius of a liquid droplet nucleating in a gaseous phase should vary with time as $t^{1/2}$, instead of $t^{1/3}$ in the absence of hydrodynamic effects. At this early stage however these subtle effects can be neglected and it is sufficient to say that unstable growth in one-component monolayers spread on a liquid subphase is akin to three-dimensional growth in alloys.

Apart from the present work and Miller's work, we know of at least two other experiments in the literature which deal with diffusion-limited aggregation in two dimensions. Hurd and Schaefer [32] have studied the aggregation of silica microspheres deposited at an air-water interface and obtained stringy structures. They surmised that unexpected anisotropic repulsive electrostatic interactions were forcing the particles to aggregate in a bead-necklace mode. The measured Hausdorff exponent had a very low 1.25 value. On the other hand, Beysens and Knöbler [33] have studied the growth of breath figures, formed by condensing supersaturated water vapor onto cold hydrophobic surfaces. To explain the growth rate of the liquid droplets, they propose a mechanism in which embryos of critical radius first condense on the glass and then diffuse to the growing droplet. However, the surface tension is large and imposes the drops to stay spherical at all times. No arborescent structures have been observed.

The advantage of the present experiment is that the interaction forces are purely isotropic and that the surface tension can be continuously tuned down to a zero value. Consequently, the experimental situation is close to the theoretical conditions. It is therefore satisfying to obtian ramified structures which appear to be self-similar over a reasonable spatial range and have an Hausdorff exponent of 1.80 ± 0.1 in fair agreement with the predicted value of 1.7. Quantitative experiments on the growth rate and dynamics of the structures are in progress.

Conclusions

We are fully aware that the present experiments leave many questions unanswered. More experiments will be required in order to understand the physical mechanism for the "bridges" observed during the compression of the circular liquid condensed domains. Similarly, the conditions for the appearance of the fingered patterns will have to be quantified in terms of compression rate, temperature range, surface pressure values. However, it remains that it is the first time, to the best of our knowledge, that the surface tension of the two coexisting phases during a nucleation and growth process can be continuously varied [34]. As a result of this fine tuning, we can observe the change from a stable front (at high line tension) to an unstable front (at low line tension). The instability is governed by a diffusion-limited process, as predicted by Mullins and Sekerka in their original paper. We believe that the thermal effects due to the generation of latent heat during the phase transformation of pure substances play no role in our experiments. Our system is therefore original. Its novelty is due to 1) the choice of a dye with a molecular structure close to that of the matrix, therefore reducing the ordinary impurity effects, and 2) the use of Langmuir monolayers floating on a large aqueous subphase which provides an efficient thermal bath and kills temperature gradients in the plane of the Langmuir films.

Acknowledgements

We gratefully acknowledge helpful discussions with F. Brochard, C.M. Knobler, C. Caroli, and P. G. de Gennes. We also thank G. Daccord for the writing of the algorithm leading to the calculation of the correlation function shown in Fig. 8.

References

1. Langmuir I (1917) J Am Chem Soc 39:1848
2. Gaines GL (1966) Insoluble Monolayers at Liquid-Gas Interfaces. Interscience Publishers, Wiley, New York
3. Nagle JF (1980) Ann Rev Phys Chem 31:157
4. Hawkins GA, Benedek GB (1974) Phys Rev Lett 32:524; Kim MW, Cannell DS (1976) Phys Rev A13:411; Pallas NR, Pethica BA (1986) The liquid-vapour transition in monolayers of n-

pentadecanoic acid at the air-water interface. preprint

5. Middleton SR, Iwahashi M, Pallas NR, Pethica BA (1984) Proc Roy Soc Lond A396:143
6. Rasing Th, Shen YR, Kim MW, Grubb S (1985) Phys Rev Lett 55:2903
7. Dutta P, Peng JB, Lin B, Ketterson JB, Prakash M, Georgopoulos P, Ehrlich S (1987) Phys Rev Lett 58:2228
8. Kjaer K, Als-Nielsen J, Helm CA, Laxhuber LA, Möhwald H (1987) Phys Rev Lett 58:2224
9. von Tscharner V, McConnell HH (1981) Biophys J 36:409,421
10. Lösche M, Sackmann E, Möhwald H (1983) Ber Bunsenges Phys Chem 87:848
11. Moore B, Knobler CM, Broseta D, Rondelez F (1986) J Chem Soc Faraday Trans 2, 82:1753
12. Mullins WW, Sekerka RF (1963) J Appl Phys 34:323
13. Daccord G, Nittmann J, Stanley HE (1986) Phys Rev Lett 56:336
14. Witten TA, Sander LM (1981) Phys Rev Lett 47:1400
15. Gunton JD, San Miguel M, Sahni PS (1983) In: Domb C, Lebowitz J (eds) Phase transitions and critical phenomena. Vol 8
16. Dimitrov DS, Panaiotov II, Richmond P, Ter-Minassian Saraga L (1978) J Colloid Interface Sci 65:483
17. McConnell HM, Tamm LK, Weis RM (1984) Proc Natl Acad Sci, USA 81:3249
18. Fischer A, Lösche M, Möhwald H, Sackmann E (1984) J Physique Lett 45:L-785
19. Rondelez F, Baret JF, Suresh KA, Knobler CM (1988) In: Velarde M (ed) Proceedings of the 2nd international conference on Physico-Chemical-Hydrodynamics, NATO ASI Series 174:857, Plenum Press, New York
20. Honjo H, Ohta S, Sawada Y (1985) Phys Rev Lett 55:841
21. Martin O, Goldenfeld N (1987) Phys Rev 35A:1382
22. Langer JS (1980) Rev Mod Phys 52:1
23. Widom B (1976) In: Domb C, Green MS (eds) Phase transition and critical phenomena. Academic Press, New York, Vol 2, p 79
24. Chuoke RL, van Meurs P, van der Poel CJ (1959) J Petrol Tech 11:64
25. Paterson L (1985) Phys Fluids 28:26
26. Woodruff DP (1973) The solid-liquid interface. Cambridge Solid-state Sci Series, Cambridge University Press, Cambridge, p 80
27. Langer JS, Müller-Krumbhaar H (1977) J Crystal Growth 42:11; and Acta Met 26:1681; Langer JS (1986) For a recent review of the literature, Physica 140A:44
28. Miller A, Knoll W, Möhwald H (1986) Phys Rev Lett 56:2633
29. Fischer A, Sackmann E (1986) J Coll Interf Sci 112:1
30. Knoll W, private communication; Rothenhäusler B, Knoll W (1988) Nature 332:615
31. De Gennes PG (1985) Compt Rend Acad Sci, Paris 300:831
32. Hurd AJ, Schaefer DW (1985) Phys Rev Lett 54:1043
33. Beysens D, Knobler CM (1986) Phys Rev Lett 57:1433
34. A brief account of the same experiments can also be found in Suresh KA, Nittmann J, Rondelez F (1988) Europhys Lett 6:437

Received March 1989;
accepted April, 1989

Authors' address:

F. Rondelez
Université Pierre et Marie Curie
Laboratoire "Structure et Réactivité aux Interfaces"
4 Place Jussieu
F-75231 Paris Cedex 05, France

Progress in Colloid & Polymer Science

Progr Colloid Polym Sci 79:194–201 (1989)

Experimental and theoretical aspects on cluster size distribution of latex particles flocculating in presence of electrolytes and water soluble polymers

R. Varoqui and E. Pefferkorn

Institut Charles Sadron, Strasbourg, France

Abstract: The size distribution of negatively charged polystyrene latex particles flocculated in the presence of 0.15 M NaCl or in the presence of poly(4-vinylpyridine) was measured using an automatic particle counter. The time evolutions of the size distribution are well described at large time by the formulas:

$$c(g, t) = t^{-2z} \psi(gt^{-z})$$

$$N(t) \sim t^{-z},$$

$c(g, t)$ being the number of flocs containing g primary colloids at time t, and $N(t)$ is the total number of flocs of any size, z is a scaling exponent and the function ψ does not depend explicitly on time. These laws are in agreement with the theoretical predictions based either on Smoluchovski's equation assuming a dynamic scaling argument, or on Monte-Carlo simulations on a three-dimensional lattice. If the flocculation occurs in the presence of an excess of electrolyte, z is equal to 1; however if P4VP is the flocculating agent, the value of z is related to the polymer concentration. The kinetics are discussed on the basis of the structure of the collision frequency in Smoluchovski's equations:

$$K(g, n) = \frac{kT}{3\eta} (R_g + R_n)(D_g + D_n).$$

$K(g, n)$ defines a rate constant. The total number of collisions of g- and n-sized flocs is $K(g, n) C_g C_n, C_g C_n$ being the number of these flocs per unit volume. In this equation R_g, R_n and D_g, D_n are, respectively, the radius of gyration, and the diffusion coefficients of g- and n-sized flocs. This simple expression holds well for an excess of electrolyte situation or at a polymer concentration were flocculation proceeds at a fast rate; Expression of $K(g, n)$ ensures $z = 1$. At low and large polymer concentrations, we have $z < 1$, which is interpreted on the basis of a model of the colloid interface in the presence of adsorbed polymer.

Key words: Polystyrene latex, water soluble polymers, poly(4-vinylpyridine), flocculation, size distribution, scaling exponents.

Introduction

There is a considerable interest, both practical and fundamental, in the mechanism of polymers on the destabilization of colloidal dispersions. The flocculation of colloids is thought to occur by bonding of different particles by parts of the polymer coil [1–3]. In the present study, we report results on the flocculation of charged polystyrene latex particles in presence of poly(4-vinylpyridine). We shall yield information on

the particle size distribution at different time intervals in the flocculation process. Recently, considerable theoretical progress has been achieved in understanding the dynamics of cluster-cluster aggregation and we shall interpret the time-dependent cluster size distribution with the aid of the new scaling theory [4, 5]. The kinetic process of the colloid aggregation in presence of electrolyte is also reported and the size distribution is compared to the one recorded in the presence of polymer. From the theory and experiments a fairly comprehensive model of the structure of colloid/polymer interface at different polymer concentration emerges.

Materials and methods

Latex particles

Latex of spherical shape and narrow size distribution was polymerized under emulsifier-free conditions by using potassium persulfate as free radical initiator. It was converted to the Na^+-form by ion-exchange. Dimensions were the following: $\bar{D}_v = 840$ nm ($\bar{D}_v$ is the average Stockes diameter obtained from the determination of the diffusion coefficient by quasi-elastic light scattering); $\bar{D}_n = 860$ nm and $\bar{D}_w = 866$ nm ($\bar{D}_n$ and $\bar{D}_w$ are, respectively, the number and the weight-averaged mean diameter, both calculated from microscopic observation). The latex as provided differed in the nature of the surface charged chemical groups. The origin of the charge at the surface of the latex comes from the sulfate surface groups SO_4^-. However, during the polymerization oxydation of SO_4^- occurs and during aging hydrolysis of SO_4^- occurs as well, so that the surface of the latex bears carboxylate COO^- and hydroxyl OH groups.

Polymer

Poly(4-vinylpyridine) (P4VP) of average molecular weight $M_w = 3.4 \times 10^5$ was prepared by free-radical polymerization in methanol, using α,α'-azobis(isobutyronitrile) as a free radical initiator. Purification proceeded through precipitation of the reaction mixture in water. After redissolution in a 1:1 dioxane/water mixture, the polymer was freeze-dried and used without fractionation.

Flocculation experiments

The latex suspension was ultrasonicated prior to use; flocculation was started by adding to a polymer solution (5 cm^3), adjusted to pH 3.5, a latex suspension (20 cm^3) at identical pH, so that the final composition in latex was $1.6 \ 10^{-2}$ wt %. As soon as the latex suspension was added to the polymer solution in a 30 cm^3-measuring cylinder, the cylinder was closed and its content mixed by gentle inversion, repeated several times. The dispersion was then allowed to stand at a temperature of 18 °C, with a fixed latex concentration of $1.6 \ 10^{-2}$ wt %. The flocculation experiments in presence of electrolyte were carried out by adding aqueous NaCl solution at pH 3.5 and 18 °C to a stable suspension at the same pH, so that the final composition was again 1.6×10^{-2} wt % in 0.15 M NaCl. Taking into account the dimensions and the surface charge of the colloid, 0.15 M NaCl corresponds to a situation of excess electrolyte which ensures diffusion-limited strong aggregation.

Size distribution determination

Small samples of approximately 1 ml of suspension were removed at intervals with a 3 mm-bore needle at a low rate corresponding to a shear of 10 s^{-1}, diluted 10^2 times with a 0.15 M NaCl aqueous solution and analyzed by the Coulter technique to obtain the number $c(g, t)$ of aggregates comprising a number g of associated primary particles at time t. By the dilution, the flocculation is stopped and the occurence of coincidence (simultaneous passage of two or more aggregates through the aperture of the cell) becomes negligibly small. The 16-channel TA II Coulter Counter with a variable threshold adapter and an aperture of 50 μm was used. All technical information concerning the determination of the distribution curve by the Coulter Counter can be found in [6]. The use of the Coulter Counter to study the aggregation kinetics of dispersions is described in a significant number of reports [7–9]. This technique has been shown to be an accurate and reproducible method for following the coagulation kinetics of polymer latices [10].

Electrophoretic mobility

The electrophoretic mobility of the latex in presence of the polymer was measured in 10^{-2} M NaCl aqueous suspensions with a Rank Brother Mark II apparatus.

Theory

The classical understanding of aggregation kinetics is given by the Smoluchovski theory [11], according to the assumptions that the collisions are binary and the fluctuations in density are sufficiently small to induce random collisions. The increase per unit time of the number of clusters of size g is given by the following equation:

$$\frac{\partial c(g,t)}{\partial t} = (1/2)\int_0^g K(g-n,n)\, c(g-n,t)\, c(n,t)\, dn$$
$$- c(g,t)\int_0^\infty c(n,t)\, K(g,n)\, dn. \qquad (1)$$

In the righthand side term, the first integral represents the gain in g-sized clusters due to collisions between $g-n$ and n-sized clusters, while the second integral represents the loss of g-sized clusters resulting from collisions between g- and any sized clusters. The rate constant K has been assumed constant by Smoluchovski, and the solution at large time takes the following form [4]:

$$c(g, t) = \frac{16\, t^{-2}}{K^2 N_0} \exp\left(-\frac{4g}{KN_0 t}\right), \qquad (2)$$

N_0 being the number of the total primary colloids. More recently, the dynamics of the diffusion limited model of cluster-cluster aggregation was carried out by Monte-Carlo simulations in which the species move with a size-dependent diffusivity [12]. The following scaling relationship was found:

$$c(g, t) \sim t^{-\omega}\, g^{-\tau}\, \psi(gt^{-z}). \qquad (3)$$

The scaling exponents ω, τ and z are related:

$$\omega = (2 - \tau)\, z \tag{4}$$

and the cut-off function $\psi(x) \simeq 1$ for $x \ll 1$ and $\psi(x) \ll 1$ for $x \gg 1$. For mass independent cluster diffusion coefficient, τ was found equal to 1.3 and ω was found equal to 2.1 – 2.3. In a more realistic model, the diffusion coefficient is choosen as:

$$D_g \sim g^y. \tag{5}$$

It was shown that for $y < 0.5$, τ is zero, and Eq. (3) then becomes:

$$c(g,\, t) \sim \tau^{-2z}\, \psi(g t^{-z}). \tag{6}$$

Moreover, z was found equal to 1 if $y = 1/D$, D being the Hausdorf dimension of the cluster which have a fractal geometry ($D = 1.75$–1.8 for three-dimensional cluster-cluster aggregation [13, 14]). Equation (6) is exactly the result derived by Swift and Friedlander [15] and Lushnikov [16], who used the following scaling transformations:

$$c(g/g_0,\, t) = g_0^2\, c(g, g_0^{1-\lambda} t) \tag{7}$$

$$K\left[g_0\,(g - n),\, g_0\, n\right] = g_0^\lambda\, K\left[g - n,\, n\right] \tag{8}$$

g_0 is any constant greater than 1, and the kernel K is homogeneous of degree λ. Equation (7) embodies the idea of a "self preserved" distribution put forward by Friedlander [15]. As shown by Lushnikov [16], the distribution then takes the form (6) with $z = 1/1 - \lambda$. Assuming a kernel K of the form:

$$K(g,\, n) = a(g^\nu + n^\nu)(g^\gamma + n^\gamma), \tag{9}$$

a being a constant, and supposing the diffusion coefficient $D_g \sim g^y$ to be inversely proportional to the radius of gyration, i. e., $\nu = -y = 1/D$, then λ is equal to 0 and z is equal to 1. Moreover, since the total number of clusters which will be denominated $N(t)$ scales like t^{-z}, we are lead to the important theoretical relationships:

$$c(g,t)N^{-2}(t) = \psi[g\, N(t)] \tag{10}$$

$$N(t) \sim t^{-1}. \tag{11}$$

We tested Eqs. (10), (11), and in the following we shall show that under some simple experimental conditions, Eqs. (10), (11) are indeed well satisfied. However, in general, although Eq. (10) still holds, the scaling exponent of $N(t)$ is not equal to 1, which signifies that one cannot expect K to be given by the simple form (9).

Results and discussion

In Figs. 1 and 2 are reported the particle size distribution at different times, respectively, in presence of NaCl and P4VP. In Fig. 3 is reported $Ln[c(g, t)N^{-2}(t)]$ as a function of $Ln[gN(t)]$ for $t > 85$ min for the flocculation of the latex in presence of 0.15 M NaCl. All the data fall well on a single curve as predicted by Eq. (10). The same representation holds also when the flocculation is carried out using P4VP as flocculating agent. This is seen in Fig. 4 where curves are represented for two polymer concentrations of $12.5\ 10^{-4}$ g and $5\ 10^{-4}$ g

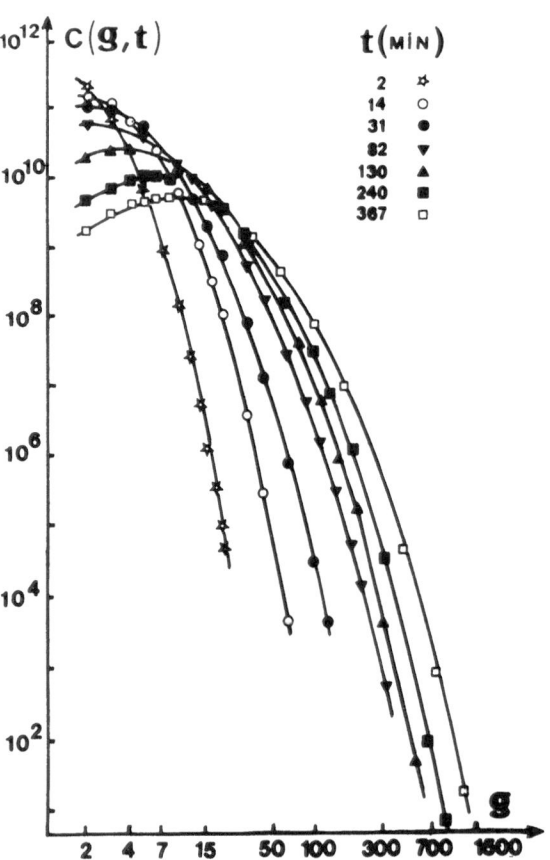

Fig. 1. Particle size distribution at different times for the latex flocculated with 0.15 M NaCl

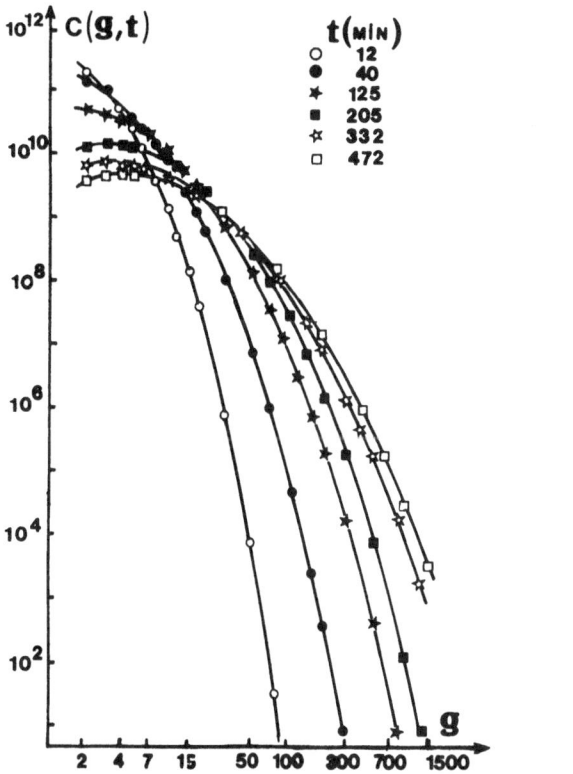

Fig. 2. Particle size distribution at different times for the latex flocculated with P4VP at pH 3.5, and polymer concentration of 1.25 10^{-3} g per g of latex

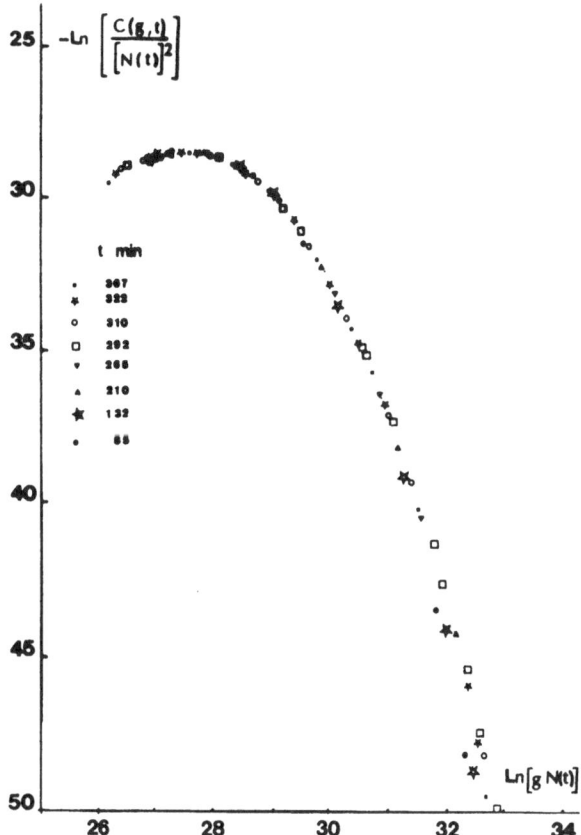

Fig. 3. Size distribution represented according to Eq. (10) for $t \geqslant 85$ min, for the flocculation of the latex in the presence of 0.15 M NaCl

per g of latex. It has been emphasized that the regime predicted by Eq. (10) is an asymptotic regime, i. e., it holds only after a certain time lag which is approximatively 150 min for the case (a) reported in Fig. 4. This time is needed to reach a situation where the distribution can be represented by a smooth function and as the self-similar morphology (embodied in Eq. (9) with the dimensions scaling like g^{ν}) on account of the finite dimensions of the primary particles, is only reached after a certain time lag.

The dependence of $\mathrm{Ln}[N(t)]$ on $\mathrm{Ln}\, t$ is represented in Fig. 5. After a time period which is of the order of 30 min, the data can be fitted by a straight line with a slope of -1, in agreement with Eq. (11). The polymer concentrations for which this behavior is verified correspond to maximum flocculation efficiency, which, as shown in Fig. 6, is in a concentration range of 10^{-3} to $2\, 10^{-3}$ g polymer per g latex. In Fig. 6, the fraction φ' of flocculated colloid:

$$\varphi' = \sum_{g=2} g\, c(g, t) \Big/ \sum_{g=1} g\, c(g, t) \qquad (12)$$

exhibits a typical broad maximum ($\varphi'_{\mathrm{MAX}} = 1$), which was also observed in the analysis of the turbidity of the supernatant or from the sedimentation mass, when the flocculation was pursued up to macroscopic phase separation (in the present investigation, the largest flocs have at maximum about 1 400 associated primary particles, and the suspension remains homogeneous without phase separation during the flocculation period).

A more systematic study (not reported here), was undertaken recently to characterize the flocculation process when $\varphi' < 1$ values were recorded at small and large polymer concentrations. The conclusions are the following: the representation in terms of the reduced variable $gN(t)$ and $c(g,t)N^{-2}(t)$ is valid at all concentrations provided large time values as considered (cf., for instance, to curve (b) in Fig. 4). However, one important difference emerges with respect to the value of the scaling coefficient z of $N(t)$ which becomes smaller than unity (cf. to polymer concentration of

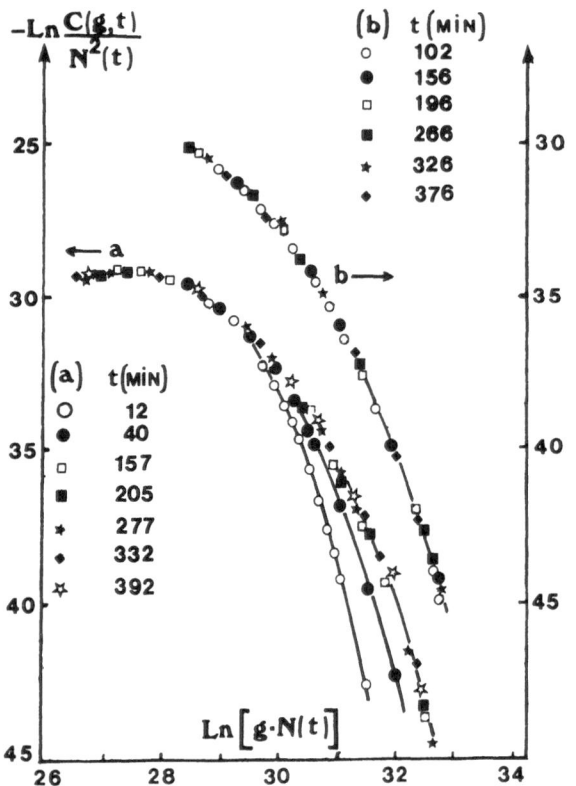

Fig. 4. Size distribution represented according to Eq. (10) for the flocculation of the latex at pH 3.5, for polymer concentration of 12.5 10⁻⁴ g (a) and 5 10⁻⁴ g (b) per g of latex

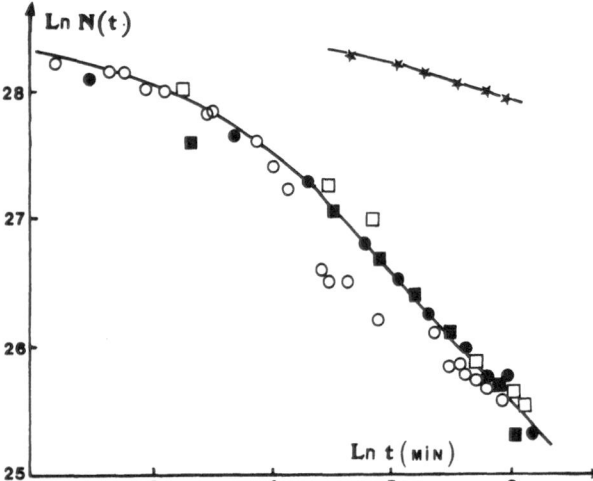

Fig. 5. Total number of clusters represented as a function of time. Flocculation in presence of different polymer concentrations (g polymer/g latex) at pH 3.5: (★) 5 10⁻⁴; (□) 10⁻³; (●) 12.5 10⁻⁴; (■) 17.5 10⁻⁴; (○) flocculation in presence of 0.15 M NaCl at pH 3.5

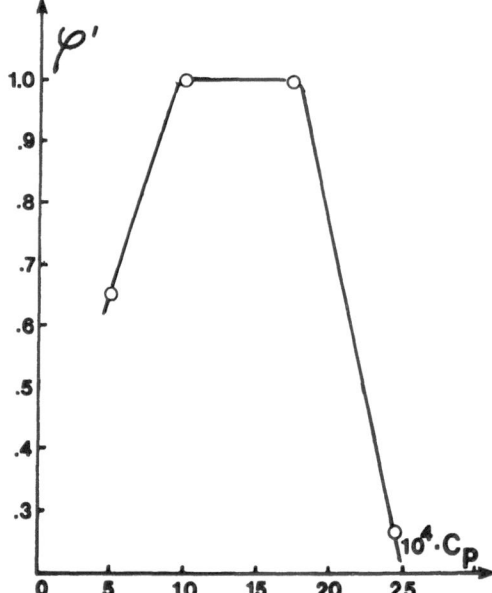

Fig. 6. Volume fraction φ' of associated particles as a function of the polymer concentration at a flocculation time of 400 min

5 10⁻⁴ g reported in Fig. 5). This trend seems to be general. Referring to the previous theoretical discussion, $z < 1$ signifies that λ becomes different from zero. Equation (9), which represents the kernel K as an homogeneous function of degree zero, does therefore not apply at low and high polymer concentrations. In order to explain this, we have to return to the significance of the maximum reported in Fig. 6, which was analyzed repeatedly using simple models in other reports [17–19].

Adsorbed polymers are viewed as a succession of monomer trains in the attractive zone of the surface with loops extending into the solution phase over distances of the order of several hundred angstroems. At low polymer concentrations, the amount of polymer per unit area of latex surface is small and the polymer can be viewed as a "flat carpet" with respect to the Debye-Hückel length $\varkappa^{-1}$, which represents the range of the electrostatic repulsions between the charged latices, (Fig. 7, situation (i)). Bridging of colloids by polymer loops is therefore unlikely. On the other hand, at large polymer concentrations the surface is thickly, densely populated; one therefore expects $L > \varkappa^{-1}$ (Fig. 7, situation (ii)). However, the dense polymer layers refuse to interpenetrate each other because of the classical monomer-monomer excluded volume effect [20]. For the maximum of flocculation, one

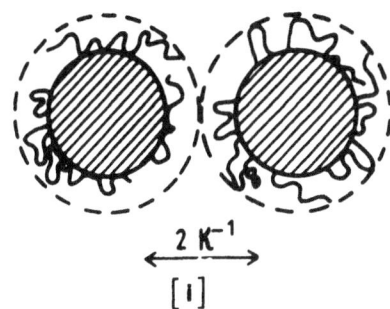

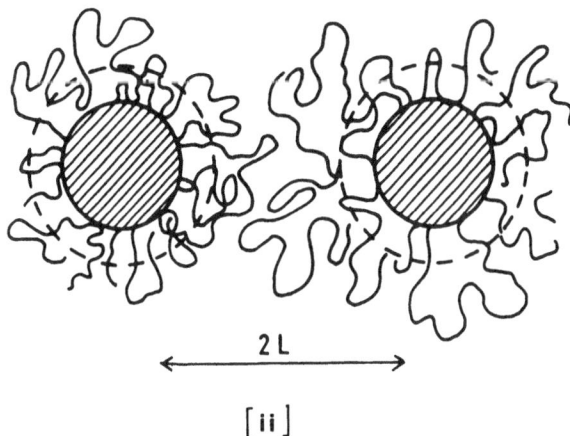

Fig. 7. Schematic representation of the colloid surface in presence of noncharged polymer at respectively low and large polymer concentration

expects the loop layer to be roughly equal to the Debye-Hückel length. At the corresponding polymer concentration, a large fraction of binary collisions leads to efficient binding and one expects therefore the

form (8) to be correct, with $\lambda = 0$ or $z = -1$. When the flocculation rate at low and high polymer concentrations slows down more and more, we are lead to multiply the r.h.s. of Eq. (9) by a probability coefficient which is proportional to a Boltzmann factor:

$$p \propto \exp\left(-\Delta G_e/kT\right) \quad \text{small polymer concentration} \quad (12)$$

$$p \propto \exp\left(-\Delta G_p/kT\right) \quad \text{large polymer concentration} \quad (13)$$

ΔG_e and ΔG_p are the free energies of electrostatic interactions and polymer-polymer interactions. The situation depicted in Fig. 7 is the one usually adopted when considering the adsorption of non-charged polymers. When P4VP is used as flocculent, the situation must be somewhat different because P4VP at pH 3.5 has approximatively 10 % of its pyridine groups in the pyridinium form. Therefore, ion pairing between NH^+ and SO_4^- groups being expected, we are probably faced with the situation described schematically in Fig. 8. At small polymer concentration, (situation (i) in Fig. 8), there is room on the latex surface for further adsorption and the latex/polymer complex has a net negative charge. The loop layer thickness should be exceedingly small, since all positive charges on the adsorbed P4VP are ion-paired, the 90 % remaining hydrophobic parts of the polymer collapse onto the surface through hydrophobic styrene-pyridine bonds. Practically, the flocculation should proceed similar to the slow coagulation of negatively charged colloids in the absence of added electrolyte. The collision frequency must be multiplied by a probability factor

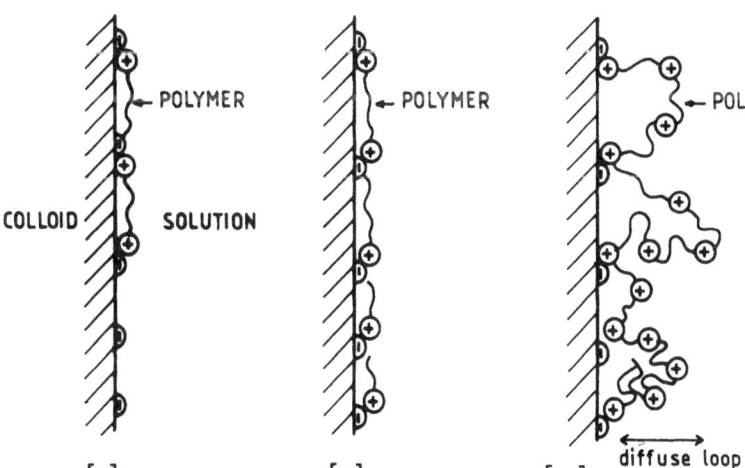

Fig. 8. Schematic of the colloid surface in the presence of P4VP at pH 3.5: (i) low polymer concentration, (ii) electroneutral colloid/polymer complex, (iii) large polymer concentration

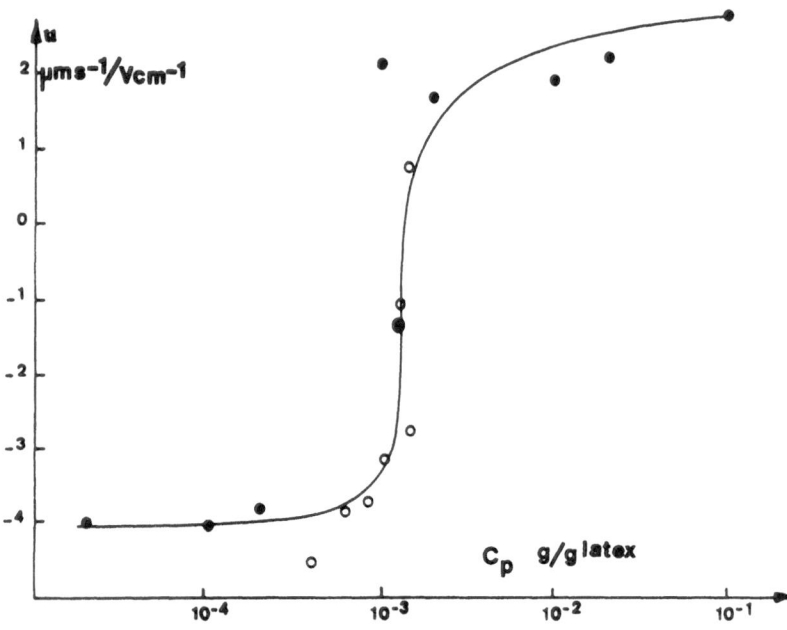

Fig. 9. Electrophoretic mobility of the isolated latex particle in presence of P4VP as a function of the polymer concentration (g polymer/g latex)

which involves the ΔG_e free energy. At an optimum polymer concentration corresponding to $\varphi' = 1$, the colloid/polymer complex is electroneutral (situation referred as (ii) in Fig. 8), and the flocculation should proceed as for charged colloids in the presence of an excess of electrolytes. The behavior of $N(t)$ with t is then exactly similar in both cases as seen in Fig. 5. At large polymer concentrations the situation (iii) in Fig. 8 is reached. After ion-pairing is completed, a further increase in adsorption is accompanied by the formation of extended charged loops. The loop layer has polyelectrolyte characteristics and both ΔG_e and ΔG_p energies becomes involved in the calculation of K.

In Fig. 9, we reported the electrophoretic mobility of the latex in the presence of P4VP as a function of the polymer concentration. A shift from negative to positive values of the electrophoretic mobility is observed at $C_p = 10^{-3}$ g polymer/g latex. This critical concentration corresponds to $\varphi' = 1$ (maximum flocculation) reported in Fig. 6. The trend of the latex/polymer electrophoretic mobility as a function of the polymer concentration indeed substantiates the schematic representation in Fig. 8 of the colloid interface in the presence of polymer.

Conclusion

We developed in the present study an experimental investigation on the aggregation kinetics of latex colloids flocculating in the presence of electrolytes and flexible water soluble polymers. The size distribution of the flocs was determined and their time evolution was analyzed in light of the theories based on "Monte Carlo" simulations or on the concept of a "self preserved" distribution, when size and time are properly rescaled. A very good agreement was found between the experimental data and the theory. In particular, in presence of an excess of electrolyte or an optimum polymer concentration, the scaling exponents have a straightforward interpretation. However, at small and large polymer concentrations, the structure of the colloid/polymer interface is not "suited" for easy bridging of colloids through polymer loops, and only a fraction of the binary encounters must be successful. This model was invoked to explain the particular value of the scaling exponents.

Acknowledgements

We thank Drs. C. Pichot and G. Graillat of the Laboratoire des Matériaux Organiques (Lyon) for kindly providing the latices. This work was performed under the auspices and with the financial support of the Programme Interdisciplinaire sur la Recherche de l'Energie et des Matières Premières (PIRSEM) of the CNRS in the theme ARC "Flocculation".

References

1. Napper DH (1983) Polymeric Stabilization of Colloid Dispersions. Colloid Science, Academic Press, New York

2. Vincent B (1982) Adv Colloid Interface Sci 4:193
3. De Gennes PG (1981) Macromolecules 14:1637; (1982) ibid 15:492
4. Jullien R, Botet R (1987) Aggregation and Fractal Aggregates. World Scientific, Singapore
5. Family F, Landau DP (eds) (1984) Kinetics of Aggregation and Gelation. North-Holland
6. Pefferkorn E, Pichot C, Varoqui R (1988) J de Physique, France 49:983
7. Higuchi WI, Okada R, Stelter GA, Lemberger AP (1963) J Pharm Sci 52:49
8. Ho NFH, Higuchi NI (1967) J Pharm Sci 56:148
9. Matthews BA, Rhodes CT (1968) J Colloid and Interface Sci 28:71
10. Matthews BA, Rhodes CT (1970) J Colloid and Interface Sci 32:339
11. Smoluchovski MV (1916) Phys Z 17:58
12. Meakin P, Vicsek T, Family F (1985) Phys Rev B 31:364
13. Kolb M, Botet R, Jullien R (1983) Phys Rev Lett 51:1123
14. Meakin P (1983) Phys Rev Lett 51:1119
15. Swift DL, Friedlander SK (1964) J Colloid Sci 19:621
16. Lushnikov AA (1973) J Colloid Interface Sci 45:549
17. Mabire F, Audebert R, Quivoron C (1984) J Colloid Interface Sci 97:120
18. Mabire F (1981) Thesis Université P et M Curie, Paris
19. Howard GJ, Leung WM (1981) Colloid Polym Sci 259:1031
20. Flory PJ (1953) Principles of Polymers Chemistry. Cornell University Press, Ithaca

Received November 1988;
accepted February 1989

Authors' address:

R. Varoqui
Institut Charles Sadron
6, Rue Boussingault
F-67083 Strasbourg Cedex, France

Progress in Colloid & Polymer Science   Progr Colloid Polym Sci 79:202–207 (1989)

D. Structure and stability of colloids

Aggregation and adsorption behavior in nonionic surfactant/oil/water systems

R. Aveyard, B. P. Binks, S. Clark, and P. D. I. Fletcher

School of Chemistry, University of Hull, Hull, England

Abstract: We present phase boundaries, viscosity, light scattering and interfacial tension data for microemulsion systems containing alkane, water and the nonionic surfactant $C_{12}E_5$. We interpret the solubilisation phase boundaries in terms of the spontaneous curvature of the surfactant monolayer and a critical concentration of surfactant required for microemulsion droplet formation. The viscosity behavior of both the oil-in-water (O/W) and water-in-oil (W/O) microemulsions is consistent with the microemulsion droplets being close to spherical and only weakly interacting at the solubilisation boundaries. The measured droplet sizes are proportional to the molar ratio of dispersed component to surfactant within the droplets. The light scattering data indicate that inter-droplet interactions are smallest at the solubilisation boundary. Planar oil/water interfacial tensions are found to be proportional to the inverse of the square of the equilibrium droplet radius. Values of the rigidity constant $(K + \bar{K}/2)$ of the surfactant monolayer estimated from these data are found to be of the order of kT.

Key words: Nonionic surfactant, microemulsion, interfacial tension, structure.

1. Introduction

There is continuing interest in the relationship between the microstructure of bulk phases containing surfactants and the properties of the adsorbed monolayer. In particular, the spontaneous curvature and rigidity of surfactant monolayers are important in determining the microemulsion droplet sizes and the interfacial tensions in Winsor types I and II multiphase equilibria in which microemulsion phases co-exist with excess oil or water [1].

Microemulsions stabilised by surfactants of the polyoxyethylene type have been investigated previously, particularly by Shinoda et al. [2], Kahlweit et al. [3], Ravey et al. [4], Aveyard et al. [5], and Kizling and Stenius [6]. However, there are insufficient systematic data for droplet sizes and interfacial tensions required for understanding the relationship between them. In this paper, we report microemulsion phase boundaries, structural data and interfacial tensions for dodecylpentaoxyethylene glycol ether ($C_{12}E_5$)/alkane/water mixtures. Using these detailed and systematic data, it is possible to relate the equilibrium microemulsion droplet radii in Winsor I and II systems and the measured interfacial tensions between the bulk phases as has been shown previously for a range of systems stabilised by ionic surfactants [7, 8]. The present results provide a first estimate of the rigidity of a nonionic surfactant monolayer at an oil/water interface.

2. Experimental

The surfactant $C_{12}E_5$ was a pure (> 99 % by glc) sample supplied by Nikko. Measured cloud points and phase boundaries were in good agreement with previous data [2, 9]. Water was distilled and passed through a Milli-Q reagent water system. *n*-Heptane (Fisons HPLC grade) was passed over alumina prior to use and *n*-tetradecane (Fluka puriss grade) was used as supplied.

Phase boundaries were determined by weighing the samples into thermostatted, tightly-stoppered flasks and noting the temperatures corresponding to the onset of permanent turbidity. It was necessary to ensure the samples were mixed thoroughly during this determination and that slow rates (approx. 0.02 °C per min) of temperature change were used. Boundary positions were reproducible for both raising and lowering the temperature.

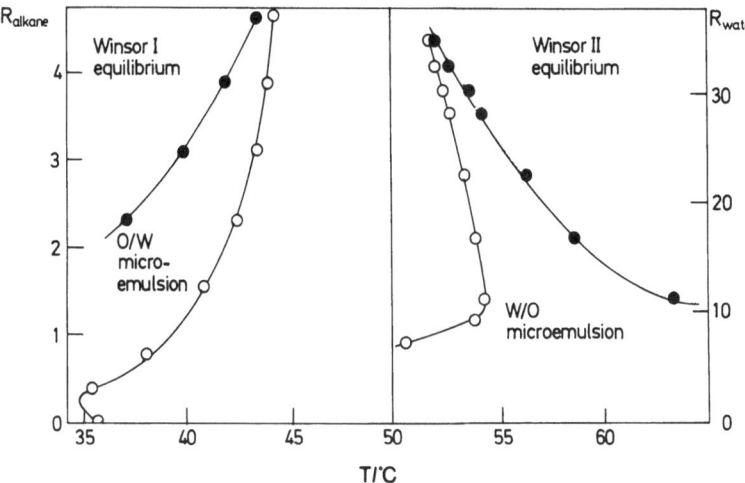

Fig. 1. Variation of R_{alkane} and R_{water} with temperature for a weight ratio of $C_{12}E_5$ to continuous solvent of 1:10. The oil is tetradecane

Relative viscosities were determined using an Ubbelohde viscometer which was thermostatted within 0.05 °C of the desired temperature. A Malvern PCS 100 photon correlation spectrometer equipped with a Spectrophysics model 124 B He-Ne laser was used to determine the hydrodynamic radii of the microemulsion particles as described previously [10]. Interfacial tensions were determined using a Kruss spinning drop tensiometer [10]. The measured tensions were observed to oscillate about mean values for temperatures lower than the phase inversion temperature [10] and values quoted here are mean values.

3. Results and discussion

Figure 1 shows the extent of the single-phase oil-in-water (O/W) and water-in-oil (W/O) microemulsion regions with respect to temperature for the case of tetradecane as oil. The amount of solubilised oil or water is expressed as the molar ratio of dispersed component to surfactant (R_{alkane} = [alkane]/[surfactant] and R_{water} = [water]/[surfactant]). The single-phase microemulsion regions are bounded on one side by the solubilisation curves (filled circles) corresponding to the separation of excess oil or water phases (the Winsor I and II equilibria). The remaining phase boundaries (unfilled circles) are the cloud point curve on the O/W side and the so-called "haze-point" curve for the W/O side. The solubilisation curves are thought to be determined primarily by the spontaneous curvature of the surfactant monolayer and the observed changes correspond to an increased tendency towards more negative curvature with increasing temperature, as has been noted previously for $C_{12}E_5$ [2, 3]. (Negative curvature is defined here as that in which the hydrophobic tail regions of the surfactant form the exterior surface of the particles.) The cloud and haze point curves cor-

respond to phase separation into surfactant-rich and surfactant-lean phases. These last transitions are thought to be associated with increasingly attractive interactions between microemulsion droplets [11, 12].

Phase boundaries for $C_{12}E_5$ stabilised microemulsions have been measured previously [2, 3]. We have extended these studies by making the measurements over a range of surfactant concentrations and the data can be used to calculate the concentration of surfactant within the microemulsion phase which is *not* adsorbed at the microemulsion droplet surfaces [13]. Figure 2 shows the maximum amount of solubilised oil or water plotted as a function of the surfactant concentration for various temperatures. The plots show that a minimum concentration of surfactant in oil is required before any significant amount of water is solubilised and a W/O microemulsion is formed [Fig. 2a]. The corresponding plots for the O/W microemulsions (Fig. 2b) pass close to the origin, as expected, since the critical micelle concentration in water is known to be of the order of 10^{-3} wt% [14]. For the W/O case the slopes of the plots yield the molar ratio of water to surfactant *within the microemulsion droplets* ($R(\text{drop})_{water}$ = [water]/([surfactant] − $c\mu c$)) and the intercepts on the abscissa give what we designate the critical microemulsion concentration ($c\mu c$) [15]. In Table 1 are listed values of the $c\mu c$ and $R(\text{drop})_{water}$ for W/O microemulsions with heptane or tetradecane as the oil. The behavior is similar for the two oils with the longer chain alkane system being shifted to higher temperatures. This observation is consistent with the notion that alkane penetration into surfactant chain regions is greater for smaller alkanes as has been demonstrated

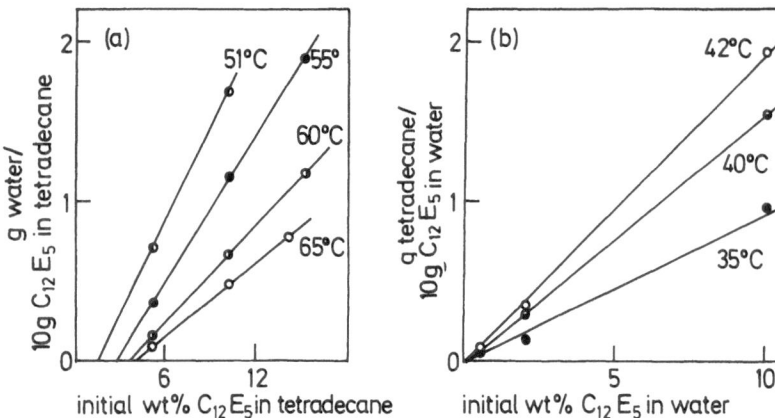

Fig. 2. Variation of maximum extent of dispersed component solubilisation with $C_{12}E_5$ concentration for (a) W/O and (b) O/W microemulsions. The oil is tetradecane

previously [16, 17]. The *cμc* values are larger for lower $R(drop)_{water}$ values which correspond to smaller droplet sizes (as will be shown later). The values determined using the solubilisation plots show good agreement with values measured from interfacial tensions as a function of concentration (Table 1). Qualitatively similar values have been reported for related systems [4, 13]. Knowledge of the *cμc* values in the W/O microemulsion phases are needed in order to be able to dilute the particles without changing their composition during the structural characterisation of these phases. (As mentioned previously, the *cμc* values in the O/W phases are negligibly small.)

We have measured the viscosities of both the O/W and W/O microemulsion phases for various extents of solubilisation and temperatures. The data were analysed according to the equation

$$\eta_{sp}/C = [\eta] + k_H [\eta]^2 C \qquad (1)$$

Table 1. Variation of $R(drop)_{water}$ and *cμc* with temperature for W/O microemulsions containing $C_{12}E_5$. The values of *cμc* shown in parentheses are determined independently from tension-concentration data [23]

alkane	temperature/°C	$R(drop)_{water}$	cμc/wt% of oil
heptane	35.0	84	1.9 (2.2)
	40.0	45	2.8
	42.0	33	(2.8)
	45.0	24	3.0
	50.0	17	3.9 (3.8)
tetradecane	51.0	44	1.4
	55.0	35	2.7
	60.0	23	3.5
	65.0	16	3.7

where C is the concentration (in g cm^{-3}) of microemulsion droplets, η_{sp} is the specific viscosity (= η (sample)/η (solvent) − 1), k_H is the Huggins coefficient, and $[\eta]$ is the intrinsic viscosity. The intrinsic viscosity is dependent on both particle shape and solvation as shown by the equation

$$[\eta] = v(V_D + \delta V_S) \qquad (2)$$

where V_D and V_S are the partial specific volumes of the droplets and solvent respectively, δ is a dimensionless quantity equal to the weight of solvent associated with a unit weight of particles and v is a shape parameter, equal to 2.5 for spheres [18]. The intrinsic viscosities are given by the intercepts of plots of η_{sp}/C vs. C as illustrated in Fig. 3 for O/W and W/O phases. At temperatures close (i.e. within 0.2 °C) to the solubilisation phase boundary in each case the intrinsic viscosities are low, the values being 5.0 ± 0.5 cm^3 g^{-1} for the O/W and 3.0 ± 0.5 cm^3 g^{-1} for the W/O microemulsions. Using Eq. (2), we calculate that the measured intrinsic viscosities for the oil microemulsion droplets close to the solubilisation phase boundary are consistent with either i) spherical particles hydrated with approximately 50 molecules of water per $C_{12}E_5$ molecule, or ii) unhydrated particles with an axial ratio of 3. For the W/O droplets the corresponding limiting values of the solvation are approximately 5 molecules of heptane per $C_{12}E_5$ (assuming spherical droplets) and axial ratios of 2 (assuming no solvation). It seems likely that the particles will be solvated to some degree and hence the droplets are probably close to spherical in shape. At temperatures further from the solubilisation boundaries the intrinsic viscosity increases moderately, whereas the slopes of the plots of η_{sp}/C against C increase sharply. This suggests that the inter-droplet

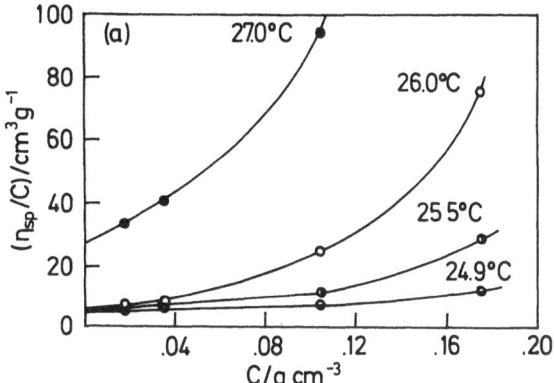

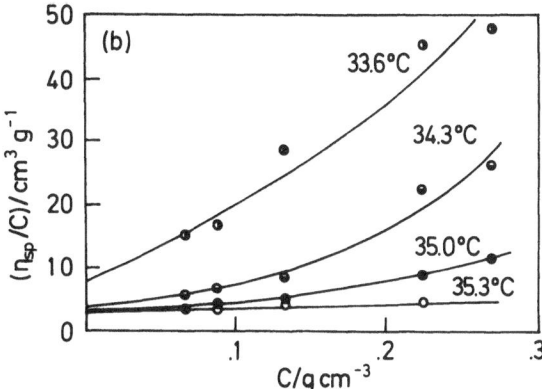

Fig. 3. Reduced viscosity plots for $C_{12}E_5$ microemulsions containing heptane: (a) O/W system, $R_{alkane} = 10$; (b) W/O system, $R(drop)_{water} = 79$

either an increase in droplet size or an increase in the attractive interactions between droplets (or both). The detailed relationship between the apparent hydrodynamic radius (as determined here) and particle interactions is complex [20]. For the purpose of this discussion, it is sufficient to note that repulsive interactions lead to a decrease of the apparent hydrodynamic radius with increasing concentration as observed for charged aqueous micelles [21]. Attractive particle interactions causing droplet clustering, as observed for W/O microemulsions close to the haze curve phase boundary, lead to an increase in the apparent r_h [22]. The concentration dependence of the apparent hydrodynamic radii for an O/W system is shown in Fig. 4(a). At a temperature close to the solubilisation boundary (22.0 °C) the concentration dependence of

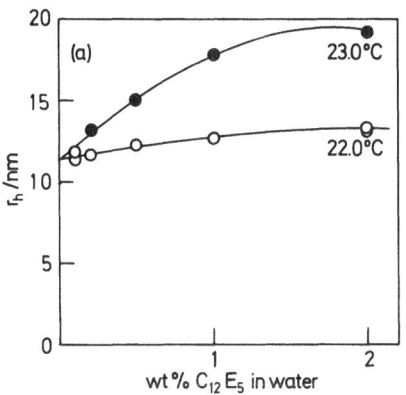

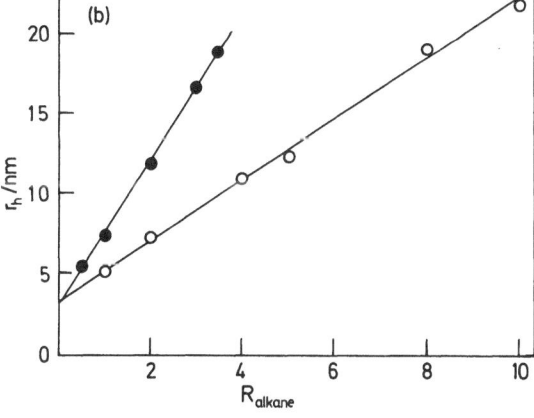

Fig. 4. Variation of apparent hydrodynamic radius with (a) $C_{12}E_5$ concentration at constant $R_{heptane} = 5.0$ and (b) with R_{alkane} for heptane droplets (unfilled circles) and tetradecane droplets (filled circles) at the solubilisation phase boundary at constant $C_{12}E_5$ concentration = 0.5 wt%; at this concentration the hydrodynamic radius is not significantly different to the value at infinite dilution

interactions increase strongly upon moving away from the solubilisation boundary for both O/W and W/O systems [19]. The viscosity behavior was found to be similar for different droplet sizes in both heptane and tetradecane-containing systems.

Spherical droplet shapes are implicitly assumed in calculating hydrodynamic radii from photon correlation spectroscopy (PCS) results. The calculated hydrodynamic radius for a sample is an apparent value which depends upon single-particle properties and inter-particle interactions. Hence, measurements must be extrapolated to infinite dilution in order to obtain the true hydrodynamic radius. Experimentally we observe, as others have [6] that the apparent hydrodynamic radius increases as one moves further away from the solubilisation phase boundary (i.e. to higher temperatures for the O/W and lower temperatures for the W/O). This increase could be due to

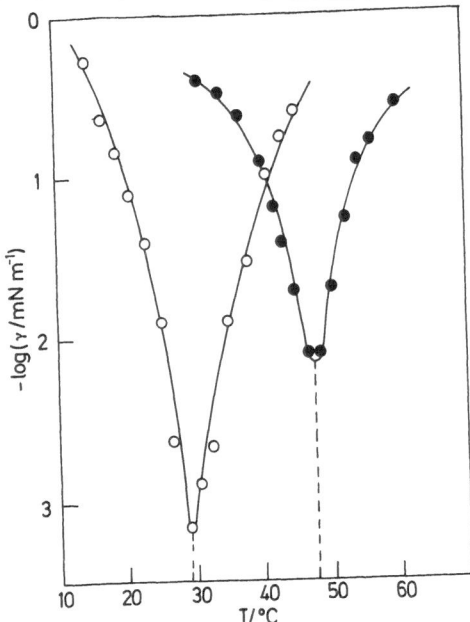

Fig. 5. Variation of the oil-water interfacial tension with temperature for heptane (unfilled circles) and tetradecane (filled circles) at an oil phase concentration of $C_{12}E_5 = 5$ wt%

the radii is small suggesting the interactions are weak. At the higher temperature (23.0 °C) the concentration dependence is increased but the intercept (corresponding to the true droplet size) remains the same. Hence, the light scattering data confirm the picture emerging from the viscosity behavior in that interdroplet interactions are weakest at the solubilisation boundary and increase as one moves from the boundary; droplet sizes are little affected.

Assuming monodisperse spherical droplets, simple geometry gives the droplet composition dependence of the hydrodynamic radii r_h as

$$r_h = (3\, R(\text{drop})\, V_{\text{mol}}/A_S) + t. \tag{3}$$

$R(\text{drop})$ is the molar ratio of dispersed component (oil or water) to surfactant in the droplets (i.e. $R(\text{drop}) = $ [dispersed component]/([surfactant]-$c\mu c$)), V_{mol} is the molecular volume of the dispersed component and A_S is the area occupied by a single molecule of surfactant at the interface between the droplet core and the surfactant monolayer. The length t is the thickness of the surfactant monolayer and contains a contribution from any entrapped solvent. Figure 4(b) shows plots of r_h vs. $R(\text{drop})$ for O/W systems with heptane and

tetradecane at temperatures very close to the solubilisation limit and at droplet volume fractions sufficiently low such that interaction effects can be neglected. The plots indicate the data are consistent with the simple model behavior predicted by Eq. (3). The measured values of the variance were in the range 0.05–0.1, indicating a reasonably low degree of polydispersity. Values of A_S were determined as 0.39 ± 0.04 and 0.29 ± 0.03 nm^2 for the heptane and tetradecane droplets, respectively. It is plausible to infer from these results that heptane penetrates the alkyl tail region of the surfactant monolayer to a greater extent than the tetradecane. Preliminary results for the W/O systems indicate the structural properties of the water droplets are qualitatively similar.

Figure 5 shows the variation of the interfacial tension y between the upper and lower phases of the Winsor I, III, and II systems as a function of temperature for heptane and tetradecane. The tensions were determined at surfactant concentrations within the total system such that the $c\mu c$ values in both the oil and water phases are exceeded. Tensions remain constant above the $c\mu c$ up to oil phase concentrations of at least 15 wt% [23]. The tensions pass through a minimum value at the phase inversion temperature (PIT), as has been observed and discussed previously [2, 3, 5].

The interfacial tension is related to the free energy difference of the interfacial film at the planar interface separating the bulk phases and at the curved surfaces of the microemulsion droplets. This energy difference contains contributions from the energy required to bend the monolayer and energies associated with dispersing the droplets within the medium and interdroplet interactions [24]. For the systems considered here we know that the tension is unaffected by changes in the volume fraction of droplets within the microemulsion phase (see earlier). Furthermore, the size of droplets formed at equilibrium in the two-phase systems of microemulsions with excess phases is independent of the droplet volume fraction. This behavior implies that the bending energy contribution dominates the overall free energy difference. Under these conditions the formation of an excess water or oil phase occurs when the surfactant monolayer curvature in the microemulsion droplet phase is virtually equal to the spontaneous film curvature. For this situation the relationship between the tension and droplet radius [25] is of the form

$$y = \text{constant}/r^2. \tag{4}$$

The constant in Eq. (4) is thought to be equal to $(2K + \bar{K})$ [8]. K is the rigidity constant of the film and is the energy required to bend a unit area of the film by a unit amount of curvature away from the spontaneous curvature. $\bar{K}$ is the Gaussian curvature elastic modulus [26]. Strictly speaking, Eq. (4) is accurate only in the limit where the film thickness is negligible in comparison with the radius. This condition is not met for the smaller microemulsion droplets and a compromise value of the droplet radius is taken here as being equal to the droplet core radius plus half the thickness of the surfactant film.

The interfacial tensions presented here scale approximately with (droplet radius)$^{-2}$ as predicted by Eq. (4), thus allowing an estimate of the constant to be made. We find $(K + \bar{K}/2)$ is approximately $1\ kT$ for the heptane system and $2\ kT$ for the tetradecane system. These values compare with values in the range $0.4 - 1.6\ kT$ calculated in this way [23] for microemulsion systems containing ionic surfactants with or without cosurfactant. Values of K measured directly by ellipsometry for various microemulsion systems are of similar magnitude [8, 27].

This work is part of an ongoing program to investigate the effects of surfactant molecular structure on film and bulk properties is such systems. It is premature to compare surfactant film rigidities directly owing to complexities introduced by the different film compositions and the differing length scales of observation [24]. However, it is tempting to speculate that the relatively high values for the nonionic systems may be a consequence of the absence of electrostatic repulsions between adjacent surfactant molecules within the film. Continuing studies of this type are expected to enable better-founded generalisations to be made concerning the dependence of monolayer rigidity on surfactant molecular structure.

References

1. Langevin D, Guest D, Meunier J (1986) Colloids and Surf 19:159
2. Shinoda K, Kunieda H, Arai T, Saijo H (1984) J Phys Chem 88:5126
3. Kahlweit M et al (1987) J Colloid Interface Sci 118:436
4. Ravey JC, Buzier M, Picot C (1984) J Colloid Interface Sci 97:9
5. Aveyard R, Lawless TA (1986) J Chem Soc Faraday Trans I 82:2951
6. Kizling J, Stenius P (1987) J Colloid Interface Sci 118:482
7. Aveyard R, Binks BP, Lawless TA, Mead J (1988) Can J Chem 66:3031
8. Binks BP, Meunier J, Abillon O, Langevin D (1989) Langmuir 5:415
9. Mulley BA (1967) In: Schick MJ (ed) Nonionic Surfactants. Marcel Dekker, New York, 1
10. Aveyard R, Binks BP, Lawless TA, Mead J (1985) J Chem Soc Faraday Trans I 81:2155
11. Corti M, Degiorgio V (1985) Phys Rev Lett 55:2005
12. Hou M-J, Shah DO (1987) Langmuir 3:1086
13. Johnson KA, Shah DO (1986) In: Mittal KL, Botherel P (eds) Surfactants in Solution. Plenum Press, New York, 6, p 1441
14. Mukerjee P, Mysels KJ (1970) Critical Micelle Concentrations of Aqueous Surfactant Systems. NSRDS-NBS 36
15. Aveyard R, Binks BP, Clark S, Fletcher PDI (1989) Chem Technol and Biotechnol, in press
16. Aveyard R, Binks BP, Mead J (1986) J Chem Soc Faraday Trans I 82:1755
17. Mukherjee S, Miller CA, Fort T (1983) J Colloid Interface Sci 91:223
18. Tanford C (1961) Physical Chemistry of Macromolecules, Wiley, New York
19. Russel WB (1984) J Chem Soc Faraday Trans I 80:31
20. Cebula DJ, Ottewill RH, Ralston J, Pusey P (1981) J Chem Soc Faraday Trans I 77:2585
21. Missel PJ, Mazor NA, Benedek GB, Carey MC (1983) J Phys Chem 87:1264
22. Brunetti S, Roux D, Bellocq AM, Fourche G, Botherel P (1983) J Phys Chem 87:1026
23. Aveyard R, Binks BP, Fletcher PDI (1989) Langmuir, in press (manuscript LA 8900420)
24. Cates ME, Andelman D, Safran SA, Roux D (1988) Langmuir 4:802
25. de Gennes PG, Taupin C (1982) J Phys Chem 86:2294
26. Helfrich W (1985) J Physique 46:1263
27. Meunier J (1985) J Phys Lett 46:1005

Received October, 1988;
accepted April, 1989

Authors' address:

Dr. P. Fletcher
School of Chemistry
University of Hull
Hull HU6 7RX, England

Progress in Colloid & Polymer Science

Progr Colloid Polym Sci 79:208–213 (1989)

Characteristic sizes, film rigidity and interfacial tensions in microemulsion systems

B. P. Binks, J. Meunier, and D. Langevin

Laboratoire de Physique de l'Ecole Normale Superieure, Paris, France

Abstract: Studies of the structure of microemulsions and of Winsor-type systems have emphasized the importance of the properties of the interfacial film through the parameters of spontaneous curvature and of the film curvature elastic modulus. We discuss here the correlations between measured dispersion sizes, surfactant film curvature elasticity (K), and interfacial tensions in several multiphase microemulsion systems. Sizes L are in agreement with simple random-space filling models, the area per surfactant molecule in the film remaining constant. For small values of K, bicontinuous structures are found when the spontaneous curvature of the film is small: the dispersion size is then close to the persistence length of the film. For larger K, more ordered structures appear. Interfacial tensions correlate only approximately with kT/L^2.

Key words: Rigidity, spontaneous curvature, microemulsion, interfacial tension, bicontinuous.

Introduction and theoretical basis

Microemulsions are dispersions of oil and water stabilized by surfactant molecules [1]. In many respects, they are small scale versions of emulsions. Indeed, they are frequently droplet-type dispersions, either of oil in water (o/w) or of water in oil (w/o). However, the droplet sizes are very small, typically 100 Å, about 100-times smaller than typical emulsion drop sizes. For this reason, microemulsions are relatively transparent. Unlike emulsions, microemulsions are thermodynamically stable due to the very low interfacial tensions ($< 10^{-2}$ mN m^{-1}) between oil and water microdomains.

The kind of dispersion obtained depends in part on the spontaneous curvature C_0 of the surfactant layer. Like in emulsions, it bends spontaneously towards the medium where the surfactant molecule is more soluble (Bancroft's rule). This has been rationalized in terms of surfactant molecular geometry by Mitchell and Ninham [2]. The surfactant molecule is replaced by a truncated cone. If the polar part is more bulky than the hydrophobic part, the interface will curve spontaneously towards water and o/w structures will be favored. By convention, C_0 is positive. If the polar

part is less bulky than the hydrophobic part, the interface will curve in the opposite direction and w/o structures will form ($C_0 < 0$). If the surfactant has the shape of a cylinder or if a surfactant for which $C_0 > 0$ and a cosurfactant for which $C_0 < 0$ are mixed, zero mean spontaneous curvature can be obtained. This probably explains why continuous inversion is possible by changing the relative proportions of oil and water in an oil-water-surfactant-alcohol mixture. The alcohol partition coefficient between the surfactant film and oil and water bulk phases can be adjusted to vary C_0 continuously from positive to negative values.

The microemulsion structure is determined from an energy balance between several terms: interfacial energy, bending energy of the surfactant monolayers, dispersion entropy, and interactions between structures [3]. The role of the bending term appears to be dominant in systems where a microemulsion coexists with excess oil and/or water (Winsor equilibria). This term as introduced by Helfrich may be written [4]

$$F_c = \frac{K}{2}(C_1 + C_2 - 2C_0)^2 + \frac{\bar{K}}{2}C_1 C_2 \text{ per unit area}$$

(1)

where K is the bending elastic modulus (or rigidity), $\bar{K}$ the saddle-splay bending constant, and C_1 and C_2 are the local principal curvatures of the layer.

In other dispersed systems, the interfacial energy term is much larger than the others. For microemulsions, however, the interfacial tension Γ between oil and water microdomains is ultralow and so the corresponding energy is small. This means that the area per surfactant molecule Σ cannot vary very much because the surface pressure of the surfactant film π cannot vary either: $\gamma = \gamma' - \pi_{\Sigma}$ (γ' = bare oil-water interfacial tension) [3]. Consequently, the droplet radius R depends mainly on composition

$$R = \frac{3}{C_S} \frac{\phi}{\Sigma} \qquad (2)$$

where ϕ is the volume fraction of dispersed phase and C_S is the number of surfactant molecules per unit volume.

When the surfactant concentration is decreased (by dilution with oil or water), R increases until it reaches the optimum radius R_0 ($= 1/C_0$). Above this point, the dispersed phase that cannot be solubilized into larger droplets is rejected and forms an excess phase [5]. The corresponding phase equilibria are called Winsor I (o/w microemulsion in equilibrium with excess oil) and Winsor II (w/o microemulsion in equilibrium with excess water).

If the spontaneous curvature is close to zero, the surfactant layer has a tendency to become planar and to promote lamellar order. If Γ and $\bar{K}$ are neglected and the bending energy is comparable to the thermal energy, one can evaluate the effect of thermal fluctuations of the film. The result is that the film is strongly wrinkled at scales larger than the persistence length [3]

$$\xi_K = a \exp\left(2\pi K/kT\right) \qquad (3)$$

where a is a molecular length. For $K \approx 10–100\, kT$ as for phospholipid bilayers, ξ_K is macroscopic and the surfactant layer is flat over large distances. If the surfactant layers contain short-chain alcohols which introduce disorder in the film, K can decrease to about kT and ξ_K to 100 Å: long-range order is destroyed and the system is macroscopically disordered (bicontinuous microemulsion). Simple geometrical space filling models [3, 6] allow us to calculate the mean diameter of the oil and water microdomains

$$\xi = \frac{6\,\phi_0\,\phi_W}{C_S\,\Sigma} \qquad (4)$$

where ϕ_0 and ϕ_W are, respectively, the oil and water volume fractions (the surfactant volume fraction assumed to be negligible). The bending constant K depends on the scale at which the surfactant layer deformations are produced [7]. If the scale is larger than ξ_K, the layer is already very rough and easier to bend than a flat layer. From Eq. (4) it is seen that when C_S decreases, ξ increases until it reaches a value of the order of ξ_K. Then the system cannot accommodate more oil and water in the microemulsion and phase separates into a microemulsion and both excess oil and water (Winsor III). When C_S increases, the models also explain the formation of lamellar phases. Indeed, ξ decreases and K and ξ_K both increase.

A particularly important problem is the origin of the ultralow interfacial tension γ (at the macroscopic interface) between the microemulsion and the excess phases. It is now well established that in most cases γ is low as a result of monolayer adsorption [8, 9]. Like the free energy of the microemulsion phase, γ contains contributions from the dispersion entropy and curvature energy. Assuming the latter is dominant and Γ and $\bar{K} = 0$ for $R = R_0$, one obtains for droplet microemulsions [3]

$$\gamma = \gamma_c = \frac{2\,K}{R^2} \qquad (5)$$

where γ_c is the energy cost (per unit area) to unbend the surfactant film. The above value of γ is independent of surfactant concentration (or ϕ), as observed experimentally [10, 11]. In some systems, the bending energy is not very large and is comparable to the dispersion entropy. As discussed by Israelachvili [12], $\gamma = \gamma_c + \gamma_e$ where

$$\gamma_e = \frac{-kT}{4\pi R^2} \ln \alpha \phi \qquad (6)$$

α is a constant depending on the approximations used. This equation accounts for the γ vs. ϕ variation reported recently in a pure surfactant system [13].

In all cases, γ is very small and so thermal fluctuations are very large. This roughness can be measured using ellipsometry [14]. The short wavelength fluctuations probed have an amplitude which is dependent on K. If γ is known, K can be deduced. The problem of the scale variation of K [15] can be taken into account and renormalization of both K and y is needed in order to interpret the measurements [16]. We have found

$$\xi_K = a \exp [3.254 \, K(a)/kT] \qquad (7)$$

where $K(a)$ is the rigidity at scale a.

Summarizing, the theories predict that there is a correlation between the behavior of the macroscopic and microscopic interfaces in Winsor equilibria. We present K values in different systems and correlate them with corresponding size and tension data.

Experimental results

We have studied four systems containing five components: oil–water–surfactant–alcohol–salt. Two anionic surfactants, sodium dodecyl sulphate (SDS) and sodium hexadecyl benzene sulphonate (SHBS), and two cationic surfactants, dodecyltrimethyl ammonium bromide (DTAB) and hexadecyltrimethyl ammonium bromide (CTAB) have been used. The addition of *n*-butanol was necessary for microemulsion formation. The compositions of the systems are (in wt. %):

toluene	47%	dodecane	38.19%
brine	47%	brine	56.93%
SDS or DTAB	2%	SHBS or CTAB	1.66%
butanol	4%	butanol	3.32%

The brine is an aqueous solution of NaCl for SHBS/SDS and of NaBr for DTAB/CTAB. The fifth system contains equal volumes of heptane and aqueous NaCl containing 0.7 wt. % AOT (sodium bis-2 ethylhexyl sulfosuccinate). Increasing the salt concentration (S) screens electrostatic interactions between surfactant headgroups and reduces C_0. In this way a continuous structural evolution from o/w to w/o systems can be obtained. At each salinity, the size of the structural elements is the largest possible (maximum swelling: R_0 for droplets, ξ_K for bicontinuous structures) because the microemulsions coexist with excess oil and/or water. All equilibrations and measurements were at 20 °C.

The variation of the interfacial tensions between the microemulsion and the excess phases, measured using surface laser light scattering [8], is shown in Fig. 1 for four systems (CTAB behaved similarly to DTAB); they are all very low, the lowest corresponding to the AOT system and the largest to the DTAB one.

The bending elasticities have been measured using ellipsometry. In the three-phase domain the interfaces between excess phases have been studied, while in the two-phase domains the studied interface was one between the excess phase and a diluted microemulsion phase. Measured and renormalized values K and $K(a)$ for $a = 5$ Å are given in Table 1; they are roughly independent of the salinity [14, 17]. Since they do not differ very much it follows from Eqs. (5) and (6) that to explain the tension differences, the largest droplet sizes in the Winsor I and Winsor II regions must be found in the SHBS system, the smallest in the DTAB one. The droplet size has been measured by us in SDS and SHBS systems using bulk light scattering techniques [8, 18], and in DTAB and CTAB systems with small-angle x-ray techniques [19]. Aveyard et al. [11, 20] report sizes for AOT microemulsions using a combination of techniques. Sizes in bicontinuous microemulsions were measured by small-angle x-ray and neutron scattering for SDS by Auvray et al. [21], and de Geyer and Tabony [22], and in SHBS and DTAB/CTAB systems by us [18, 19]. A typical x-ray spectrum for a middle-phase microemulsion of DTAB-containing equivalent amounts of oil and water is shown in Fig. 2 a. At low q, a maximum in intensity is observed. Although very broad, the peak is evidence that there is a well defined length scale in the microstructure. The peak has been attributed by some authors to a form-factor feature [23], and by others to a structure-factor feature [15]. Empirically $\xi \approx \pi/q_{max}$. At higher q values, the spectrum follows Porod's law $I(q) \propto q^{-4}$. When $q \xi \ll 1$, then the flat portions of the oil-water interfaces are seen. Figure 2b illustrates such behavior, evidencing directly the existence of well defined surfactant layers. The departure of the asymptotic regime in the intermediate q range leads to a minimum of $q^4 I$ which can be taken as an estimate of the

Table 1. Measured and renormalized bending constants, measured dispersion sizes ξ in bicontinuous microemulsions, and calculated persistence lengths for the studied systems

System	K/kT	$K(a)/kT$	ξ (Å)	ξ_K (Å)
CTAB	0.40	0.55	92	63
DTAB	0.40	0.55	95	63
SDS	0.65	1.00	230	375
SHBS	0.40	0.86	400	175
AOT	1.10	1.65	birefringent phase	2250

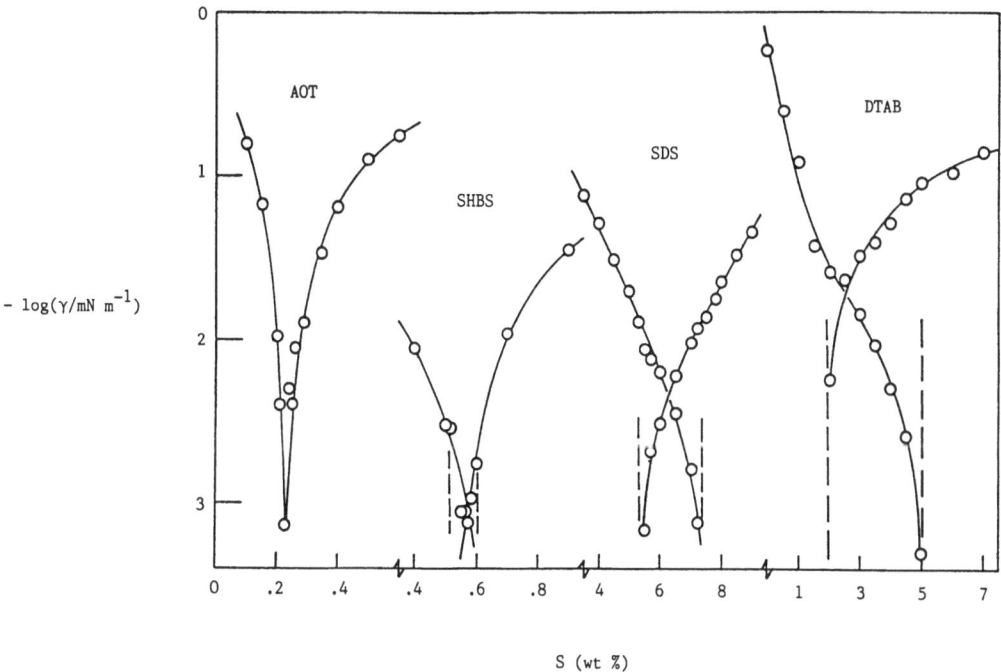

Fig. 1. Interfacial tensions between the microemulsion and the excess oil phase (decreasing values) and the microemulsion and the excess water phase (increasing values) vs. salinity for the different systems. Dashed lines indicate the extent of the three-phase region. *Ordinate:* $-\log (\gamma/\mathrm{mN\ m^{-1}})$; *abscissa:* S (wt. %)

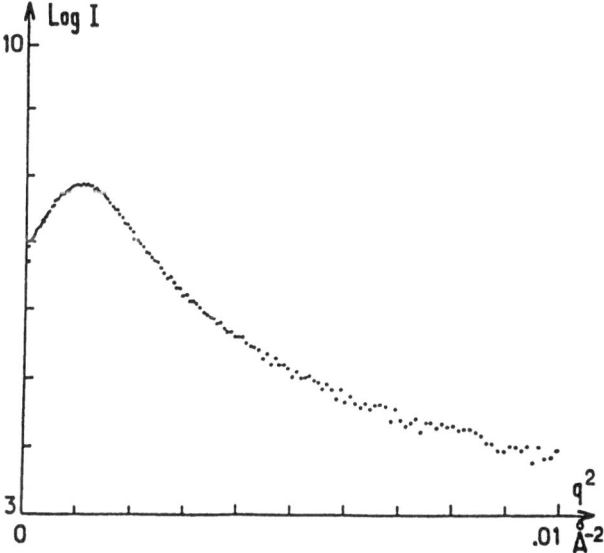

Fig. 2a. X-ray spectrum for a bicontinuous microemulsion – DTAB system, $S = 2.5$. *Ordinate:* log I; *abscissa:* q^2 (Å^{-2})

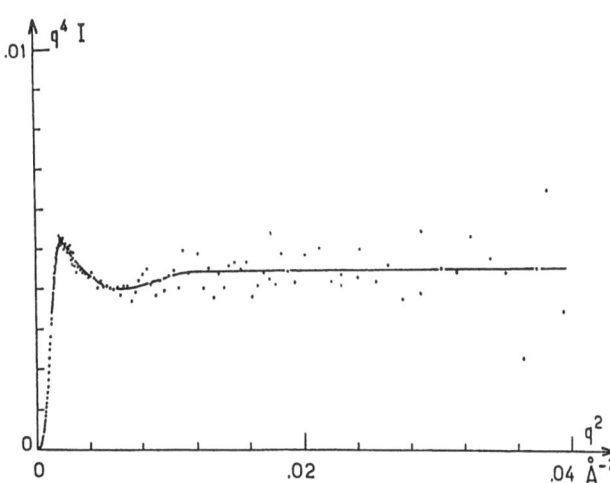

Fig. 2b. Porod's plot for the DTAB microemulsion, $S = 2.5$. The line is a guide for the eye. $q^4\ I$ units are arbitrary. *Ordinate:* $q^4\ I$; *abscissa:* q^2 (Å^{-2})

distance at which the interface begins to bend, i.e., as an estimation of ξ: $\xi = a/q$ min. Auvray et al. [21] found $a \approx 2\pi$ by calculating ξ from Eq. (4). In other systems a can be larger [19]; the differences in a values may reflect differences in microstructure.

The complete data are presented as a summary in Fig. 3, showing the variation of dispersion sizes with oil and water volume fractions. The abscissae indicate increasing salinity, the ordinates normalized sizes $\bar{L} = LC_S \Sigma/6$. For droplets, $L = 2R$ and for middle-phases $L = \xi = \pi/q_{max}$. The figure shows again that droplet structures are compatible with the data in the WI and WII regions (Eq. (2)). The length ξ varies as $\phi_0 \phi_W$, confirming that the structure is bicontinuous in WIII regions (Eq. (4)).

From the experimental data, we have made a comparison between ξ and ξ_K (calculated using renormalized $K(a)$ values and Eq. (7)). The results are given in Table 1. Again the data are in qualitative agreement with theoretical predictions: the ratio ξ/ξ_K is of the order of unity although it varies from one system to another. As expected, the smaller values of ξ_K (< 500 Å) lead to bicontinuous phases. For AOT, the larger ξ_K value helps explain the formation of an organized phase, possibly lamellar and of the highest rigidity. In the cationic systems, the small sizes and high tensions are consistent with a relatively low rigidity. The comparison of the K values between SDS and SHBS systems is more surprising. However, it must be recalled that i) SHBS films contain three alcohol molecules per surfactant, whereas SDS ones contain only one, ii) the saddle-splay bending constant $\bar{K}$ has been neglected but may be an important parameter related to surfactant molecular geometry, iii) the role of interactions between surfactant layers is not taken into account in the calculation of ξ_K.

We can test the theories for interfacial tensions with our results. As seen earlier, tensions y are expected to be of the order kT/L^2. At minimum tension, since the spontaneous curvature is zero, y is expected to arise from entropy contributions to the free energy alone. This does not seem to be the case since $y \xi^2/kT$ depends on the system: 0.74 for CTAB, 0.38 for DTAB, 0.65 for SDS, and 0.35 for SHBS. In the two-phase region, y contains two kinds of terms: y_e and y_c. The second term is probably responsible for the variation of yR^2/kT with salinity and surfactant nature. To test this, we performed a numerical comparison; the results are given in Table 2. It can be seen that for DTAB and CTAB, since the curvature energy is very small, the entropic contribution is no longer negligible

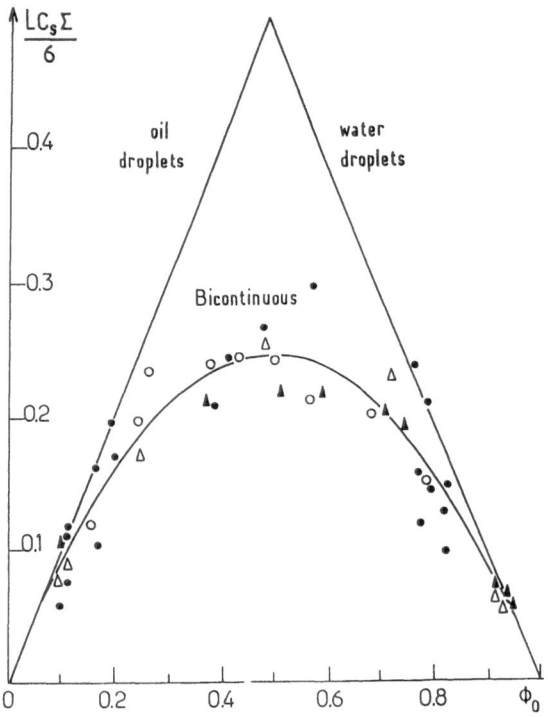

Fig. 3. Reduced characteristic size ($L = 2R$ or ξ) vs. oil-volume fraction for different Winsor microemulsions. Lines are theoretical: triangle for droplets, parabola for bicontinuous structures. $\triangle$: CTAB, $\blacktriangle$: DTAB, $\bullet$: SDS, $\circ$: SHBS. *Ordinate:* $LC_S \Sigma/6$; *abscissa:* ϕ_0

as in SDS and SHBS systems. However, in all systems, serious discrepancies arise between y and $(y_c + y_e)$. This may arise from oversimplifications in the theories. Further improvement of these theories and possibly new experimental developments (measurement of Γ and $\bar{K}$) are needed to achieve a complete understanding of the problem.

Conclusions

We measured the surfactant film bending elasticity in a variety of oil-water-surfactant systems. We made a correlation between this elasticity and the microstructure of the microemulsion phases and the interfacial tensions between the microemulsion and the excess oil or/and water phases. In the AOT system, the film rigidity is large and the phases containing equal amounts of oil and water are not microemulsions but have lamellar structures. For the other systems K is of the order of kT; the persistence length of the surfactant layer being microscopic, when the spontaneous curvature is small, the lamellae break and give rise to a

Table 2. Measured radii and interfacial tensions for droplet microemulsions, and calculated γ_c (Eq. (5)) and γ_e (Eq. (6)), $\alpha = 1$) contributions to the tension

System	S (wt. %)	ϕ	R (Å)	γ (mN m^{-1})	γ_c (mN m^{-1})	γ_e (mN m^{-1})
CTAB	2	.091	52	.348	.165	.029
	4	.110	61	.129	.120	.019
	24	.075	46	.111	.210	.039
	30	.070	42	.147	.252	.049
DTAB	1	.100	64	.120	.108	.018
	6	.076	51	.105	.171	.032
	7	.065	54	.138	.152	.030
	9	.053	42	.134	.252	.053
SDS	3.5	.114	85	.075	.112	.0097
	5	.165	130	.020	.048	.0034
	8	.206	180	.023	.025	.0032
	10	.169	125	.075	.052	.0037
SHBS	0.40	.170	110	.0089	.058	.0047
	0.52	.280	250	.0029	.011	.0006
	0.70	.290	190	.010	.019	.0011
	0.90	.180	130	.035	.041	.0033

bicontinuous structure. The structural sizes are in good agreement with those calculated assuming the area per surfactant is constant. The experiments also show that at maximum solubilization (where excess oil and water separate) the size ξ is dependent on the K value, and close to the persistence length. Although the agreement with the theories is qualitatively good, quantitative predictions can differ by factors as large as 2 or 3 and this probably arises from oversimplification of the theories.

Acknowledgements

Dr. B. P. Binks thanks the Royal Society for the award of a European Postdoctoral Fellowship. This work has been supported by the CNRS PIRSEM (Greco Microemulsions), and the CEE (contract EN3COO18F).

References

1. Rosano HL, Clausse M (1987) Microemulsion Systems. Surf Sci Ser 24, Dekker, New York
2. Mitchell DJ, Ninham BW (1981) J Chem Soc Faraday Trans II 77:601
3. de Gennes PG, Taupin C (1982) J Phys Chem 86:2294
4. Helfrich W (1973) Z Naturforsch C28:693
5. Safran SA, Turkevich LA (1983) Phys Rev Lett 50:1930
6. Talmon Y, Prager S (1978) J Chem Phys 69:2984
7. Meunier J, Langevin D, Boccara N (1987) Physics of Amphiphilic Layers. Springer Verlag, Berlin
8. Cazabat AM, Langevin D, Meunier J, Pouchelon A (1982) Adv Colloid Interface Sci 126:175
9. Aveyard R, Binks BP, Mead J (1987) J Chem Soc Faraday Trans I 83:2347
10. Pouchelon A, Chatenay D, Meunier J, Langevin D (1981) J Colloid Interface Sci 82:418
11. Aveyard R, Binks BP, Clark S, Mead J (1986) J Chem Soc Faraday Trans I 82:125
12. Israelachvili J (1987) Surfactants in Solution. Vol 4, Plenum Press, New York, p 3
13. Fletcher PDI (1987) Chem Phys Lett 141:357
14. Meunier J (1985) J Phys Lett 46:L1005
15. Cates ME, Andelman D, Safran SA, Roux D (1988) Langmuir 4:802
16. Meunier J (1987) J Phys Paris 46:1819
17. Binks BP, Meunier J, Abillon O, Langevin D (1989) Langmuir 5:415
18. Guest D, Langevin D (1986) J Colloid Interface Sci 112:208
19. Abillon O, Binks BP, Otero C, Langevin D, Ober R (1988) J Phys Chem 92:4411
20. Aveyard R, Binks BP, Lawless TA, Mead J (1988) Can J Chem 66:3031
21. Auvray L, Cotton JP, Ober R, Taupin C (1984) J Phys Paris 45:913
22. de Geyer A, Tabony J (1985) Chem Phys Lett 113:83
23. Zemb TN, Hyde ST, Derian PJ, Barnes I, Ninham BW (1987) Europhys Lett 4:561

Received November, 1988;
accepted January, 1989

Authors' address:

Dr. B. P. Binks
School of Chemistry
University of Hull
Hull HU6 7RX, United Kingdom

Progress in Colloid & Polymer Science   Progr Colloid Polym Sci 79:214–217 (1989)

Different spin probe positions related to structural changes of nonionic microemulsions

C. T. Cazianis and A. Xenakis

Institute of Biological Research, The National Hellenic Research Foundation, Athens, Greece

Abstract: The ESR spectra of iodoacetamide, 5-doxyl stearic acid and FDNB derivative spin labels, in nonionic microemulsion systems, formulated with tetraethyleneglycoldodecylether in decane were measured in various amounts of water and surfactant content. At certain microemulsion compositions, the high field line of the ESR spectrum, splits into a doublet, suggesting that the spin probe locates in two different positions, i. e., the water core and the micromembrane which separates the aqueous from the oil phase. The method used for this study is based on i) the partition of spin labels between the microphases of the system expressed as the ratio of the high field peaks (H_{-1}/H_{-1}^*) corresponding to the two different spin states; ii) the mobility of the spin labels in the dispersed phase expressed in terms of the degree of anisotropy (H_1/H_o); and iii) the polarity of the microenvironment of the label expressed in terms of the hyperfine coupling constant a_n. The variation of these spectral parameters is discussed in connection with the corresponding structural state of the microemulsions.

Key words: Microemulsions, ESR spectroscopy, nonionic systems.

Introduction

Spin labelling is a spectroscopic method which can produce important information on a microenvironmental level, about conformation, mobility, and polarity in biological systems (membranes, proteins) [1, 2], as well as in physicochemical systems (micelles or microemulsions) [3].

Certain microemulsions have been studied as model systems of biomembranes by several techniques [4]. However, an ionic microemulsion formulated with water/AOT/Heptane has been studied by the spin label method [5] and the ESR spectra obtained in this study suggested that at low temperature, the spin probe locates in two positions, namely, the water pool and the shell formed by the surfactant molecules.

The intent of this work was to examine the potential use of nonionic microemulsions as model systems of biomembranes.

The results presented here indicate that, under certain conditions, nonionic microemulsions behave as biological membranes [6] at room temperature.

The ESR spectra of iodoacetamide (IA), 1-fluoro-2,4-dinitrobenzene derivative (FDNB), and 5-doxyl stearic acid (DX) spin labels were measured in nonionic microemulsion systems consisting of water/tetraethyleneglycoldodecylether ($C_{12}EO_4$)/decane. Analysis of these spectra shows the existence of different spin states, depending on the exchange rate of the probe between the water core and the micromembrane separating the latter phase from the continuous oil phase. The fact that the rate of exchange depends upon the physical characteristics of the spin label as well as the composition of the microemulsions, may be useful in approaching the conditions needed for using microemulsions as model systems of biomembranes.

Expermimental

Materials

The iodoacetamide (IA) spin label: 4-(2-iodoacetamide) 2,2,6,6-tetramethylpiperidinyloxyl was a product of Synvar. 5-Doxyl stearic acid (DX) spin label was a product of Sigma,

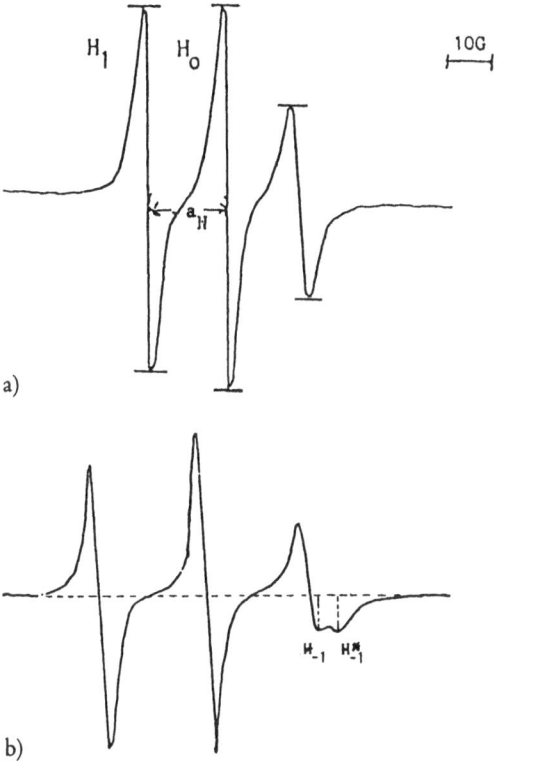

a)

b)

Fig. 1. ESR Spectra of a) iodoacetamide spin label, b) FDNB analogue spin label recorded in a microemulsion with $R = [H_2O]/[C_{12}EO_4] = 15$. The concentration of the surfactant was 20 % w/w in decane. The concentration of the probe was $4 \cdot 10^{-5}$ M

while the spin-labelled fluoro dinitrobenzene derivative (FDNB): 4-(2,4,-dinitro-5-fluorophenoxyl)-2,2,6,6-tetramethyl-1-piperidinyloxyl was synthesized as described elsewhere [7]. Tetraethyleneglycoldodecylether ($C_{12}EO_4$) was obtained from Nikko Chemicals Co. Ltd., n-decane was a product of Merck, Darmstadt, F.R.G., and twice-distilled water was used.

Methods

Nonionic microemulsion systems were prepared by adding the appropriate amounts of water in stock solutions of different concentrations of $C_{12}EO_4$ in decane. The total amount of water added was calculated to give the desired values of the molar ratio $[H_2O]/[C_{12}EO_4] = R$ for the different surfactant contents. Special care was taken to keep the samples at constant temperature (22 °C).

The spin probes were dissolved in a concentration of about $5 \cdot 10^{-5}$–10^{-4} M in each sample.

ESR spectra were recorded on a Bruker ER 200D spectrometer system operating at X-band, equipped with a variable temperature unit. The samples were contained in an E-248 cell. Typical settings were: field set, 3 471 G; scan range, 100 G; time constant, 1s; scan time, 16 min; microwave power, 31 nW; microwave frequency, 9.76 GHz; modulation amplitude, 2.5 G.

Results

The behavior of iodoacetamide, FDNB and 5-doxyl stearic acid spin probes was examined in nonionic microemulsion systems, consisting of water/n-decane/$C_{12}EO_4$. Thus, the ESR spectra of these probes were recorded in microemulsion systems with various amounts of water (as expressed by the hydration ratio of the reversed micelles, $R = [H_2O]/[C_{12}EO_4]$) and different surfactant content (ranging from 5 to 25 %).

Figures 1a and 1b show the two representative types of the ESR spectra recorded in the microemulsion systems used in this study. Figure 1a, presents a typical one-spin state spectrum obtained when iodoacetamide spin-label was used. The ESR spectra of this probe did not show any qualitative change throughout the entire range of compositions of the microemulsion systems. In contrast, the type of spectrum 1b shows the existence of two different spin probe positions and was obtained when FDNB and 5-doxyl stearic acid spin labels were used in microemulsions of specific compositions (detergent content: 10 %, $R > 8$). The possibility of observing this latter type of spectrum, where the different number of spin probe positions are revealed, depends on the rate of exchange of the particular spin label between the different microphases of the system and it has been observed in a biological membrane system [6].

The characteristic splitting of the high field line into a doublet (lines H_{-1}, H^*_{-1}) is interpreted as the sum of two spectra: one arises from probes in aqueous solution (H^*_{-1}) and the other from labels located in the micromembrane which separates the aqueous from the oil phase. The partition of the spin probe between the two microphases is expressed as the ratio H_{-1}/H^*_{-1}. Figure 2 presents the variation of this ratio for FDNB spin label as a function of R for various surfactant concentrations. It is shown that for a specific surfactant content the H_{-1}/H^*_{-1} ratio decreases with increasing R. In addition, this parameter (H_{-1}/H^*_{-1}) presents a significant variation which is abrupt between $R = 8$ to 15. The above observation is in agreement with the variation of the mobility parameter of the FDNB (Fig. 3) and 5-doxyl (Fig. 4) spin labels in the dispersed phase expressed in terms of the ratio of the low field to the center line (H_1/H_o) vs. R in nonionic microemulsions of various surfactant concentrations. Namely, the mobilities of FDNB and 5-doxyl stearic acid spin labels exhibit abrupt changes in slope at the characteristic R values between 8 and 15. It should be pointed

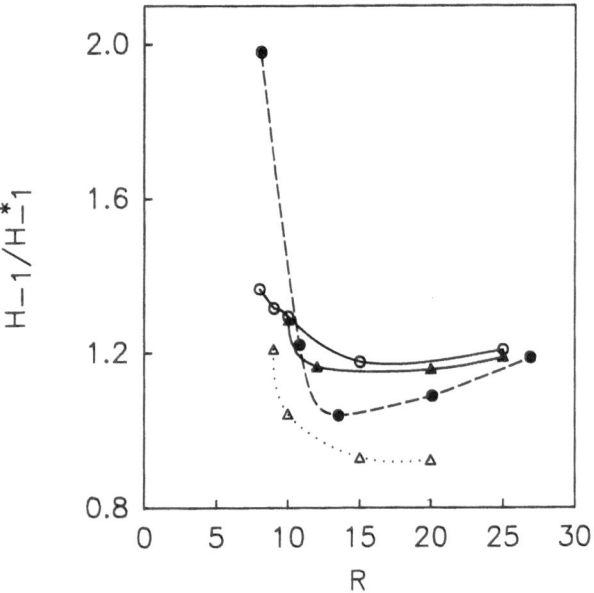

Fig. 2. Variation of the ratio of the two high field lines H_{-1}/H^*_{-1} vs. R, for various surfactant concentrations. a) (O—O): 10%; b) (●—●): 15%; c) (△—△): 20%; d) (▲—▲): 25% (FDNB spin label)

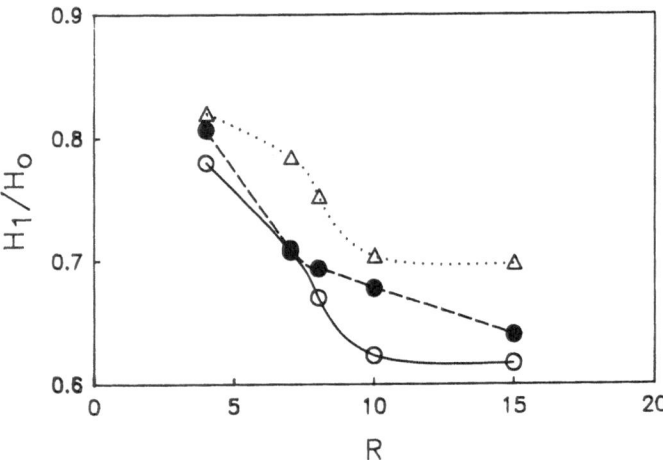

Fig. 4. Variation of the mobility of 5-doxyl stearic acid expressed as the ratio H_1/H_o vs. R for various surfactant concentrations a) O—O: 10%; b) ●—●: 15%; c) △—△: 20%

out that in microemulsion systems with surfactant content of 5% or below this value, the abrupt change was not observed.

Contrary to the above observations, iodoacetamide spin label, a rather small and hydrophilic molecule, shows behavior different from those of FDNB and 5-

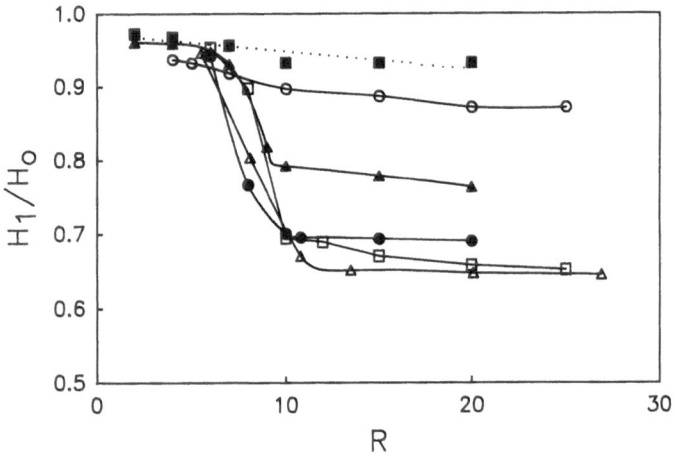

Fig. 3. Variation of the mobility of the spin labels expressed as the ratio H_1/H_o vs. R, for various surfactant concentrations. a) (O—O): 5%; b) (●—●): 10%; c) (△—△): 15%; d) (▲—▲): 20%; e) (□—□): 25% (FDNB spin label). (■·····■) correspond to the iodoacetamide spin label

doxyl labels, which are concidered hydrophobic and amphiphilic molecules, respectively. More specifically, Fig. 3 shows that the mobility parameter of IA remains almost invariable throughout the entire range of R values studied. This means that this probe spends more time in the water phase than in the micromembrane.

In addition, the fact that this probe is tumbling with a very rapid rate in water, with a correlation time, $\tau_c = 7.9 \times 10^{-11}$ s, compared to those of FDNB, $\tau_c = 1.6 \times 10^{-10}$ s, and 5-doxyl, $\tau_c = 1.7 \times 10^{-10}$ s (calculated as described elsewhere [9,10]), could explain the one-spin probe spectrum of Fig. 1a obtained throughout the entire range of compositions of the microemulsion systems studied.

In conclusion, these observations suggest that structural changes occur within the nonionic microemulsion systems for the particular water contents ($R = 8$ to 15). Namely, the beginning of the appeerence of the splitting of the high field line on one hand and the abrupt change of the mobility parameter on the other, correspond to the onset of structural changes, while the plateau parts of the plots of Figs. 2, 3, and 4, correspond to the completion of these structural changes from shapeless aggregates to reversed micelles (8).

References

1. Swartz HM, Bolton JR, Borg DC (eds) (1972) Biological Applications of Electron Spin Resonance. Wiley Interscience, New York

2. Berliner LJ (ed) (1976) Spin Labelling: Theory and Applications. Academic Press, New York
3. Taupin C, Drolaitzky M (1987) In: Zana R (ed) Surfactant Solutions: New Methods of Investigation, Vol 22. Marcel Dekker, Inc, New York
4. Fendler JH, Fendler EJ (1975) Catalysis in Micellar and Macromolecular Systems. Academic Press, New York
5. Yoshioka H (1983) J Colloid Interface Sci 95:81
6. Hubbell WL, McConnell HM (1968) Proc Natl Acad Sci 61:12
7. Cazianis CT, Sotiroudis TG, Evangelopoulos AE (1980) Biochem Biophys Acta 621:117
8. Ravey JC, Buzier M, Oberthur A (1987) Prog Colloid Polym Sci 73:113
9. Martinie J, Michon J, and Rassat A (1975) J Amer Chem Soc 97:1818
10. Griffith OH, Cornell DW, and McConnell HM (1965) J Chem Phys 43:2909

Received November, 1988;
accepted January, 1989

Authors' address:

C. T. Cazianis
Institute of Biological Research
The National Hellenic Research Foundation
48, Vas Constantinou Ave.
11635 Athens, Greece

Progress in Colloid & Polymer Science   Progr Colloid Polym Sci 79:218–225 (1989)

Perfluoropolyether microemulsions

A. Chittofrati, D. Lenti, A. Sanguineti, M. Visca, C. M. C. Gambi[1]), D. Senatra[1]), and Z. Zhou[2])

Montefluos Spa, Colloid Laboratory, Spinetta Marengo, Italy
[1]) University of Florence, Department of Physics, Florence, Italy
[2]) On leave from Hubei University, People's Republic of China

Abstract: In the present paper a systematic investigation is presented of the phase behavior of perfluoropolyether (PFPE) surfactant, PFPE oil, water and isopropyl alcohol mixtures. Monophasic regions of isotropic, transparent, and fluid samples are identified in the thermal range 20° ÷ 60 °C. Preliminary data by light-scattering investigation support the hypothesis of the presence of O/W and W/O structural aggregates.

Key words: Perfluoropolyether (PFPE) polymers, PFPE microemulsions, phase diagrams on PFPE mixtures, light scattering on PFPE mixtures, DSC analysis on a PFPE surfactant.

Introduction

The aim of this work is to understand whether microemulsion's formation can take place by mixing perfluoropolyether (PFPE) surfactant, PFPE oil and water in both the presence and the absence of a short-chain hydrocarbon alcohol. The perfluoropolyethers (PFPE) used are fluorinated polymers produced by Montefluos (Milan, Italy) of general structure:

$$CF_3-[(O-CF_2-CF)_n-(O-CF_2)_m]-O-$$
$$\underset{CF_3}{|} \qquad \qquad n/m = 20 \div 40$$

The properties of PFPE polymers, both oil and surfactant, and of their mixtures with water and alcohol have been reported in [1]. PFPE polymers have the main properties of fluorocarbons (thermal, chemical, and biological inertness, high permeability to gases, low compatibility with aqueous and hydrocarbon-based systems). The rigidity of the fluorocarbon chain (due to the bulky fluorine atoms) which allows fluorocarbons to be liquid only at very low molecular weights, is reduced in PFPE polymers by the etheral bridges of the polymer chains (C–O–C). The liquid state in PFPE is allowed even at molecular weights as high as 10,000 or more; the thermal range of the liquid state is also increased.

As fluorocarbon microemulsions have been prepared in the presence of fluorinated surfactants, for example as blood substitutes [2, 3], PFPE microemulsions, if they exist, could be prepared in wider molecular weight and temperature ranges.

Because the knowledge of the phase diagram is a fundamental basis to understand surfactant systems, this paper mainly focuses on the phase diagram investigation which has been performed on the binary, ternary and quaternary mixtures of PFPE surfactant, PFPE oil, water and isopropyl alcohol, in the thermal range 20° ÷ 60 °C; isopropyl alcohol has been chosen for its ability to sharply increase the surfactant solubility. The phase transition temperatures of the surfactant have been evidenced by differential scanning calorimetry in parallel with the visual observation of the surfactant properties in the range 20° ÷ 80 °C. Some surface tension and interfacial tension measurements as well as structural investigation by quasi-elastic light scattering and light intensity measurements have also been carried out.

Experimental

Materials

PFPE polymers are prepared industrially by U.V. photoinitiated oxidation of perfluoropropene at temperatures in the range − 40° ÷

— 80 °C [4]. As an intermediate of the production process, functional products are obtained, characterized by a single carboxyl end group. The products are polydispersed; practically monodispersed fractions are obtained, by distillation under reduced pressure.

The PFPE surfactant and the PFPE oil used in this work have the same R_f group of general formula reported in the introduction.

The PFPE surfactant R_f—CF_2—$COO^- NH_4^+$, of narrow molecular weight distribution (95 % by gaschromatographic analysis), has a molecular weight of 710; R_f has 3 ÷ 4 monomer units. After the neutralization of the PFPE acid (R_f—CF_2—$COOH$) with aqueous ammonia, the surfactant was dried at 60 °C under vacuum for several days. The water content of the surfactant, as determined by Karl Fisher titration in anhydrous methanol is 0.28 % by weight. From now on, the latter water content will affect the surfactant concentration values. The surfactant, which is highly hygroscopic, was stored under vacuum.

The PFPE oil R_f—CF_3 (narrow molecular weight distribution) has a molecular weight of 800, thus R_f has 4 monomer units. The oil is a transparent liquid of density 1.8 g/cm³, index of refraction 1.282 and viscosity 0.0684 poises; its ability to solubilize air is 26 cm³ of air every 100 cm³ of oil; the solubility in water is 14 10⁻⁶ w/w.

Isopropyl alcohol of RPE grade (Carlo Erba, Italy) and double distilled or Millipore Milli-Q water have been used.

Methods

a) Surface and interfacial tension were determined at 25 °C by the Du Nouy ring method on a Kruss 110 tensiometer.

b) Differential scanning calorimetry (DSC). The study of the thermal properties of the PFPE surfactant was performed with DSC by means of a Mettler TA 3000 thermal analyzer equipped with a DSC-30 low temperature cell. The heat flow rate (dQ/dt) vs temperature was recorded at constant pressure during the controlled heating of previously frozen surfactant samples (DSC-ENDO). The temperature rate (dT/dt) of 4 °C/min was used in the DSC-ENDO measurements of samples frozen with a 2 °C/min scan speed. The temperature interval investigated extends from −130° to +100 °C.

c) Quasi-elastic light scattering analysis (QELS) was performed on a Brookhaven apparatus (BI − 200SM goniometer plus BI − 2030 AT correlator and computer with 128 channels) with a time resolution of 0.1 µs. The light source is a vertically polarized Argon ion laser (Spectra Physics series 2000, $\lambda = 514.5$ nm) steadily kept at 0.1 W power. A calibration procedure was carried out on a dilute monodispersed suspension of polystyrene latex spheres. The data were analyzed using the cumulant technique (software provided by Brookhaven Instrument Co.). The correlation function is expanded about an average linewidth $\bar{\Gamma}$ as a polynomial in the sample time with cumulants as parameters to be fitted [5,6]. The expansion is stopped at the second moment result and a weighted, least-squares technique is applied to the second order polynomial to determine the constants and their standard deviations. The so-called average mutual diffusion coefficient $\bar{D}$ is related to $\bar{\Gamma}$ by the relation $\bar{D} = \bar{\Gamma}/K^2$ where $K = (4\pi/\lambda) n \sin(\theta/2)$ is the scattering wave vector, n the index of refraction of the sample and θ the scattering angle. The deviation of the experimental curve from a single exponential decay is usually given by $\mu_2/\bar{\Gamma}^2$ (where μ_2 is the second moment of the distribution) and called "degree of polydispersity" [6]. In some cases multi-exponential fittings were done by a non-negatively constrained least squares method (software provided by Brookhaven Instrument Co.). Dust-free samples were obtained by filtration

through Millipore filters (0.2 ÷ 0.45 µm-pore size). For samples exhibiting high scattering power, where the scattering due to dust is relatively low with respect to the scattering of the sample itself, measurements were carried out before and after filtration, to test if a variation was induced on the sample by the filtration procedure. No significant difference in the measured variables was observed, thus the filtration procedure seems not to modify the samples.

d) Light Intensity Measurements. The intensity of the scattered beam has been measured as total counts of photon (accumulated during one experimental run) divided by the experimental duration in seconds over identical scattering volumes for all the samples. The statistical error is the standard deviation over several measurements. Angular intensity measurements on two Rayleigh scatterers (benzene and toluene) have been done in the range 45° ÷ 150 °C; the measured values (corrected for dead time, dark count, and scattering volume effects) are identical, with a deviation between the values of less than 3 %.

Results and discussion

Surface and interfacial tension measurements

The surface tension of PFPE oil is 20 dynes/cm at $T = 25$ °C. The interfacial tension PFPE oil/water is 40 dynes/cm; the addition of PFPE surfactant decreases the interfacial tension to few dynes/cm. Furthermore, a critical micellar concentration C.M.C. = 3 10⁻⁵ mole/liter has been measured and the surface tension of $\simeq 15 \div 20$ dynes/cm has been found for the water — surfactant solution at concentrations higher than the C.M.C. value.

DSC Analysis

The DSC analysis was done on the PFPE surfactant. The thermal spectra (endotherms a–b and c) are reported in Fig. 1. The thermal curve (a) corresponds to the very first experimental run made; curves (b) and (c) are successive measurements on the same sample. The only remarkable difference consists in the disappearence of the thermal event at $T \sim 50$ °C in the DSC spectra (b) and (c). The latter behavior seems to be characteristic of the given surfactant as it results from visual observation of this component in the thermal range 20° ÷ 80 °C. The surfactant is white and solid for $T < 40$ °C; its aspect changes gradually from white to transparent in the range $40° \leq T \leq 50$ °C but it remains solid in this range. At $T \simeq 50$ °C the surfactant becomes a transparent not birefringent fluid of very high viscosity; the viscosity decreases slowly for a temperature increase towards 80 °C. During following runs between 20 °C and 80 °C the surfactant remains transparent solid, or transparent liquid of high viscosity, depending on the temperature value. The heating

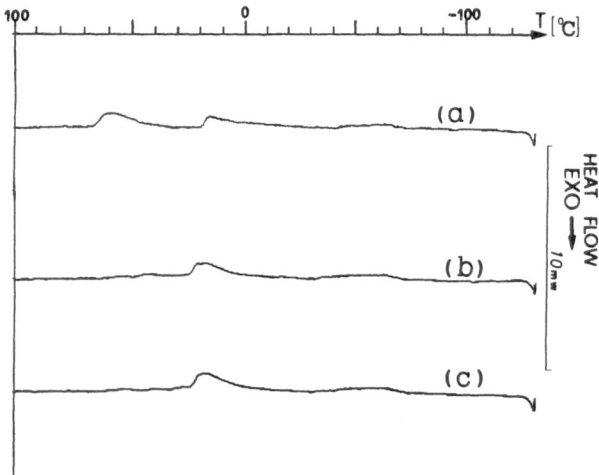

Fig. 1. DSC recordings of the PFPE surfactant: three successive measurements of the same sample are shown

at $T > 50\,°C$ seems to be necessary to obtain an homogeneous structure from the initial complex organization of the surfactant.

Phase diagram

For each binary, ternary, and quaternary phase diagrams of PFPE surfactant, PFPE oil, water, and isopropyl alcohol mixtures of given composition were prepared at room temperature and put into glass tubes (closed by teflon plugs). Each sample was stabilized at $80\,°C$ in order to enhance the solubility before starting the phase diagram investigation. At the end of the stabilization time the number of phases was counted and the behavior of each phase described after examination by visual observation and between two crossed polarizers. Phase equilibrium was considered as achieved when no further change appeared with time. The stabilization time, which strongly depends on the composition and/or nature of the samples, was more than 5 h in the range $40° ÷ 60\,°C$ and more than 10 h in the interval $20° ÷ 40\,°C$. The thermal range investigated is $20° ÷ 60\,°C$.

The following symbols and abbreviations have been used throughout: C is the concentration of one component over the total sample amount in w/w. For the phase diagrams: 1ϕ corresponds to monophasic, isotropic, homogeneous and transparent domains; 2ϕ corresponds to biphasic transparent domains. S: surfactant, O: oil, W: water and A: alcohol. If not otherwise specified, a 10% composition step was used to prepare the samples for phase diagram investigation.

The binary phase diagrams of surfactant − water, surfactant − oil, and oil − water systems are reported in Fig. 2. The following informations can be deduced:

a) The surfactant − water system gives birefringent phases for any composition at any temperature tested;

b) Surfactant is insoluble in oil for $T < 30\,°C$ for any composition tested; the samples are transparent solid homogeneous not birefringent for an oil concentration $C < 30\%$, while for $C \geq 30\%$ two phases are observed, the lower is white − turbid and the upper is transparent − fluid (see Fig. 2, PP' line). For temperatures higher than $30\,°C$ solubility is achieved. Surfactant − oil samples belonging to the monophasic domain are characterized by differences in the fluidity value; at a given temperature the fluidity increases gradually from the surfactant rich corner to the oil rich corner.

c) As shown in the experimental section, the water solubility in oil is too low to be reported in the phase diagram scale.

The surfactant − alcohol and the oil − alcohol systems have been also studied in the range $20° ÷ 60\,°C$. Isopropyl alcohol solubilizes surfactant at any concentration higher than 5% and practically at any temperature tested. However it is insoluble in oil. We recall that isopropyl alcohol is soluble in water in any proportion.

The ternary phase diagram of the surfactant − oil − water system is reported in Fig. 3 where the monophasic domains at three main temperatures are shown. The monophasic domain (transparent not birefringent

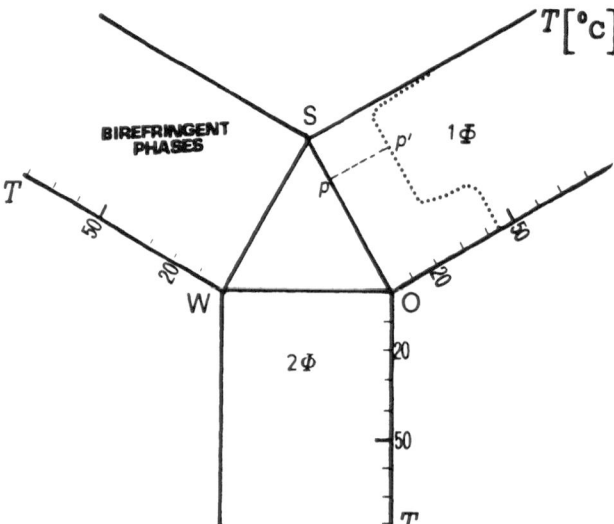

Fig. 2. Binary phase diagrams investigation of the S-W, W-O, and O-S mixtures

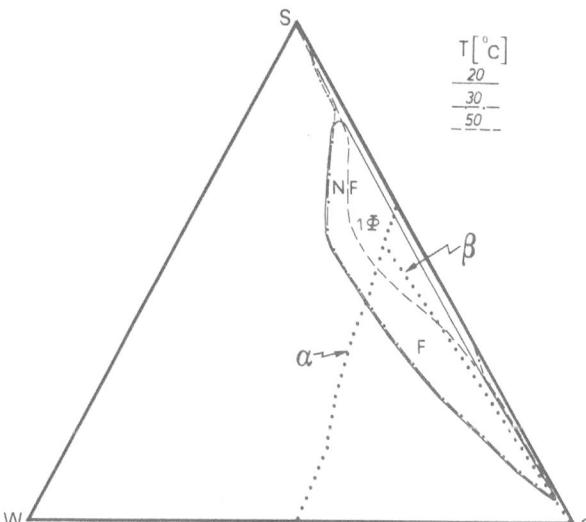

Fig. 3. Ternary phase diagram of the system S-W-O mixture. Monophasic regions are shown as a function of temperature. α- and β-lines identify excluded regions on the basis of geometrical considerations. NF and F symbols identify not-fluid samples, respectively

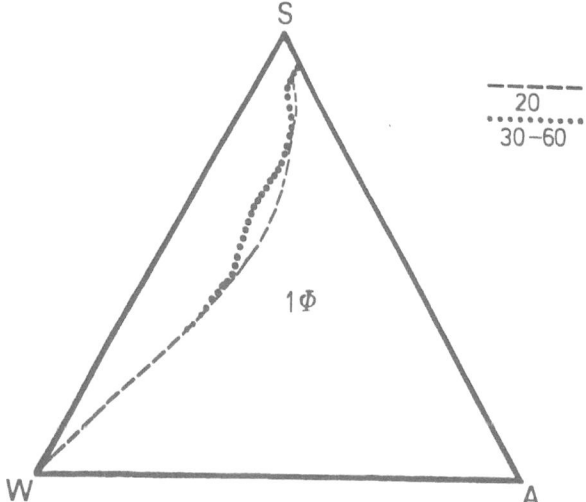

Fig. 4. Ternary phase diagram of the S-W-A mixture. Monophasic regions are shown as a function of temperature

samples) is close to the oil — surfactant side at all the temperatures in the range 20° ÷ 60 °C. At $T = 20$ °C the domain is a loop, along the S–O side; the maximum water amount of the domain is ~ 17%. At $T = 30$ °C the loop remains the same as before but it opens towards the O–S side in the range 5 ÷ 60% of oil. At $T = 50$ °C the domain becomes thinner (the maximum water amount is 12%) and wider along the O–S side. The sample viscosity tends to decrease upon approaching the oil rich corner. Therefore poorly fluid samples (*NF*) as well as fluid samples (*F*) are observed (see Fig. 3).

In Fig. 4 the ternary phase diagram of the surfactant — water — alcohol system is reported. The monophasic domain is very large for all the temperatures tested. The minimum alcohol amount which solubilizes any proportion of the surfactant — water mixture is about 20%. This phase diagram shows the great ability of the alcohol to solubilize both surfactant and surfactant plus water birefringent phases suggesting that the alcohol may affect the interfacial film curvatures.

In Fig. 5 pseudoternary phase diagrams of the surfactant — alcohol — oil — water system are reported. The surfactant plus alcohol mixture is considered as a pseudocomponent. Four main S/(S + A) ratios were studied, namely: 0.6, 0.7, 0.8, and 0.9 (w/w). They correspond respectively to the molar ratios 0.13, 0.19, 0.34, and 0.76. The thermal range studied is always 20° ÷ 60 °C. For the lower ratio S/(S + A) = 0.6 the

monophasic region is narrow, strictly close to the water side and independent of the temperature. At S/(S + A) = 0.7 the monophasic domain enlarges from the water side towards the oil corner to cover regions having oil content always lower than 20%. At S/(S + A) = 0.8 the monophasic domain changes drastically; the solubility towards the water side is decreased; however the solubility towards the oil side is increased. The domain enlarges for temperature increase. At S/(S + A) = 0.9 the monophasic domain is very narrow and close to the oil side; a temperature increase, increases the solubility along the same side.

From the investigation of binary and ternary phase diagrams (Figs. 2 and 3), the following main considerations can be made:

a) At $T < 30$ °C the surfactant is insoluble in water as well as in oil; however the larger monophasic domain is exhibited by the W–S–O system.

b) At T > 30 °C, against a temperature increase, the surfactant becomes more and more soluble in oil and the monophasic domain correspondingly spreads towards the O–S side. The increase of the surfactant solubility in oil decreases the ability of the S–O mixture to solubilize water.

From the previous considerations we can deduce that, at T < 30 °C, some structural organization of the components water, surfactant, and oil could explain the formation of the wide monophasic domain. Furthermore, from composition considerations, only water in oil structures can be hypothesized for this domain. However, we cannot a priori exclude a simple

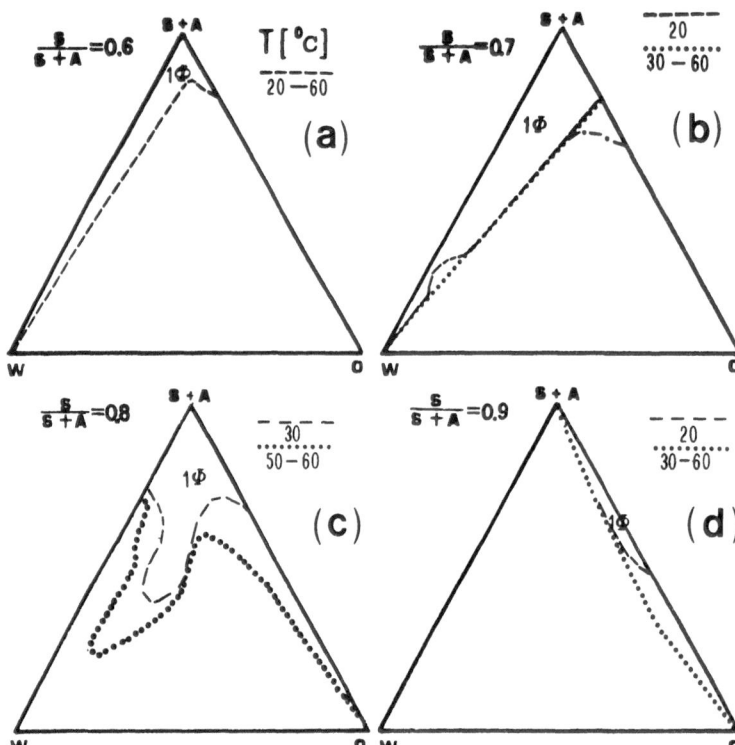

Fig. 5. Pseudoternary phase diagrams of the mixture S–A–W–O for different S/(S + A) ratios (% w/w). The temperature dependence of monophasic domain is also shown

cosolubilization of the components. In the case that water in oil droplets exist, it is reasonable to suppose the water inside the core and the surfactant at the interface while the oil is the continuous medium. From geometric considerations [7], two excluded regions for w/o structures should be taken into account, one due to the close-packing limit and the other due to the minimum surface-limit available for the polar-head area of the surfactant, which can be roughly assumed at around 50 Å^2, as deduced from surface pressure-area curves [8]. The close-packing limit of hard spheres, 0.64, can be expressed as: $\phi = (V_w + V_s)/(V_o + V_w + V_s)$ where V are the volumes of water, surfactant, and oil, respectively. In Fig. 3 the α line has been calculated for $\phi = 0.64$, assuming for the density of surfactant and oil the value 1.8 g/cm^3 because the two types of molecules are very similar (the excluded region falls in the upper part of the triangle). The polar-head area limit originates the β line which excludes the region very close to the S–O side. If σ is the surface polar-head area, R the droplet radius, l the surfactant length, V_s^* the surfactant molar volume, and N_A the Avogadro number, then:

$$\sigma = \left(\frac{3V_s^*}{N_A}\right)\left(\frac{(R-l)^2}{R^3 - (R-l)^3}\right).$$

The last equation implies that under the hypothesis of constant l, the points of constant radius fall in the same locus as those of constant σ, the lower limit of σ being the area of the surfactant polar head. The l value of 13 Å is found from molecular weight and density values supposing a 50 Å^2 polar-head area. It is worth noting that the region identified through the geometrical evaluation agrees with the one found experimentally on the actual system for what concerns the low-viscosity domain (F) of isotropic fluid and transparent samples. We have to point out that the assumption that each component is insoluble in the others (which is fundamental for the geometrical model hypotheses) is very rough in our case, expecially because at oil concentrations less than 30%, the binary mixture S–O is an homogeneous transparent solid, at the visual observation, which induces us to think that some kind of interaction is possible between oil and surfactant. In order to detect whether structures exist in the monophasic domain of the S–W–O system, light-scattering investigations have been performed. As the index of refraction of the oil and of the surfactant tail must be approximatively the same, (because the two molecules have an identical R_f group), we expect a continuous medium charcterized by an index of refraction of 1.282 and a dispersed phase (water) of index of refraction 1.333.

Table 1. Sample composition in % w/w

	S1	S2	S3	S4	S5	S6	S7
S	47.80	9.54	45.93	28.24	18.65	41.78	9.44
O	49.98	90.01	44.06	64.78	74.13	51.37	88.98
W	2.25	0.45	10.01	6.98	7.22	1.97	0.45
A	–	–	–	–	–	4.88	1.14

The effect of the isopropyl alcohol addition to the ternary system S–O–W has been investigated specifically with the aim of finding water in oil structures. For ionic surfactants it is a general result that short chain alcohols are too soluble in the aqueous phase to be useful as cosurfactant [9]. For example, isopropanol is insoluble in sodium dodecyl sulfate and the monophasic domain of the ternary W–A–S occupies the region close to the W–A side. Since isopropanol solubilizes the PFPE surfactant quite well (see Fig. 4), we have carried out a preliminary investigation to test if isopropanol favors the formation of w/o, o/w structures or both, independently on the cosurfactant or cosolvent effect of the alcohol itself. From the phase diagram of Fig. 5 it is seen that the monophasic domain belongs either to the region accessible to w/o or to o/w structures; the above geometrical considerations can also apply to an o/w system [7]. It follows that light scattering investigation may offer some evidence for the presence of structures. Such a result would strongly support that a transition from w/o curvature to the o/w one, takes place due to the presence of the alcohol.

QELS and light intensity measurements

A preliminary QELS analysis has been carried out on ternary and quaternary systems with the aim of obtaining structural information [10, 11]. The composition of the samples investigated is reported in Table 1. For all the samples, the autocorrelation function of the scattered intensity has been measured at different angles in the angular range $45\,°C \div 150\,°C$. Constant $\bar{\Gamma}/K^2$ values have been detected; therefore the scattering process is diffusive and the mutual diffusion coefficient can be espressed by $\bar{D} = \bar{\Gamma}/2K^2$ (homodyne detection). As an example of the angular dependence, the $\bar{D}(\theta)$ values for sample S_2 at $T=35\,°C$ are reported in Table 2. For the evaluation of the scattering wave vectors of all the samples analyzed, the index of refraction used was that of the oil (1.282) because the samples are mainly composed of oil. In fact, all the n values estimated by the refractivity formula [12] were found to differ less than 1% from the n value of oil. In Table 3 the $\bar{D}$ values at $\theta = 90\,°$ are reported for all the samples investigated. Furthermore, in Table 3, $\mu^2/\bar{\Gamma}^2$ and the intensity values are also given. For all the samples the I values have been detected at $\theta = 90\,°$. No sharp peak has been observed in the $I(\theta)$ dependence; further investigation on the $I(\theta)$ dependence will be the object of a future work. The $\bar{D}$ and I values reported in Table 3 are averages over several measurements; the standard deviations are also given. All the samples have been studied in cylindrical cells of identical diameter (12 mm). For sample S_4, which has an intermediate scattering power value between S_1 and S_5, a cylindrical cell of 26 mm diameter has been also used to test if multiple scattering affects the measurements. As the $\bar{D}$ values are identical, within the experimental error, it is reasonable to deduce that multiple scattering does not affect the results.

The oil-sample alone was also studied; no autocorrelation function could be found in the limit of resolution of the apparatus, and the intensity was practically undetectable.

Because oil by itself does not scatter light, the oil-rich samples (S_2, S_4 and S_5) scatter light strongly, therefore, some structural organization must be inside the S_2, S_4 and S_5 samples due to the presence of water and surfactant. In fact, the I values of S_2 S_4, and S_5 samples are the highest values detected (see Table 3). Correspondingly, $\bar{D}$ values in the range $1 \div 0.37 \times 10^{-8}$ cm^2/s have been obtained. In the frame of ternary samples, the S_1 and S_3 ones with about a 50% oil to surfactant ratio and a scattering power less than the S_2,

Table 2. Angular dependence of the mutual diffusion coefficient for S2 at $T = 35\,°C$

θ	45°	60°	90°	120°	135°	150°
$\bar{D}(\theta) \times 10^8$ cm^2/s	1.23 ± 0.08	1.11 ± 0.09	1.19 ± 0.08	1.24 ± 0.06	1.22 ± 0.03	1.20 ± 0.03

Table 3. Light-scattering data

	S1	S2	S3	S4	S5	S6	S7
$T(°C)$	25	35	25	25	25	25	–
$\bar{D}(\theta) \times 10^7$ cm^2/s	0.84 ± 0.01	1.19 ± 0.08	0.86 ± 0.02	0.57 ± 0.01	0.37 ± 0.01	1.19 ± 0.02	–
$\mu_2/\bar{\Gamma}^2$	0.09	0.11	0.08	0.05	0.09	0.08	–
I(a.u.)	8.70 ± 0.1	120 ± 1	17.6 ± 0.1	45.3 ± 0.1	149 ± 1.2	12.4 ± 0.1	–
$T(°C)$	50	50	–	25	–	50	50
$\bar{D}(\theta)10^7$ cm^2/s	2.14 ± 0.16	0.42 ± 0.02	–	$0.58^a) \pm 0.02$	–	2.82 ± 0.12	1.08 ± 0.02
$\mu_2/\bar{\Gamma}^2$	0.20	0.13	–	–	–	0.13	0.04
I(a.u.)	8.20 ± 0.05	85.9 ± 1.4	–	–	–	10.6 ± 0.1	41.4 ± 0.7

a) cell diameter 26 mm

S_4, and S_5 samples, a $\bar{D}$ value falls within the previous range.

In summary, all the ternary samples of Table 3 appear to possess a structural organization. As the water content of all the samples is lower than 10 % we can hypothesize that the samples are of water-in-oil-type; a simple cosolubilization of the components can therefore be excluded. The degree of polydispersity, always lower than 10 % at $T = 25$ °C and lower than 20 % at all the temperatures tested, may support the interpretation that the samples are practically monodispersed. In fact, even for the highest $\mu_2/\bar{\Gamma}^2$ value (20 %), a single-peak distribution was obtained by the NNLS technique.

Light-scattering measurements were also done on samples of the monophasic regions of the phase diagrams b, c, and d of Fig. 5. For the phase diagrams b and c (samples close to the W–(S + A) side), the autocorrelation functions display multiexponential decays and the intensity values are lower than those of samples of Table 3. However, a structural organization is exhibited as QELS signals have been detected. A quantitative investigation will be carried out in future work. On the basis of considerations about the relative proportions between the components, it is reasonable to expect o/w structures. For the phase diagram d of Fig. 5 two samples have been investigated (S_6 and S_7 of Table 1). $\bar{D}$, I and $\mu_2/\bar{\Gamma}^2$ values comparable with those of the ternary samples studied have been obtained. On the basis of geometric considerations, w/o structures are expected in the latter case.

All the samples have been studied at 25 °C if monophasic and transparent at that temperature; in case of phase separation as in the S_2 and S_7 samples, a study has been carried out a $T = 50$ °C. For comparison few other samples (S_1 and S_6) have also been studied at $T =$ 50 °C. In S_1 and S_6 samples $\bar{D}$ increases with temperature while in S_2 it decreases.

Conclusion

The hypothesis that microemulsions form by mixing, in given proportions, PFPE surfactant, PFPE oil and water (in presence or absence of isopropyl alcohol) is strongly supported by the phase diagram and light-scattering results. Typical samples (isotropic, transparent, and homogeneous) belonging to the monophasic regions of ternary and quaternary phase diagrams exhibit a structural organization. At the present state of the research, quantitative considerations about the dimensions of the structures, their geometry, as well as the interactions between the structural aggregates cannot be made. However some preliminary considerations can be put forward by comparing the findings of the studied samples which belong to the ternary phase diagram. For such samples, aggregates of water – surfactant type can be hypothesized as previously discussed. Furthermore, as the $\bar{D}$ value of the samples S_3, S_4, and S_5 decreases towards the oil-rich corner and the fluidity of the samples increases towards the same corner, we can reasonably think that the interactions between the aggregates are quite low in the S_5 sample. Thus we can give a rough estimate of the hydrodynamic radius R_H of the aggregates, assuming the $\bar{D}$ value close to the mutual diffusion coefficient at zero concentration of the dispersed phase. Using the Stokes-Einstein formula [10] and also assuming for the viscosity of the continuous medium that of the oil, an R_H value of 86 Å can be estimated, a value which is typical of microemulsion systems.

References

1. Chittofrati A, Lenti D, Sanguineti A, Visca M, Gambi CMC, Senatra D, Zhou Z, in press, Colloids & Surfaces, X Chemistry of Interfaces Conference, S Benedetto del Tronto, May 1988
2. Ceschin C, Roques J, Malet-Marrins MC, Lattes A (1985) J Chem Tech Biotechnol 35A:78
3. Selve C, Castro B, Lempoel P, Mathis G, Gartiser T, Delpuech JJ (1983) Tetrahedron 39:131
4. Sianesi D, Pasetti A, Fontanelli R, Bernardi GC, Caporiccio G (1973) Chim Ind 55:208
5. Koppel DE (1972) J Chem Phys 57(11):4814
6. Brown JC, Pusey PN, Dietz R (1975) J Chem Phys 62(3):1136
7. Biais J, Bothorel P, Clin B, Lalanne P (1981) J Colloid Interface Sci 80(1):136
8. Caporiccio G, Burzio F, Carniselli G, Biancardi V (1984) J Colloid Interface Sci 98(1):202
9. Friberg SE (1985) J Dispersion Sci and Technology 6(3):317
10. Guest D, Langevin D (1986) J Colloid Interface Sci 112:209
11. Chang NJ, Kaler EW (1986) Langmuir 2:184
12. Born M, Wolf E (1983) In: Principles of Optics. Pergamon Press

Received October, 1988;
accepted April, 1989

Authors' address:

C.M.C. Gambi
Dipartimento di Fisica
L.E. Fermi 2
I-50125 Florence, Italy

Progress in Colloid & Polymer Science   Progr Colloid Polym Sci 79:226–232 (1989)

Experimental evidence for bicontinuous structures in L_3 phases

D. Gazeau[1]), A. M. Bellocq[1]), D. Roux[1]), and T. Zemb[2])

[1]) Centre de Recherche Paul Pascal (CNRS), Domaine Universitaire, Talence, France
[2]) C.E.N. Saclay, Département de Physico-chimie, Gif sur Yvette, France

Abstract: Anomalous flow birefringent phases, sometimes designated as L_3, have been identified in both the water-rich and the oil-rich parts of the phase diagram of the water(-NaCl)-dodecane-pentanol-SDS system. The presence of these phases appear to be associated with that of swollen lamellar phases. Conductivity and neutron-scattering results provide evidence that the structure of these phases consists of a highly connected, sponge-like, random bilayer-continuous surface. The surfactant surface separates, depending upon the system, into either two water-continuous domains or two oil-continuous domains. The data are consistent with a recent theoretical model.

Key words: L_3 phase, bicontinuous structure, phase diagram, electrical conductivity, neutron scattering, fluctuating membranes.

1. Introduction

Surfactants in solution give rise to a large variety of structures such as micellar, lamellar, hexagonal, or cubic [1, 2]. These different structures can be understood as phases of surfaces [3], since surfactant aggregation builds surfaces which organize themselves in order to accomodate energetic and entropic contributions to the free energy.

Recently, a lot of interest has been focused on phases made of very flexible surfactant films. These systems have the property of having a low bending constant ($\varkappa \sim k_B T$) and can make either isotropic liquid phases (microemulsions) or lamellar phases (lyotropic smectics). In both cases, phases with very low concentration of surfactant can be prepared, leading to large characteristic distances (100–1000 Å). Thermal fluctuations have been demonstrated to be essential for explaining the stability of these phases [4–10]. In the interesting case of dilute lamellar phases experimental results have demonstrated that the very large dilutions that can be obtained are due to repulsive interactions entropically driven [7–10]. Indeed constrained thermal undulations of the membranes are responsible for the existence of a long-range repulsion between neighboring membranes. On theoretical

basis, it has been proposed [11] that dilute lamellar phases can only be stable when the repeating distance between membranes remains less than the persistence length of the film (ξ_k). For large dilutions the lamellar phase is expected to melt in a phase of connected random surfaces stabilized by entropy [11]. In a random mixing approximation, the structure of this phase is that of a sponge: a random surface made of a bilayer film separates two identical continuous phases made of the solvent. This phase has to be related to the regular bicontinuous microemulsion phase which is made of a mixing of water and oil separated with a random monolayer film. The main difference being that the sponge phase is composed of only one type of solvent water *or* oil instead of water *and* oil. One should notice that this phase has a different symmetry than the regular isotropic liquid micellar phase. Indeed, in this phase, the average mean curvature at the midplane of the bilayers is equal to zero and the two continuous phases are identical. One expects a phase transition toward an asymetric phase where there is less "inside" than "outside" [11, 12]. This asymetric phase can be seen as a phase of connected vesicles going continuously toward a phase of single isolated vesicles.

In experimental phase diagrams there exists effectively what has been called an anomalous isotropic

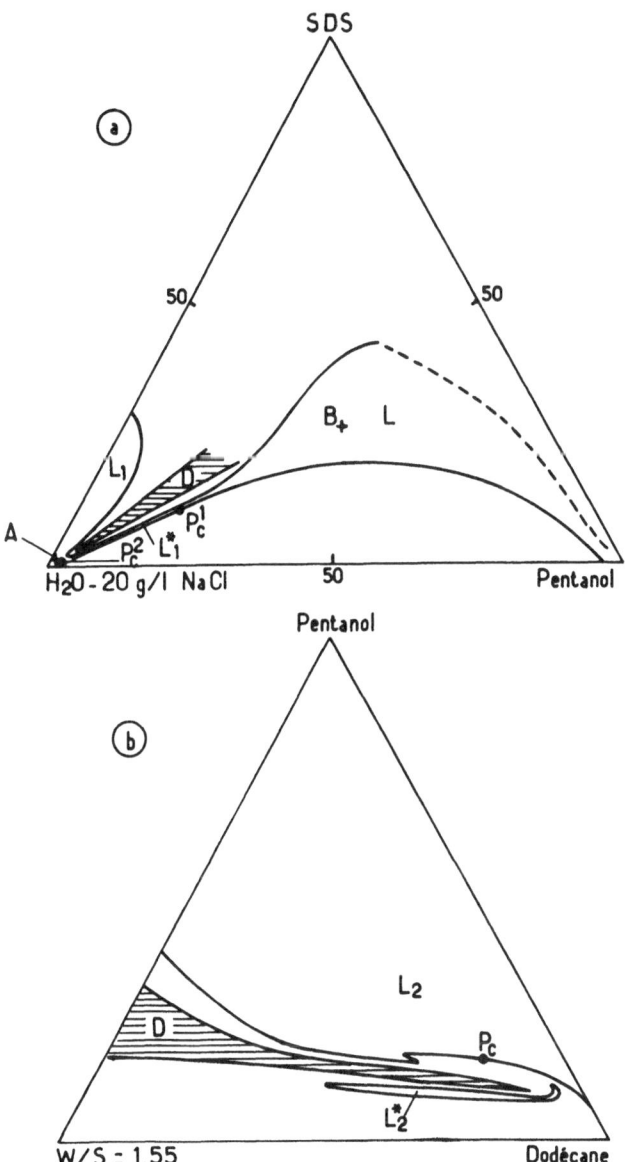

Fig. 1. a) Section at the salinity 20 g/l NaCl of the phase diagram of system I ($T = 25\,°C$). b) Section at constant water over SDS ratio ($H_2O/SDS = 1.55$) of the phase diagram of system II ($T = 25\,°C$)

aqueous and oily L^* phases made of water(NaCl)-dodecane-pentanol-sodium dodecylsulfate (SDS). Our data suggest that their structure consists of locally surfactant bilayers joined together into a multiply-connected random surface. The surfactant surface separates either two aqueous continuous domains or two oil continuous domains. Our experimental results mainly confirm, and generalize on different systems, the results obtained by Porte et al. [17], and are consistent with the model proposed by Cates et al. [11].

2. Phase diagrams

The aqueous L^* mixtures studied in this paper and noted hereafter L_1^* are made of water-NaCl-pentanol-SDS (system I) and the oily ones, noted L_2^*, consist of water-dodecane-pentanol-SDS (system II). The phase diagrams of these two quaternary mixtures have been published [16, 19]. We will present here the main features of these phase diagrams. At P and T constant, the phase diagram of a quaternary mixture may be represented in a tetrahedron. Figure 1 shows two pseudoternary sections of the tetrahedron obtained by keeping constant the ratio between two components. In Fig. 1a, the salinity is fixed at 20 g NaCl/lH$_2$O; in Fig. 1b, the water over surfactant ratio is held constant and is equal to 1.55 (expressed in weight). Comparison of these two diagrams shows strong similarities. In both diagrams one identifies, in addition to the isotropic micellar phases L_1 or L_2 and to the lamellar phase D, a flow birefringent phase L_1^* or L_2^*. The flow birefringence increases with the solvent, brine, or oil content. In Fig. 1b, the L_2 and L_2^* domains are separated by a coexistence region; in contrast, in Fig. 1a a single isotropic domain, continuous from the brine corner to the alcohol corner is observed. In the salted system, the L_1 region is connected on one side to the L_1 phase and on the other side to the alcohol-rich phase.

Whatever the mixture, the change from L to L^* is obtained by varying the alcohol concentration or by dilution of the lamellar phase. The investigation of the complete diagrams provides evidence that the existence of the L_1^* and L_2^* phases is associated with the presence of the nearby swollen lamellar phase D. Each region L_1^* and L_2^* extends well into the dilute region and occurs at surfactant concentrations ranging between 1 and 15 wt % of SDS. Both regions L_1^* and L_2^* are very narrow and limited by straight lines, indicating a constant alcohol/SDS ratio $R_{A/S}$ in the interfacial film. Expressed as the molar ratio, $R_{A/S}$ is equal to 5 in L_1^* and only 1.3 in L_2^*. All these special features suggest

phase (L^*) or the L_3^* phase in the vicinity of dilute lamellar phases [13–17]. Based on measurements of diffusion coefficient Lindman et al. have suggested that for nonionic surfactants [18] the phase could be made of very large discs. In a recent study [17], Porte et al. suggested that this phase (for one quaternary water-rich system) may consist of connected surfaces organized in a dilute bicontinuous phase.

In this paper we present results of conductivity and small angle neutron scattering (SANS) studies of both

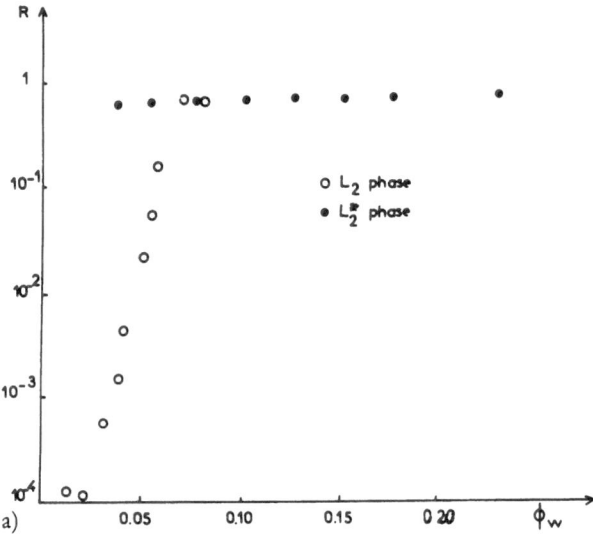

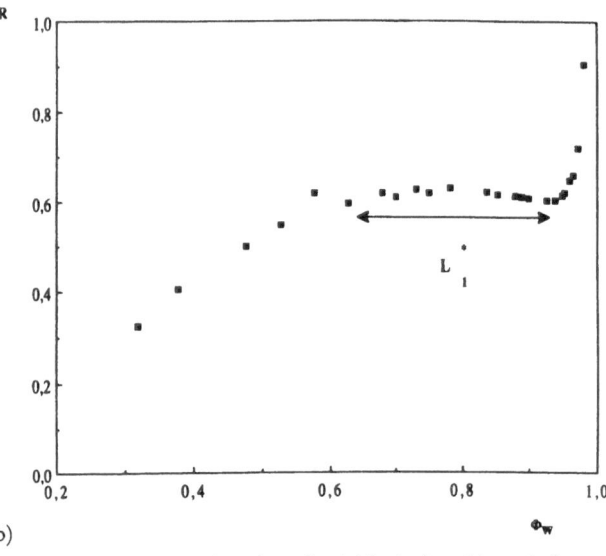

Fig. 2. a) Ratio of the reduced conductivities in the L_2^* (or L_2) phase and in the reference mixture (mixture water, SDS in the ratio 1.55) vs. ϕ_w. b) Ratio of the reduced conductivities in the L_1^* phase and the reference solution vs. ϕ_w

that the structure of the phases L_1^* and L_2^* can be investigated as the function of the dilution.

3. Conductivity data

Figure 2a presents the reduced conductivity of the L_2^* phase relative to the reduced conductivity of the pure water/SDS mixture in the ratio 1.55 as a function of the water volume fraction along a dilution path (see Fig. 1a). On the same graph are also shown conductiv-

ity measurements made in the droplet phase L_2. In the L_2^* phase, the reduced conductivity is remarkably constant and can be extrapolated to 0.63 at low concentration. It can be noticed in particular that for a water-volume fraction as low as 4 % the conductivity is three orders of magnitude higher than that of the droplet phase.

For the oil-less system, conductivity measurements have been made along the line AB of Fig. 1a. In Fig. 2b is reported the ϕ_w-dependence of the ratio R between the reduced conductivity σ/ϕ_w measured along the line AB and that of corresponding reference solutions of SDS in brine. In a reference solution, the ratio between the SDS and brine concentrations is the same as in the sample considered. Figure 2b exhibits three distinct regimes: an initial increase of R with ϕ_w, then a plateau of practically constant value ($R = 0.62 \pm 0.02$) in the range of ϕ_w comprised between 0.60 and 0.95, and finally a rapid increase of R up to 1 above $\phi_w = 0.95$. The region where the plateau is observed corresponds to the region of existence of the L_1^* phase.

The high values of σ measured in the L_2^* phase clearly demonstrate that in this phase water forms a continuous channel. Figures 2a and 2b show that in the regions L_1^* and L_2^* the mobility of ions is constant and it is reduced to approximately 2/3 the value measured in the reference solutions. Such a value is expected for diffusion within randomly distributed parallel planar confinements [20] and also for the obstruction produced by oblate aggregates [21]. This value is also consistent with the tortuosity coefficient expected for a random distribution of insulating platelets in a conducting medium [22]. Then conductivity data suggest that the L_2^* phase is a solution in oil of connected bilayers of SDS and alcohol containing water and that the L_1^* phase is a solution in water of connected bilayers of SDS and alcohol. In order to precisely determine the structure, we performed neutron-scattering experiments.

4. Sans data

Neutron-scattering experiments were performed on the spectrometer PAXY (Laboratoire Léon Brillouin, Saclay, France). The sample detector distances were set to cover the q-range between $q_{min} = 6.10^{-3}$ Å^{-1} and $q_{max} = 0.6$ Å^{-1}. This range is obtained with two overlapping positions for the detector. In the samples investigated, water was replaced by heavy water. This replacement leads to a slight shift of the phase diagram shown in Fig. 1b towards the low alco-

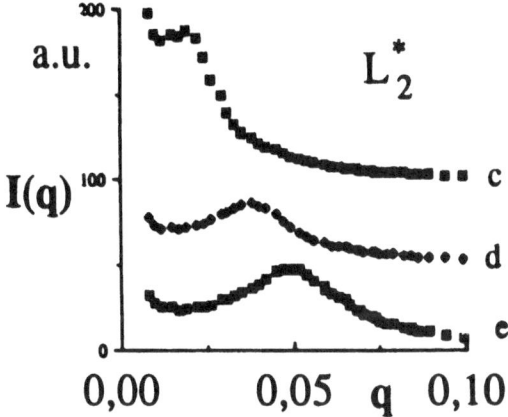

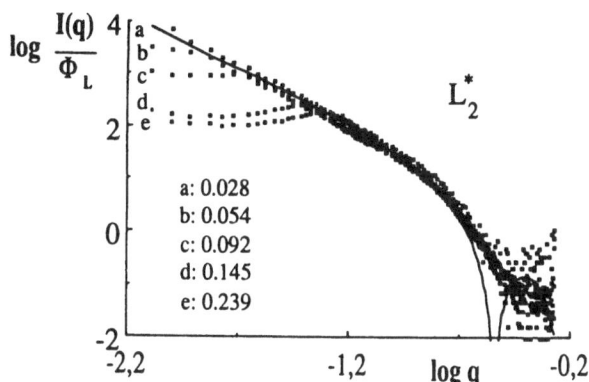

a: 0.028
b: 0.054
c: 0.092
d: 0.145
e: 0.239

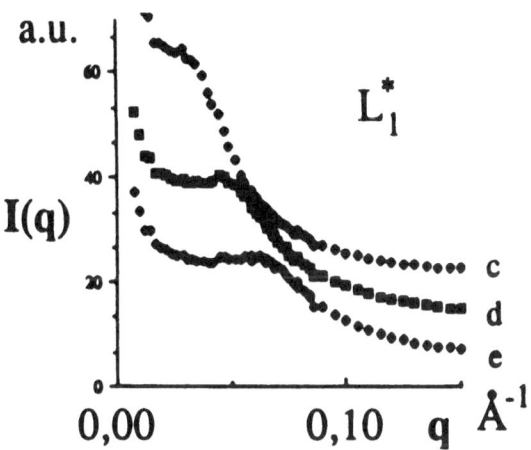

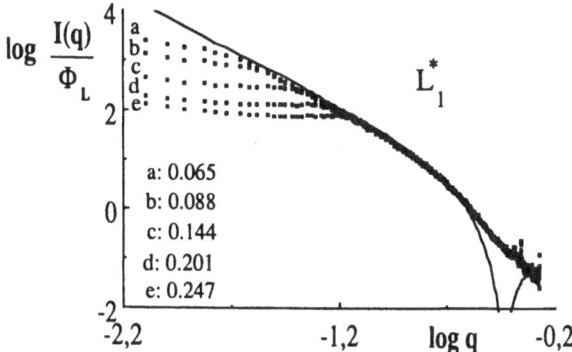

a: 0.065
b: 0.088
c: 0.144
d: 0.201
e: 0.247

Fig. 4. Log-log plot of the reduced intensity (I/ϕ) vs. q for various L_1^* and L_2^* solutions of different ϕ_L. Full line: calculated spectra for a non-interacting dispersion of lamellae ($\delta = 18.5$ Å for the L_1^* phase; $\delta = 21$ Å for the L_2^* phase; $\Delta\varrho = 6.3 \cdot 10^{10}$ cm^{-2})

Fig. 3. Experimental SANS spectra for L_1^* and L_2^* solutions of different volume fractions of bilayers ϕ_L. The concentrations are given in Fig. 4

hol content. Absolute intensity scaled measurements were performed for seven L_1^* samples (ϕ_L ranging from 0.065 to 0.659 of Fig. 1a and seven L_2^* samples (ϕ_L ranging from 0.028 to 0.239). Figure 3 shows two series of typical spectra corresponding respectively to the oil and brine dilution. All the spectra, but especially the most dilute sample of L_2^*, exhibit a correlation peak which shifts towards the low q range as the oil or brine concentration increases. Figure 4 shows the reduced absolute scattering $L_n \; I/\phi_L$ for all the samples of the two series (L_1^* and L_2^*). ϕ_L is the volumic fraction of lamellae. In the L_2^* phase ϕ_L comprises the polar part of the interfacial film, i.e., the water layer plus the polar head of the surfactant molecules. For the L_1^* phase, ϕ_L

contains the aliphatic chains of the surfactant and alcohol molecules. The scattering for a random dispersion of lamella is also shown as a solid line in Fig. 4.

$$\frac{I(q)}{\phi_L} = \frac{2\Pi}{q^2} \, \Delta\varrho^2 \, \delta \left(\frac{\sin q \; \delta/2}{q \; \delta/2}\right)^2 \qquad (1)$$

where $\Delta\varrho$ is the excess of scattering length, and δ the thickness of the lamella [23].

Clearly for q vectors larger than the peak position (q_m) all the curves merge in a unique curve identical to what is expected for the scattering of lamella. One should notice that this comparison has been done with no fitting parameters since all the measurements have been done in absolute units and that the thickness of the bilayers are obtained from the Bragg peak in the lamellar phases [7, 8]. The peak position q_m corresponds to the breakdown of the scattering dominated by the form factor of lamella. For $q < q_m$ the scattering

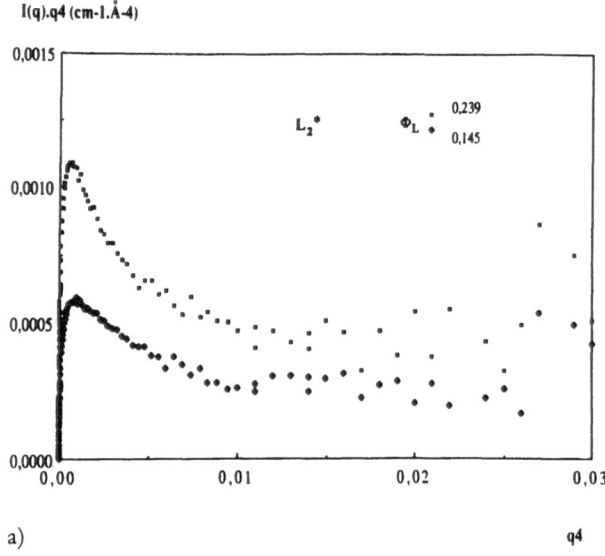

a)

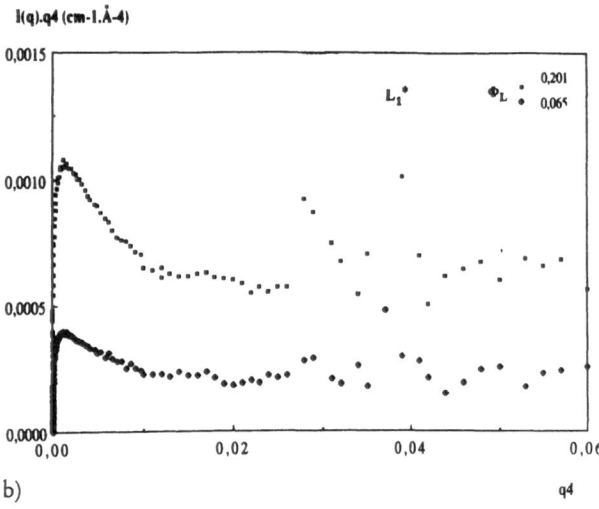

b)

Fig. 5. Iq^4 vs. q^4. Solutions L_1^* and L_2^*

is expected to be dominated by correlations between pieces of lamella and obviously depends strongly on the volume fraction. The large q $(q > q_m)$ scattering is dominated by the form factor of a lamella and can be divided in two regimes.

i) For $q > q_1 = 2\Pi/\delta$ (δ is the thickness of the bilayer) the scattering corresponds to the Porod law ($I(q) = 2\Pi \Sigma \Delta\varrho^2/q^4$; Σ is the interface area per unit volume [23]. Since along a dilution line the thickness of the bilayer is supposed to be constant, the breakdown of the Porod law should arrive for the same value q_1,

whatever the dilution. This is more accurately illustrated in Fig. 5 where $I \cdot q^4$ is plotted against q^4 for two L_2^* samples corresponding to $\phi_L = 0.145$ and 0.239 (Fig. 5a) and two L_1^* samples $\phi_L = 0.065$ and 0.201 (Fig. 5b). For $q > q_1$ the curves are constant (Porod regime) and the breakdown arrives for the same value of q_1. The value of δ obtained that way ($\delta = 2\Pi/q_1$) is respectively, $\delta = 21$ Å for the L_2^* phase, and $\delta = 19$ Å for the L_1^* phase. The coefficient of the Porod law is proportional to the total amount of surface and this allows also us to calculate δ. Using the Porod coefficient one obtains $\delta' = 23$ Å for the L_2^* phase and $\delta' = 17$ Å for the L_1^* phase, values consistent with the breakdown of the Porod law.

ii) For $q_m < q < q_1$ the scattering is dominated by the form factor of a flat lamella (formula 1) and this gives us a third method for determining the thickness of the bilayers. This is illustrated by a plot of $\ln[I(q) \cdot q^2]$ as a function of q^2, the slope allows us to determine the value of δ. One obtains $\delta'' = 21.3$ Å for the L_2^* phase and $\delta'' = 18$ Å for the L_1^* phase.

The large q analysis of the scattering clearly demonstrates that the scattering is dominated by the form factor of flat platelets and allows us to determine, using three independent methods, the thickness of the platelets. The comparison of the measured thicknesses of the bilayer with those measured in the nearby lamellar phases [7, 8] confirms that the bilayer film is of the same nature as that in the lamellar phase ($\delta_1 = 19$ Å and $\delta_2 = 21$ Å).

The way the peak position (q_m) changes with the dilution should give us indications on the nature of the correlations between platelets. Taking into account correlations in the model proposed by Cates et al. [11], the peak position should be at $2\Pi/\xi$, where ξ is the characteristic size due to random mixing [24]. In this model $\xi = \alpha \cdot \delta/\phi_S$ ($\alpha = z/4$ where z is the number of nearest neighbors in the model, $z = 6$ for the cubic lattice chosen and $\alpha = 3/2$). Figure 7 shows a plot of the characteristic distance ξ obtained from the peak position ($\xi = 2\Pi/q_m$) as a function of the inverse volume fraction of bilayers ϕ_L for the two systems studied. Clearly the behavior is linear and the slope corresponds respectively to 29.5 Å (L_2^*) and 30.5 Å (L_1^*). Taking the values of δ previously measured we get $\alpha = 1.4$ (L_2^*) and $\alpha = 1.6$ (L_1^*), in both cases the value is close to the 1.5 expected in the model of [11]. The value $\alpha = 1.5$ of the model of [11] is model-dependent, but the order of magnitude of α seems to be realistic since a Voronoi tesselation will give $\alpha \cong 1.46$.

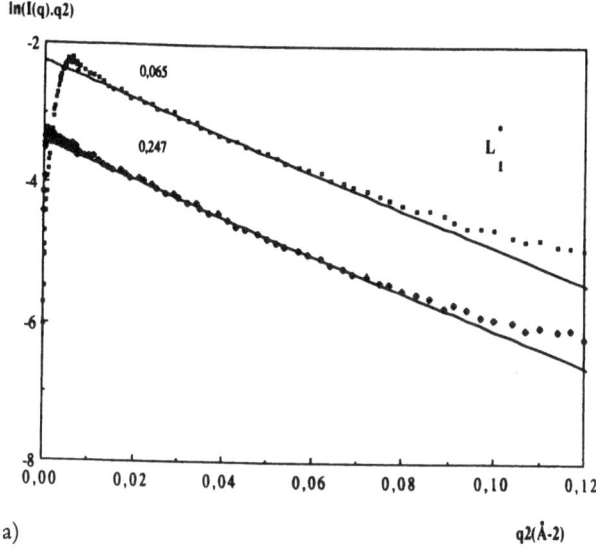

a)

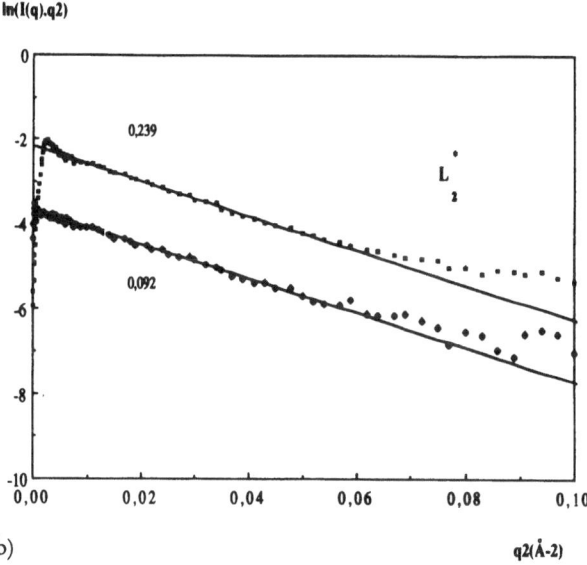

b)

Fig. 6. $\ln I(q)q^2$ vs. q^2. Solutions L_1^* and L_2^*

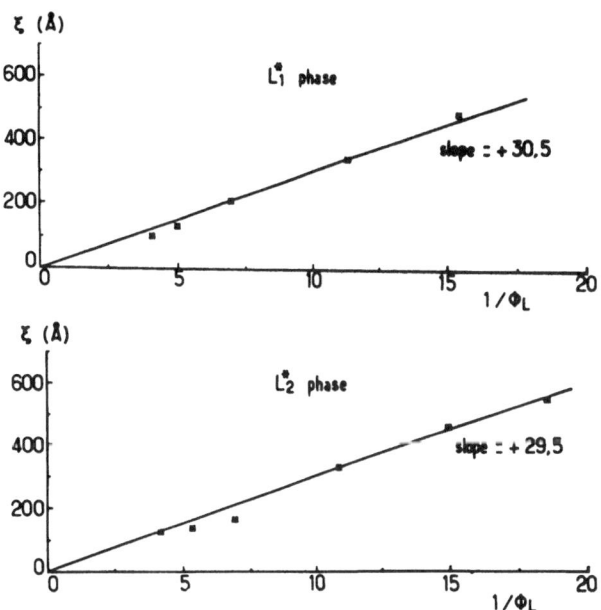

Fig. 7. Bilayer volumic fraction dependence of the distance ξ for the solutions L_1^* and L_2^*

measurements obtained by Porte et al. [17] and theoretical predictions by Cates et al. [11].

The stability of this phase against ordered phases (lamellar or cubic) is probably due to two terms. The liquid nature of this phase indicates that entropy is dominating as suggested by Cates et al. [11]. However, as it has been pointed out by Porte et al. [25] and more recently by Anderson et al. [26], an elastic term favoring Gaussian curvature is necessary to explain the phase diagram behavior.

Acknowledgement

The authors would like to acknowledge A. Brulet and L. Auvray for very helpful assistance on PAXY (LLB). Moreover, we benefited from many interesting discussions with G. Porte, M. Cates, F. Nallet, and C. Safinya.

5. Discussion

Conductivity and neutron-scattering data obtained in the L^* phase are clearly consistent with a structure of randomly connected lamellae consisting of bilayers of surfactant which separate two solvent continuous domains (water or oil depending on the system).

This phase is isotropic and liquid-like (no long-range order) and is then distinguishable from the well-known cubic phases of connected surfaces (plumber's nightmare) [1–3, 12]. These results confirm previous

References

1. Ekwald P (1975) In: Brown GH (ed) Advances in Liquid Crystals, Vol 1. Acad Press, New York, p1
2. Winsor PA (1954) Solvent Properties of Amphiphilic Compounds. Butterworths Scientific Publications, London
3. Charvolin J (1985) J de Phys 46:C3-173
4. De Gennes PG, Taupin C (1982) J Phys Chem 86:2294-2304
5. Safran SA, Roux D, Cates ME, Andelman P (1986) Phys Rev Lett 57:491-494

6. Andelman D, Cates ME, Roux D, Safran SA (1987) J Chem Phys 7229-7241
7. Safinya CR, Roux D, Smith GS, Sinha SK, Dunon P, Clark NA, Bellocq AM (1986) Phys Rev Lett 57:2718
8. Roux D, Safinya CR (1988) J Phys France 49:307
9. Larche F, Appell J, Porte G, Bassereau P, Marignan J (1986) Phys Rev Lett 56:1700
10. Nallet F, Roux D, Prost J (1989) Phys Rev Lett 62:276
11. Cates ME, Roux D, Andelman D, Milner ST, Safran SA (1988) Europhys Letters 5:733
12. Huse DA, Leibler S (1988) J Phys France 49:605-621
13. Benton WJ, Miller CA (1983) J Phys Chem 87:4981
14. Lang J, Morgan RD (1980) J Chem Phys 73:5849
15. Fontell K (1975) In: Colloidal Dispersions and Micellar Behavior. American Chem Society Washington DC, ACS Symp Ser n° 9, p 270
16. Bellocq AM, Roux D (1987) In: Friberg SE, Bothorel P (eds) Microemulsions: structure and dynamics. CRC Press, p 33
17. Porte G, Marignan J, Bassereau P, May R (1988) J Phys France 49:511-519
18. Nilsson PG, Lindman B (1984) J Phys Chem 88:4764-4769
19. Guerin G, Bellocq AM (1988) J Phys Chem 92:2550
20. Lindblom G, Larsson K, Johansson L, Fontell K, Forsen S (1979) J Amer Chem Soc 101:5465
21. Jonsson B, Wennerstrom H, Nilsson PG, Linse P (1986) Colloid Polym Sci 264:77
22. Landauer R (1977) In: Garland JC, Tanner DB (eds) Electric Transport and Optical Properties of Inhomogeneous Media. Am Inst of Physics, p1
23. Glatter O, Kratky O (1982) Small Angle X-ray Scattering. Acad Press
24. Milner ST, Safran SA, Andelman D, Cates ME, Roux D (1988) J Phys France 491:1065
25. Porte G, Appell J, Bassereau P, Marignan J (1988) Communication to the Second European Colloid and Interface Society Conference, Arcachon, France, September; (1989) J Phys France 50:1335
26. Anderson D, Wennerstrom H, Olsson U (1989) J Phys Chem 93:4243

Received December, 1988;
accepted April, 1989

Authors' address:

A. M. Bellocq
CRPP CNRS
Domaine Universitaire
F-33405 Talence Cedex, France

Progress in Colloid & Polymer Science   Progr Colloid Polym Sci 79:233–238 (1989)

Small-angle neutron scattering experiments of micellar solutions under shear

J. Kalus[1]), H. Hoffmann[2]), and P. Lindner[3])

[1]) Lehrstuhl für Experimentalphysik I der Universität Bayreuth, Bayreuth, F.R.G.
[2]) Lehrstuhl für Physikalische Chemie I der Universität Bayreuth, Bayreuth, F.R.G.
[3]) Insitut Laue-Langevin, Grenoble, France

Abstract: Rodlike micelles formed by the surfactants N-hexadecyloctyldimethylammo-niumbromide (C16-C8DAB) and tetradecyltrimethylammoniumsalicylate (TTMA-Sal) were aligned by shear gradients Γ up to 2000 s^{-1}. Small-angle neutron diffraction patterns (SANS) were recorded for different shear gradients. The quiescent solutions show a peak in the scattering intensity distribution having its origin in particle interaction. With increasing shear gradient the scattering intensity becomes very anisotropic. The scattering data are explained on the basis of alignment of rods, taking into account an effective pair correlation function. Both surfactants show evidence for the phenomenon of a shear induced phase transition. Transient SANS measurements on a time scale of 100 ms show two relaxation times for C16-C8DAB.

Key words: Micelles, phase transition, relaxation, rotational diffusion.

1. Introduction

Information on the orientational distribution of ani-sotropic micellar particles can be obtained from small-angle neutron diffraction (SANS) data with two-dimensional data acquisition [1]. In our experiments the micellar particles are immersed in heavy water and are aligned by a shear gradient Γ, applied to the system [2]. The shape of the micelles of both substances we examined; N-hexadecyloctyldimethylammonium-bromide (C16-C8DAB) [3] and tetradecyltrimethyl-ammoniumsalicylate (TTMA-Sal), was found to be rodlike. The radii are $R = 1.93 \pm 0.03$ nm and 1.94 ± 0.07 nm respectively, as determined by SANS meas-urements. Information of the length L was more diffi-cult to obtain. The neutron diffraction measurements [3] give evidence that L, as compared with the radius R, is large, probably around 21 nm for C16-C8DAB. For TTMA-Sal no estimate about L is available at the moment from SANS. Electric birefringence measure-ments gave an orientation time of 3 µs which corre-sponds to a length of about 40 nm. The micelles are charged. The charges are due to the dissociation of the negatively charged bromide and salicylate ions respec-tively. The degree of dissociation α is about 0.23 for

C16-C8DAB, as determined from conductivity meas-urements [3] and 0.1 for TTMA-Sal.

The change of the anisotropic diffraction pattern with increasing shear rates enables us to analyze both the structural and the dynamical behavior of the micelles. Theoretical work on the distribution of the axes of rodlike particles immersed in a sheared liquid is available [4, 5] and can be used to describe the aniso-tropic diffraction pattern. However, the theory does not take into account the interaction between micelles. The interaction itself depends in a complicated man-ner on the distribution function f of the rod axes. When the rods carry electric charges, a strong nearest-neighbor order will be established in the systems. The nearest-neighbor order shows up as a correlation peak in the scattering experiment. The structure factor $S(Q \rightarrow 0)$ in such systems can be very low. Q is the magni-tude of the scattering vector of the radiation ($Q = 4\pi \sin(\Theta/2)/\lambda$). λ and Θ are the neutron wavelength and the scattering angle, resepectively. λ is typically around 1 nm. At present, it is very difficult to calculate the orientational distribution function of interacting rods. Schneider et al. [6] made calculations for a sim-plified interaction in which the charge of the rods was

was concentrated at certain positions along the rods. These calculations show that neighboring rods try to arrange perpendicular to each other.

The evaluation of the exact shape and size of aniso-metric micellar aggregates at high concentration becomes very difficult because of the presence of the correlation peak mentioned above. The major dimension of the micelles is usually hidden under this correlation paek. This situation can become even more complicated by the presence of two or more different structures which could be in equilibrium with each other. In this case it seems attractive to try to orient one type of the structure while leaving the other type in an isotropic unoriented or in a weakly oriented state. This would provide additional information for the evaluation and determination of the structure. In previous papers [2, 7] we have shown that an additional information can indeed be obtained when the rodlike micelles are aligned by shear. It seems possible, for instance, to extract the lengths of the rods from these experiments even in the presence of a strong correlation peak.

In the present investigations we carried out SANS measurements of rodlike micelles. in a sheared solution showing a correlation peak.

We present in this paper experimental results that give an overview of what can be observed in SANS scattering experiments under shear. A theoretical evaluation is not included in this paper and, therefore, is only indicated in a very brief version.

Our paper is organized as follows. In section 2, 3, and 4 we describe the sample preparation, the shear- and the small angle scattering-apparatus. Section 5 is dedicated to the experimental results. In section 6 we mention briefly theoretical calculations and their fits to the experimental data. A semiquantitative description and conclusions are given in the last section.

2. Sample preparation

C16-C8DAB ($C_{16}H_{33}C_8H_{17}N(CH_3)_2Br$) was received as a gift from the Hoechst company, Gendorf, FRG. The compound has a melting point of $146 \pm 1\,°C$. The purity was checked by surface tension measurements. Solutions did not give a minimum of the surface tension at the cmc of 0.17 mM/l. A 50 mM solution dissolved in D_2O was examined at 25°C. At this concentration an isotropic phase is found, in which rodlike micelles were shown to exist. This was verified by SANS and SAXS measurements [3]. At a higher concentration of about 80 mM/l a two-phase region begins where one of the phases shows a liquid crystalline lamellar structure and the other is still isotropic [3]. At the choosen concentration three electric birefringence signals with time constants which are up to four orders of magnitude different were observed ($\sim 5\,\mu s$, $\sim 500\,\mu s$,

and $\sim 50\,ms$). The birefringence Δn, plotted against shear rate Γ, shows an anomaly as shown in Fig. 1. The angle of extinction χ is a steady function of the shear rate Γ, see Fig. 2, but does not follow the usual theoretically expected behavior [8]. More details about the phase diagram can be found in [3, 8].

TTMA-Sal ($C_{14}H_{29}N(CH_3)_3$-(C_6H_4OHCOO)) was prepared by ion exchange procedure from TTMA-Cl and Na-salicylate solutions or from a solution of the hydroxide of the surfactant TTMA-OH and salicylic acid. The hydroxides can be obtained by titration of the chloride with AgOH. Both methods of preparation gave identical results. TTMA-Sal shows, above a threshold value of the shear rate Γ, a shear-induced structure (see [9] and references cited in this paper). We studied a 5 mM solution dissolved in D_2O at 25°C.

3. The shear apparatus

Alignment of the rods was made with two different shear apparatus described in detail in [2, 10]. The shear rate Γ was varied between 0 and $2000\,s^{-1}$.

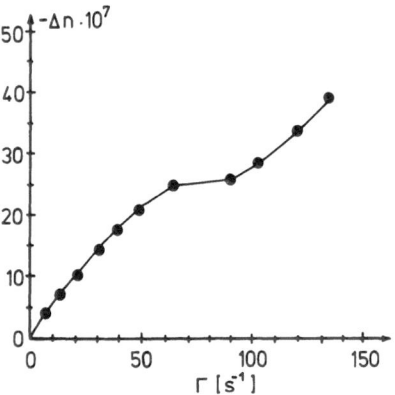

Fig. 1. The size of birefringence Δn as a function of the shear rate Γ for C16-C8DAB at 25°C

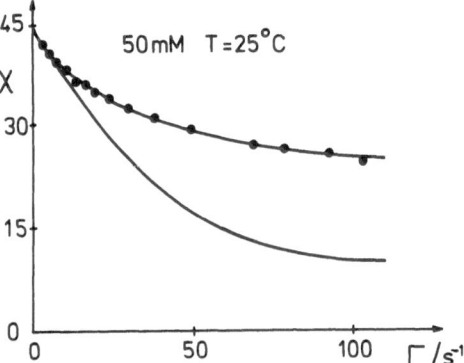

Fig. 2. The extinction angle χ as a function of the shear rate Γ for C16-C8DAB at 25°C. The curve joining the experimental points is a guide to the eye. The lower solid line is expected by theory, taking into account the slope of the experimental curve for $\Gamma \to 0$

4. The small-angle scattering apparatus

Experiments were carried out at the neutron small-angle diffraction instruments D 11 at the Institute Laue-Langevin in Grenoble with neutrons of a wavelength $\lambda = 1.0$ nm and 0.5 nm respectively. The wavelength resolution is typically $\Delta\lambda/\lambda = 0.09$ fwhm [11]. The measurements showed that it was possible to obtain extremely anisotropic scattering patterns even at low shear gradients. Temperature during the measurements was kept constant at 25 ± 0.5 °C.

5. Experimental results

Figures 3 and 4 show contour plots of the scattering intensities for TTMA-Sal. The Γ values are 0 and 100 s^{-1}, respectively. The anisotropy increases, as we know, steadily with increasing shear gradient, at least up to $\Gamma = 380$ s^{-1}. The smooth curve in Fig. 4 is due to a fit procedure explained below. There is an indicaiton of a correlation peak around $Q = 0.08$ nm^{-1}. The disturbance near $Q = 0$ stems from the beam stop.

Figures 5, 6, and 7 show isometric plots for C16-C8DAB of the scattering intensities at low scattering angles. The Γ values are 0, 200, and 2 000 s^{-1}, respectively. An isometric plot for higher scattering angles and $\Gamma = 2\,000$ s^{-1} is shown in Fig. 8. The sharp peak,

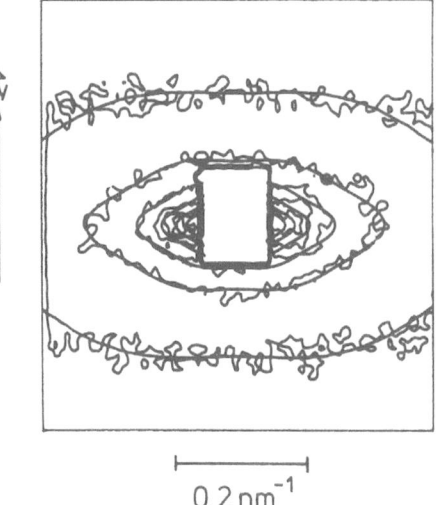

$\vdash\!\!\!\!\longrightarrow\!\!\!\!\dashv$
0.2 nm^{-1}

Fig. 4. Same as in Fig. 3, but for $\Gamma = 100$ s^{-1}. The smooth solid line is due to a fit described in the text. The same contour lines as in Fig. 3 are shown plus addtional lines at the intensities 900 and 1 075, which were ab-absent in Fig. 3. The direction of the veloscity vector $\vec{w}$ of the sheared solution is indicated. The gradient of $\vec{w}$ is a vector perpendicular to the paper plane

denoted SP in Figs. 6-8 keeps its shape, but increases in intensity with increasing Γ. This sharp peak shows a threshold behavior and appears somewhere between $\Gamma = 50$ and 100 s^{-1}. The magnitude of the momentum transfers Q of the sharp correlation peaks of Fig. 8 are $Q_1 = 0.284 \pm 0.001$ nm^{-1} and of a second one at $Q_2 = 0.510 \pm 0.05$ nm^{-1}, respectively. It turns out that

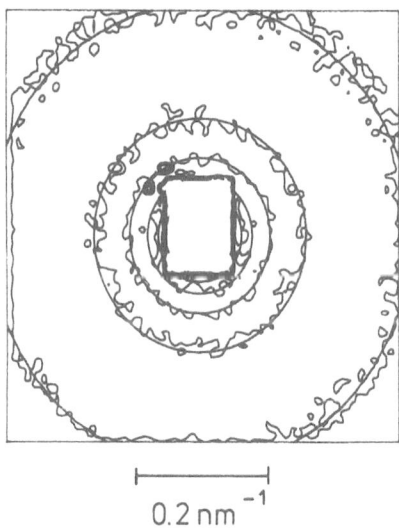

$\vdash\!\!\!\!\longrightarrow\!\!\!\!\dashv$
0.2 nm^{-1}

Fig. 3. Contour map of the scattering intensity of TTMA-Sal for the shear gradient $\Gamma = 0$ s^{-1}. The smooth solid line is due to a fit described in the text. The intensities increase to the center of the figure and are 200, 375, 550, and 725. The disturbance in the middle is due to the beam stop. The momentum transfers are given in units of nm^{-1}. $\vec{Q}$, the scattering vector, is in the paper plane

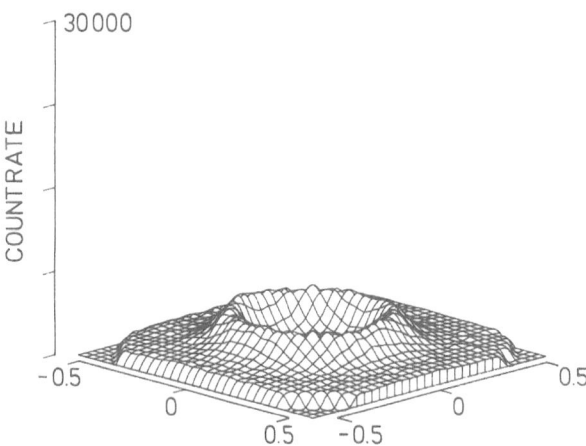

Fig. 5. Isometric plot of the scattered intensity of C16-C8DAB with a shear gradient of $\Gamma = 0$ s^{-1}. The momentum transfer is given in units of nm^{-1}

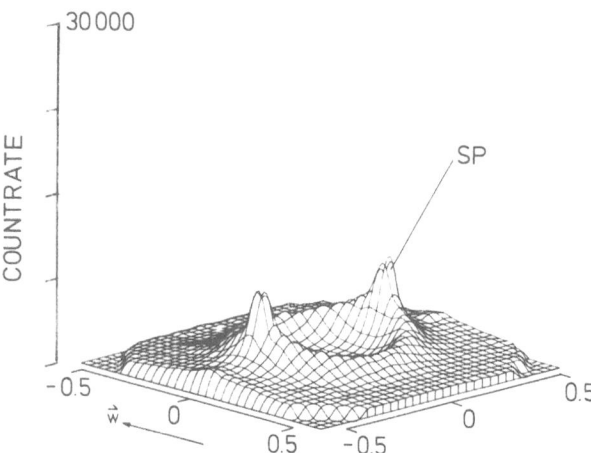

Fig. 6. Same as Fig. 5, but for $\Gamma = 200\ \mathrm{s}^{-1}$. The direction of the velocity vector $\vec{w}$ of the sheared solution is indicated. The gradient of $\vec{w}$ is a vector parallel to the countrate axes

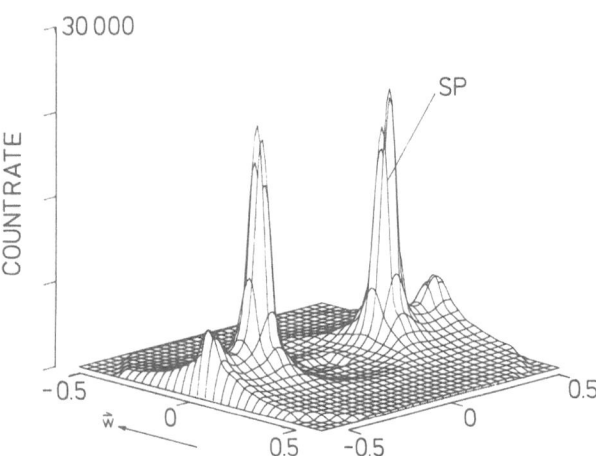

Fig. 7. Same as in Fig. 5, but for $\Gamma = 2\,000\ \mathrm{s}^{-1}$

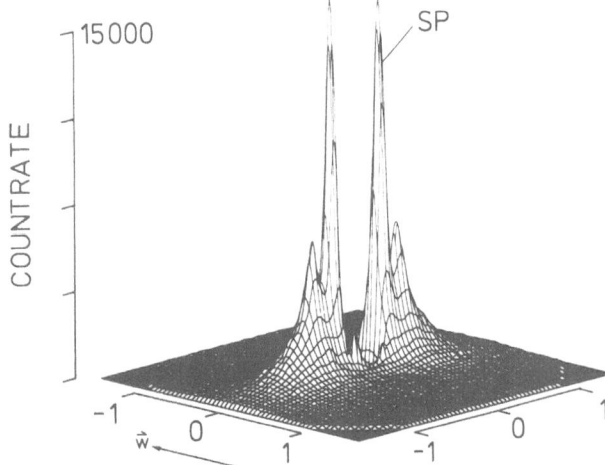

Fig. 8. Same as in Fig. 5, but for $\Gamma = 2\,000\ \mathrm{s}^{-1}$ and for higher scattering angle

$Q_2/Q_1 = 1{,}80$ is near $\sqrt{3}$, which could be an indication that a hexagonal structure of parallel rodlike micelles is formed under shear. The Q-value of the rim of the ringlike intensity structure in Figs. 5–7 is equal to Q_1 and stays constant for all Γ-values.

The fact that the sharp peak SP stays constant in shape as a function of the shear gradients indicates that above $\Gamma = 50\ \mathrm{s}^{-1}$ a subdivision of the intensity distribution into two parts is possible. The first part shows a ringlike structure, showing a weak anisotropy at higher shear rates. A second part, giving rise to the sharp peak SP, superimposed on the ringlike structure, indicates a high degree of alignment, even at low values of Γ. We call the micelles responsible for these two parts in the scattering pattern "type I micelles" and "type II micelles", respectively. To elucidate the behavior of the system further, we performed a time-dependent SANS-measurement with a time resolution of 100 ms. In that experiment Γ was kept constant at a value of $300\ \mathrm{s}^{-1}$ for times $t < 0$ and suddenly reduced at $t = 0$ (within a time smaller than 100 ms) to zero. We measured the anisotropic scattering intensity in 32 subsequent time channels, each having a time width of 100 ms. The technique is described in more detail in [2]. Figures 9 and 10 show contour plots for times $t < 0$ and for time $t = 3.2$ s. A clear relaxation of the peaked structure is seen. For a simple presentation of this behavior we plotted in Fig. 11 the peak intensity as a function of time and fitted two exponentials to this curve.

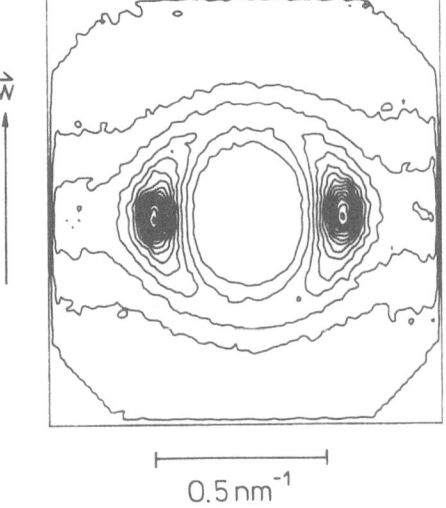

Fig. 9. Contour plot for C16-C8DAB at $\Gamma = 300\ \mathrm{s}^{-1}$ and time $t < 0$ for the transient scattering experiment. For more details see text. The intensities are 200, 400, 600, and so on. The moment transfers are given in units of nm^{-1}

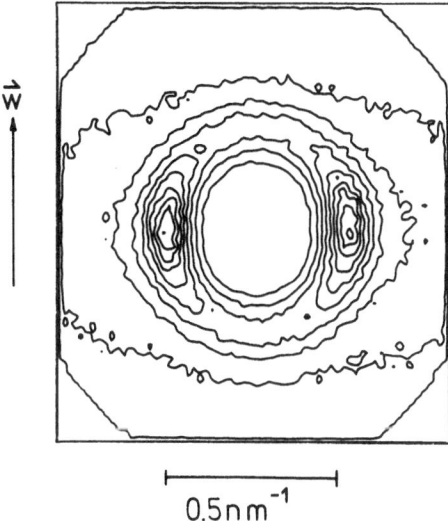

$\vec{W}$

$\overline{0.5\,nm^{-1}}$

Fig. 10. Same as in Fig. 9, but 3.2 s after the sudden decrease of $\Gamma =$ 300 s^{-1} to zero

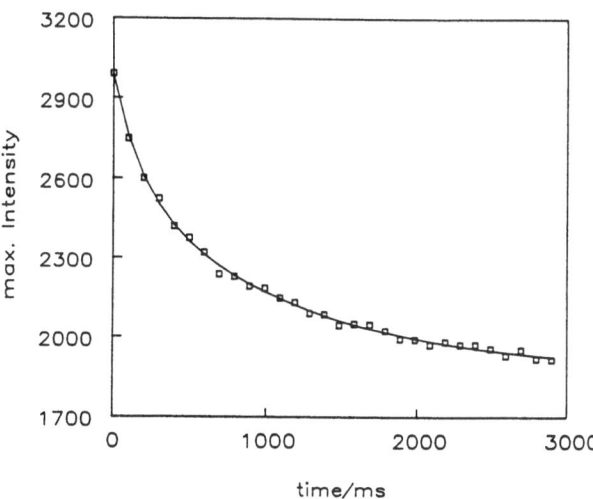

Fig. 11. The time dependence of peak intensity of C16-C8DAB for the transient experiment

6. The model calculations

The model used to describe the anisotropic scattering intensity $I(\vec{Q})$ of an ensemble of identical, centrosymmetric particles measured in a small-angle scattering experiment is described elsewhere [12]. With the

introduction of severe simplifications $I(\vec{Q})$ can be written as

$$I(\vec{Q}) = C\{\langle F^2(\vec{Q})\rangle + \varrho\langle F(Q)\rangle^2 \int_0^\infty \cos(\vec{Q}\vec{r})$$

$$\cdot [P(\vec{r}) - 1] \cdot d\vec{r}\}$$

C incorporates the intensity of the neutron beam, the solid angle, the detector efficiency and the shape of the sample. $F(\vec{Q})$ is the form factor of the micelle depending on the scattering vector, the orientation, the length L and the radius R. $\langle F(\vec{Q})\rangle = \int F(\vec{Q}, \vec{u}) \cdot f(\vec{u})\,d\vec{u}$ is the mean form factor and $f(\vec{u})$ is the orientational distribution function of the rod axes. $\vec{u}$ is a vector specifying the orientation of the particles in space. In our case, $\vec{u}$ gives the orientation of the rod axis. ϱ is the number density and $P(\vec{r})$ is the pair correlation between the center of gravities of the micelles. We solved numerically an equation of motion of rodlike micelles in a shear gradient, giving the orientational distribution function $f(\vec{u})$ of the rod axes, with a method described in [2]. There remains only one free parameters in the equation of motion, the rotational diffusion coefficient D of the particles.

For TTMA-Sal we did not succeed to fit this model with reasonable accuracy to the measured curves at $\Gamma = 100$ s^{-1}. Only by introducing a second species of long rodlike micelles which are much more aligned and which are in equilibrium with micelles, showing no or only a weak alignment the fit became quite reasonable, as shown by the smooth solid lines in Fig. 4. One should note that the fit in Fig. 4 includes, apart from an overall intensity factor, only two free fit parameters, the rotational diffusion coefficient D ($D \sim$ 1.4 s^{-1}) and a second intensity parameter which describes the relative amount of the not aligned micelles ($\sim 50\%$).

For C16-C8DAB we show in Fig. 12 fitted D-values which are related to the type I micelles. The increase of these values indicates, that a change in length or in a length distribution or a shear induced change in shape or a change in the interaction between the micelles or a combination of all these possibilities might be responsible for this fact. The intensity of the sharp peak SP can be analyzed and shows, as mentioned, an increase with increasing shear gradient. This behavior, beginning at a threshold value of Γ somewhere between 50 s and 100 s^{-1}, is shown in Fig. 13. A closer analysis shows that this peak is associated with a shear-induced phase transition in which micelles of type I are transformed into micelles of type II. The

The result of the fit is also shown in Fig. 11. The values of the two relaxation times are $\tau_1 = 0.18 \pm 0.03$ s and $\tau_2 = 1.1 \pm 0.1$ s.

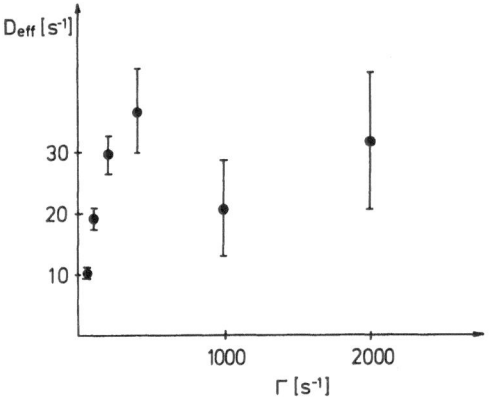

Fig. 12. The rotational diffusion coefficient D as a function of the shear rate Γ for C16-C8DAB

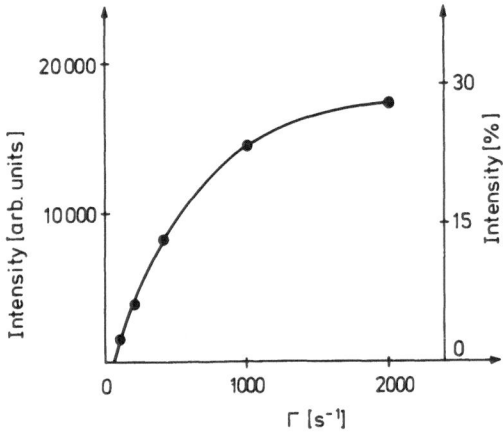

Fig. 13. The intensity of the SANS pattern related to type II micelles

tpye-II micelles are organized quite differently from type-I micelles. There are strong indications that type-II micelles form a hexagonal two-dimensional lattice. We present more details in a forthcoming paper [8].

Acknowledgements

We thank the Institute Laue-Langevin for providing the neutron beam facilities and to W. Grießl for his assistance in computer work. This work has been supported by the Bundesministerium für Forschung and Technology Grant No. 03-KA1BAY-0.

References

1. Schmatz W, Springer T, Schelten J, Ibel K (1974) J Appl Cryst 7:96–116
2. Herbst L, Hoffmann H, Kalus J, Thurn H (1985) Neutron scattering in the nineties. IAEA, Vienna, pp 501–506
3. Neubauer G, Hoffmann H, Kalus J, Schwandner B (1986) Chemical Physics 110:247–253
4. Boeder P (1932) Z Physik 75:258–281
5. Herbst L, Hoffmann H, Kalus J, Thurn H, Ibel K, May RP (1986) Chemical Physics 103:437–445
6. Schneider J, Karrer D, Dhont JKG, Klein R (1987) J Chem Phys 87:3008–3015
7. Neubauer G, Herbst W, Hoffmann H, Ibel K, Kalus J (1988) Material Science Forum 27/28:147–150
8. Kalus J, Hoffmann H, Chen S-H, Lindner P (1988) J Phys Chem 93:4267–4276
9. Wunderlich I, Hoffmann H, Rehage H (1987) Rheol Acta 26:532–542
10. Lindner P, Oberthür RC (1984) Revue Phys Appl 19:759–763
11. Ibel K (1976) J Appl Cryst 9:296–309
12. Kalus J, Hoffmann H (1987) J Chem Phys 87:714–722

Received October, 1988;
accepted March, 1989

Authors' address:

Prof. Dr. J. Kalus
Lehrstuhl für Experimentalphysik I
Universität Bayreuth
Postfach 10 12 51
D-8580 Bayreuth, F.R.G.

Progress in Colloid & Polymer Science   Progr Colloid Polym Sci 79:239–243 (1989)

Polymerization of cetyltrimethylammonium methacrylate direct micelles

B. Lerebours[1])[2]), B. Perly[2]), and M. P. Pileni[1])[2])

[1]) Université de Paris VI, Laboratoire de Structure et Réactivité aux Interfaces, Paris, and
[2]) Centre d'Etudes Nucléaires de Saclay, Départment de Physico-chimie, Gif-sur-Yvette, France

Abstract: By means of ion exchange chromatography, a new surfactant was synthesized. Its purity and characteristics, molecular and micellar, were investigated. The originality of this surfactant lies in its counterion, which is a polymerizable group. The polymerization of such direct micellar solution did not lead to precipitation. The use of different techniques like light scattering, NMR, fluorescence spectroscopy, and photoelectron transfer reactions allow us to describe the polymerized aggregates as cylindrical entities which remain in size and shape, even under dilution, below the critical micellar concentration of the unpolymerized micelles.

Key words: Counterion exchange, direct micelle polymerization, properties of polymerized aggreates.

Introduction

Micelles and microemulsions are dynamic aggregates which regenerate themselves permanently. Two lifetimes were defined to describe the dynamic of micelles: One, short and attributed to the mean exchange time of surfactant monomer between micelles and bulk. The other, longer, defines the mean time after which the micelle is totally regenerated [1]. This dynamic confers to the micelles a limited stability for guest molecules [2].

The challenging aim of the following study was to polymerize micelles without perturbing the thermodynamic equilibrium (monomer in micelles $\langle = \rangle$ monomer in the bulk) which would lead to a precipitation. Moreover, we have to face the difficulty of the high curvature of the micelle, a limiting factor to the topological alignment of the polymerizable groups (and soforth) to the growth of the polymer. To preserve the dynamic of monomers, the polymerization of micelles by means of covalent bonds in the lipidic [3] or head group parts [4] of the micelle have to be excluded.

We chose to polymerize the counterions [5]. For this purpose we exchange the bromide counterion of a surfactant by a methycrylate ion.

The formation and characterization of this surfactant and its characteristics after polymerization are presented.

Material and methods

The surfactants cetyltrimethylammonium bromide and chloride (CTAB, CTAC; Eastman Kodak) were recrystallized from acetone. The methacrylic acid was used just after bidistillation under reduced pressure (12 mm Hg).

The resin used for the ionic exchange was a AG1-X2 (Biorad). Azoisobutyronitrile (AIBN) and Pyren butyric acid (PBA) were used as supplied (Aldrich).

The zinc tetramethylpyridylporphyrin (ZnTMPyP⁴⁺, 4 Cl⁻) and the dioctylviologens were synthezised (departement of Physico-chimie, Saclay) as previously described [6].

The surface tension experiments were carried out on a Krüss digital tensiometer K10.

The light scattering apparatus and set-up was described previously [7]. For static light scattering experiments, the scattered light (546 nm) was measured at 90° of the incident beam. The intensity measurements were compared to a benzene solution. Specific refractive index increments were measured using a Rayleigh interferometer. The data from the static light scatttering experiments were extrapolated at the cmc.

The auto-correlation treatment of the dynamic light scattering measurements was obtained by a Malvern auto-correlator.
^{1}H NMR spectra were recorded on a Bruker WM500 spectrometer. The fluorescence emission and excitation spectra were measured on a Perkin-Elmer LS5.

Photoelectron-transfer experiments were performed with a conventional flash photolysis apparatus at 560 nm [8].

Results and discussion

Synthesis of the surfactant

The bromide ion of CTAB surfactant is exchanged on a AG1-X2 chromatography column by a methacrylate ion.

Firstly, the resin should be washed with NaOH to exchange its chloride ions with hydroxide ions. Then the resin is charged with methacrylic acid; this acid has to be freshly bidistilled under 12 mm Hg to eliminate the polymerization inhibitor agents. Then the CTAB was passed through the column and the cetyltrimethylammonium methacrylate CTAM was collected and liophylized. Water washes should be performed at each step until a neutral pH is reached in order to eliminate the non-adsorbed material.

Characterization

Three methods were used to test the purity of the surfactant and the yield of ionic exchange.

a) The Br^- titration of the CTAM solution by $AgNO_{3-}$ and bromide selective electrode confirms the absence of bromide (detection limit 10^{-7} M).

b) The vinyl bond (3026, 1640 cm^{-1}), the carboxylic group (1562 cm^{-1}), the acrylate group (1231 cm^{-1}), the presence of a CH_2 group in a CH_3 group (2957 cm^{-1}) were obtained by IR Fourier transform spectra of solid CTAM in nujol or voltalef.

c) Using ^1H NMR, it is possible to determine the yield of counterion exchange by comparing the relative integration of the peaks due, partly, to the protons of the methycrylate group and partly to the protons of the cetyltrimethylammonium group (see Fig. 1). This method confirms the presence of one methacrylate for each surfactant.

Micellar properties

From binary solutions (surfactant-water) at concentration close to the critical micellar concentration (cmc), small spherical micelles usually form [9]. The macroscopic properties of these solutions can be strongly modified by different parameters:

— addition of salt [10]
— increase of detergent concentration [11]
— variation of the counterion [12], of the polar head group [13], or of the length of the alkyl chains [14].

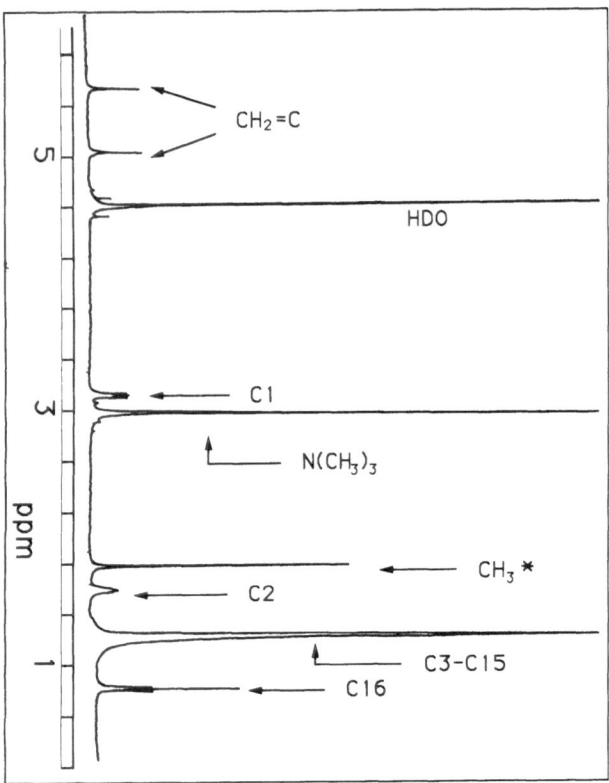

Fig. 1. ^1H NMR spectrum of a $2 \cdot 10^{-2}$ M CTAM solution in D_2O. The comparison of the integration peak leads to $(M^-)/(CTA^+) = 0.9 \pm 0.1$

The shape of the aggregates can be different when changing the counterion. For example, CTAC micelles over a wide range of concentration present a spherical shape, but CTAB micelles evolve from spherical entities to cylindrical ones when raising the concentration [15].

For these reasons it is essential to determine precisely all the characteristics of this new surfactant CTAM.

The critical micellar concentration was determined by surface tension measurements, ring method, and gave the value of $9.8 \cdot 10^{-4}$ M. The dynamic light scattering measurements allowed measurement of the hydrodynamic radius of the diffractants. Above the cmc we obtained a value of 30 Å and none below. By static light scattering the aggregation number could be calculated. It was found equal to 106 above the cmc and nul below. To these values is added a theoretical sphericity limit value of the aggregation number nl defined by simple geometrical model. This calculation is based on the following hypothesis: the limit of the spherical micelle radius has to be equal or inferior to the extend-

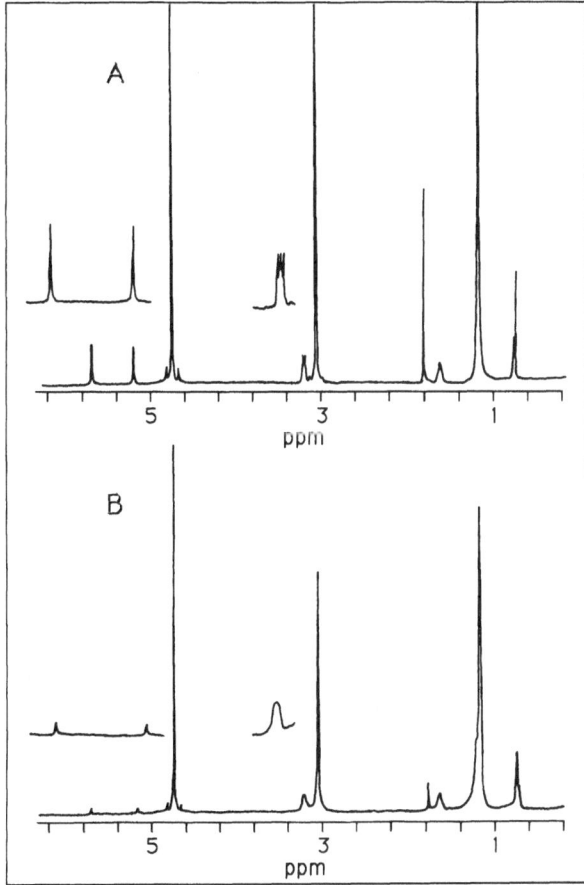

Fig. 2. ¹H NMR spectra of a $2 \cdot 10^{-2}$ M CTAM solution before (A) and after polymerization (B). The comparison of the integration peak leads to $(M^-)/(CZA^+) = 0.9 \pm 0.1$ for (A) and $(M^-)/(CTA^+) = 0.15 \pm 0.1$ for (B)

ed length of the alkyl chain. The volume of one extended alkyl chain monomer [16] compared to the limit volume of the spherical micelle gives the aggregation number limit; in the case of CTAM nl is equal to 111.

The good agreement between experimental and theoretical n values lead to the conclusion that the CTAM micelles are spherical and close to the limit of sphericity.

Polymerization modes

A micellar CTAM solution, previously degassed by argon bubbling or freeze thaw cycles, is polymerized using different techniques:

— UV irradiation of 3 ml samples with a 500 W lamp during 1–5 h with a water filter for thermal protection.

— Gamma irradiation of different doses of several tenths of krad.

— Thermally initiated reaction by AIBN 1 % w/w during 12 h–3 days at 85 °C.

These different techniques give the same data. The samples obtained are limpid or slightly bluish.

Characterization of the polymer

A $2 \cdot 10^{-2}$ M CTAM solution was polymerized using one of the three different techniques developed above.

The polymerization process was followed by ¹H NMR spectroscopy. Figure 2 shows the CTAM spectra before and after polymerization. The decrease of the vinyl protons peaks intensity and of the methacrylate methyl group with polymerization is observed. Polymerization of the counterion leads to unobservable lines due to excessive broadening. This also affects the micelle itself to a lesser degree. The motion of the polar head is reduced as shown by the broadening of the ¹H lines arising from the first CH_2 group at 3.1 ppm. Conversly, all other methylene groups still experience local motion which are weakly affected by polymerization of the counterions and experience no or few line broadening. This can be taken as a proof of the encapsulation of the micelle by the polymerized counterions.

Characterization of the polymerized aggregates

Using the surface tension apparatus, the polymerized CTAM samples presented no cmc. The light scattering techniques were used and the results are as follows. We call sample A a polymerized sample of $2 \cdot 10^{-2}$ M CTAM. The hydrodynamic radius found for sample A was 540 Å; after dilution of the sample A (CTAM $= 10^{-4}$ M) the radius was 470 Å. The aggregation number of the sample A, before and after dilution, was found equal to 12 000. No sensitive change of size and aggregation number appears when diluting a CTAM micelle polymerized sample to a concentration 10-times lower than the CTAM cmc. Conversely, the same study on a non-polymerized sample gives no value for the hydrodynamic radius and aggregation number after dilution. After dilution in the first case the aggregates can still be detected; in the second case only the free monomers are present.

Using Perrin's formula, Tanford [17] deduced that a cylindric entity of 30 Å radius having a hydrodynamic radius of 500 Å corresponds to an aggregate of 15 000 monomers (mean surface per polar head 40 Å). For a

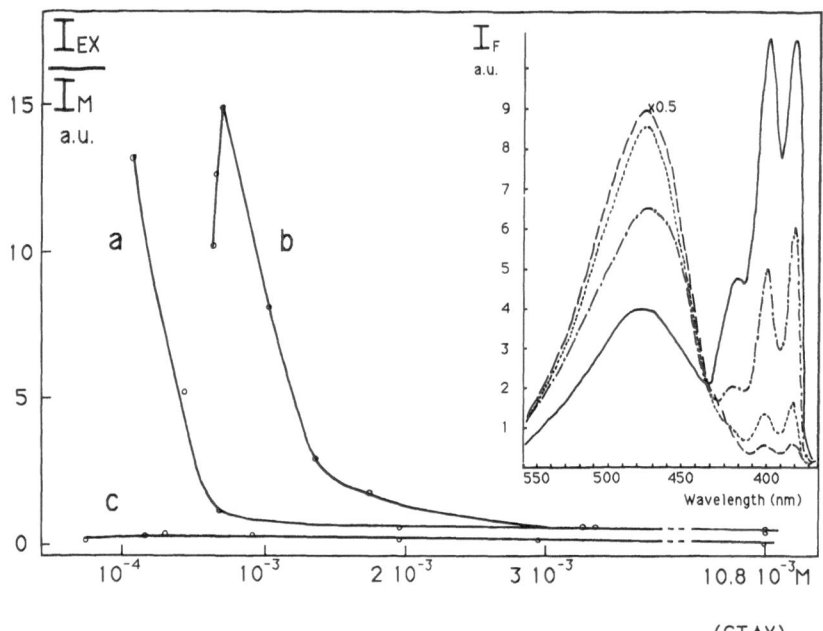

Fig. 3. Excimer over monomer fluorescence intensities of 10^{-6} M of pyren butyric acid with dilution of surfactants. a) CTAC; b) CTAM; c) polymerized CTAM. Insert: Fluorescence emission spectra of 10^{-6} M of pyren butyric acid in various CTAM concentration. $(--) = 5 \cdot 10^{-4}$ M; $(---) = 1.2 \cdot 10^{-3}$ M; $(-\cdots-) = 2 \cdot 10^{-3}$ M; $(\underline{\quad}) = 3 \cdot 10^{-3}$ M

spherical item a same hydrodynamic radius of 500 Å would lead to an aggregation number of 195 000.

We conclude that after polymerization the CTAM micelles samples are composed of cylindrical aggregates of high molecular weight which maintain their shape and size under dilution.

To clarify the concept of polymerized items and decide if they can be described as polyelectrolytes organized in statistical pelots or as micelles with a charged interface and an hydrophobic core, we used the properties of an interfacial fluorescent probe.

Interface effect

The pyren butyric acid (PBA) is an interfacial fluorescent probe with fluorescent emission maxima at 370 nm for its monomer form and 480 nm for its excimer form. Such a probe is partially soluble in water and forms dimers leading to the excimer fluorescence emission in such medium. In alcohol solution, PBA is well solubilized and the fluorescence emission of the monomer is observed. Using such a probe it has been previously shown that in CTAC micellar solution, the fluorescence of the monomer is obtained [18], whereas by dilution below the cmc, the fluorescence of the excimer is observed. This is interpreted in terms of solubilization of PBA at the micellar interface. Similar experiments were performed in CTAM micellar solutions and in CTAM polymerized aggregates.

Figure 3 (insert) shows the monomer fluorescence spectrum observed above the cmc of CTAM and that of the excimer below the cmc. The 10^{-6} PBA fluorescence is recorded at various CTAM concentrations by progressive dilutions. As it was observed previously with CTAC, the excimer emission appears at the cmc.

The same experiment performed with CTAM polymerized aggregates shows that all along the dilution process only monomer fluorescence was observed. This unchanged fluorescence behavior obtained after dilution indicates that PBA remains at an interface, even below the concentration corresponding to the cmc of its unpolymerized equivalent.

Hydrophobic effect

Photoelectron transfer reaction have been intensively studied between Zn porphyrins water-soluble and different viologens [19]. We showed previously that the presence of micelles like CTAC increases the lifetime of the products of this phototransfer reaction by incorporation of one of them (long chain reduced viologens) in the hydrophobic core of the micelle [20]. The kinetic order of the disparition reaction of these reduced viologens was found to be first order in micellar systems and second order or diffusion controlled in water and in CTAC solutions below the cmc.

Similar experiments were performed using CTAM micellar solutions and polymeric aggregates. The results obtained with CTAM conformed to those obtained with CTAC. Above the cmc the reaction followed a first order and the halftime $T_{1/2}$ of the reduced viologen was equal to 13 ms; below the cmc the reaction kinetic order was 2 and the $T_{1/2}$ was equal to 0.8 ms. Using CTAM polymerized aggregates, the back reaction was found to be partially delayed. This delay can be explained in terms of entrance of reduced viologen inside the polymeric aggregate. The incorporation is less effective with polymerized aggregates than with unpolymerized ones. The half-lifetime of reduced viologen, (7 ms) and the kinetic order of the reaction (1) were totally unchanged by 100-times dilution of the polymerized aggregates. We can conclude that a hydrophobic zone is still available. However, the polymethacrylate net surrounding the aggregate could constrain the incorporation of the reduced viologens.

Conclusion

In this work we investigated a new surfactant whose characteristic is its polymerizable counterion. In aqueous solution this surfactant forms spherical micelles which are close to the limit of sphericity.

We could polymerize these counterions and follow the polymerization yield by ^{1}H NMR. It showed that the polymerization was complete. Such a solution was investigated by physico-chemical techniques and reactivity reactions. From reactivity reactions we know that the diluted polymerized aggregates conserve interfaces and hydrophobic character. This is an indirect proof of the existence of polymerized micelles. These polymerized aggregates are large and their size and shape remain upon dilution.

References

1. Aniansson GEA (1985) Progr Colloid Polym Sci 70:2
2. Lerebours B, Chevalier Y, Pileni MP (1985) Chem Phys Lett 117:89
3. Hamid SM, Sherrington DC (1987) Polymer 28:325
4. Hamid SM, Sherrington DC (1987) Polymer 28:332
5. Lerebours B, Perly B, Pileni MP (1988) Chem Phys Lett 147:503
6. Chevalier S (1982) Ph D Thesis
7. Hayoun M (1982) Note CEA N-2283
8. Pileni MP (1978) J Chim Phys 75:32
9. Lindman B, Wennerstrom H (1980) In: Mukerjee P, Mysels KJ (eds) (1971) Critical Micelle Concentration Of Aqueous Surfactant Systems NSRDS-NBS 36 Washington, DS: US Governement Printing Office, Top Curr Chem 87:1; Wennerstrom H, Lindmann B (1979) In: Micelles, Physical Chemistry of Surfactant Association. Phys Rep 52:1
10. Appell J, Porte G (1983) J Physique Lett 44:689
11. Hayter JB, Penfold J (1983) Colloid Polym Sci 261:1022
12. Jansson M, Stilbs P (1985) J Phys Chem 89:4868
13. Stigter D (1974) J Phys Chem 78:2480
14. Lianos P, Zana R (1983) J Phys Chem 87:1289
15. Ruckenstein E, Nagarajan R (1976) J Colloid Interface Sci 57:388
16. Tanford C (1972) J Phys Chem 76:3020
17. Tanford C (1974) J Phys Chem 78:2469
18. Gregoritch SJ, Thomas JK (1980) J Phys Chem 84:1491
19. Chevalier S, Lerebours B, Pileni MP (1984) J Photochem 27:301
20. Lerebours B, Chevalier Y, Pileni MP (1985) Chem Phys Lett 117:89

Received November, 1988;
accepted January, 1989

Authors' address:

M. P. Pileni
Centre d'Etudes Nucléaires de Saclay
Département de Physico-chimie
BP 121
F-91191 Gif-sur-Yvette Cedex, France

Progress in Colloid & Polymer Science   Progr Colloid Polym Sci 79:244–248 (1989)

Microviscosities in alkane/surfactant ionic micelles

A. Malliaris

N.R.C. "Demokritos", Athens, Greece

Abstract: Some common aliphatic hydrocarbons were solubilized in aqueous ionic surfactant micelles. The effect of these additives on η_{mic}, the microviscosity in the pallisade region of the micelle, was studied. It was found, by means of fluorescence quenching, that η_{mic} is independent of the nature and quantity of the additive. These findings are discussed in terms of diffusion kinetics and the validity of the Einstein-Stokes equation in micelles.

Key words: Microviscosity, micelle, fluorescence quenching, solubilization, diffusion.

Introduction

Micellar systems, along with several other microheterogeneous media, have been extensively studied during the last decade by fluorescence methods [1]. In particular, transient fluorescence has been proved to be one of the most powerful techniques available for the extraction of micellar microproperties [2]. Thus, mean micellar aggregation numbers, intra and intermicellar kinetics, micelle/water exchange rates of additives, etc., have been determined from computer analysis of the time-resolved quenched fluorescence decay of micelle-bound fluorophors [3]. Among these micellar parameters, the rate constant of the intramicellar fluorescence quenching, k_q, is of particular interest, because of its relevence to diffusion-controlled chemical reactions.

In a recent publication we discussed our fluorescence probing results about the effect of n-alkanes upon the micellization of aqueous ionic surfactants [4]. In the present article we focus our attention specifically to the intermicellar fluorescence quenching rate constant, and through it to the microviscosities in the interior of the aggregates. The results are examined in relation to diffusion controlled reactions in microaggregates, and the validity of the Einstein-Stokes equation in compartmentalized systems where the molecular dimensions are comparable to the dimensions of the space available for diffusion.

Experimental

The origin and purification of the ionic surfactants used in this study, i. e., sodium dodecyl sulfate (SDS), tetradecyltrimethylammonium chloride (TTAC), and cetyltrimethylammonium chloride (CTAC), as well as cetylpyridinium chloride (CPC) used as the fluorescence quencher, and pyrene (Py) used as the fluorophor, have been described elsewhere [4]. Recall however that both CPC and Py do not exchange with the intermicellar aqueous phase within the fluorescence lifetime of the fluorophor ($\tau_o = 300-400$ ns) [5]. Instead, they remain associated with their host micelles for times much longer than τ_o (exchange time for CPC $> 10^{-5}$ s [5], and for Py $> 10^{-4}$ s [6]), therefore they are treated as immobile reactants [7, 8]. For the collection of fluorescence decay data the single photon counting method [9] was employed. Finally, the molecular volumes V_m of the alkanes were calculated from densities and molecular weights, while critical micelle concentrations (CMC) of the surfactants were taken from the literature [10], and were assumed to be unaltered by the solubilization of the alkanes [5].

Results and discussion

Under the conditions of the present experiments, i. e., when the reactants are immobile, the micellar parameters are given by Eqs. (1–3), where k_o is the unquenched fluorescence life

$$k_o = A_2 \tag{1}$$

$$[Q]/[M] = A_3 \tag{2}$$

$$k_q = A_4 \tag{3}$$

time of the fluorophor in the micelle, $[Q]$ and $[M]$ are the molar concentrations of the quencher and that of the micelles, respectively, and k_q is the rate constant of the intramicellar fluorescence quenching. A_2, A_3, and A_4 represent time-independent fitting parameters, related to the time dependence of the fluorescence intensity by Eq. (4) [6,11]. I_o is

$$I_t = I_o \exp\{-A_2 t - A_3[1 - \exp(-A_4 t)]\} \qquad (5)$$

the maximum fluorescence intensity at the time of the flash ($t = 0$), and I_t is the intensity at some later time t.

N_s and N_A are, respectively, the mean surfactant aggregation number, and the number of solubilized n-alkane molecules per micelle. N_s can be estimated from eqs. (2) and (6) where $[C_s]$ is the total surfactant concentration

$$N_s = ([C_s] - [CMC])/[M] \qquad (6)$$

and N_A is calculated from Eq. (7)

$$N_A = [C_A]/[M] \qquad (7)$$

on the assumption that the added n-alkane concentration $[C_A]$, is totally distributed in the micellar pseudophase. Finally, R_m, the radius of the mixed micelle containing N_s surfactants and N_A additive alkanes, is given by Eq. (8). Note that the

$$R_m = \{3[N_s(27.4 + 26.9\, n_c) + N_A V_m]/4\pi\}^{1/3} \qquad (8)$$

expression $27.4 + 26.9\, n_c$ gives the volume of the aliphatic chain of a surfactant with n_c carbon atoms on its chain [12]. It is evident that here we have assumed that the micellar core includes even the very first CH_2 unit which is located immediately next to the ionic head of the surfactant. All the micellar parameters obtained in this study are listed in Table 1.

Fluorescence quenching is a typical diffusion-controlled process since usually the rate determining step is the diffusional approach of the excited fluorophor and the quencher [13], as shown in Eq. (9). The mutual approach and

$$F^* + Q \; - - - \; [F^*Q] \; - - - \; F + Q \qquad (9)$$

separation of F^* and Q are described by the corresponding rate constants, k_D and k_{-D}. The presense of the solvent has the implication that after $F\cdot$ and Q have

an encounter they are trapped in a "cage" $[F^*Q]$, where they undergo numerous collisions between each other before they either interact or fly apart. In the present case k_{-D} is practically negligible and all encounters lead to quenching. Such bimolecular processes, when electrostatic effects are ignored are expected to proceed with a rate constant calculated from the simple Smoluchowski theory as expressed by Eq. (10) [14–16]. N is Avogardo's

$$k_q = \pi N R_{F^*Q} D_{F^*Q}/250 \qquad (10)$$

number, R_{F^*Q} is the so-called encounter radius which measures the separation of F^* and Q at which quenching occurs, and $D_{F^*Q} = D_{F^*} + D_Q$ is the relative diffusion coefficient. On the other hand, values of the diffusion coefficient D of a spherical molecule having diameter R and diffusing with constant velocity in a continuous medium with viscosity n can be estimated by means of the Einstein-Stokes equation shown below.

$$D = KT/6R\pi n. \qquad (11)$$

K is the Boltzmann constant and T the absolute temperature. Finally, a combination of Eqs. (10) and (11) produces Eq. (12), which gives k_q in terms of the

$$k_q = [NKT/1500\, n]\,[(R_{F^*} + R_Q)^2/R_{F^*}R_Q] \qquad (12)$$

dimensions of the colliding species and the viscosity of the medium. Introducing typical values in Eq. (12), e. g., $R = 3$–$4 \cdot 10^{-8}$ cm and $n = 0.01$ Poise, we obtain values of k_q, at room temperature, of the order of 10^{10}–10^{11} M^{-1} s^{-1}. This is the so-called diffusion-controlled limit of k_q, and it is about the upper limit of the rate constant for most reactions occurring upon molecular collision.

Equation (12) indicates that the fluorescence quenching rate constant is inversely proportional to the viscosity of the medium. Although it has been proved that this is true in the vast majority of ordinary solutions, there are nevertheless many occasions where Einstein-Stokes behavior is not observed. Most of these deviations arise either from ambiguities in the molecular dimensions [17], or occur in viscous solutions [18]. As far as micelles are concerned, and in view of their restricted space available for molecular motion, the validity of the Einstein-Stokes equation is not self-evident. In the following we discuss, on the basis of our fluorescence quenching data, the applicability of the diffusion equation to intramicellar processes.

Table 1. Micellar parameters for alkane/surfactant mixed aggregates

X add. mol. f.	N_s	N_A	Rm Å	k_q (micel.) $\times 10^{-7}$ s^{-1}	$[Q]$ $\times 10^2$	k_q (homogen.) $\times 10^{-7}$ M^{-1} s^{-1}	η cP
n-Hexane/0.4 M SDS							
0	107	0	20.8	2.73	6.28	4.35	14
0.051	112	6	21.3	2.70	6.01	4.49	14
0.113	117	15	21.9	2.65	5.71	4.64	14
0.176	131	28	23.2	2.61	5.14	5.08	12
0.228	142	42	24.1	2.33	4.79	4.86	13
n-Octane/0.4 M SDS							
0.043	112	5	21.3	2.35	6.01	3.91	16
0.086	127	12	22.5	2.11	5.44	3.88	16
0.134	142	22	23.7	1.76	4.94	3.56	18
n-Decane/0.4 M SDS							
0.027	109	3	21.1	2.36	6.12	3.86	16
0.041	118	5	21.7	2.18	5.81	3.75	17
0.096	132	14	22.9	1.94	5.26	3.68	17
n-Hexane/0.4 M TTAC							
0	85	0	20.2	1.85	5.06	3.66	17
0.082	89	8	20.8	1.72	4.80	3.58	18
0.153	94	17	21.5	1.56	4.53	3.45	18
0.194	100	24	22.1	1.43	4.31	3.32	19
n-Octane/0.4 M TTAC							
0.071	92	7	21.1	1.54	4.68	3.29	19
0.133	104	16	22.3	1.44	4.24	3.39	19
0.18	114	25	23.3	1.24	3.92	3.16	20
n-Decane/0.4 M TTAC							
0.052	91	5	20.9	1.53	4.76	3.21	20
0.109	106	13	22.4	1.28	4.21	3.04	21
n-Hexane/0.4 M CTAC							
0	120	0	23.6	1.01	3.83	2.64	24
0.078	119	10	23.8	1.06	3.77	2.81	23
0.132	125	19	24.5	0.98	3.58	2.74	23
0.177	135	29	25.3	0.91	3.38	2.69	24
0.215	142	39	26.0	0.92	3.21	2.86	24
n-Octane/0.4 M CTAC							
0.059	128	8	24.4	0.97	3.61	2.69	24
0.149	154	27	26.5	0.81	3.10	2.61	24
0.192	160	38	27.1	0.82	2.98	2.75	23
n-Decane/0.4 M CTAC							
0.044	130	6	24.5	0.93	3.58	2.60	25
0.09	142	14	25.5	0.87	3.33	2.61	24
0.128	157	23	26.7	0.78	3.06	2.55	25
0.162	171	33	27.7	0.73	2.86	2.55	25

See text for explanation of symbols and other details.

The intramicellar fluorescence quenching has always been assumed to follow pseudo-first order kinetics, in spite of its bimolecular nature. This assumption has also been proved mathematically from the solution of Fick's equation under the appropriate boundary conditions [19–22]. Therefore, the values of k_q (micel.) listed in Table 1 are expressed in units of s^{-1}. Assuming that both the fluorophor Py and the quencher CPC, reside in the region of the micelle/water interface, we can estimate the effective quencher concentration in the micelle. Thus, taking as thickness of the interfacial volume within which the reactants are confined, 5 A for TTAC and CTAC, and 4 A for SDS micelles, we found the quencher concentrations $[Q]$ listed in Table 1 for the various micelles. Note that the thickness of the micelle/water interface was taken equal to the length of the ionic head, which in turn was estimated from the corresponding bond lengths [23]. Furthermore, knowing the quencher concentration in the micelle, and the pseudo-first order intramicellar fluorescence quenching rate constant, a second order rate constant can be calculated corresponding to an assumed homogeneous solution which would exhibit the same diffusion characteristics as the micellar interface, i. e., the same viscosity. Values of this rate constant are also listed in Table 1 as k_q (homogen.) in units of $M^{-1}s^{-1}$. On the other hand, using ordinary Stern-Volmer plots of the fluorescence life time of Py in ethanol and in the presence of varying amount of CPC, we found $k_q = 6.5 \cdot 10^9 \ M^{-1}s^{-1}$, for this homogeneous solution with $n = 1.0$ cP. Finally, assuming the Einstein-Stokes equation to be valid in micellar systems, we were able to determine from Eq. (13) the microviscosity in the micellar site

$$\eta_{mic} = \eta \ (\text{ethanol}) \ k_q (\text{ethanol})/k_q (\text{homogen.}) \quad (13)$$

where the fluorescence quenching occurs. Values of micellar microviscosities obtained by means of Eq. (14) are listed in Table 1 and indicate that the pallisade region of a micelle is about 15- to 25-times more viscous than ethanol. Recall that such high viscosities have been previously reported from measurements of the excimer/monomer fluorescence intensity ratio of dipyrenylpropane [24–26]. Also, in earlier studies of fluorescence polarization it was fond that the microviscosity in micelles of TTABr and CTABr, at the site of solubilization of the fluorophor, was between ca. 20 and 30 cP depending on the nature of the probe (2-methylanthracene or perylene) [27]. It is concluded therefore that the Einstein-Stokes equation can be applied in the study of micellar aggregates, and consequently the microviscosity in the pallisade region of the micelle η_{mic}, is expected to be inversely proportional to k_q. On the basis of these arguments, values of η_{mic} were calculated from intramicellar fluorescence quenching rate constants, and are listed in Table 1. Note that recently some preliminary results were published that also indicate the validity of Eq. (12) in micelles and vesicles [16].

Examination of the values of the microviscosities shown in Table 1, reveals two main conclusions. First, the microviscosity in the micelles of the quaternary

ammonium salts is definitely higher ($\eta = 17–25$ cP) than that in the micelles of SDS ($\eta = 12–18$ cP), and second, the addition of alkanes does not seem to affect the micellar microviscosity. For the first conclusion, i. e. the fact that η is higher in micelles of TTAC and CTAC than in SDS, the explanation lies in the well-known specific interaction between Py and the $^+N(CH_3)_3$ head group [28]. It has been shown that pyrene associates weakly with the quaternary group, and therefore its mobility in the micellar interfacial region, where pyrene is known to reside, is impeded. A consequence of this interaction is the fact that the diffusion of Py molecules becomes slower in TTAC and CTAC compared to its diffusion in SDS micelles. Therefore k_q and η_{mic}, measured via k_q, turn out to be lower in the micelles with the interaction than in the ones without it, as it is seen in Table 1. It is interesting that the interaction between pyrene and $^+N(CH_3)_3$ is also demonstrated in the way Py dissolves in these micelles. Thus, in TTAC and CTAC the solubilization is very easy and occupation numbers of pyrene per micelle equal to one or more are readily obtained. On the contrary, incorporation of Py in micelles of SDS is quite difficult, it needs continuous stirring and heating, and the occupation number is never more than ca. 0.6–0.8.

The second conclusion mentioned above, i. e., that η_{mic} does not depend on the nature and mole fraction of the added alkane, has some implications concerning the site of solubilization of these additives. Thus, it is well established that pyrene resides mostly in the palisade region of its host micelle, where also one expects to find the fluorescence quenching pyridinium group of the quencher CPC used here. Therefore, the microviscosity measured by means of the k_q refers to this particular micellar region. Since, the solubilization of the additive alkane molecules has no effect on the microviscosity, it is concluded that the addition of these molecules does not change the composition of the palisade region. This indicates that either alkane molecules stay away from the palisade region, or that they go there by replacing surfactant aliphatic chains and therefore retain a constant polarity in this region. Note that the constancy of the polarity in the palisade region has been proved from measurements of the ratio of the first to the third fluorescence peak of pyrene. This ratio, which constitutes a reliable index for the local polarity [29], did not change when alkanes were solubilized in the micelles. Because the aliphatic chains of the surfactants used in this study cannot be distinguished from the added alkane molecules, it is very difficult to decide if the additives go in the micellar core or if they just push the tails of the folded surfactant aliphatic chains into the core and they take their place close to the interface. In either case the environment in the palisade region will not be changed and therefore the microviscosity, micropolarity, and all other microproperties will remain unaffected by the solubilization.

References

1. Thomas JK (1987) J Phys Chem 91:267
2. Malliaris A (1988) Intern Rev Phys Chem 7:95
3. Boens N, Malliaris A, Van der Auweraer M, Luo H, De Schryver F (1988) Chem Phys 121:199
4. Malliaris A (1987) J Phys Chem 91:6511
5. Malliaris A, Lang J, Zana R (1986) J Chem Soc, Faraday Trans 1 82:109
6. Infelta PP, Gratzel M, Thomas JK (1974) J Phys Chem 78:190
7. Infelta PP (1979) Chem Phys Lett 61:88
8. Malliaris A (1987) Adv Colloid Interface Sci 27:153
9. O'Connor DV, Phillips D (1984) Time-correlated Single Photon Counting. Academic Press, New York
10. Mukerjee P, Mysels KJ (1971) Critical Micelle Concentrations of Aqueous Surfactant Systems. NSRDS-NBS 36
11. Van der Auweraer M, Dederen C, Palmans-Windels C, De Schryver FC (1982) J Amer Chem Soc 104:1800
12. Tanford C (1972) J Phys Chem 76:3020
13. Hague DN (1971) Fast Reactions. Wiley-Interscience, New York
14. Smoluchchowski MV (1917) Physik Chim 92:124
15. Noyes RM (1961) Progr React Kinet 1:129
16. Olea AF, Thomas JK (1988) J Amer Chem Soc 110:4494
17. Edward JT (1970) J Chem Ed 47:261
18. Dainton FS, Henry MS, Pilling M, Spencer PC (1977) J Chem Soc, Faraday trans 1 73:243
19. Hatlee MD, Kozak JJ, Rothenberger G, Infelta PP, Gratzel M (1980) J Phys Chem 84:1508
20. Hatlee MD, Kozak JJ (1980) J Chem Phys 72:4358
21. Hatlee MD, Kozak JJ (1981) J Chem Phys 74:1098
22. Hatlee MD, Kozak JJ (1981) J Chem Phys 74:5627
23. Pauling L (1960) The Nature of the Chemical Bond. 3rd Edition, Cornell University Press, Ithaca, New York
24. Lianos P, Lang J, Strazielle C, Zana R (1982) J Phys Chem 86:1019
25. Turley WD, Offen HW (1985) J Phys Chem 89:2933
26. Turley WD, Offen HW (1986) J Phys Chem 90:1967
27. Shinitzky M, Dianoux A-C, Gilter G, Weber G (1971) Biochemistry 10:2106
28. Viaene K, Verbeeck A, Gelade E, De Schryver FC (1986) Langmuir 2:456
29. Thomas JK (1980) Acc Chem Res 80:283

Received October, 1988;
accepted March, 1989

Author's address:

A. Malliaris
N.R.C. "Demokritos"
GR-Athens 153 10, Greece

Progress in Colloid & Polymer Science

Progr Colloid Polym Sci 79:249–256 (1989)

On certain solved and unsolved problems with water/PDMS/surfactant systems

A. Messier, G. Schorsch, J. Rouviere[1]), and L. Tenebre[1])

Rhone-Poulenc, Courbevoie, France
[1]) Laboratoire de Physico-chimie des Systèmes polyphasés, L.A. 330, U.S.T.L., Montpellier, France

Abstract: Most of the properties of the polydimethylsiloxane (PDMS) oils can be deduced from the structure of the linear polymeric chain of the macromolecule, including flexibility and stability of the Si-O-Si backbone, and hydrophobicity of the methyl substituents of the silicon atom.

But in fact, the advantages of pure PDMS for industrial application are limited. For the major uses, it is always necessary to either disperse variable amounts of silica in PDMS or to emulsify PDMS in water.

The interfacial properties of the water/PDMS/surfactant system were therefore investigated according to phase diagrams, surface tension, and ellipsometric measurements (carried out at CNRS Montpellier), and x-ray and ESR measurements (carried out at Collège de France, Paris).

The relationship between the amphiphilic behavior of the PDMS chain and the practical advantages of the PDMS/water system will be discussed.

Key words: Silicones, nonionic surfactant, interfacial film, emulsion, lamellar phase.

Introduction

This molecule is original, compared with other polymers for several reasons:

1. The mobility of the PDMS chain due both to:

the flexibility of the chain resulting from a large valence angle Si—O—Si: 140 °C, a great interatomic distance Si—O, and an absence of substituents on the O atom —

and to the small interactions between the chains due to the apolar behavior of the methyl groups on the Si atoms.

This mobility is confirmed by a very low glass transition temperature $T_g = -123\,°C$.

The consequences are fluidity of the products, spreading and film forming properties, gas solubility and permeability.

2. The hydrophobicity of the chain due to the apolar methyl groups

This entails incompatibility with hydrophilic products; they are used as demolding agents, and for paper treatment for stickers.

Silicone oils are used in their raw form or in emulsion; in the later case it is essential thing to find a compromise between stability (for storage and handling) and coalescence (for film forming).

Terminology is the following [1]:

Dimethylsiloxanes:

· M_2:

· M_2D:

· M_2D_2:

· M_2D_n:

· D_3:

· D_4:

· D_n:

Dimethylsilanols:

· D_2OH_2:

· D_3OH_2:

· D_nOH_2:

1. Phase diagrams

Phase diagrams of the ternary system water-oil-surfactant system have a practical importance: 1) during formulation; they indicate the bi-phase and the monophase regions which are met when preparing the emulsions, and 2) during application where, starting from a ternary mixture, they are formed when water is eliminated.

Examples: 1) *PDMSiloxane-surfactant-water (see Fig. 1). The surfactant is a mixture of two nonionic surfactants: an octylphenol with apolar head consisting of

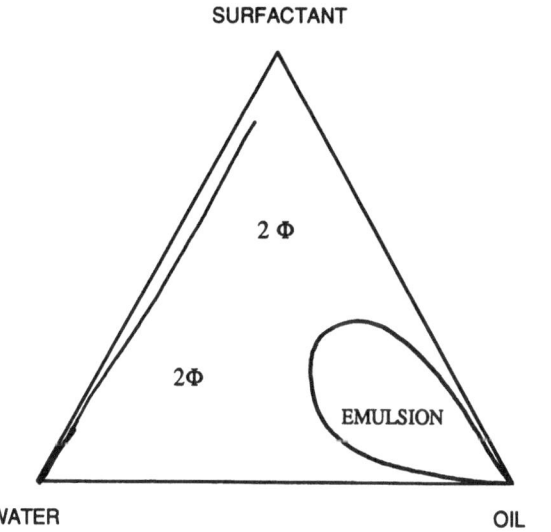

Fig. 1. Phase diagram of a ternary system PDMS-nonionic surfactant-water

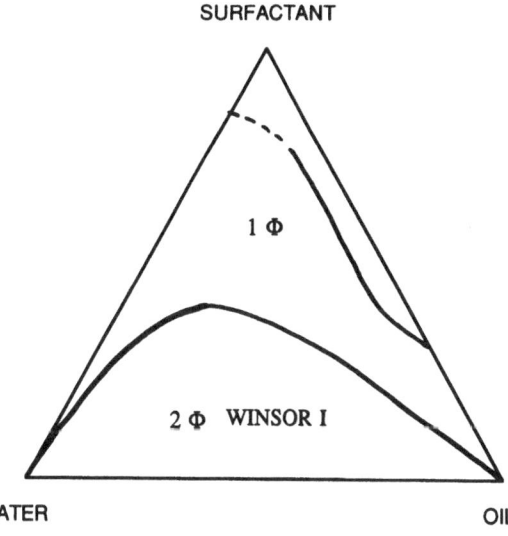

Fig. 3. Phase diagram of a ternary system PDMSilanol-nonionic surfactant-water

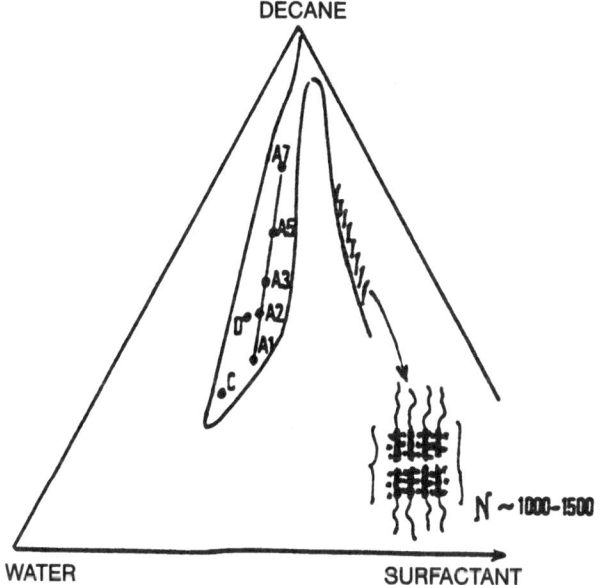

Fig. 2. Phase diagram of a ternary system decane-$C_{12}EO_4$-water (from [2])

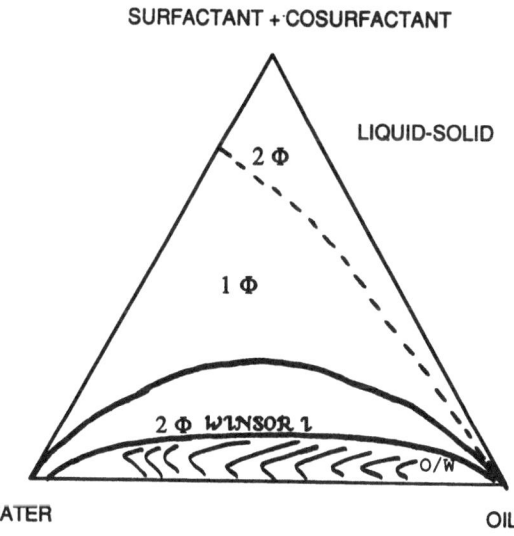

Fig. 4. Phase diagram of a pseudo-ternary system oil-surfactant-cosurfactant-water

10 ethylene oxydes and a nonylphenol with a polar head of 20 ethylene oxydes. The polydimethylsiloxane is $M_2D_{\langle 46 \rangle}$.

2) *Decane-surfactant-water (see Fig. 2) [2]. The surfactant consists of an alkyl chain of 12 carbon atoms and a polar head of 4 ethylene oxydes.

They are an important biphase region of Winsor I type, and a total emulsion region, with high oil con-

tent. These diagrams show the same hydrophobic character of PDMS as alkane.

3) *PDMSilanols-surfactant-water (see Fig. 3). The nonionic surfactant is the same as in Fig. 1. The mass of the PDMSilanol is about $D_{\langle 4 \rangle}OH_2$.

The phase diagram is similar to diagrams obtained with pseudo-ternary mixtures: oil-surfactant-cosurfactant-water (see Fig. 4, $C_{10}H_{22}$-OBS-C_5OH-H_2O).

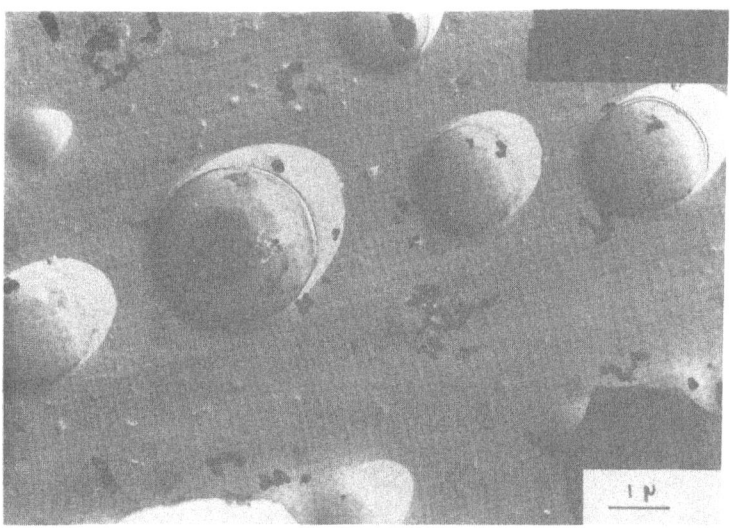

Fig. 5. and 6. Pictures of emulsions of silicone in water, taken by electron microscopy

One can see a monophase region, and a biphase region, Winsor I type, where oil-in-water emulsion can be prepared. This analogy is due to hydroxyl end groups which confer a cosurfactant character to polydimethylsilanols. These diagrams are important for industrial processing of emulsions.

The first stage is to prepare a water-surfactant mixture. As the surfactant content is high, the mixture falls into the field of liquid crystals [3]. Then silicone oil is added. Finally, water is added for dilution. Some pictures are given in Figs. 5 and 6, taken by electron microscopy.

In the final emulsion (water content, 80 %), one can see oil droplets which are covered by several randomly arranged layers of surfactant. In the picture (the emulsion has been concentrated after a centrifugation), we are able to see a plan view of the broken layers. All the droplets appear to be covered with a flaccid, slightly adhesive coat.

Referring to the first stage of emulsion processing, we wanted to see how a water surfactant system in the liquid crystal field evolves when silicone oil or alkane oil is added. When oil is added to a surfactant-water hexagonal phase, we observed no swelling. When oil is

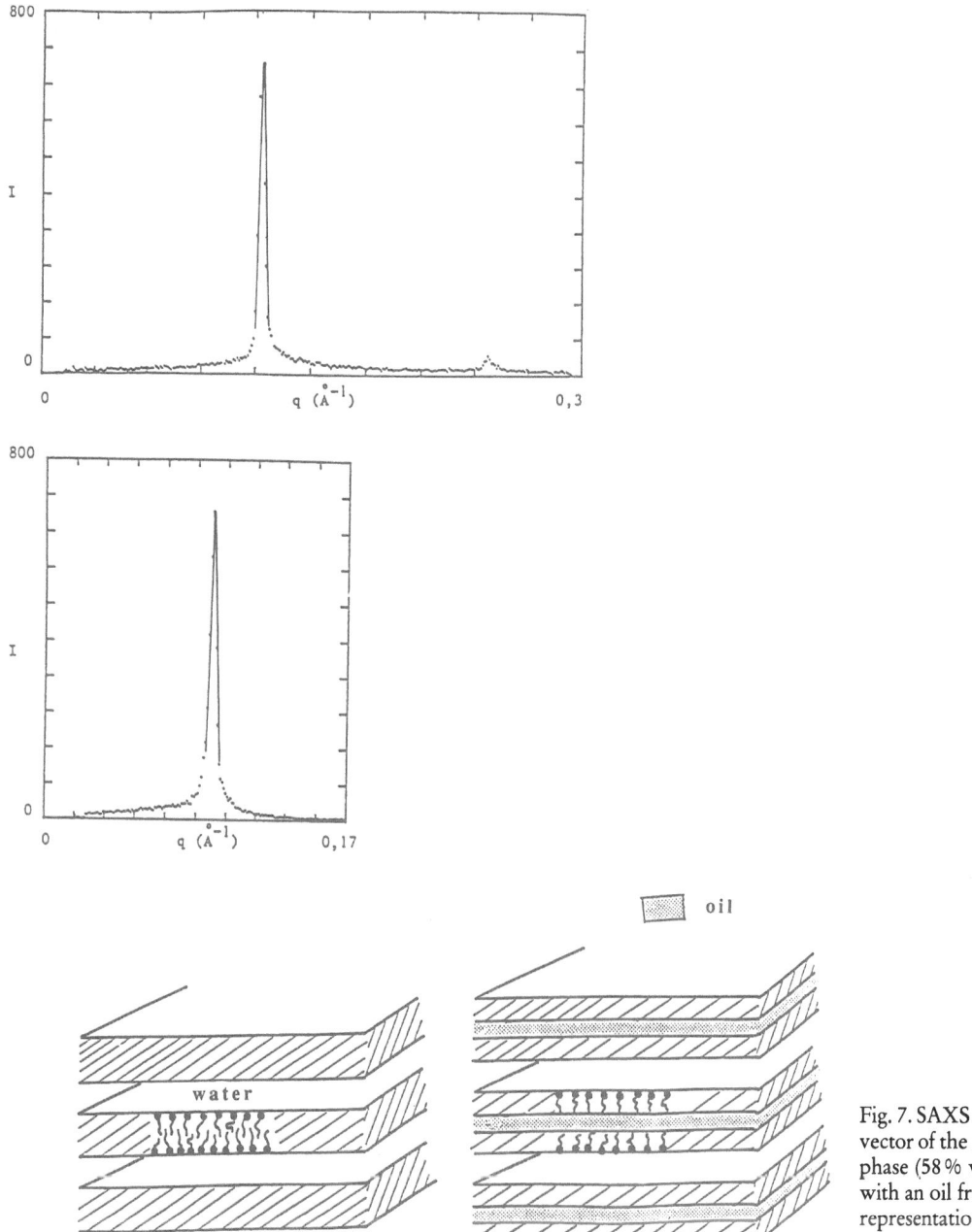

Fig. 7. SAXS intensity vs. the diffusion vector of the surfactant-water lamellar phase (58% w/w) $L\alpha$ (a) of $L\alpha + M_2$ with an oil fraction $\phi_0 = 0.18$ and (b), representation of the initial and swollen lamellar phases (c)

added to a surfactant-water lamellar phase, we observed an apparent swelling, which we investigated with x-ray scattering.

For polydimethylsiloxanes and polydimethylsilanols we can distinguish two different kinds of behavior, depending on the chain length. Several spectra of the scattered intensity vs. the diffusion vector are reported in Figs. 7–9. The initial lamellar phase $L\alpha$ is obtained with a nonionic surfactant in solution in water (58% w/w). This surfactant is "C_nEO_m" type, where n is 9 and 11, and m is about 5. The first spectrum is the surfactant-water lamellar phase one. In Fig. 7 the intensity is scattered by a sample $L\alpha + M_2$, with an oil fraction $\phi_0 = 0.18$. The lamellar phase is swollen

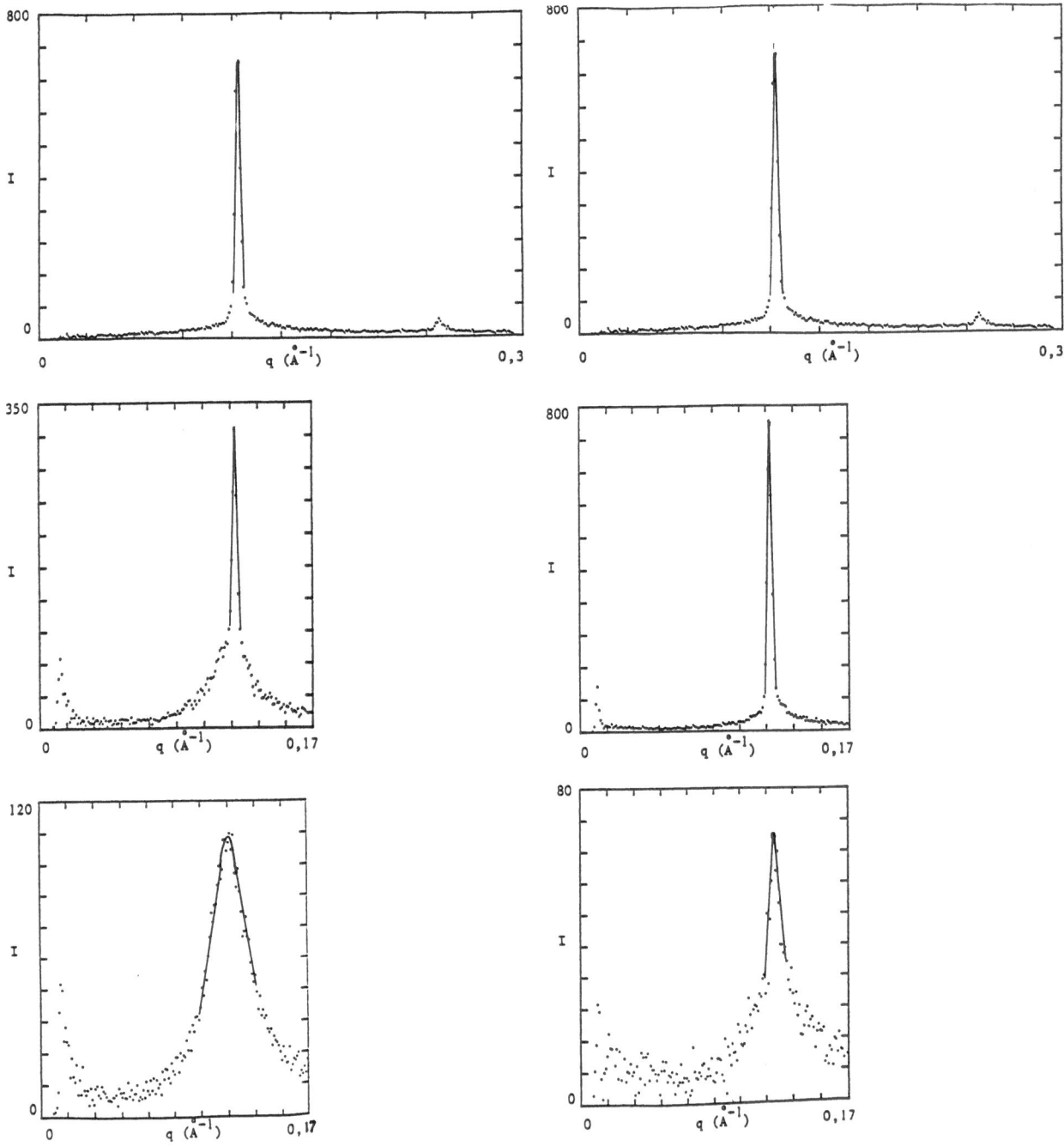

Fig. 8. SAXS intensity vs. the diffusion vector of the surfactant-water lamellar phase (58 % w/w) $L\alpha$ (a) of $L\alpha + M_2D_{\langle 15\rangle}$ with $\phi_0 = 0.23$ (b), and $\phi_0 = 0.6$ (c)

Fig. 9. SAXS intensity vs. the diffusion vector of the surfactant-water lamellar phase (58 % w/w) $L\alpha$ (a) of $L\alpha + M_2D_{\langle 350\rangle}$ with $\phi_0 = 0.34$ (b), and $\phi_0 = 0.64$ (c)

with oil, and one can verify that the surfactant area per molecule is unchanged.

In Figs. 8 and 9 are represented the following spectra: $-L\alpha + M_2D_{\langle 15\rangle}$, with $\phi_0 = 0.23$ and $\phi_0 = 0.6$; $-L\alpha + M_2D_{\langle 350\rangle}$, with $\phi_0 = 0.34$ and $\phi_0 = 0.64$.

In both cases there is no change in the position of the diffraction peak, but only in its width when the oil fraction is large enough. The oil does not swell the lamellar phase, but disperses it in small crystals with the initial structure. Like alkanes, short chains of PDMS ($M_w <$

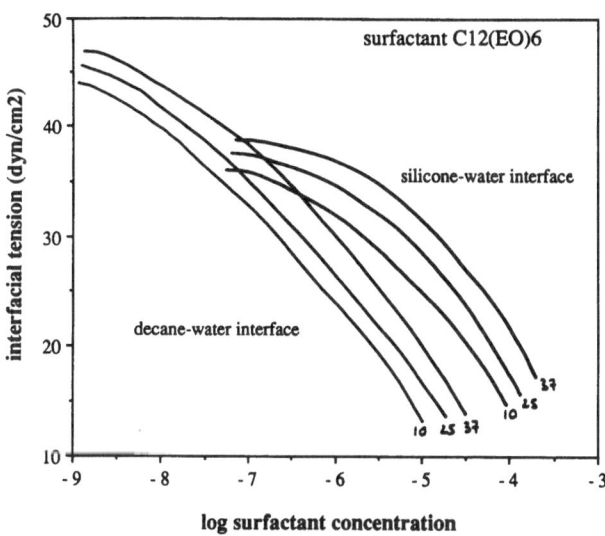

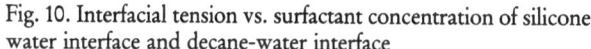

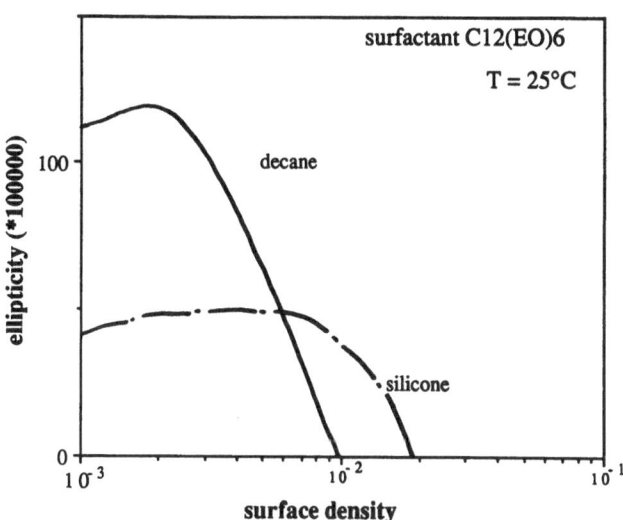

Fig. 10. Interfacial tension vs. surfactant concentration of silicone-water interface and decane-water interface

Fig. 11. Ellipticity vs. surface density of silicone and decane surfaces

500) make a lamellar system swell (the surfactant area per molecule being unchanged), while polymer-like long chains disperse a lamellar system into lamellar crystals with the original structure: there is no swelling and the reticular distance remain constant. The size of these crystals is decreased when more oil is added and the width of the diffraction peak is larger.

The influence of the chemical identity of the end group of the chains is not important in these two kinds of behavior. Its influence on the state of the interfacial film for short chains is very important. We studied the interfacial film in swollen lamellar phase, i.e., intial lamellar phase + short chains of PDMS with electron spin resonance of labelled surfactants, e.g. water-sur-factant-oil.

This technique consists of introducing a spin-labelled surfactant molecule in the interfacial film and detecting its movements by electron spin resonance. This local technique gives information about the immediate environment of the labelled molecule. Order parameter can be determined directly on the spectra. It is practically the same with heptane, silo-xane, as in the original water surfactant lamellar phase. It is smaller with silanol. With a computer simulation, the film undulation angles in the lamellae can be calculated.

The film undulations are unchanged with siloxanes or alkanes (about 16°), but are increased with silanols (about 26°); this proves the existence of an interaction of the hydroxyl end groups with the films.

Concerning the interaction of linear silicone oils with lamellar phases (water + nonionic surfactant), two parameters seem to be important: the length of silicone chains and the chemical identity of the end groups of these chains [4].

2. The water-oil interface studied by measuring both interfacial tension and ellipticity when putting surfactant molecules at the interface

Referring to the polar nature of the Si—O bond, we investigated the true hydrophobic nature of PDMS.

These techniques allow us to calculate the quantity of absorbed molecules, the orientation of these molecules in relation to the surface, the principal conformations, and the structure of the solvent at the interface. Some experimental results are given in Figs. 10 and 11.

After the first results we noticed that the surfactant was embedded in an upright fashion into the alkane-water interface. With short silicones, we found it to be only partly compatible with the surfactant at the polar end.

3. Conclusion

The following questions remain:

1) Spreading: what is the chain conformation at air-water interface?

The first experiment to deposit short oligomers of PDMS into the air-water interface showed the amphiphilic character of siloxanes. The case of long chains of siloxanes and silanols has yet to be studied.

2) Coalescence of emulsion: For purposes of application, we must know what is the exact mechanism, and what are the mechanical properties of the interfacial film (cohesion, adhesion)?

Acknowledgements

This work is the result of a team effort carried out by a scientific group financed by C.N.R.S. and Rhone-Poulenc.

References

1. Noll (1968) Chemistry and Technology of Silicones. Academic Press
2. Mittal KL, Lindman B (eds) (1984) Surfactants in Solution. Plenum Press
3. Mitchell DJ, Tiddy GJT, Waring L, Bostock T, McDonald MP (1983) J Chem Soc Faraday Trnas I 79:975
4. Messier A (1986) Thèse Université Paris VI

Received October, 1988;
accepted April, 1989

Authors' address:

A. Messier
Rhone-Poulenc
25, quai Paul Doumer
F-92408 Courbevoie Cedex, France

Progress in Colloid & Polymer Science   Progr Colloid Polym Sci 79:257–262 (1989)

Interfacial charges manifestations: Kerr and dielectric relaxation studies in a microemulsion system

M. Paillette and N. Belhadj-Tahar[1])

Groupe de Physique des Solides de l'Ecole Normale Supérieure, Paris, France
[1]) Laboratoire Dispositifs Infrarouges et Microondes (CNRS ID 035450), Université Pierre et Marie Curie, Paris, France

Abstract: Dielectric relaxation (10^2–10^9 Hz), transient- and phase (1–10^7 Hz) electric birefringences were carried out cojointly on a W/O ternary microemulsion system from water, benzene, and BHDC as surfactant, for different ratios w_o: (water)/(BHDC) and volume fractions ϕ values. From the Kerr-effect approaches it is possible to distinguish the existence of dimers, trimers, and an unexpected sign-reversal of the Kerr constant as w_o decreases, whatever the ϕ values. Preliminary analysis of the dielectric measurements evidences a frequency power law of the conductivity below 10^3 Hz. The hypothesis of a quasi *dc* (i. e., dispersive) transport charge process is proposed and credits the possible presence of temporal self-similar clusters. Four marked dielectric relaxations have been found. The contributions of the Maxwell-Wagner interfacial polarization, the space charge manifestations via a surface resistivity and a double-layer polarization, as well as a possible molecular relaxation process are proposed.

Key words: Dispersed system, dielectric loss and relaxations, electro-optical effects.

1. Introduction

A microemulsion is a stable, thermodynamically isotropic single liquid phase composed of water, oil, and a small quantity of a surface-active agent or surfactant.

A large number of experimental approaches such as static- and dynamic light scattering, electric conductivity, and photobleaching recovery techniques has been interpreted to provide information on droplet size and interactions between microemulsion droplets [1–3]. The influence of attractive interactions between droplets induces a clustering with formation of anisotropic structures [4].

Three different experimental methods in conjugation, sensitive to the auto- and different types cross-correlation terms were used. They are:
– in the frequency domain: dielectric relaxation spectroscopy (10^2–10^9 Hz) [5] and phase electric birefingence (PEB) (1–10^7 Hz) [6];
– in the time domain: the transient electric birefringence (TEB) ($3 \cdot 10^{-4}$–$8 \cdot 10^{-8}$ s) [7–8].

They offer a sensitive means and a rather elegant picture of the early stages of clustering in the low volume fraction range ($\phi \leq 0.04$) [9].

The present work is divided into three sections. The first section is devoted to the material and the different experimental methods performed. The second section shows qualitatively the different results of the droplet aggregation using Kerr effect. The third section discusses the presence of charge aggregates via a quasi-*dc* (i. e., dispersive) charge transport process. It also elucidates the new interfacial structure of the isolated droplets via the different relaxations obtained from the dielectric spectroscopy.

2. Materials and experimental

a) Microemulsion systems

The ternary microemulsion is a mixture of water in benzene (as oil) with benzyl-hexadecyl-dimethyl ammonium chloride (BHDC) as the surfactant. All chemicals were obtained commercially and used without further purification except water (Millipore freshly bidistilled, resistivity $\simeq 18$ M$\Omega \cdot$ cm).

The droplet radius is fixed by the water to surfactant ratio $w_o =$ [H₂O]/[BHDC] given by weight.

Five series of samples were prepared with different w_o. For each series, the droplet volume fraction ϕ was fixed about (4 ± 1)%. The size of the droplets and the parameter of the interactions between them are listed in (Table 1) [1].

The mean features of these results are that R_H increases with w_o and the negative α values corresponding to attractive interactions between droplets increases with increasing w_o. They are the origin of the very rapid clustering effect with a rate limited solely by the time taken for cluster collision to occur via Brownian diffusion. The chosen ϕ values are well below the percolation threshold ϕ_{wp} ($\approx$ 10–11%) of the electrical conductivity [3].

b) Experimental techniques carried out at room temperature

The dielectric relaxation experiments permit to yield the frequency dependence of the complex permittivity $\varepsilon^*(\omega)$ of the dispersed system, i. e., the values of the real ε' and imaginary parts ε'', respectively. Capacitance and conductance measurements were carried out over the frequency range of 100 Hz–13 MHz by means of the Impedance Analyzer model HP 4192A and over the range of 1 MHz to 1 GHz with the model HP 4191A. The measuring cells were i) a coaxial capacitor (C_{air} = 17.5 pF) [10²–10⁷ Hz], and ii) a cylinder cell [10⁶–10⁹ Hz], accurately calibrated by benzene as [10].

The transient electric birefringence consists in applying an electric field E rectangular pulse and measuring both the steady state electric birefringence Δn expressed by means of the Kerr constant $B(0)$ such as:

$$\Delta n(0) = B(0) \, \lambda \, E^2$$

where λ is the probe light wavelength and the time rise and decay functions $\Delta n(t)$ usually equivalent to a superposition of exponentials with characteristic times τ_i. Measurements were preformed by means of an electric-pulse generator Cober 605 P and analyzed with a digital storage oscilloscope-computer system [11].

The phase electric birefingence yields the frequency dependence of the complex Kerr constant $B^*(2\omega)$ at twice the frequency of the applied sinusoidal electric field $E(\omega)$. This new linear and differential method (carbon disulfide as reference) providing the magnitude of the modulus $|B(2\omega)|$ and the phase $\varphi(2\omega)$ has been previously described elsewhere [6]. The frequency bandwidth depends on that of the lock-in amplifier detection (1 Hz–50 MHz).

Table 1. Values of the hydrodynamic radii (R_H) of the droplets and of first virial coefficients (α) for different water to surfactant ratio (w_o)

w_o (wt)	R_H (Å)	α
0.33	36	− 11
0.5	41	− 12
0.66	47	− 13
1	55	− 14
1.1	61	− 19

To avoid heating and spurious effects all these measurements necessitate low excitation strength, and thus offer a very sensitive detection.

3. From the two Kerr effect experimental approaches

The conjugation (for the first time) of these tow modes from the same materials evidences two major features depending on w_o and ϕ values.

a) Presence of dimers, trimers, and larger aggregates

In the very low concentration range the spherical [11–13] isotropically polarisable droplets do not give any orientational contributions except a possible electric field-induced deformation [14]. No $B \sim \phi$ dependence has been detected.

As ϕ increases (≤ 0.04) the alignment and/or orientational contributions expected from anisotropic polarizable clusters, detected from TEB and PEB correspond to the Kerr constant $B(t)$ and $B(2\omega)$ respectively [11].

$$B(t) \equiv B(2\omega) = f \cdot v \cdot A\phi^m$$

where f is a shape-dependent factor associated with the electrical and optical polarizabilities, v is the droplet volume, and $A\phi^m$ is the volume fraction of clusters with $m = 2$ for dimers, $m = 3$ for trimers, etc.

Part a of Fig. 1 (log-log scale) illustrates clearly the manifestation of the presence of dimers ($B \sim \phi^2$) and trimers ($B \sim \phi^3$) up to $\phi \sim 0.04$. A satisfactory agreement between the two experimental approaches is noted.

Part b of Fig. 1 plots the log-frequency dependence of the imaginary part $B''(2\phi)$ of $B^*(2\omega)$ relative to the Kerr constant of carbon disulfide at w_o constant and ϕ variable [12].

Two relaxations that increasingly resolve as ϕ decreases are observed. The low frequency one is ascribed to the presence of large aggregates ($\tau_i = 1$–3 µs) whose relative number increases with ϕ and the efficiency of the aggregation. This frequency domain of relaxation (50–100 kHz) corresponds to the dielectric relaxation domain of the inner core counter-ions double-layer manifestation. The high-frequency relaxation corresponds to the alignment of the dimers. The relaxation time fits satisfactorily with the value calculated from a dumb-bell-like pair of droplets model [15].

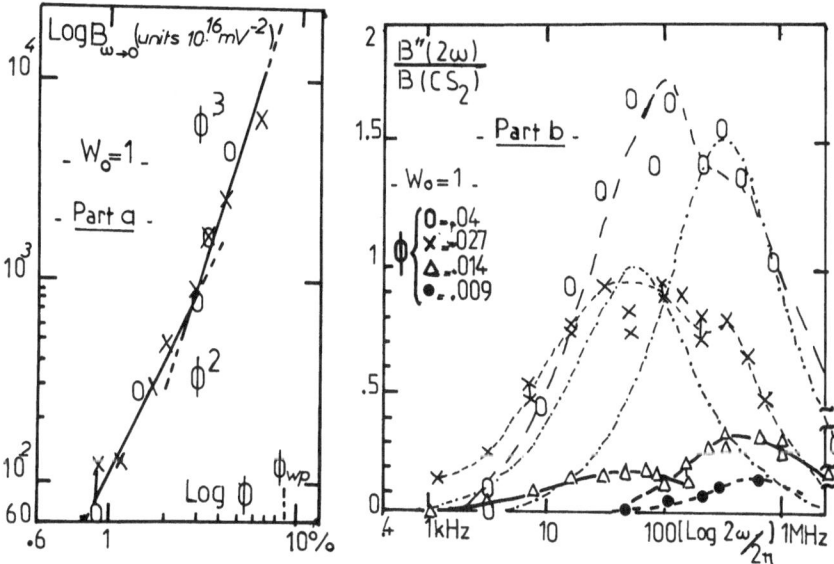

Fig. 1. a) Log-log plot of the Kerr constant B vs. volume fraction ϕ (ϕ_{wp}: percolation threshold) from PEB ($\times$) and TEB (O) measurements. b) Log-frequency dependence of the imaginary part $B''(2\omega)$ of the Kerr constant relative to carbone disulfide at $w_o = 1$ and ϕ variable. The lines are a guide for the eye, except the dot-dash line which corresponds to the Fourier transform of the short and long times decay functions measured by TEB [11]

b) The unexpected sign reversal of the Kerr constant as w_o decreases, whatever the ϕ values

The two forms of experimental approaches confirm the occurrence of this effect. The physical situation observed from TEB described elsewhere, shows that B scales roughly as ϕ^2 in the whole concentration range and thus indicates the dimer contribution [11, 15].

Parts a and b of Fig. 2 illustrate, respectively, the behaviors of the modulus $|B(2\omega)|$ and the phase $\varphi(2\omega)$ from PEB measurements of $B^*(2\omega)$ vs. the log frequency as w_o varies from $1(B^* > 0)$ up to 0.33 ($B^* < 0$), whatever the ϕ values. (Further measurements are underway to yield a complete description as a function of w_o.)

As described by the theoretical approach, this situation implies two conditions [16]:

– the existence of a permanent dipole moment, and
– a negative value of the optical polarizability anisotropy; the two moments, induced and permanent, are not directed along the same axis as i. e. the symmetry axis of the dimer.

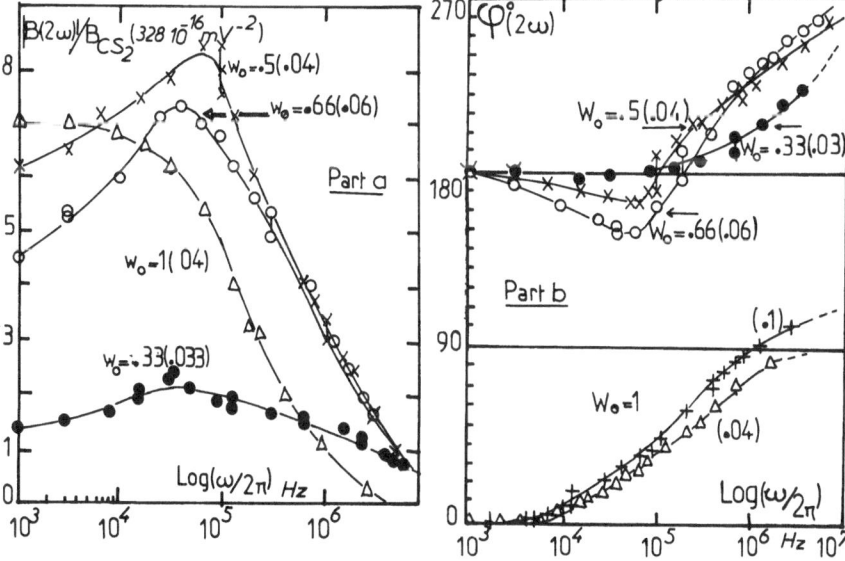

Fig. 2. Log-frequency dependences of the modulus $|B(2\omega)|$ relative to $B(CS_2)$ a), and the phase $\varphi(2\omega)$ b) of the complex Kerr constant $B^*(2\omega)$ as w_o varies [1–0.33], from PEB measurements

A two-body effect in the form of a local deformation of the interfaces of colliding droplets has been attempted [11, 17].

To yield further information about the polarization of the interfaces, the mobility of charges on the surfaces, and in the atmosphere surrounding the droplets, dielectric relaxation spectroscopy of this system has been carried out.

4. From dielectric relaxation spectroscopy

The interpretation and diffusion of these results are in progress [18]. Two major features are evidenced.

a) Experimental evidence of temporal self-similar clusters at very low frequencies

Below 1 kHz, the *ac* dielectric dispersion reveals the existence of a quasi *dc* (i.e., dispersive) transport charge process between clusters; the real and imaginary part of $\varepsilon^*(\omega)$ increases as frequency decreases [19].

The essential difference between this process and the *dc* conduction is that $\varepsilon^*(\omega)$ obeys a power-law frequency response that follows the Jonscher approach [19]:

$$\varepsilon^*(\omega) = \varepsilon_\infty + a(i\omega)^{n-1}$$
$$= \varepsilon_\infty + a\omega^{n-1}\left(\cos(1-n)\frac{\pi}{2} + i\sin(1-n)\frac{\pi}{2}\right)$$

$$(1)$$

where a is a constant and the exponent n defines a measure of the stuctural order between clusters [20].

In the present case, as n approaches zero ($4-8 \cdot 10^{-4}$ for $\phi \simeq 0.04$) this should signify that the aggregates are non-interacting and the charges move freely. A number of significant properties confirm our hypothesis:

α) Below 1 kHz, contrary to the previous assertions, Fig. 3 shows a frequency dependence of the conductivity σ where two behaviors are visible [21]. First, for $\omega < \omega_{B_z} = \dfrac{\varkappa_{B_z}}{\varepsilon_{B_z}\varepsilon_v}$, the characteristic angular frequency of the insulating phase, where $\varkappa_{B_z}, \varepsilon_{B_z}$ are, respectively, the conductivity and the permittivity of benzene, and ε_v is the absolute vacuum permittivity, σ is frequency-independent ($\sigma(\omega < \omega_{B_z}) = \sigma_{dc}$). Second, for $\omega_{B_z} < \omega < \omega_w = \dfrac{\varkappa_w}{\varepsilon_w \varepsilon_v}$ the characteristic angular frequency of the conducting phase (water), σ scales as $\omega^n \cdot (\sigma(\omega_{B_z} < \omega < \omega_w) \sim \omega^n)$.

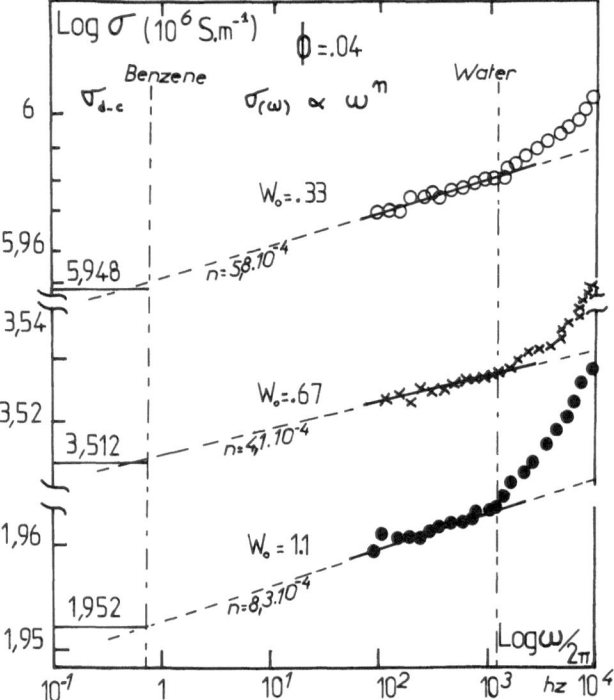

Fig. 3. The logarithm plots of the conductivity σ (in the low-frequency range) vs. the frequency at w_o variable and ϕ constant (0.04)

β) In this frequency range the impedance $Z(\omega)$ scales as $N_w^{-0.66}$, where N_w is the number of water pole at constant ϕ, whatever w_o.

γ) Relation (1) implies that the loss angle $\Delta(\omega)$ is frequency-independent, that is:

$$\tan \Delta(\omega) = \frac{\varepsilon''(\omega)}{\varepsilon'(\omega) - \varepsilon_\infty} = \frac{1}{2}\pi(1-n).$$

This behavior is reminiscent of the frequency dependence of $\varepsilon^*(\omega)$ at the percolation threshold ϕ_{wp} of a conductor-insulator mixture [22]. All these results represent macroscopic averages taken over samples which are large compared to the average correlation radius ξ for the clusters. The samples look *homogeneous*: the cluster are *temporally self-similar* [23]. The conduction mechanism is questionable.

b) Maxwell-Wagner interfacial polarization and molecular relaxation of the isolated droplets

Data analysis leads to evidence of the multiple dielectric relaxation processes of isolated droplets whose interpretations are in progress [18].

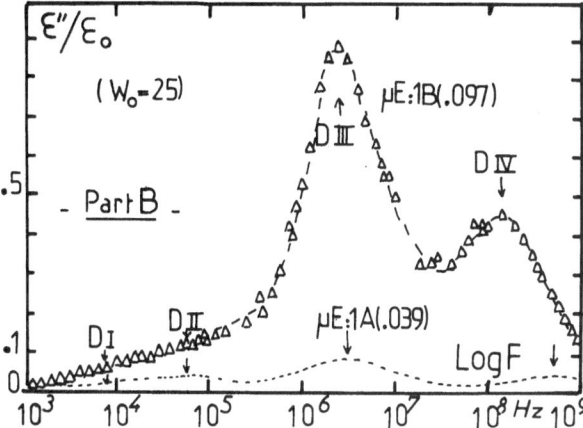

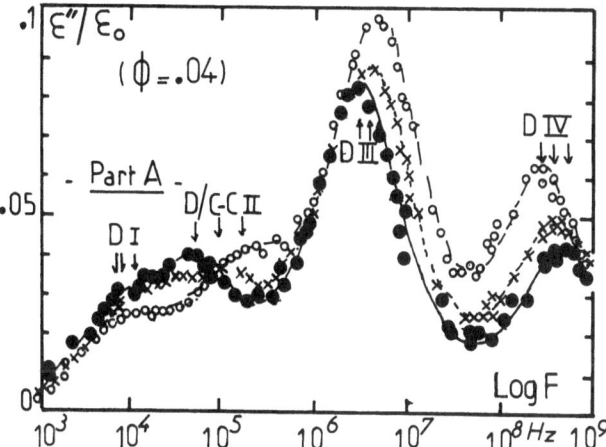

Fig. 4. A) The spectrum of the imaginary part $\varepsilon''(\omega)$ of the relative complex permittivity $\varepsilon^*(\omega)/\varepsilon_v$ as a function of log frequency at w_o variable (ϕ constant). ●: $w_o = 0.33$; ×: $w_o = 0.5$; O: $w_o = 0.66$. B) Comparison of the spectrum of the imaginary part ε'' as a function of log frequency at w_o constant and ϕ variable (1 A: 0.039; 1 B: 0.097)

Figure 4 (Part A) shows the form of the frequency spectrum of $\varepsilon''(\omega)$ of the systems at w_o variable and ϕ constant (0.04). Four relaxation domains labelled I, II, III, IV, respectively, are clearly visible.

The Debye-type relaxation (I) (characteristic relaxation time τ_I independent of R_w) corresponds to the manifestation of the conducting particles dispersed in the insulating phase [24].

The relaxation II (from Debye to Cole-Cole types as w_o decreases) seems to correspond to the surface polarization of the counterion (Cl$^-$) associated to a thermal diffusive process ($\tau_{II} \sim R_w^2$) bound to the presence of an inner core double-layer [25]. The stronger, well-resolved relaxation III is attributed to the mani-

festation of the surface conductivity of a layer around the droplets ($\tau_{III} \sim R_w$) [26]. The detailed analysis in progress leads to numerous parameters.

The origin of the fourth relaxation domain appears more controversial. Figure 4 (Part B) illustrates the frequency spectrum of ε'' (ε'' vs. log f) of the series $w_o = 1.1$ and ϕ variable (0.04–0.10) [27]. We observe that the intensity of relaxation III increases quicker than the ϕ ratio values and that the relaxation time τ_{IV} is translated towards the low-frequency side. It seems consistent to a mechanism of ionization of the surface layer which increases the electrolytic nature of the water in the inner core, leading to a possible relaxation.

In summary, this set of sensitive experimental approaches over many decades of frequencies leads to new results about the electrical polarization of this microemulsion system. A new dynamic area seems open to the interpretation of these results, for example: the role of the attractive interactions, the mobilities of the charges on the surfaces and around them.

The potential usefulness of these measurements in microemulsions and its extension in colloid and polymer systems must be outlined.

Acknowledgements

I am grateful to Prof. A. Fourrier-Lamer for lending the equipment for the dielectric measurements, and to Dr. J. C. Badot for very useful discussions. I also thank Profs. A. M. Cazabat and G. Mayer for fruitful discussions and valuable advice.

References

1. Chatenay D, Urbach W, Cazabat AM, Langevin D (1985) Phys Rev Lett 54:2253–2256
2. Cazabat AM, Langevin D (1981) J Chem Phys 74:3148–3158
3. Dvolaitzky M, Guyot M, Lagües M, Le Pesant JP, Ober R, Sauterey C, Taupin C (1978) J Chem Phys 69:3279–3288; Lagües M (1979) J Phys Lett 40:L 331–333
4. Ober M, Taupin C (1980) J Phys Chem 84:2418–2422; Cazabat AM, Chatenay D, Langevin D, Meunier J (1982) Faraday Disc Chem Soc 76:291–303; Cazabat AM, Chatenay D, Guering P, Urbach W, Langevin D, Meunier J (1987) In: Rosano HL (ed) Microemulsion Systems. M Dekker, New York
5. Eicke HF, Shepperd JCW (1974) Helv Chem Acta 57:1951–1963; Chou SI, Shah DO (1981) J Phys Chem 85:1480–1485; Peyrelasse J, Boned C (1985) J Phys Chem 89:370–379
6. Paillette M (1982) Opt Comm 41:140–144
7. Benoit H (1951) Ann Phys 6:561–609
8. Eicke HF, Markovic Z (1981) J Colloid Interface Sci 79:151–158; (1985) ibid 85:198–204; Hilficker R, Thomas RH (1985) Chem Phys Lett 120:272–275
9. Guering PH, Cazabat AM (1983) J Phys Lett 44:L 601–607
10. Belhadj-Tahar N, Fourrier-Lamer A (1986) IEEE MTT 34:346–350

11. Guering PH, Cazabat AM, Paillette M Meunier J (1986) Proc VI Intern Conf on Int and Coll Sci, New Dehli, in press
12. Guering PH, Cazabat AM, Paillette M (1986) Europhys Lett 2:953–960
13. Paillette M, Guering PH, Cazabat AM (1986) Opt Comm 60:244–250
14. Mayer G (1984) Opt Comm 52:215–220
15. Guering PH (1985) Thesis, Orsay, unpublished
16. Thurston GB, Bowling ID (1969) J Colloid Interface Sci 30:34–35
17. Discussion with Mayer G (1985) University Paris VI
18. Paillette M, to be published
19. Jonscher AK (1983) In: Dielectric Relaxations in Liquids. Chelsea Dielectrics Press, London
20. Dissado A, Rowe RC, Haidar A, Hill RM (1987) J Colloid Interface Sci 117:310–324; (1988) ibid 122:354–366
21. Bhattacharya S, Stokes JP, Kim MW, Huang JS (1985) Phys Rev Lett 55:1884–1887; Kim MW, Huang JS (1986) Phys Rev A 34:719–722
22. van Dijk MA, Casteleyn G, Joosten JGH, Levine YK (1986) J Chem Phys 85:626–631
23. Gefen Y, Aharong A, Alexander S (1983) Phys Rev Lett 50:77–80
24. Maxwell JC (1891) In: Treatise of Electricity and Magnetism. 3rd ed, Clarendon Press, Oxford; Wagner KW (1914) Arch Electrotech 2:371
25. Schwarz G (1962) J Phys Chem 66:2636–2642
26. O'Konski CT (1960) J Phys Chem 64:605–619
27. Cole RH, Delbos G, Winsor IV P, Bose TK, Moreau JM (1985) J Phys Chem 89:3338–3343

Received October, 1988;
accepted April, 1989

Authors' address:

M. Paillette
Groupe de Physique des Solides de l'Ecole Normale Supérieure
Laboratoire Associe du Centre National de la Recherche Scientifique
a l'Ecole Normale Superieure et a l'Universite Paris VII
(UA 17) Tour 23, 4e étage
2 place Jussieu
75251 Paris Cedex 05, France

Progress in Colloid & Polymer Science

Progr Colloid Polym Sci 79:263–269 (1989)

Percolation phenomenon in waterless microemulsions

J. Peyrelasse, C. Boned, and Z. Saidi

Université de Pau et des Pays de l'Adour, Centre Universitaire de Recherche Scientifique, Laboratoire de Physique des Matériaux Industriels, Pau, France

Abstract: A waterless microemulsion (glycerol/AOT/isooctane; molar ratio [glycerol]/
[AOT] = 3.2) has been studied at $T = 25\,°C$ as a function of the (glyerol + AOT) ϕ
volume fraction. The measured properties are electric conductivity, dynamic viscosity,
and dielectric relaxation (time domain spectroscopy).

An increase of conductivity σ and dynamic viscosity η are observed as the volume
fraction ϕ increases. The quantities $d(\ln \sigma)/d\phi$ and $d(\ln \eta)/d\phi$ go through a maximum, as
well as static permittivity ε_s. At the same time, the characteristic frequency ν_R of dielectric
relaxation goes through a minimum. The dielectric relaxation may be fitted by a generalized Davidson-Cole distribution of relaxation times.

The results are discussed within the framework of the theory of percolation. The
minimum of ν_R at the percolation threshold, and the maximum of ε_s, confirms more specifically the theoretical prediction in the domain of dielectric relaxation. The analogies
with microemulsions of the water/AOT/oil type are also discussed.

Key words: Waterless microemulsion, percolation, conductivity, dielectric relaxation,
viscosity.

Introduction

Microemulsions are systems which have attracted
considerable research interest over the last few years as
a result of their fundamental interest as well as their
practical applications. A microemulsion is a macroscopically monophasic system, which is generally the
result of mutual solubilization of water and oil. This
mutual solubility requires the addition of one or more
surface agents. A number of systems require the presence of alcohol for their formation and stability. The
resulting systems have four or five components (after
addition of brine) which makes analyzing the experimental results awkward. But it is not always necessary
to add a cosurfactant to obtain a microemulsion. It has
been known for some time that sodium bis(2-ethyl-
hexyl) sulfosuccinate (AOT) enables a high degree of
solubilization of water in oil without having to add another constituent. The result is a ternary system,
which is easier to study. Surfactants other than AOT
also exhibit this property. Similarly, it is possible to
form detergentless microemulsion. For example [1]

the water/2-propanol/n-hexane system. It is also possible [2] to prepare microemulsions with liquids other
than water such as formamide, glycerol, etc. Here the
choice of a suitable surfactant enables ternary systems
to be obtained: for example the methanol/AOT/cyc-
lohexane system [3] was one of the first to be indicated
in the literature. Waterless microemulsions are increasingly being subjected to systematic studies [2–10].

For water microemulsions in oil, when the water
volume fraction or the temperature increases, rapid
growth of the electric conductivity (of the order of several magnitudes) can be observed in certain cases [11–
20]. Several studies [16, 17, 19] indicate that this is a
dynamic percolation phenomenon [13, 21] involving
the formation of clusters of water droplets in a continuous medium. When the droplets are sufficiently
close to each other effective transfer of charge carriers
may take place [22]. Similarly, the dielectric behavior
of these systems can also be interpreted with reference
to an analog percolation model [16, 17, 23–26].
Finally, as far as dynamic viscosity is concerned the

phenomenon of percolation has been indicated very recently for ternary systems with AOT [27, 28], and it has also been mentioned [29] in a study of dynamic viscosity in the course of the Winsor I → III → II transition as a function of salt content.

For waterless microemulsions the percolation phenomenon has been indicated [14] for electric conductivity. In this paper we present a more comprehensive report concerning not only the study of stationary conductivity, but also that of the dielectric relaxation and dynamic viscosity of a ternary waterless microemulsion (glycerol/AOT/isooctane).

Experimental techniques

1. System studied

The samples were prepared with AOT (Fluka AG purum), isooctane (Fluka AG puriss) and glycerol (Prolabo Rectapur). The molar ration $n = $ [glycerol]/[AOT] was maintained constant at 3.2. As the microemulsions studied were of the glycerol in isooctane type, we can define the dispersed matter volume fraction (glycerol + AOT) as the ratio:

$$\phi = \frac{V_{\text{glycerol}} + V_{\text{AOT}}}{V_{\text{glycerol}} + V_{\text{AOT}} + V_{\text{isooctane}}}$$

in which V_i represent the volumes of the constituent i. Finally all the measurements were carried out at $T = 25.0 \pm 0.1\,°C$. We should point out here that the neighboring glycerol/AOT/heptane microemulsions have been studied [6] by light scattering. The resulting systems consist of discrete spherical droplets of glycerol stabilized by the surfactant. Droplet size is independent of temperature and depends primarily on the molar ratio n.

2. Experimental apparatus

The stationary electric conductivity σ was measured using a Wayne Kerr B 331 precision bridge operating at 1 592 Hz. The measurement cell is of Phillips-Mullard type and was calibrated with aqueous solutions of KCl. The relative uncertainty for the electric conductivity is estimated at 0.1%.

Kinematic viscosity measurements were made according to the capillary viscosimeter method (Viscotimer Lauda S/1). The density was measured using an automatic DMA 45 Anton Paar KG densitometer. Density measurements were carried out with an absolute uncertainty less than 10^{-3} g/cm³. Relative uncertainty of dynamic viscosity η can be estimated at less than 0.5%.

To measure dielectric relaxation we relied on a time domain spectroscopy method. The development of our apparatus has been reported in two technical articles [30, 31]. It consists of a Tektronix WP 1200 system with a reflectometer and a digital processing oscilloscope, all controlled by computer. Accurate determination of the complex permittivity ε^* is possible between 2 MHz and 8 GHz as long as $\sigma < 10^{-2}$ Sm^{-1}. Using alcohols with well known properties, we checked that the relative uncertainty of the real part ε' and the imaginary part ε'' ($\varepsilon^* = \varepsilon' - j\varepsilon''; j^2 = -1$) is 3%. Low frequency

study of complex permittivity was carried out with a Hewlett Packard 4192 A impedancemeter, also controlled by computer.

Experimental results

1. Dielectric properties

a) General considerations: For all the samples studied we observed that the microemulsions considered showed dielectric relaxation in addition to electrical conductivity. The complex permittivity $\varepsilon^*(v)$ (v: frequency of the electrical field applied) can be written:

$$\varepsilon^*(v) = \varepsilon_R^*(v) - j\,\frac{\sigma}{2\pi\,\varepsilon_o\,v}$$

in which $\varepsilon_R^*(v) = \varepsilon'(v) - j\,\varepsilon_R''(v)$ represents the relaxation term, σ conductivity, and ε_o the dielectric permittivity of a vacuum.

b) Study of the dielectric relaxation of our microemulsion: as an example Fig. 1 represents the variations of ε' and ε_R'' as a function of v for a sample with a volume fraction $\phi = 0.22$. Figure 2 represents the variations of ε_R'' as a function of ε' for the same sample. It is observed that the latter curve is not an arc of a circle and that the values of $d\varepsilon_R''/d\varepsilon'$ obtained for $v \to 0$ and for $v \to \infty$ are different. Curves with the same shape are obtained [32] when the dielectric properties of pure glycerol are studied. They can be represented by the generalized Davidson and Cole relationship:

$$\varepsilon_R^*(v) = \varepsilon_d + \frac{\varepsilon_s - \varepsilon_d}{\left[1 + \left(j \cdot \dfrac{v}{v_R}\right)^\alpha\right]^\beta} \qquad (1)$$

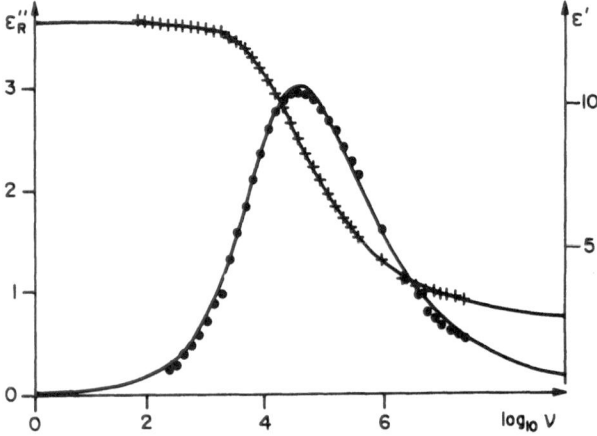

Fig. 1. Variations of ε' and ε_R'' vs. v (Hz) $\phi = 0.22$. + (ε') and ● (ε_R''): experimental points; ——: theoretical curve calculated using Eq. (1)

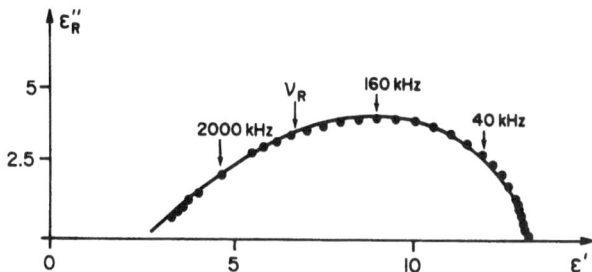

Fig. 2. Plot of $\varepsilon_R'' = f(\varepsilon')$. The frequency is expressed in kHz. $\phi = 0.22$. ●: Experimental points; ——: theoretical curve calculated using Eq. (1)

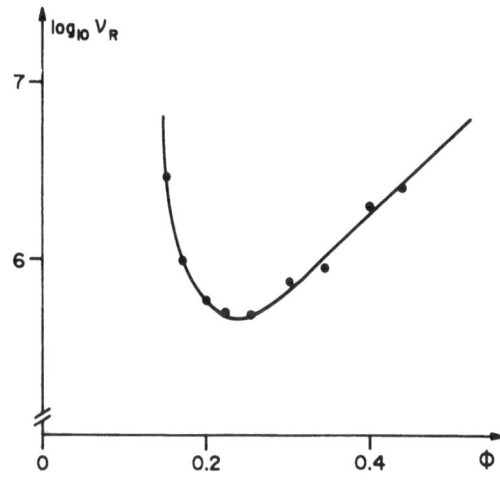

Fig. 4. Variation of $\log_{10} \nu_R$ vs. $\phi (\nu_R$ in Hz)

in which ε_s and ε_d are values of $\varepsilon_R^*(\nu)$ when $\nu \to 0$ and $\nu \to \infty$, α and β are parameters characterizing the distribution of relaxation times and ν_R is the characteristic relaxation frequency.

We determined by numerical analysis the values of $\varepsilon_d, \varepsilon_s, \nu_R, \alpha$, and β which give the best fit for experimental curves. Figures 1 and 2 plot the experimental points and the curves calculated from Eq. (1). It can be seen that there is a highly satisfactory fit.

Figure 3 represents the variations of ε_s vs. ϕ. This curve moves through a flattened maximum in the neighborhood of $\phi = 0.30$. Figure 4 shows variations of $\log_{10} \nu_R$ with ϕ. It can be seen that ν_R presents a minimum for $\phi = 0.24$. Variations of ε_d with ϕ have no noteworthy particularity. It can be observed, as is usual in the case of dispersed systems, that ε_d is a monotone function (here increasing) of the volume fraction ϕ. The parameters α and β do not vary significantly; their mean values are 0.85 and 0.53, respectively.

c) Study of conductivity: Fig. 5 represents the variations of conductivity vs. the volume fraction. It will be noted that σ remains very low while ϕ is less than 0.20. Above $\phi = 0.25$ conductivity increases very rapidly.

2. Study of viscosity

We have represented in Fig. 6 the variations of the dynamic viscosity η as a function of the volume frac-

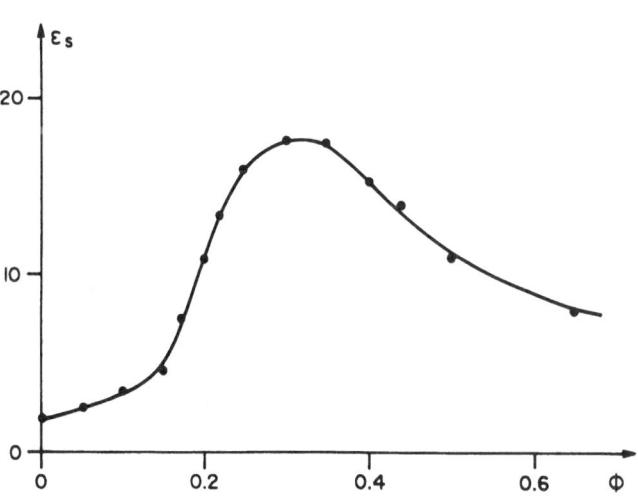

Fig. 3. Variation of ε_s vs. ϕ

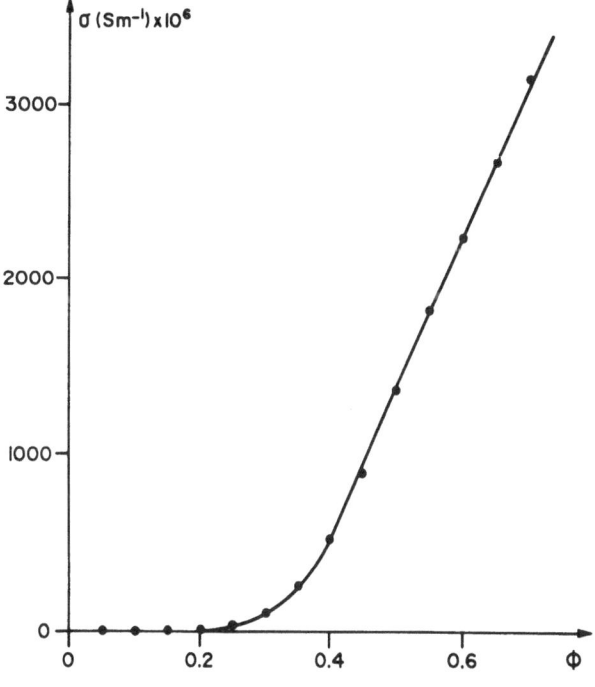

Fig. 5. Variation of σ vs. ϕ

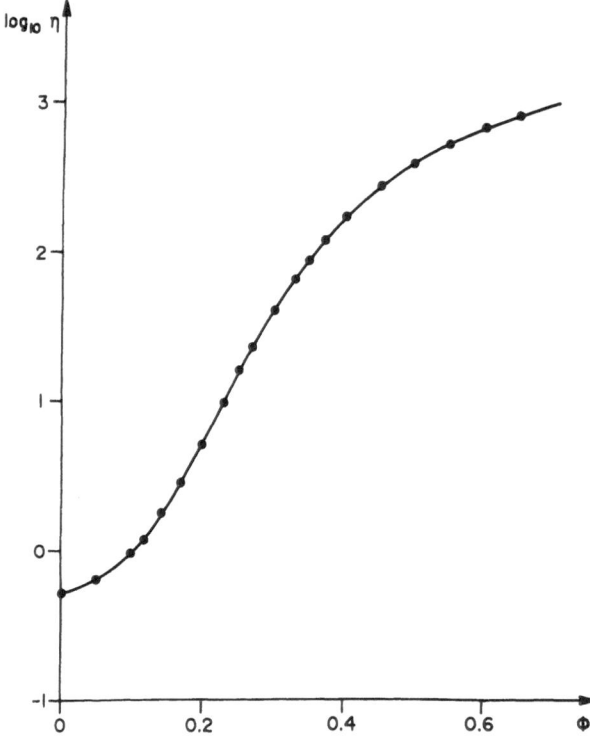

Fig. 6. Variation of $\eta(cP)$ vs. ϕ

tion ϕ. It can be seen that it varies very considerably and that for $\phi = 0.7$ it has a higher value than pure glycerol.

Percolation — a brief summary

The determination of the complex permittivity of heterogeneous systems is a well-known problem which has attracted much research. From the most general point of view the complex permittivity ε^* of a heterogeneous binary system must be accounted for by a relation of the form $\varepsilon^* = G(\varepsilon_1^*, \varepsilon_2^*, \phi, p_k)$, where ε_1^* and ε_2^* are the complex permittivities of constituents 1 and 2, ϕ the volume fraction of constituent 1, and p_k represents parameters which enable the function G to contain all the information on the geometry of the dispersion and on the interactions taking place within the system. The models by which the function G can be obtained are most often based on the effective medium or mean-field theories. They generally yield good results when the interactions in the mixture are weak, which is often the case at low volume fractions of one of the components, and as long as the dispersion can be

considered as macroscopically homogeneous. On the other hand when the dispersed particles can no longer be considered as isolated, in other words when clusters of varying sizes form, the conventional models no longer apply and the concept of percolation can successfully be used.

We have already indicated in the introduction that for microemulsions, percolation played a part in the description of conductivity. The relations concerning percolation for conductivity are particular cases of a more general relation concerning complex permittivity [33, 34]:

$$\frac{\varepsilon^*}{\varepsilon_1^*} = |\phi - \phi_c|^\mu \, f\left[\frac{\varepsilon_2^*/\varepsilon_1^*}{|\phi - \phi_c|^{(\mu + s)}}\right] \qquad (2)$$

where the exponents μ and s are positives. The function $f(z)$ where z is a complex variable satisfies the following asymptotic forms:

$$\phi > \phi_c \quad |z| \ll 1 \quad f(z) = C_1 + C_1' z$$

$$\phi < \phi_c \quad |z| \ll 1 \quad f(z) = C_2 z$$

$$|z| \gg 1 \quad \forall \phi \qquad f(z) = C \, z^{\frac{\mu}{\mu + s}}.$$

If one assumes that the two constituents are dielectric conductors of static permittivity ε_{1s} and ε_{2s} and of conductivity σ_1 and σ_2 we have:

$$\varepsilon_1^* = \varepsilon_{1s} - j \frac{\sigma_1}{2\pi \varepsilon_0 \nu} \qquad \varepsilon_2^* = \varepsilon_{2s} - j \frac{\sigma_2}{2\pi \varepsilon_0 \nu}.$$

The system presents a static permittivity ε_s and an overall conductivity σ. We will determine these two quantities for the following particular cases:

a) $\phi < \phi_c \ |z| \ll 1$ we obtain:

$$\varepsilon_s = C_2 \varepsilon_{2s} (\phi_c - \phi)^{-s} \qquad (3)$$

$$\sigma = C_2 \sigma_2 (\phi_c - \phi)^{-s} \qquad (4)$$

b) $\phi > \phi_c \ |z| \ll 1$ we obtain:

$$\varepsilon_s = C_1' \varepsilon_{2s} (\phi - \phi_c)^{-s} \left(1 + \frac{C_1}{C_1'} \frac{\varepsilon_{1s}}{\varepsilon_{2s}} (\phi - \phi_c)^{(\mu + s)}\right)$$

$$\sigma = C_1 \sigma_1 (\phi - \phi_c)^\mu \left(1 + \frac{C_1'}{C_1} \frac{\sigma_2}{\sigma_1} (\phi - \phi_c)^{-(\mu + s)}\right).$$

As $\mu + s$ is positive and $\varepsilon_{1s}/\varepsilon_{2s}$ does not tend towards infinity, when ϕ is reasonably close to ϕ_c we obtain:

$$\varepsilon_s = C_1' \varepsilon_{2s} (\phi - \phi_c)^{-s}. \tag{5}$$

Thus, when close to ϕ_c, ε_s varies according to $|\phi - \phi_c|^{-s}$ to either side of ϕ_c. Moreover, if $\sigma_2/\sigma_1 \ll 1$ (for example with a perfect insulator $\sigma_2 = 0$), close to the percolation threshold one can still find:

$$\frac{\sigma_2/\sigma_1}{|\phi - \phi_c|^{(\mu+s)}} \ll 1 \quad \text{then} \quad \sigma = C_1 \sigma_1 (\phi - \phi_c)^{\mu}. \tag{6}$$

Finally, according to suggestion made by Bergman, the following was proposed [35, 36] for the characteristic frequency:

$$\nu_R \alpha |\phi - \phi_c|^{(\mu+s)} \tag{7}$$

which is only valid when ϕ is near ϕ_c. So, when $\phi \to \phi_c$, then $\nu_R \to 0$.

In reality a maximum of $\dfrac{1}{\sigma} \dfrac{d\sigma}{d\phi}$, along with a maximum of ε_s are observed [24]. When a percolation phenomenon is present, it might be thought that the viscosity will also present particular variations, and this was observed for microemulsions with water [27–29, 37, 38]. In particular, the quantity $\dfrac{1}{\eta} \dfrac{d\eta}{d\phi}$ moves through a maximum at the percolation threshold. It should be noted at this point that Eqs. (4) and (6) describing conductivity, and Eqs. (3) and (5) for static permittivity are only valid if $|z| \ll 1$, in other words, if $\sigma_2/\sigma_1 \ll 1$ and if ϕ is close to ϕ_c. In particular, it is no longer possible to use all of these laws at infinite dilution ($\phi \to 0$) and at unit concentration ($\phi \to 1$). Moreover they cannot be utilized in the immediate vicinity of ϕ_c where no discontinuity of σ or ε_s should be observed. For example, as regards conductivity, a continuous transition between the values predicted by Eqs. (4) and (6) can be observed. The transition interval can be defined [33] by $\Delta = \left(\dfrac{\sigma_2}{\sigma_1}\right)^{\frac{1}{\mu+s}}$. When $\sigma_2/\sigma_1 \ll 1$, Δ is small and the equations can account for variations of σ vs. ϕ even when very close to the percolation threshold. This is the case for the water/AOT/undecane system [20].

As for viscosity, the connection with the theory of percolation has been less widely developed and stud-

ied. However by analogy with the previous equations we can write [28]:

$$\eta = C_2'' \eta_2 (\phi_c - \phi)^{-s'} \quad \text{if } \phi < \phi_c \tag{8}$$

$$\eta = C_1'' \eta_1 (\phi - \phi_c)^{\mu'} \quad \text{if } \phi > \phi_c \tag{9}$$

where s' and μ' are positive scaling exponents.

Discussion

As we have already pointed out, a great deal of research has been done on conductivity of microemulsions of water in oil. In this case the conductivity ratio $\dfrac{\sigma_2}{\sigma_1} = \dfrac{\sigma_{\text{oil}}}{\sigma_{\text{water}}}$ is much smaller than 1. The percolation Eqs. (4) and (6) are applicable and give a highly satisfactory fit with experimental results [20]. The dynamic aspect of the theory of percolation in microemulsions is expressed by the values [11, 21] of the exponents $\mu = 1.94$ and $s = 1.2$. Note also that the percolation threshold value that can be obtained through numerical analysis from Eqs. (4) and (6) coïncides with the position of the very marked maximum that appears on the curve $\dfrac{1}{\sigma} \dfrac{d\sigma}{d\phi}$ vs. ϕ. In the case of the microemulsions of glycerol in isooctane that we are concerned with here, the ratio $\dfrac{\sigma_2}{\sigma_1} = \dfrac{\sigma_{\text{isooctane}}}{\sigma_{\text{glycerol}}}$ does not satisfy $\sigma_2/\sigma_1 \ll 1$. Figure 7 shows that $\dfrac{1}{\sigma} \dfrac{d\sigma}{d\phi}$ presents a maximum (for $\phi = 0.25$), in the same way as in the previous case. This corresponds to the percolation threshold, but Eqs. (4) and (6) can no longer be used to provide a quantitative description, since the extent of the transition interval Δ is too large.

By extrapolating the theoretical conditions concerning conductivity to the case of dynamic viscosity, one might suppose that Eqs. (8) and (9) will only be valid if $\dfrac{\eta_2}{\eta_1} \ll 1$. In the case of water in oil microemulsions this condition is not met, but however $\dfrac{1}{\eta} \dfrac{d\eta}{d\phi}$ does present a maximum, which corresponds to the percolation threshold [28]. In the case of glycerol in isooctane microemulsions, this would a priori appear to be a more favorable case since $\dfrac{\eta_2}{\eta_1} \approx 5.10^{-4}$ at $T = 25\,^{\circ}\text{C}$. Figure 8 represents the curves $\eta^{-1/1.2} (\phi < \phi_c)$ and $\eta^{1/1.94} (\phi > \phi_c)$. One obtains two straight lines which cross over one the ϕ-axis at $\phi_c = 0.25$ (in Fig. 8

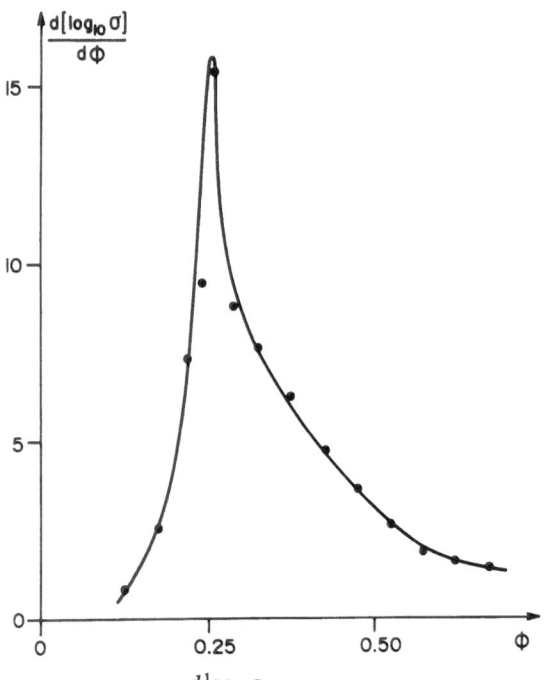

Fig. 7. Variation of $\dfrac{d \log_{10} \sigma}{d\phi}$ vs. ϕ

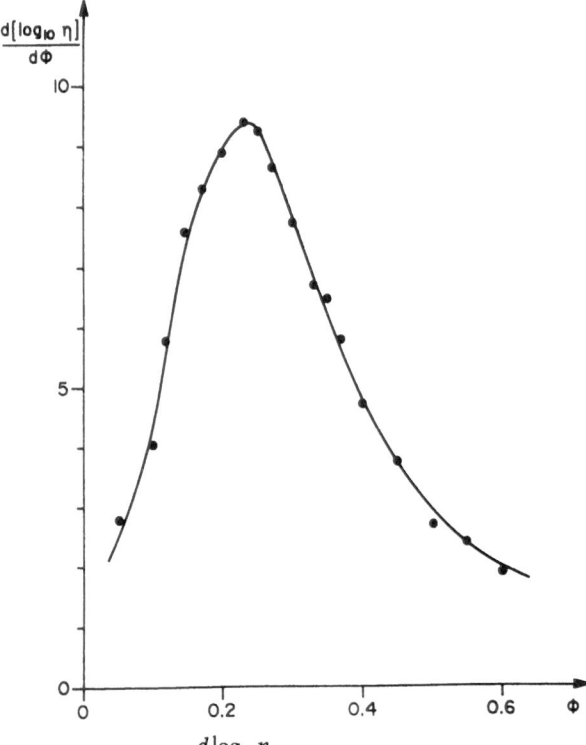

Fig. 9. Variations of $\dfrac{d \log_{10} \eta}{d\phi}$ vs. ϕ

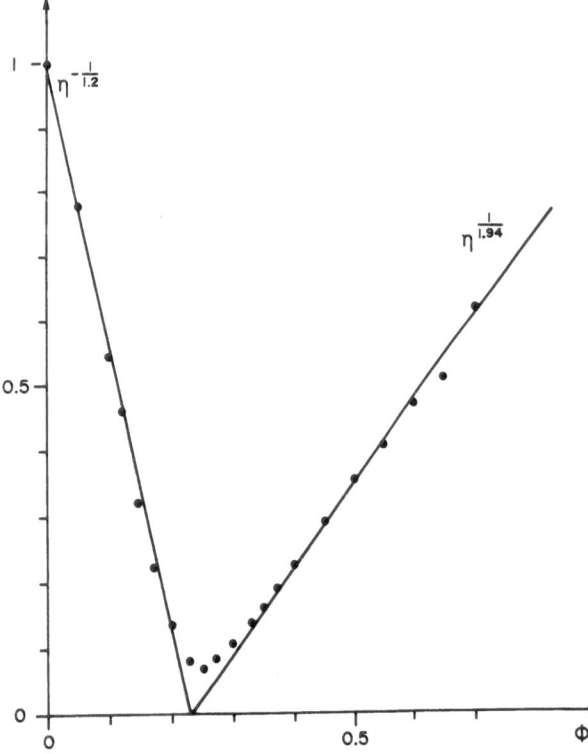

Fig. 8. Variations of $\eta^{-1/1.2} (\phi < \phi_c)$ and $\eta^{1/1.94} (\phi > \phi_c)$ vs. ϕ (normalized to 1 for $\phi = 0$ and $\phi = 1$)

the curves have been normalized to 1 for $\phi = 0$ and $\phi = 1$). Figure 9 plots the variations of $\dfrac{1}{\eta} \dfrac{d\eta}{d\phi}$ vs. ϕ. The previous analysis shows that Eqs. (8) and (9) can be applied in this case (since the transition interval Δ associated with viscosity is small). Moreover, it shows that the scaling exponents s' and μ' of viscosity are those which appear in dynamic models of percolation in microemulsions. This clearly shows the general character of the phenomenon of dynamic percolation.

As for the dielectric aspect (Eqs. (3), (5), and (7)), as Δ (associated with dielectric constant) is high, it is impossible to carry out a quantitative determination. However, the equations give a good qualitative representation, since ε_s moves through a maximum (Eqs. (3) and (5)) and ν_R moves through a minimum (Eq. (7)), as ϕ increases. The maximum for ε_s is located at $\phi = 0.30$, a higher value than that corresponding to the maxima for $\dfrac{1}{\sigma} \dfrac{d\sigma}{d\phi}, \dfrac{1}{\eta} \dfrac{d\eta}{d\phi}$, or the minimum for ν_R, which are at $\phi = 0.24$. In the case of water in oil microemulsions such a shift has already been observed [16, 24], and interpreted as a "hopping" phenomenon.

Conclusion

As for water and oil type microemulsions, the behavior of ternary waterless microemulsions can be interpreted, at least in this case, by means of a percolation phenomenon which corresponds to a particular behavior of electrical conductivity, dynamic viscosity, and dielectric relaxation. There are close analogies between the properties of waterless microemulsions and those of waterbased microemulsions. It should be stressed that the scaling exponents are the same in both cases. Moreover, using these systems, one can test the theoretical predictions (maximum of ε_s and minimum of ν_R) relative to the behavior of dielectric relaxation in the presence of a percolation phenomenon.

Finally, this preliminary study will have to be followed up by the determination of the "line of threshold" in the ternary diagram, which will allow a more complete discussion as a function of interactions [24].

References

1. Smith GJ, Donelan CE, Barden RE (1977) J Colloid Interface Sci 60:488–496
2. Rico I, Lattes A (1984) Nouveau Journal de Chimie 8:429–431
3. Pouligny B, Lalanne JR, Couillaud B, Ducasse A, Sarger L (1981) Optics Communications 37:271–276
4. Rico I, Lattes A (1984) J Colloid Interface Sci 102:285–287
5. Friberg SE, Podzimek M (1984) Colloid Polym Sci 262:252–253
6. Fletcher PDI, Galal MF, Robinson BH (1984) J Chem Soc Faraday Trans I 80:3307–3314
7. Fletcher PDI, Robinson BH, Tabony J (1986) J Chem Soc Faraday Trans I 82:2311–2321
8. Das KP, Ceglie A, Lindman B (1987) J Phys Chem 91:2938–2946
9. Auvray X, Petipas C, Anthore R, Rico I, Lattes A, Ahmah-Zadeh Samii A, de Savignac A (1987) Colloid Polym Sci 265:925–932
10. Friberg SE, Liang YC (1987) Colloids and Surfaces 24:325–336
11. Lagües M, Ober R, Taupin C (1978) J Phys Lett (Paris) 39:L 487–L 491
12. Lagourette B, Peyrelasse J, Boned C, Clausse M (1979) Nature 281:60–62
13. Lagües M (1979) J Phys Lett (Paris) 40:L 331–L 333
14. Lagües M, Sauterey C (1980) J Phys Chem 84:3503–3508
15. Cazabat AM, Chatenay D, Langevin D, Meunier S (1982) Far Discuss Chem Soc 76:291–303
16. Van Dijk MA (1985) Phys Rev Lett 55:1003–1005
17. Bhattacharya S, Stokes JP, Kim MW, Huang JS (1985) Phys Rev Lett 55:1884–1887
18. Chatenay D, Urbach W, Cazabat AM, Langevin D (1985) Phys Rev Lett 54:2253–2256
19. Kim MW, Huang JS (1986) Physical Review A 34:719–722
20. Moha-Ouchane M, Peyrelasse J, Boned C (1987) Phys Rev A 35:3027–3032
21. Grest GS, Webman J, Safran SA, Bug ALR (1986) Phys Rev A 33:2842–2845
22. Safran SA, Webman J, Grest GS (1985) Phys Rev A 32:506–511
23. Van Dijk MA, Casteleijn G, Joosten JGH, Levine YK (1986) J Chem Phys 85:626–631
24. Peyrelasse J, Moha-Ouchane M, Boned C (1988) Phys Rev A 38:904–917
25. Clarkson MT, Smedley SI (1988) Phys Rev A 37:2070–2078
26. Clarkson MT (1988) Phys Rev A 37:2079–2090
27. Berg RF, Moldover MR, Huang JS (1987) J Chem Phys 87:3687–3691
28. Peyrelasse J, Moha-Ouchane M, Boned C (1988) Phys Rev A 38:4155–4161
29. Quemada D, Langevin D (1985) J Theor Appl Mech, Numero Special:201–237
30. Peyrelasse J, Boned C, Le Petit JP (1981) J Phys E 14:1002–1008
31. Boned C, Peyrelasse J (1982) J Phys E 15:534–538
32. Davidson DW, Cole RH (1951) J Chem Phys 19:1484–1490
33. Efros AL, Shklovskii BL (1976) Phys Stat Sol 76 B:475–485
34. Stroud D, Bergman J (1982) Phys Rev B 25:2061–2064
35. Benguigui L, Yacubowicz J, Narkis M (1987) J Polym Sci: Part B, Polym Phys 25:127–135
36. Benguigui L (1985) J Phys Lett (Paris) 46:L 1015–L 1021
37. Eicke HF, Hilfiker R, Holz M (1984) Helv Chim Acta 67:361–372
38. Borkovec M, Eicke HF, Hammerich H, Das Gupta B (1988) J Phys Chem 92:206–211

Received November, 1988;
accepted January, 1989

Authors' address:

Dr. J Peyrelasse and Prof. C. Boned
Laboratoire de Physique des Matériaux Industriels
Centre Universitaire de Recherche Scientifique
Avenue de l'Université
F-64000 Pau, France

Progress in Colloid & Polymer Science   Progr Colloid Polym Sci 79:270–271 (1989)

Non-aqueous silica dispersions.
Charged particle interactions studied by scattering of light

A. P. Philipse[1]) and A. Vrij

Van't Hoff Laboratory, University of Utrecht, Utrecht, The Netherlands
[1]) Present address: ECN, PO Box 1, 1755 ZG Petten, The Netherlands

Abstract: A novel type of stable colloidal dispersions in weakly-polar organic solvents will be described. The particles consist of a charged, spherical silica core (coated with a silane coupling agent) and are nearly uniform in size. An important feature of this silica is that it can be optically matched up to high particle volume fractions. It provides therefore a very appropriate model dispersion for light scattering studies (static and dynamic) of the *combined effect* of charge and high (core) volume fraction on "direct" and "hydrodynamic" interactions [1].

Key words: Silica, light scattering, colloidal, dispersion, dynamics.

A novel type of stable colloidal dispersions in weakly-polar organic solvents will be described. The particles consist of a charged, spherical silica core (coated with a silane coupling agent) and are nearly uniform in size. An important feature of this silica is that it can be optically matched up to high particle volume fractions. It provides therefore a very appropriate model dispersion for light scattering studies (static and dynamic) of the *combined effect* of charge and high (core) volume fraction on "direct" and "hydrodynamic" interactions [1].

In previous studies of charged systems (mainly aqueous dispersions of latex particles, e.g. [2]) the (core) volume fraction, ϕ, had to be kept very low in order to avoid multiple scattering effects. At low (core) volume fractions the hydrodynamic interactions become insignificant.

The synthesis was performed in two steps. Particle cores were obtained with a Stöber-type of synthesis in which tetraethoxysilane is hydrolyzed to form colloidally-sized spheres. These can be grown further in steps [3] to paucidisperse particles with the desired size. A quantitative model for the growth process was derived and tested [4]. These systems were used to study poly- and bi-dispersity effects on light scattering [3].

In a second step of the synthesis a stabilizing (protective) layer is attached to the particle surface, using the silane coupling agent: 3-methacryloxypropyl trimethyloxysilane. This leads to dispersions which are stable in a range of weakly-polar organic solvents. The optically very homogeneous particles [5] carry an electrical charge, which induces the formation of "colloidal crystals" at high volume fractions [6]. Also glass-like phases are sometimes observed.

Static and dynamic interactions were measured between the silica spheres ($a = 79$ nm) suspended in a nearly optically-matching, salt-free mixture of ethanol and toluene up to $\phi \cong 0.10$ [7]. The static structure factor, $S(K)$, is compared with theoretical calculations based on the RMSA approximation for a single-component fluid of charged macro-particles (supramolecules) [8]. Also an effective diffusion coefficient, D_e, (obtained from the initial decay of the field autocorrelation function) was determined and combined with $S(K)$ to obtain the function $H(K)$, which represents the hydrodynamic interactions. Here $K = (4\pi/\lambda) \sin (\theta/2)$ is the length of the scattering vector, and θ is the scattering angle.

From $H(K)$, obtained for the first time as far as we know for charged colloidal particles, we conclude that the long-range electrostatic repulsion between the spheres has a pronounced influence on the hydrodynamics of large-scale, collective particle motions, whereas small-scale single-particle diffusion is relatively unaffected. The autocorrelation functions show a non-single exponential decay at $K \rightarrow 0$.

References

1. Philipse AP (1987) Dissertation, University of Utrecht
2. Brown JC, Goodwin JW, Ottewill RH, Pusey PN (1976) In: Kerker M (ed) Colloid and Interface Sci IV. Academic Press, New York
3. Philipse AP, Ref 1, Ch 2 Philipse AP, Vrij A (1987) J Chem Phys 87:5634
4. Philipse AP, Ref 1 General Appendix, Philipse AP (1988) Colloid Polym Sci 266:1174
5. Philipse AP, Ref 1, Ch 4, Philipse AP, Smits C, Vrij A (1989) J Colloid Interface Sci 129:335
6. Philipse AP, Ref 1, Ch 3, Philipse AP, Vrij A (1989) J Colloid Interface Sci 128:121
7. Philipse AP, Ref 1, Ch 5, Philipse AP, Vrij A (1988) J Chem Phys 88:6459
8. Hayter JB, Penfold J (1981) J Chem Soc Faraday Trans 1, 77:1851; Hayter JB, Hansen JP (1982) Institute of Laue-Langevin Report

Received October, 1988;
accepted March, 1989

Authors' address:

A. Vrij
Van't Hoff Laboratory
University of Utrecht
Padualaan 8
NL-3584 Utrecht, The Netherlands

Progress in Colloid & Polymer Science　　　　Progr Colloid Polym Sci 79:272–278 (1989)

Fluorinated and hydrogenated nonionics in aqueous mixed systems

J. C. Ravey, A. Gherbi, and M. J. Stébé

Laboratoire de Physico-Chimie des Colloides LESOC, UA CNRS 406, Faculté des Sciences, Université de Nancy I, Vandoeuvre les Nancy, France

Abstract: In mixing nonionic fluorinated and hydrogenated surfactants, one or several types of mixed aggregated structures can be obtained. Several types of measurements have been performed and combined in order to obtain detailed information on this subject: ternary phase diagram determinations, x-ray diffraction on the liquid crystal phases, small-angle neutron scattering on the isotropic phases, and surface tension measurements on dilute aqueous solutions. It is shown how the surface tension results fully correlate the phase behaviors of these systems. At constant hydrophobic tail, according to the oxyethylene chain length the mixing may be almost ideal or markedly nonideal, although every one phase system (isotropic or liquid crystal) is actually constituted by only one type of mixed aggregate. And it is shown how the mutual oleophoby of these surfactants is enhanced by a marked dissymmetry in the hydrophilic chains but in presence of large amounts of water. The mixing is described by the regular solution formalism, making use of a polynomial representation of the interaction parameter.

Key words: Mixed micelles, nonionic surfactant, fluorinated surfactant, phase diagram, surface tension.

Introduction

Many models have appeared in the literature describing interactions of surfactants in mixed micelles [1–10, 20], in order to explain the actual behavior of the mixtures. In particular, most of the nonionic/nonionic mixed systems studied so far have been recognized to behave almost ideally [3, 10], at least if both the surfactants are perhydrogenated compounds. But when fluorocarbon and hydrocarbon surfactants are mixed, the surface adsorption and micellization properties of their solutions must also reflect the "mutual phobicity" between FC- and HC-chains [3, 9, 11–14].

From the experimental point of view, there exist several means to determine the composition of the mixed micelles; but they are in two classes according to the concentration of the solutions. In the first case, one deals directly with the micelles themselves at finite concentration [12, 17]; in the other case, the surface properties of the solutions at very dilute concentrations (near the CMC) are investigated [3, 5]. However one wonders whether both types of measurements are truly correlated. To answer that question the ternary phase diagram water/mixed surfactants should also be considered.

In this preliminary paper, surface tension-related investigations, as developed by Funasaki et al. [3, 15], will be presented for nonionic mixtures with new ethoxylated fluorinated surfactants. Emphasis will be put on the effect of the number of the oxyethylene groups. A few structural determinations will be merely mentioned (x-ray and small-angle neutron scattering). But most importantly, we want to show how the surface tension data, interpreted in terms of the regular solution theory, can be tightly correlated to the ternary phase diagram-water $C_{12}EO_6$/fluorinated surfactants, and to the structures of the aggregates present in these different "macro-phases".

Experimental

Materials

The HC-chain surfactant used throughout the present investigations was hexaethylene glycol dodecyl ether ($C_{12}EO_6$). It was purchased from Nikko Chemicals and used without further purification.

The FC-chains surfactants belong to a series of new nonionic compounds, of general formula $C_m F_{2m+1} - C_2H_4 - \Sigma - E_n$ (E stands for $-OC_2H_4-$, while Σ represents $-SC_2H_4$). The data reported in the present work correspond to $m = 6$, and $n = 2$, and $n = 7$. Their formula will be abbreviated as $R_f E_n$. Details on their synthesis have been given elsewhere [16].

Methods

The small-angle neutron scattering technique was used for determinations of the composition of the mixed micelles by using the well known contrast variation method (use of mixtures of D20/H20) [12]; their size and shape was also investigated, but will not be discussed here. Investigations were performed at the Institute Laue Langevin (Grenoble) and at the Leon Brillouin Laboratory (Saclay, France).

X-ray diffraction was carried out to determine the type of the liquid crystals: lamellar (L_a), hexagonal (H_1), and bicontinuous cubic (V_1). The Bragg spacings were measured as a function of the overall composition of the ternary one anisotropic phase systems and of the temperature (10°-60°C). Of course, these determinations concerned very concentrated systems, since it was also expected that mixed surfactant solution may form mixed crystal phase (LC) and/or two kinds of LC in mutual equilibrium, in the same way as micelles.

Surface tension

Surface tension measurements were performed by using the Wilhelmy plate method, in order to evaluate the critical micellar concentration (CMC), and as a function of the total surfactant concentration above the CMC; indeed, according to the method outlined by Funasaki [3], such measurements provide information about the composition of the mixed micelles (one or two types) in equilibrium with monomers. For that purpose, one assumes that, above the CMC, the surface tension of aqueous solutions containing a single surfactant remains constant, and that the composition of bulk solutions (monomers) and micelles containing two surfactants are a function of the surface tension alone, regardless of the amount of the micelles.

Using the pseudo-phase separation model [1], and by fitting the two surface tension curves vs., respectively, the composition of the monomers and that of the mixed micelles, it is possible to determine a best estimate of the $\beta(x)$ function, which measures the excess free energy of mixing in the mixed micelle, x being the mole fraction of fluorinated compound into the micelles.

The experiments and calculations were performed at various temperatures. For that purpose temperature was monitored between 10° and 60°C.

Results

Two systems have been investigated, which differ only in the number (n) of oxyethylene groups into the hydrophilic chain of the *fluorinated* surfactant ($n = 2$ and 7). The nonionic hydrogenated surfactant was $C_{12}E_6$, whose aqueous solutions exhibit a cloud point at about 49°-50°C for concentrations in the range 1-5 % w/w. For higher concentrations, liquid crystalline phases are formed (lamellar, cubic and hexagonal), as described in literature [22].

The ternary phase diagrams

a) System $C_{12}E_6/R_f E_2$/water:
The diagram at 20°C is shown in Fig. 1, and there is a continuous change of the phase behavior with temperature. The three classical one-isotropic-phase regions L_1, L_2, L_3, are obtained as for many of the $C_{12}E_6$/oil/water systems. Similarly, that system forms classical liquid crystals (lamellar, hexagonal and bicontinuous cubic), which are true one phase *mixed crystals* (see the discussion of x-ray data).

At very low water content (L_2 phase) these FC- and HC-chain surfactants are completely miscible. Similarly, the mixed lamellar phase (L_a) extends continuously from the water/$C_{12}E_6$ to the water/$R_f E_2$ basis, incorporating a maximum of water for comparable concentrations of each surfactant.

On the other hand, the V_1, H_1, and L_1 phases can solubilize only limited amounts of the fluorinated

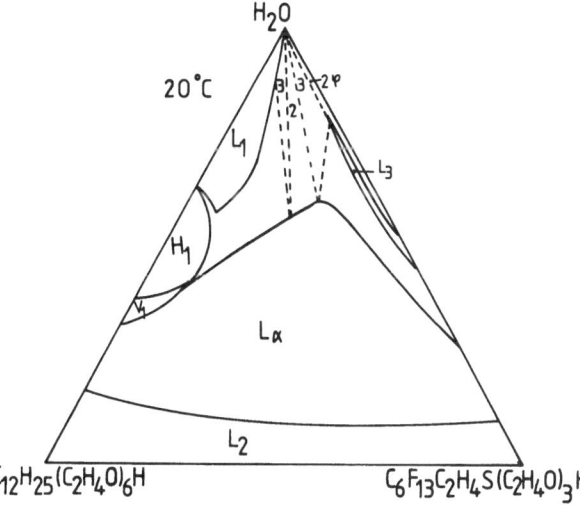

Fig. 1. Phase diagram at 20°C of the system $C_{12}E_6/R_f E_2$/water. L_1, L_2, L_3 are isotropic phases. H_1, V_1, L_a are hexagonal, cubic, lamellar phases, respectively

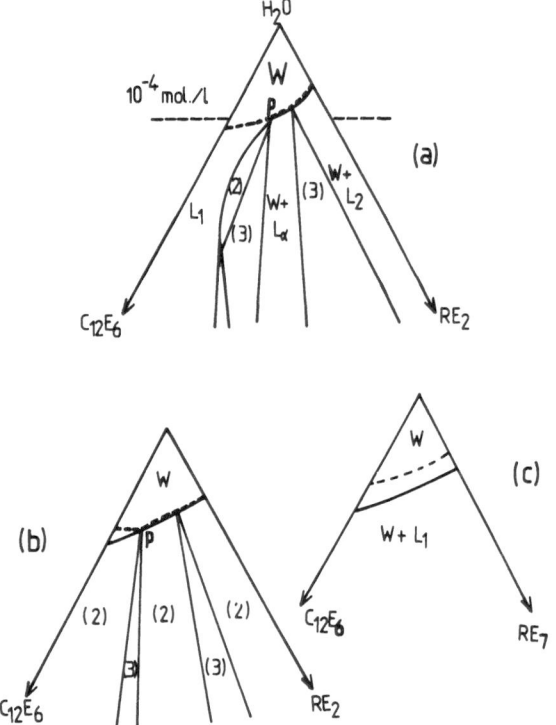

Fig. 2. Schematic representations of the water-rich part of the phase diagrams for $C_{12}E_6/R_fE_2$ at 20 °C (a) and 60 °C (b), and for $C_{12}E_6/R_fE_7$ at 60 °C (c)

compound (about 10 %). Similarly, the L_3 phase of the aqueous R_fE_2 can be swollen by the perhydrogenated amphiphile, but only for a fixed composition of the surfactant mixture. The corresponding domain is quite thin and extends towards the water corner, but

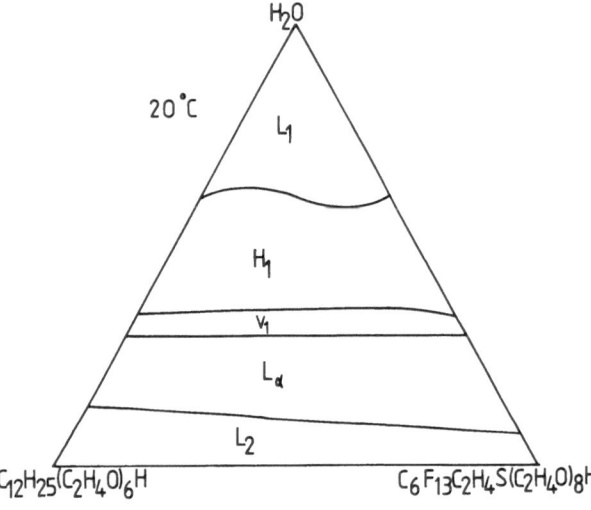

Fig. 3. Phase diagram at 20 °C for the system $C_{12}E_6/R_fE_7$/water

without a connection to it. Most important for the present investigation are the multiphasic domains, as reported in Fig. 1. A careful analysis of this phase behavior shows that L_3 may be in equilibrium with almost pure water (W), (two phase region), and with L_α (three phase region). For intermediate $C_{12}E_6/R_fE_2$ ratios, L_α may also be in equilibrium with W. For higher values of this ratio, thin three-phase and two-phase regions arise where the micellar phase L_1 is in equilibrium with L_α and/or W.

We have to emphasize that this *above W* phase actually contains *only monomers* (i.e., no aggregates nor micelles). This has been checked by measuring the surface tension of W (after centrifugation to ensure the precipitation of L_α and L_3 phases is total), and that of subsequent dilutions of W with pure water. Complementary NMR studies have confirmed such an important result. Schematically, the water-rich region may be represented by the drawings of Fig. 2. The dotted line in Fig. 2 is the CMC-line. A few degrees above the critical point of the aqueous $C_{12}E_6$, the micellar part of W actually reduces to a very small domain (Fig. 2b) which finally disappears when the surfactant-rich L_1 phase no longer exists. Meanwhile the L_3 region has merged to L_2. Such a situation reminds us of the phase behavior of water/oil/nonionic surfactant for temperatures above its HLB temperature [23, 24].

b) System $C_{12}E_6/R_fE_7$

In this case, the two surfactants have about the same hydrophilicity: the critical temperature (lowest cloud point) is about 44° and 49 °C, respectively, for R_fE_7 and $C_{12}E_6$. Quite simple phase diagrams are obtained (Fig. 3), exhibiting a remarkable symmetry whether isotropic phases (L_1, L_2) or anisotropic phases (H_1, V_1, L_α) are concerned. Clearly, such a phase behavior suggests an almost ideal mixing of the surfactants, whatever the structures in which they are commited. This symmetry persists at higher temperatures above the "common" critical temperature of this L_1 phase. This high surfactant content L_1 phase is in equilibrium with very dilute micellar pseudo-phase.

Structural results

In the present paper, they will be only briefly summarized.

X-ray diffraction

As usual, x-ray diffraction has been performed for a "precise" determination of the delineation of the liquid crystal domains. We have to emphasize that, whatever the $C_{12}E_6/R_fE_n$ ratio, the Bragg reflections were always very sharp, without any doublet. Moreover, for a given value of this ratio, the *area per polar head* (σ) of the surfactant was found *constant* with respect to the water content, till demixing occured.

And most interestingly, for lamellar phase, this σ value was found to vary perfectly linearly with the mole fraction of the fluorinated surfactant, in the whole range 0–1. And this was also obtained for R_fE_7 systems, whether H, V, L_α were concerned.

Hence, all these results clearly suggest that these crystalline phases are true mixed crystals. Of course we cannot disregard some segregation of FC- and HC chain surfactant, but only on a very short scale.

Small angle neutron scattering

Measurements have been performed for both systems in their micellar L_1 phase, at a total surfactant concentration of 6 % w/w, and for various $C_{12}E_6/R_fE_n$ ratios. Here the classical contrast variation method consists of using different H_2O/D_2O isotropic compositions for the water component. If we are dealing with a single type of mixed micelle, the neutron scattering at zero angle $I(0)$ will be exactly zero when the mean scattering length density of the particle matches the scattering length of the solvent (water) (12). This technique is made possible because H_2O and D_2O have a very different scattering length.

The experimental results shown in Fig. 4 represent the variations of $\pm \sqrt{I(0)}$ as a function of the isotropic composition of water, and various $C_{12}E_6/R_fE_7$ ratios (20 °C). A critical analysis of these data will be presented elsewhere: the main conclusion is that, given the uncertainty on the very small values obtained in the vicinity of particular H_2O/D_2O ratios, a *true* matching point is highly probable ($I(0) = 0$), whatever the $C_{12}E_6/R_fE_n$ ratio. In other words, whatever the sample in the L_1 region, there is only one type of mixed micelle, whose composition is practically identical to the overall $R_fE_n/C_{12}E_6$ composition.

Surface tension

The surface activity properties of these new fluorinated compounds have been described in a previous paper [18]: CMC, low surface tension γ, free energy of adsorption and micellization...

Now, two sets of investigations have been performed; the first concerned systems at the CMC

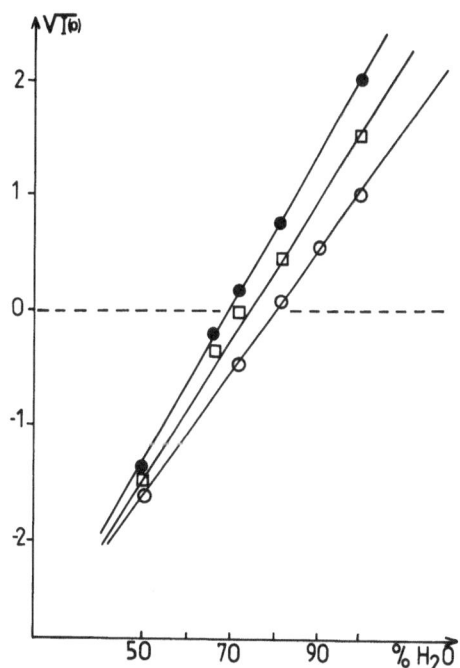

Fig. 4. Variation of $\pm \sqrt{I(0)}$ against the isotopic composition of the water component (20 °C) for different $C_{12}E_6/R_fE_7$ ratios: (●) 0.1; (□) 0.6; (○) 2.0 w/w

[determined by the classical break in the surface tension – log (concentration)]. The second concerned systems about 10-times more concentrated than the CMC (the exact value is not relevant).

From these experimental data, and according to the method developed by Funasaki [3, 15], the variations of γ and CMC as functions of the composition of the *mixed micelles* can be obtained. These data are shown in Figs. 5, 6 (20 °C).

Clearly, depending on the system, we get very different results: from a mere inspection of the CMC curves (Fig. 6a) vs. the monomer composition, we can suspect a very nonideal behavior for $C_{12}E_6/R_fE_2$ (existence of a pronounced maximum) which is at variance with the behavior of $C_{12}E_6/R_fE_7$ (Fig. 6b).

From a thermodynamic point of view, we have assumed that the mixing of the surfactants in the aggregates can be described by the regular solution formalism [1, 7, 8]. Using the experimental data, the interaction parameter β, related to the excess free energy of the mixing, can be derived by fitting the curves of Figs. 6 to theoretical calculations. We have to note that the best β value (or β (x) function, x being the mole fraction of R_fE_n in the aggregates) will result from the fitting of *both* the CMC curves, i.e., the one versus the composition of the monomers, and the one versus x,

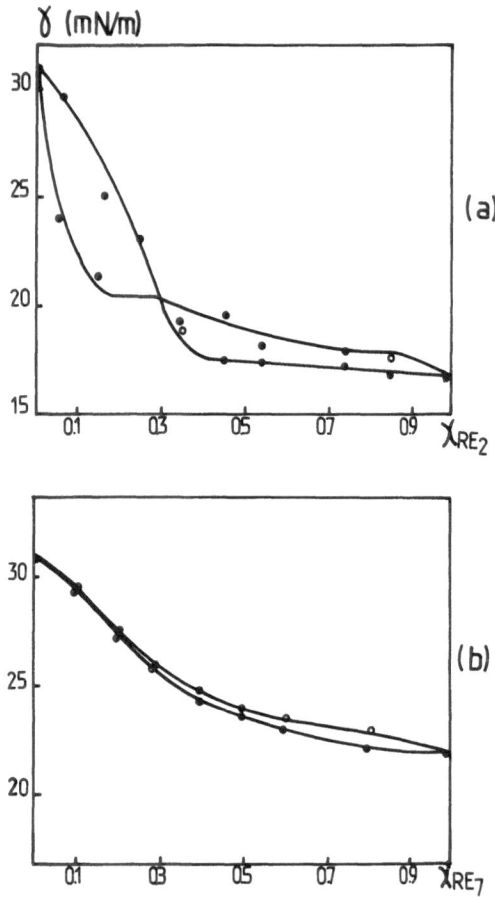

Fig. 5. Plots of surface tension against the mole fractions of fluorinated surfactant in monomers (●) and in the micelles (○). The full lines are calculated according to the regular solution formalism (see text)

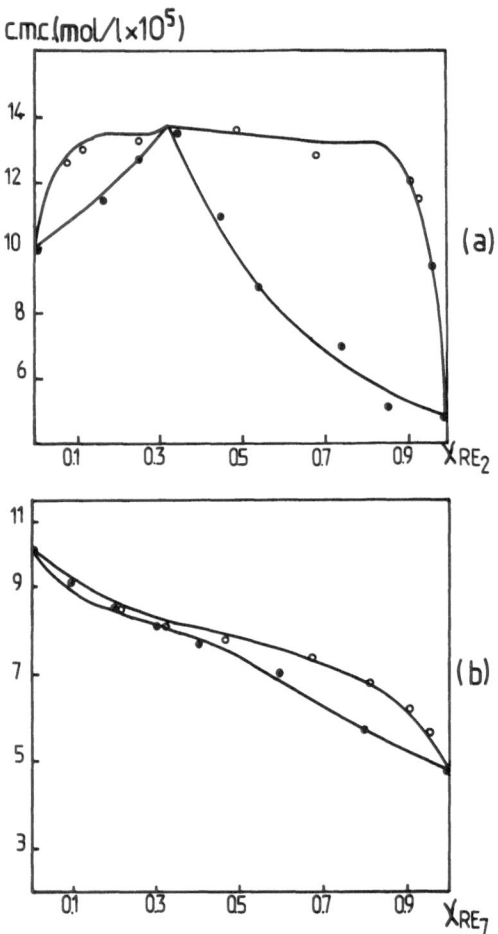

Fig. 6. Plots of the CMC against the mole fractions of fluorinated surfactant in monomers (●) and in micelles (○) for two systems at 20 °C. The full lines are calculated according to the regular solution theory

the composition of the micelles/aggregates. Such a procedure is much more stringent than a mere fitting of the CMC data as a function of the monomer mole fraction alone. This is particularly true for $C_{12}E_6/R_fE_7$ which could have been termed as ideal otherwise.

As a matter of fact, the main result is the non-constancy of the interaction parameter β, which must be considered as a function of x. Its functional form depends on the systems, the temperature, etc. Assuming the polynomial representation of Redlich-Kister [19] holds, which can formally describe all cases of nonideality:

$$\beta(x) = A + B(2x - 1) + C(2x - 1)^2 + \dots$$

we found: $A = 2.4$, $B = 0$, $C = 0.7$, and $A = 0.4$, $B = 0.6$, $C = -0.4$, respectively for $C_{12}E_6/R_fE_2$ and C_{12}/R_fE_7 (20 °C).

Now, from these theoretical values of $\beta(x)$, it is possible not only to calculate the monomer and mixed micelle composition, but also to find out the eventual existence of two types of mixed micelles (aggregates) in mutual equilibrium. In other words, the phase diagram (i.e., the regions where different types of micelles (aggregates) exist and/or coexist) can be calculated in the very dilute aqueous ternary system [1]. One example of such a diagram is shown in Fig. 7.

Discussion

Both the phase diagrams and structural studies show that, throughout the different one phase regions $(L_1, L_\alpha, H_1, V_1)$, the micelles/aggregated structures are mixed systems whose composition is (practically) identical to the corresponding overall fluorinated/

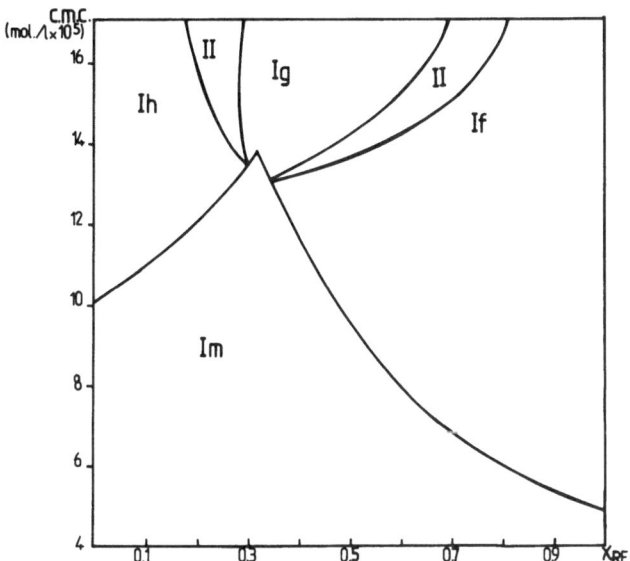

Fig. 7. Micellar pseudo-phase diagram for the system $C_{12}E_6$-R_fE_2 at 20 °C calculated according to the regular solution formalism. *Im* is the monomer region. *Ih* is the mainly hydrogenated mixed micelle region. *If* is the mainly fluorinated mixed micelle region. *Ig* is the mixed lamellar aggregate region. II are regions where two mixed aggregated structures are in mutual equilibrium

hydrogenated surfactant molar ratio (i.e., one single type of mixed structure for every overall composition). This result is valid whether R_fE_2 or R_fE_7 is concerned. However, surface tension data of the dilute aqueous solutions are quite different depending on if the oxyethylene chain of the fluorinated surfactant is shorter or is longer.

For $C_{12}E_6/R_fE_7$, $\beta(x)$ is found to be quite an asymmetric function with respect to $x = 0.5$, since it would increase from about -0.4 to $+0.6$ (in kT units at 20 °C). Therefore, as far as hydrophilic surfactants are concerned, the solubilization of the fluorinated amphiphile into the hydrocarbon chain micelles appears somewhat favored; and there is a slight reluctance for the HC-chain compound to be incorporated into fluorinated micelles. Such a dissymmetrical behavior was already noted by Mukerjee and Mysels [13] on other types of systems. At any rate, given these small values of β, we consider as a first approximation that $C_{12}E_6$ and R_fE_7 mix almost ideally, in perfect accordance with the structural results and the phase behavior.

For $C_{12}E_6/R_fE_2$ things seem much more complex. Here, $\beta(x)$ is roughly symmetrical with respect to $x = 0.5$, but varies in the range 2.4-3 (at 20 °C) (the minimum value is for $x = 0.5$). Hence this system appears

strongly nonideal, suggesting a demixing must occur between mixed aggregates.

As a matter of fact, the theoretical "phase diagram" calculated from the above $\beta(x)$ functions shows that *two* pairs of mixed aggregates may well be in separate mutual equilibrium, in presence of monomers. Indeed, above the CMC curve two two-branch delineations originate from two points on this curve. The upper branches are the classical critical demicellization concentrations [20]. Therefore, for concentrations above the CMC, by increasing the overall $R_fE_2/C_{12}E_6$ ratio, we successively meet one type of HC-chain surfactant-rich micelles, two types of HC-rich mixed micelles in mutual equilibrium, one type of mixed aggregate with comparable amounts of each surfactant, two types of FC-rich "micelles" in mutual equilibrium, and, finally, one type of fluorinated surfactant-rich aggregates.

By comparing this theoretical "phase diagram" at very high dilution to the actual one (Figs. 1, 2), we see there is a perfect (qualitative) agreement between them. The first type of mixed micelles is the classical micellar L_1 phase. But the other types of "mixed aggregates" are, respectively, the surfactant aggregates of the L_α and L_3 phases. In that case "three phases" means: two types of aggregates in presence of monomers (W phase), and "two phases" means: one type of aggregate in equilibrium with W.

Hence, it is possible to correlate the full-phase diagrams to the data derived from surface tension measurements of very dilute aqueous solutions. For these systems, "two types of mixed micelles in equilibrium" simply means "two surfactant-rich phases in equilibrium with monomers". Let us emphasize that such a conclusion results from the fact that both L_3 and L_α can be in equilibrium only with aqueous monomeric solutions. This is not the case for R_fE_7 systems, since when L_1 is still present the monomeric region must be contiguous to a micellar domain, although it is very small (Fig. 2c); in that case the mixing is almost ideal. At any rate, every one-phase region (L_1, L_α, ...) is actually constituted by one single mixed aggregate.

It is interesting to note that, whatever the systems, these fluorinated and hydrogenated surfactants are totally miscible, as are C_6F_{14} and C_6H_{14} (at least above about 20 °C). Hence the mutual oleophoby is enhanced by a marked disymmetry in the hydrophilic chains *and in the presence of large amounts of water.* This could originate from an important entropic contribution to the excess free energy of mixing, as suggested by the non-constancy of the interaction β

parameter with respect to the composition of these mixed aggregates. For the sake of comparison, let us note that Rosen [21] obtained $\beta \approx -0.4$ for analogous mixed nonionic but hydrogenated system $C_{12}E_3$-$C_{12}E_8$, and Nishikido et al. [4] found β positive for $C_{12}E_{49}$-$C_{12}E_5$.

References

1. Kamrath RF, Franses EI (1984) In: Mittal KL, Lindman B (eds) Surfactants in Solution. Plenum Press, New York 1:129-142
2. Kamrath RF, Franses EI (1986) In: Scamehorn JF (ed) Phenomena in Mixed Surfactant Systems. ACS Symposium Series 311, American Chemical Society, Washington 44-60
3. Funasaki N, Hada S (1979) J Phys Chem 83:2471-2475; Ibid (1982) 86:2504-2508; Ibid (1983) 87:342-347
4. Nishikido N, Moroi Y, Matuura R (1975) Bull Chem Soc Jpn 48:1387-1390
5. Nguyen CM, Rathman JF, Scamehorn JF (1986) J Colloid Interface Sci 112:438-446
6. Osborne-Lee IW, Schechter RS, Wade WH, Barrakat Y (1985) J Colloid Interface Sci 108:60-74
7. Shinoda K, Nomura T (1980) J Phys Chem 84:365-369
8. Holland PM (1986) In: Scamehorn JF (ed) Phenomena in Mixed Surfactant Systems. ACS Symposium Series 311, American Chemical Society, Washington 102-115
9. Asakawa T, Johten K, Miyagishi S, Nishida M (1985) Langmuir 1:347
10. Clint JH (1975) J Chem Soc Faraday Trans I 71:1327-1334
11. Zhao GX, Zuh BY (1986) In: Scamehorn JF (ed) Phenomena in Mixed Surfactant Systems. ACS Symposium Series 311, American Chemical Society, Washington 184-198
12. Burkitt SJ, Ottewill RH, Hayter JB, Ingram BT (1987) Colloid Polym Sci 265:628-636
13. Murkejee P, Mysels K (1975) In: Mittal KL (ed) Colloidal Dispersion and Micellar Behavior. ACS Symposium Series 9, American Chemical Society, Washington 239-252
14. Suzuki T, Ueno M, Meguro K (1981) J Am Oil Chem Soc 58:800-803
15. Funasaki N, Hada S, Neya S (1983) Bull Chem Soc Jpn 56:3839-3840
16. Cambon A, Delpuech JJ, Matos L, Serratrice G, Szonyi F (1986) Bull Soc Chim 6:965-970
17. Asakawa T, Miyagishi S, Nishida M (1985) J Colloid Interface Sci 104:279-281; (1987) Langmuir 3:821-827
18. Matos L, Ravey JC, Serratrice G (1989) J Colloid Interface Sci 128:341-347
19. Prausnitz JM (1969) In: Molecular Thermodynamics of Fluid Phase Equilibria. Prentice-Hall, Englewood-Cliffs, New-Jersey
20. Mysels KJ (1978) J Colloid Interface Sci 66:331-334
21. Rosen MJ, Hua XY (1982) J Colloid Interface Sci 86:164
22. Mitchell DJ, Tiddy GJT, Waring L, Bostock T, McDonald MP (1983) J Chem Soc Faraday Trans I 79:975-1000
23. Ravey JC, Buzier M In: Mittal KL (ed) Proceedings of 6th International Symposium on Surfactants in Solution, Plenum Press, New Dehli, India (in press)
24. Ravey JC, Stébé MJ (1987) Progr Colloid Polymer Sci 73:127-133

Received October, 1988;
accepted February, 1989

Authors' address:

J. C. Ravey
Lab. de Physico Chimie des Colloides LESOC
UA CNRS 406
Faculté des Sciences
Université de Nancy I
BP 239
F-54506 Vandoeuvre les Nancy Cedex, France

Progress in Colloid & Polymer Science

Progr Colloid Polym Sci 79:279–286 (1989)

The critical region of water-in-oil microemulsions: new light scattering results

J. Rouch, A. Safouane, P. Tartaglia[1]), and S. H. Chen[2])

Centre de Physique Moléculaire Optique et Hertzienne, Université de Bordeaux I et CNRS (U. A. n. 283) Talence, France
[1]) Dipartimento di Fisica, Universita' degli Studi di Roma La Sapienza, Roma, Italy
[2]) Department of Nuclear Engineering, and Center for Materials Science and Engineering, Massachusetts Institute of Technology, Cambridge, Massachusetts, U.S.A.

Abstract: A three-component microemulsion system consisting of mixtures of water, decane, and a surfactant AOT (WDA), has been studied by means of extensive light scattering measurements, including the intensity, turbidity and linewidth. The measurements have been performed in the one-phase region over a very large temperature range, along the critical and several off-critical constant microemulsion droplet volume fractions. The standard theory of critical binary fluids, complemented by the linear model equation of state, accounts very well for the intensity data in the vicinity of the lower phase separation temperature T_p using a single value for the short range correlation length, $\xi_0 = (13.5 \pm 1.5)$ Å. We have also been able to fit the dynamic light scattering data by combining a mode-coupling theory including the background effects, and the critical equation of state using a cutoff wavelength q_D^{-1} equal to the constant average diameter of microemulsion droplets. Moreover, we find clear evidence for a crossover from critical to single particle behavior in both static and dynamic light scattering data. A crossover temperature T_x has been identified for $\xi(T_x) = q_D^{-1}$ so that q_D, which can only be measured far away from T_p, plays the most important part in controlling the critical dynamics in the whole temperature range.

Key words: _M_icroemulsions, _c_ritical phenomena, _l_ight scattering.

Three-component mixtures of water, *n*-decane, and the surfactant AOT can form water-in-oil (w/o) microemulsions of well-defined droplet size determined by the molar ratio ω of water to AOT [1–3]. For a solution with $\omega = 40$, with a volume fraction of the droplets $\phi_c = 0.098$, one observes a lower consolute critical point at $T_c = 39.860$ °C. Above T_c the solution separates into two microemulsion phases of different composition but with the same ω ratio [3, 4]. Keeping ω constant and changing the amount of *n*-decane varies the volume fraction of the droplets but not their radii [3]. In this respect ω can be considered [4] to a first approximation as a field variable in the sense of Griffiths and Wheeler [5], i. e., a quantity which has the same value in the coexisting phases. Therefore, a w/o microemulsion can be regarded, close to a critical point [6, 7], as a pseudo-binary system with ϕ as the order parameter [3] (at least for small values of ϕ), having a Widom-Fisher-type [8, 9] renormalization of the critical indices, since in reality it is a ternary system [10].

The samples were kept in a cylindrical optical glass vessel of 1 cm diameter and after freezing at liquid nitrogen temperature, sealed under vacuum. Since AOT might undergo hydrolysis, the experiments were conducted very rapidly. Old sealed samples show a slight change of T_c and a systematic increase of the linewidth by 2–3 %. In order to be able to compare our results with others obtained on the same mixture, we set $\omega \approx 40$. The volume fraction of the droplets was changed by adding various amounts of *n*-decane. The coexistence curve was obtained by recording the phase separation temperature as a function of composition and its shape fitted to a power-law with an index $\beta = 0.35 \pm 0.05$, a value consistent with the Widom-Fisher renormalization of the critical indices. The experimental setup and the light photon correlation spectrometer have been described in detail in [9]. We measured the total scattered intensity, the turbidity, and the characteristic time $1/\Gamma_{exp}$ of the order parameter correlation function for the critical sample and for three off-critical solutions with values of $\phi = 0.064$, 0.129, and 0.22. All the samples were studied in a temperature range larger than 25 °C, from 15 °C to the phase transition temperature T_p and at the scattering angle of 90°. We report in Fig. 1 the data for the scattered intensity and the linewidth as a function of the distance from the critical point, for the sample at the critical value of the volume fraction.

As we noted above, SANS experiments [2, 3] have shown that the average radius of the surfactant-coated water droplet is a unique and linear function of the molar ratio ω, and that the droplet shape and size distribution does not change upon transition from one to two phases, implying that ω is a field variable. Therefore, WDA at constant pressure and ω, can be treated thermodynamically as a one-component system and then one can apply the mode-coupling theory for critical dynamics in fluid systems.

The complete treatment of the mode-coupling theory has been reported by many authors [12–14]. We want to emphasize the most important point, as far as the analysis of light scattering data is concerned: the effects of the background on the transport coefficients. Since the solution can be treated as a binary system, the hydrodynamic slow variables are the local volume fraction of the droplets, ϕ, which is proportional to the local mass fraction for small values of ϕ, the velocity field transverse with respect to the wave vector q and the local entropy. The corresponding transport coefficients are the concentration conductivity α and the shear viscosity η. The mass diffusion coefficient D can be expressed in terms of α, the number density of the droplets ϱ, and of the non-local q-dependent osmotic compressibility χ_T as $D = \alpha/(\varrho\chi_T)$. When close to the critical point, the transport coefficients can be written as a sum of a regular or background part α_B or η_B and a nonlocal frequency-dependent critical part α_C or η_C. One can set up the mode-coupling equations taking into account both parts [13]. In order to solve them self-consistently one can use the following simplifying assumptions: i) neglect the frequency dependence of the transport coefficients; ii) neglect the higher order contributions to the transport coefficients; iii) use the Ornstein-Zernike-form for the non-local susceptibility $\chi_T(q, T)$, assuming Fisher exponent η to be zero, that is,

$$\chi_T(q, T) = \frac{\chi_T}{1 + q^2\xi^2} = \frac{\chi_0}{\xi_0^2}\frac{\xi^2}{1 + q^2\xi^2} \qquad (1)$$

where χ_0 is a constant. Since in a light scattering experiment, the frequency change is negligible, the magni-

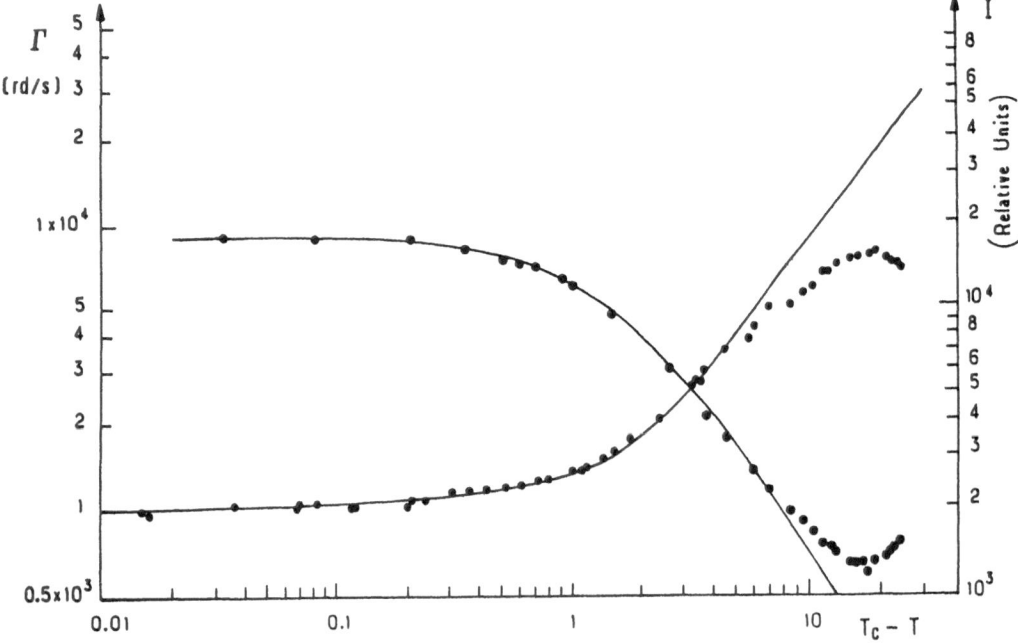

Fig. 1. A log-log plot of the total scattered intensity (curve bending downwards) and of the relaxation rate (curve bending upwards) vs. $T_c - T$ for the critical microemulsions, $\phi_c = 0.098$. The dots are the experimental results, whereas the solid lines are the best fits to the data according to the q-dependent structure factor and the mode-coupling theoretical prediction including background effects

tude of the wavevector is $q = (4 \pi n / \lambda) \sin (\theta/2)$, where n, λ, and θ are respectively the refractive index of the sample ($n = 1.410$), the wavelength in vacuum of the incoming laser beam ($\lambda = 632.8$ nm), and the scattering angle.

The decay rate of the order parameter fluctuations $\Gamma(q, T)$, the quantity directly measured by photon correlation spectroscopy, is proportional to D and can be written as the sum of two terms $\Gamma(q, T) = \Gamma_C(q, T) + \Gamma_B(q, T)$. The background part $\Gamma_B(q, T)$ and the critical part $\Gamma_C(q, T)$ can be written as

$$\Gamma_B(q, T) = q^2 \frac{\alpha_B(T)}{\varrho \chi(q, T)} = q^2 D_R(T) (1 + q^2 \xi^2);$$

$$D_B(T) = \frac{\alpha_B(T) \xi_0^2}{\varrho \chi_0 \xi^2} \tag{2}$$

$$\Gamma_C(q, T) = q^2 \frac{\alpha_C(T)}{\varrho \chi(q, T)} = q^2 D_C \frac{K(q \xi)}{(q \xi)};$$

$$D_C = \frac{k_B T}{6 \pi \eta \xi} \tag{3}$$

and $K(x)$ is the universal scaling function for the decay rate as given in terms of the scaling variable $x = q\xi$. In an analogous way one can introduce the q-dependent shear viscosity $\eta(q, T) = \eta(T) [1 + F(q\xi, T)]$, where $F(x)$ is a scaling function and $\eta(T)$ its macroscopic value $\eta(T) = \lim_{q \to 0} \eta(q, T) = \eta_C(T) + \eta_B(T)$ with

$$\eta_C(T) = \eta(T) \frac{3}{15 \pi} \int_0^\infty dy \frac{y^6}{(1 + y^2)}$$

$$\times \frac{1}{K(y, T) + \frac{D_B(T)}{D_C(T)} y^2 (1 + y^2)} \tag{4}$$

which shows clearly the dependence of the viscosity on the background of the decay rate. In the same long wavelength limit Γ gives

$$\lim_{q \to 0} \frac{\Gamma_C(q, T)}{q^2} = R \frac{k_B T}{6 \pi \eta(T) \xi};$$

$$R = \frac{2}{\pi} \int_0^\infty dy \frac{1}{(1 + y^2)(1 - F(y, T))}, \tag{5}$$

i. e., the correction factor R to the Stokes-Einstein law applied to correlated regions of size ξ in the critical sys-

tem. A set of coupled integral equation can be written for the two scaling functions $K(x)$ and $F(x)$ and can be solved numerically if the background is included, and with a self-consistent perturbation scheme when D_B/D_C is neglected. This type of calculation is valid only very close to the critical point when the scaling functions $K(x)$ and $F(x)$ become universal, and in this case one gets the value $R = 1.027$. Since the viscosity anomaly is experimentally known to be weak, a simple iterative procedure can be used to evaluate both transport coefficients. To lowest order one neglects the q dependence of the viscosity, so that $F = 0$ and K has a universal expression

$$K(x) = \frac{3}{4} \left[1 + x^2 + \left(x^3 - \frac{1}{x} \right) \tan^{-1} x \right]. \tag{6}$$

When substituted into the equation for $\eta_C(T)$ this function allows the evaluation of the critical part of the shear viscosity in terms of the background linewidth Γ_B. It must be noted that the neglect of the background prevents the integral for the critical part of the viscosity from converging. An approximate evaluation of $\eta_C(T)$ has been given by Oxtoby and Gelbart [13] in the form

$$\eta(T) = \eta_B(T) (q_D \xi)^\psi = \eta_B(T) \left(\frac{C k_B T \varrho \chi_0}{6 \pi \eta_B \alpha_B \xi_0^2} \xi \right) \tag{7}$$

where ψ is a universal exponent equal to $8/15 \pi^2 = 0.054$, and $C = 0.9$. The parameter q_D is the Debye upper cutoff wavevector for the nonlocal shear viscosity. To the same order of approximation the background decay rate $\Gamma_B(q, T)$ is given by [11]

$$\Gamma_B(q, T) = q^2 D_c \frac{3}{4} \frac{C}{q_D \xi} (1 + x^2) \frac{\eta}{\eta_R}. \tag{8}$$

In order to compare more easily the theoretical predictions with the experimental results, it is customary to express Eqs. (2) and (3) in a reduced form involving only universal quantities. We therefore define a reduced relaxation rate which is the sum of the two following contributions

$$\Gamma_C^* = \Gamma_C \frac{6 \pi \eta}{k_B T q^3 R} = \frac{K(x)}{x^3}; \quad \Gamma_B^* = \Gamma_B \frac{6 \pi \eta}{k_B T q^3 R}. \tag{9}$$

The comparison between the mode-coupling theory and the dynamic light scattering experiment is now straightforward. In fact the inverse of the experimental order parameter fluctuations correlation time $\Gamma_{\exp}$ can be reduced in the same way in order to have a dimensionless quantity. The difference $\Gamma^*_{\exp} - \Gamma^*_B$ is now to be compared with the universal function Γ^*_c given by Eq. (3). This can be easily done along the line of critical volume fraction $\phi = \phi_c$ where the long-range correlation length ξ has the usual definition $\xi = \xi_0 \varepsilon^{-\nu^*}$, and can be deduced from combined total scattered intensity and turbidity measurements [15]. However, one has to remember that, since the sample is really a ternary mixture, the critical indices γ^* and ν^* are renormalized. According to Widom [8] and Fisher [9] they are given by $\gamma^* = \gamma/(1-\alpha)$ and $\nu^* = \nu/(1-\alpha)$ where $\gamma = 1.24$, $\nu = 0.63$, and $\alpha = 0.11$, as predicted by a renormalization group calculation [14].

For off-critical mixtures ξ cannot be directly inferred from total scattered intensity measurements. In order to derive this quantity we adopt the model used in [15]. In this method one first expresses the correlation length ξ as a function of the compressibility χ_T and then calculates χ_T in terms of a scaling equation of state valid close to the critical point [16]. In order to do that we first relate it to the spatial integrals of the pair-correlation function of the two species G_{ij} ($i,j = 1,2$):

$$\chi_T = \frac{\phi(1-\phi)}{RT}$$
$$\times \left[1 + \frac{\phi(1-\phi)}{V} N_A (G_{11} + G_{22} - 2G_{12}) \right] \quad (10)$$

where V is the molar volume of the mixture, N_A the Avogadro number, and R the gas constant. Using the Ornstein-Zernike form of the correlation function, which is a good approximation, one obtains [17]

$$G_{ij} = a_{ij} \int d\vec{r} \, \frac{\exp{(-r/\xi)}}{r} = 4\pi a_{ij} \xi^2 \quad (11)$$

where the coefficients a_{ij} are constant. Since the compressibility is directly linked to the order parameter fluctuations, and putting $a = a_{11} + a_{22} - 2a_{12}$, one gets

$$\xi^2 = \frac{V}{4\pi a N_A \phi(1-\phi)} \left[\frac{k_B T N_A \chi_T}{\phi(1-\phi)} - 1 \right]. \quad (12)$$

The linear model equation of state can be used to obtain an explicit temperature and volume fraction dependence of χ_T and then of ξ. In this model [16] the critical part of the appropriate thermodynamic potential

π, the field conjugate to the order parameter h and the reduced temperature ε are expressed in terms of two parametric variables r and θ. In the case of a two-component system [18] the linear model equations can be written in the following way

$$h(r,\theta) = ar^{\beta\delta}\theta(1-\theta^2); \quad \varepsilon(r,\theta) = r(1 - \bar{b}^2\theta^2);$$

$$\phi - \phi_C = \frac{\partial \pi}{\partial h} = \pm \bar{g}\theta r^\beta \quad (13)$$

where the critical indices β, δ have their usual meaning, the parameter $\bar{g}$ is the nonuniversal amplitude of the coexistence curve, and $\bar{b}^2 = (\delta - 3)/(\delta - 2\beta\delta + 2\beta - 1)$. The compressibility is then

$$\chi_T = \frac{1}{N_A k_B T} \frac{\bar{g}}{a} \frac{r^{-\gamma}}{1 + d_2 \theta^2}; \quad d_2 = \frac{2\beta\delta - 3}{(1-2\beta)}. \quad (14)$$

By substituting Eq. (14) in (12), and assuming that for $\theta = 0$, ξ behaves as $\xi = \xi_0 \varepsilon^{-\nu}$, we get

$$\xi^2 = \xi_0^2 \frac{V}{V_c} \left[\frac{\phi_c(1-\phi_c)}{\phi(1-\phi)} \right]^2 \frac{r^{-\gamma}}{(1 + d_2\theta^2)} \quad (15)$$

where V_c is the critical molar volume of the system. By inverting the linear model equations we deduce the parametric variables r and θ in terms of the experimentally accessible variables. Then by using the mode-coupling equations and the viscosity data [19], we are able to describe completely the static and the dynamic properties of microemuslions over the whole critical regime.

We now discuss our experimental results. Far below the phase separation temperature, for $T_c - T$ ranging from 25 °C to 10 °C and for all samples, the scattered intensity first decreases and reaches a minimum around 20 °C–22 °C and then rises again (Fig. 1). One observes exactly the reverse for the decay rate (Fig. 1). In this domain we can assume that microemulsions behave like a macromolecular solution [1]. Phenomenologically it is possible to express both the scattered intensity I_s and the diffusion coefficient D in terms of the droplet-droplet structure factor $S(q)$ by [20, 21] $I_s = K S(q)$ where K is a constant factor depending on the wavelength of the incident light, the index of refraction of the microemulsion, the scattering geometry, and

$$D = \frac{k_B T}{6\pi\eta R_H S(q=0)} \quad (16)$$

where η is the shear viscosity of the solution and R_H the hydrodynamic radius of the droplets. We verified that the relaxation rate Γ has a proper q^2 dependence, such as $\Gamma = Dq^2$ in this temperature domain. Using viscosity data of Berg et al. [19], we can extract from Eq. (16) the effective hydrodynamic radius $R_H S(0)$. This quantity was found to be constant, independent of volume fraction and temperature, and equal to (83 ± 5) Å. This means that $S(0)$ is not changing appreciably in this range and can be taken approximately to be unity. This result is consistent with the findings of Kotlarchyck et al. [2, 3] from SANS measurements of the water-core radius. They gave a value for the average droplet size of (70 ± 5) Å, which is independent of concentration and very slightly decreasing on increasing temperature. It must be noted that light scattering measures a hydrodynamic radius which takes into account the viscous drag of the droplets in the oil continuous phase and this could account for the slight difference between the two determinations.

The extension of the critical domain is roughly 5 °C below T_c for the critical sample, and about 2 °C below T_p for the off-critical ones. In the critical regime, where the droplet correlation function obeys the Ornstein-Zernike-Fisher-form, the q-dependent structure factor is given by

$$S(q) = \frac{\varrho k_B T \chi_T}{(1 + q^2 \xi^2)^{1 - \eta/2}} \tag{17}$$

where the constants and the indices have been defined before. For the w/o critical microemulsions, intensity data corrected for turbidity can be quite accurately fitted to Eq. (17) with the renormalized values of γ^* and ν^* and choosing $\xi_0 = (13.3 \pm 1.5)$ Å (Fig. 1). This value of ξ_0 is significantly larger than those usually found for pure fluids or binary liquid solutions [11, 15]. However, it is of the same order of the values found for critical non-ionic micellar solutions [20], and for the same microemulsion system, (12.2 ± 2.0) Å, found by Kim and Huang [6, 23], and (11.0 ± 2.1) Å by Kotlarchyk, Chen, and Huang [24], (the water used was D_2O). The temperature range, where the fit is of good quality, can be significantly enlarged from T_c to $T_c - T \approx$ 10 °C provided one subtracts a small constant background from the data. This background represents roughly one-half of the total intensity at the minimum at 22 °C and accounts for 4 % of the intensity close to the critical point. From the physical point of view it represents the contribution of the single particle non-critical scattering to the total scattering.

For off-critical line samples Eq. (17) is still valid but χ_T and ξ must be calculated using the linear model equation of state. Figure 2 shows the results of this calculation for the correlation length for three values of the volume fraction. The scattered intensity is now proportional to the factor $x^2/(1 + x^2)$. It is worth

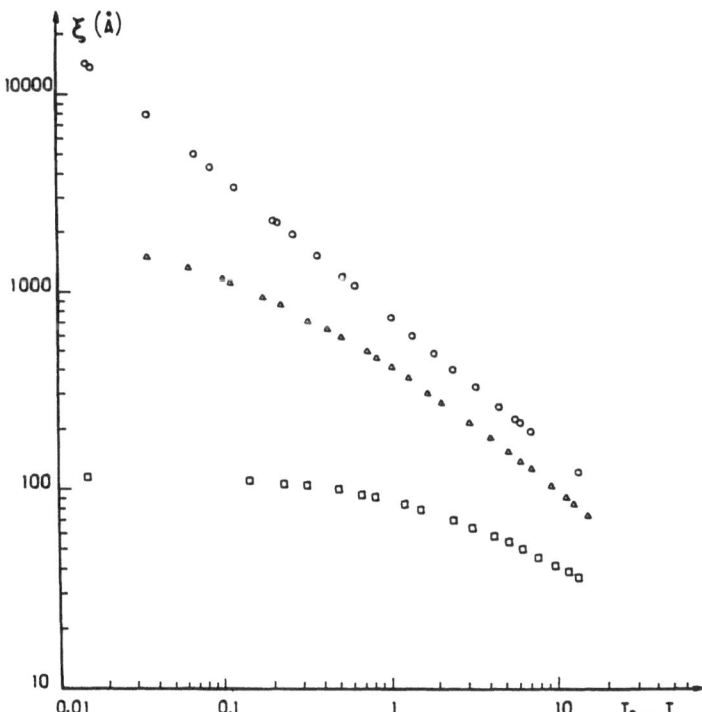

Fig. 2. A log-log plot of the correlation length vs. $T_p - T$ for three volume fractions. Open circles: critical volume fraction $\phi_c = 0.098$; open squares: $\phi = 0.22$; open triangles: $\phi = 0.064$. The computation has been made using the linear model equation of state

noting that in the hydrodynamic regime, for $x \ll 1$, one no longer observes the strong $\varepsilon^{-\gamma^*}$ divergence of the osmotic compressibility. Besides, since χ_T remains finite at the phase transition temperature $T_p \neq T_c$, one does not observe a dramatic increase of the scattered intensity close to T_p. Our experimental data can be fitted to Eq. (17) with the same value of $\xi_0 = (13.3 \pm 1.5)$ Å only if a constant background is first subtracted from the intensity values. However, since off-critical mixtures are poor scatterers, even close to T_p, this background, which is of the same order as for the critical sample, can now account for 20% of the scattered intensity close to T_p. All the intensity data corrected in this way, including the critical and the off-critical

samples, are plotted in Fig. 3 as a function of the scaled variable x in order to show the scaling behavior predicted by Eq. (17) with $\eta = 0$. A very good agreement between the theory and the experiment is observed.

Let us discuss now the experimental results for the dynamical behavior of the system. In order to calculate the background relaxation rate Γ_B^*, the Debye cutoff q_D must be known. In the case of a critical binary mixture one can deduce q_D from viscosity data by extrapolating to the critical concentration data taken for off-critical mixtures [11]. For w/o microemulsions the situation is not so clear, since the viscosity behaves in a complicated way as a function of the volume fraction. We shall therefore use an approach similar to

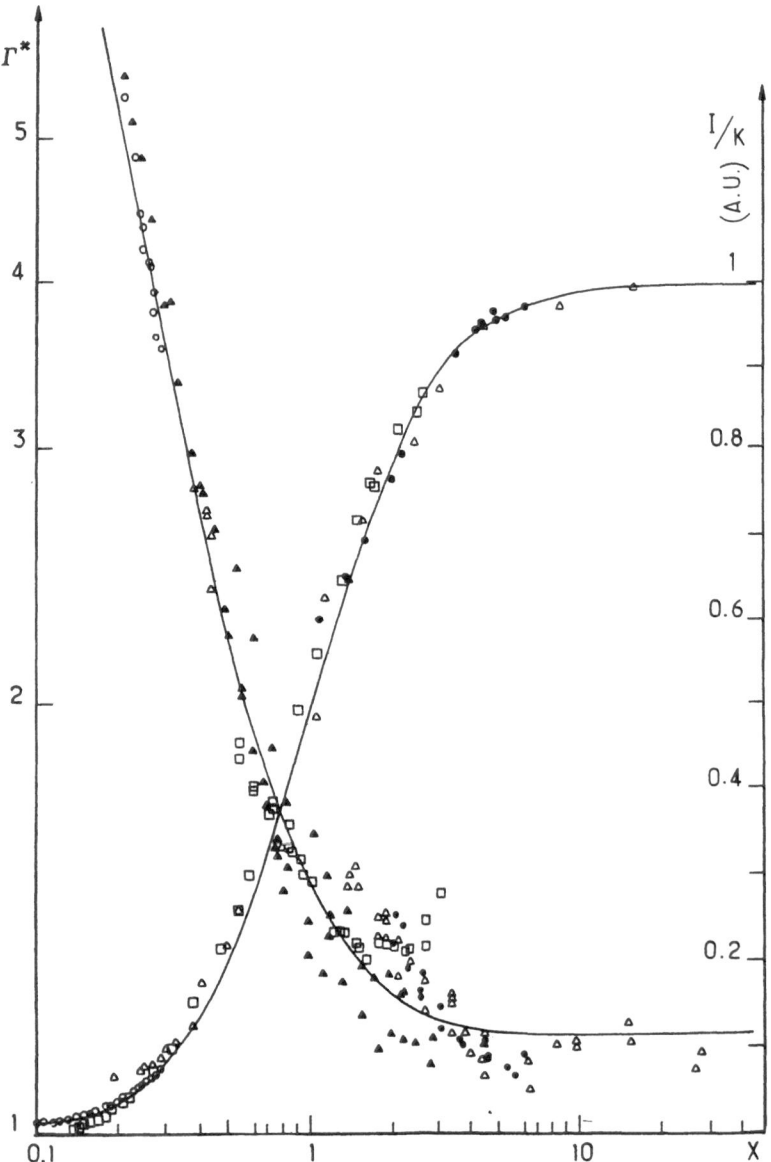

Fig. 3. Semilog plot of the scaled scattered intensity I/K (increasing curve) and log-log plot of the reduced relaxation rate Γ^* (decreasing curve) as a function of $x = q\xi$. Open triangle: critical volume fraction $\phi_c = 0.098$; open circles: $\phi = 0.22$; solid circles: $\phi = 0.064$; squares: $\phi = 0.129$. The solid triangles refers to the Huang and Kim data [6]. The solid lines are the predictions of the expressions for the structure factor and the mode-coupling theory for the linewidth

Fisher's [25], where q_D is assumed to be the size of the droplets. Since $R_H S(0) \approx 83$ Å we get $q_D = 0.60 \times 10^6$ cm^{-1}. The best fit of our linewidth data in the critical regime up to $x = 30$ is obtained by using $q_D = (0.61 \pm 0.05) \times 10^6$ cm^{-1} (Fig. 1). These two values are extremely close to one another. With $q_D = (0.65 \pm 0.05) \times 10^6$ cm^{-1}, we can fit Huang and Kim data [6, 23], both in the hydrodynamic and in the critical regime for x ranging from 0.2 to 2. The numerical value of q_D^{-1} we obtained for w/o microemulsions, i. e., $q_D^{-1} \approx 160$ Å, is very large compared to the same parameter deduced for binary or ternary critical mixtures [11, 15], i. e., $q_D^{-1} \approx (5 \div 10)$ Å. This implies that the background Γ_B^* given by Eq. (9) is very large in our system. In fact, in the critical regime where Γ_B^* varies slowly with x, the background accounts for 20 % of the decay rate. This result explains the fact that the universal amplitude factor $R = 1.2$ was necessary to fit the experimental data when the background effect was not properly accounted for [6, 23]. In the latter case its contribution reflects in the value of R, which should increase with the size of the droplets, all the rest of the parameters being fixed to their proper values. In the present work R is set to its theoretical mode-coupling value of 1.027.

We turn now to the off-critical solutions. The same value $q_D = (0.61 \pm 0.05) \times 10^6$ cm^{-1} very well fits the relaxation rates along iso-ϕ lines ending at the coexistence curve and close to the phase transition temperature. It is worth noting that in this case the linear model equations have to be solved in order to calculate ξ. For the different solutions x was ranging from 0.2 to 8. However, we must remark that for the most concentrated sample, $\phi = 0.22$, we have to choose $q_D = 1.8 \times 10^6$ cm^{-1}. This value, although fairly larger than the preceding one, is still of the order of the inverse of the diameter of the droplets at this concentration [24]. In Fig. 3 the reduced decay rate Γ^* for the critical and the off-critical iso-ϕ lines is plotted agisnt x together with the results of Huang and Kim [23] and the theory. The agreement is very good for x ranging from 0.2 to 30.

Let us go back now to the low temperature domain. With the above numerical values of ξ_0, q_D and the critical indices γ^* and ν^*, we obtain $q_D \xi(T_X) = 1$ for $\Delta T = 10\,°C$ below T_c, i. e., around 29 °C, which corresponds to the crossover temperature. If we move roughly 5 °C above this temperature, the sample obeys the usual critical laws, whereas moving 5 °C below it the previously strongly divergent quantities like the osmotic compressibility or the correlation time of the order parameter fluctuations remain constant. This result explains why (using the standard theories of critical phenomena) it is not possible to account for our experimental data in this intermediate temperature domain, where the dynamics of the system are not governed by the scaling variable x but by the temperature through the long range correlation length ξ.

We must note at this point that all the fits between the experimental data and the theoretical expressions have been performed assuming the renormalized values of the critical indices, as we stressed many times earlier. The attempts to fit the data using the universal critical indices lead us to much poorer results. We must conclude that the system behaves only approximately as a binary critical mixture and the renormalized indices are reminiscent of the real nature of the system, i. e., a three-component system.

In conclusion we stress that we observed a twofold effect of the background, namely a static and a dynamic one. The first one is the contribution of single particle scattering, important only for the scattered intensity. The second one comes from the non-critical contribution to the transport coefficients which enters in the mode-coupling dynamics. The important fact is that only the parameter q_D, which can be measured far from the critical temperature, controls both effects since it is related to the size of the droplets in the case of single particle scattering, and to the upper momentum cutoff in the mode-coupling theory.

References

1. Graciaa A, Lachaise J, Letamendia L, Chabrat P, Rouch J, Vaucamps C, Bourrel M, Chambu C (1977) J Phys Lett (Paris) 38:L-253; (1978) J Phys Lett (Paris) 39:L-235
2. Kotlarchyk M, Chen SH, Huang JS (1982) J Phys Chem 86:3273
3. Kotlarchyk M, Chen SH, Huang JS, Kim MW (1983) Phys Rev A 28:508; (1984) Phys Rev A 29:2054 and references therein
4. Belloq AM, Roux D (1984) Phys Rev Lett 52:1895; Roux D (1984) Thèse d'Etat, Université de Bordeaux I; Honorat P (1984) Thèse de 3 cycles, Université de Bordeaux I
5. Griffiths RB, Wheeler JC (1970) Phys Rev A 2:1047
6. Huang JS, Kim MW (1981) Phys Rev Lett 47:1462
7. Honorat P, Roux D, Bellocq AM (1984) J Phys Lett 45:L-961
8. Widom B (1967) J Chem Phys 46:3324
9. Fisher ME (1968) Phys Rev 176:257; Fisher ME, Sceney PE (1970) Phys Rev A 2:825
10. Rouch J, Safouane A, Tartaglia P, Chen SH (1988) Phys Rev A 37:4995
11. Chen SH, Lai CC, Rouch J, Tartaglia P (1983) Phys Rev A 27:1086
12. Kawasaki K (1970) Ann Phys (NY) 61:1; Lo SM, Kawasaki K (1973) Phys Rev A 8:2176
13. Oxtoby DW, Gelbart WM (1974) J Chem Phys 61:2957
14. Hohenberg PC, Halperin BI (1977) Rev Mod Phys 49:435

15. Rouch J, Tartaglia P, Chen SH (1988) Phys Rev A 37:3046
16. Schofield P (1969) Phys Rev Lett 22:606
17. D'Arrigo G, Mistura L, Tartaglia P (1977) J Chem Phys 66:80
18. Leung SS, Griffiths RB (1973) Phys Rev A 8:2670
19. Berg RF, Moldover MR, Huang JS (1987) J Chem Phys 87:3687
20. Berne BJ, Pecora R (1976) Dynamic Light Scattering. J Wiley, New York
21. Ackerson B (1976) J Chem Phys 64:242
22. Corti M, Degiorgio V (1985) Phys Rev Lett 55:2005 and references therein
23. Kim MW, Huang JS (1982) Phys Rev B 26:2073
24. Kotlarchyck M, Chen SH, Huang JS (1984) Phys Rev A 29:2054
25. Fisher ME (1986) Phys Rev Lett 57:1911

Received October, 1988;
accepted February, 1989

Authors' address:

Prof. J. Rouch
Centre de Physique Moléculaire Optique et Hertzienne
Université de Bordeaux I
351 Cours de la Liberation
F-33405 Talence Cedex, France

Progress in Colloid & Polymer Science   Progr Colloid Polym Sci 79:287–292 (1989)

Influence of the stabilizing coating on the rate of crystallization of colloidal systems

C. Smits[1]), W. J. Briels[2]), J. K. G. Dhont[1]) and H. N. W. Lekkerkerker[1])

[1]) Van't Hoff Laboratory, University of Utrecht, Utrecht, The Netherlands
[2]) Technical University Twente, Chemical Physical Laboratory, Enschede, The Netherlands

Abstract: We report visual observations and preliminary light scattering experiments on the crystallization process in several colloidal systems. The main difference between the colloidal systems studied here is the range of their pair-interaction potential which is varied by changing the stabilizing coating of the particles. This range is found to have a pronounced influence on the rate of crystallization.

Key words: Colloidal crystals, phase behavior, light scattering.

1. Introduction

One of the most remarkable phenomena exhibited by concentrated suspensions of monodisperse spherical colloidal particles is the spontaneous transition from a fluid-like structure to a crystalline arrangement of the particles. This ordering was first observed by Williams and Smith [1] in centrifuged pellets of a purified virus suspension. Because the spacing of the particles is of the order of optical wavelengths, light is separated into colours when scattered giving rise to beautiful iridescence; hence the name, Tipula Iridescent Virus (TIV), given to the virus by Smith and Williams [2]. Subsequent to the discovery of this phenomenon in the TIV system it was also observed in dispersions of synthetic colloids notably in deionized dispersions of polymer latex particles [3, 4]. In fact most experiments in this area have been performed on quite dilute suspensions of these charge stabilized colloids exhibiting long range repulsive interactions [5].

More recently the phenomenon of crystalline ordering has also been observed in dispersions of sterically stabilized particles which have a steep repulsive interaction [6, 7]. From computer simulations on crystallization in atomic systems it is clear that the harshness of the repulsive part of the pair potential has a pronounced effect on the process of homogeneous nucleation and thereby on the rate of crystal growth [8]. Both the time lapse for the onset of nucleation and the time required for the completion of the nucleation

process are longer for steeper repulsive potentials. In view of the fact that the structure in colloidal dispersions may be treated in the same way as in simple liquids it appears reasonable to expect that the trend found for atomic liquids also applies to colloidal systems. In fact colloidal systems offer unique possibilities to investigate the influence of the range of the repulsive potential since by suitable modification of the surface properties of the particles the repulsive interaction can be varied more or less continuously.

In this study we describe first observations of the crystallization process of colloidal silica particles with different (steric) stabilization layers resulting in different ranges of the repulsive interaction.

Section 2 of this paper consists of a brief description of the systems used in this study. Then in Section 3 we describe our observations on each system and in Section 4 we describe preliminary light scattering experiments. Finally we make some concluding remarks in Section 5.

2. Materials

The colloidal particles used here consist of an amorphous silica core, synthesized according to Stöber's method [9] which were coated with different types of stabilization layers. The following four types of dispersions with increasing range of the repulsive interactions were studied.

The first dispersion is stearyl silica. The surface silanol groups of the amorphous silica particles are esterified with octadecyl alcohol

[10]. The hydrodynamic radius of the particles is $a = 160$ nm while the thickness of the coating is $d \cong 2$ nm. Extensive experimental work has led to the conclusion that these particles (nearly perfectly) behave as hard spheres [11].

The second dispersion is called PIB/1-silica. These particles are coated with terminally attached polyisobutene chains [12] with number averaged molecular weight $M_n = 1300$. In this case $a = 170$ nm and $d = 4$–5 nm.

The third silica is called PIB/2-silica. The difference with the previous system is that now the terminally attached polyisobutene chains have a number averaged molecular weight $M_n = 13000$. This results in $d = 30$–50 nm and $a = 210$ nm.

The fourth silica is referred to as TPM-silica. It consists of silica particles coated with γ-methacryloxy-propyltrimethoxysilane [13]. For this suspension $a = 160$ nm and $d = 2$–4 nm. The range of the repulsive interaction potential is large due to residual surface charges. In ethanol-toluene mixtures the additional range of the interaction due to these surface charges is of the order of 100 nm [14].

The range of the repulsive interaction clearly increases in going from stearyl silica to TPM-silica.

3. Observations of crystallization phenomena with the naked eye

In this section we present some characteristic observations on the crystallization process in the four systems described in the previous section. The rate of crystallization depends markedly on the type of system. On this basis the four systems studied here can be divided into two groups. On the one hand we have stearyl silica and PIB/1-silica. Here crystallization only occurs after (slow) sedimentation under gravity. On the other hand PIB/2-silica and TPM-silica crystallize rapidly whenever the appropriate concentration is achieved. We start with a description of our observations on the first group of systems.

Allowing a dilute (say 10 per cent volume fraction) dispersion of stearyl silica in cyclohexane to settle under gravity a sediment is formed in a few days. At this

stage no crystallization is apparent. However after a period ranging somewhere from one to three months crystallites appear at the interface between the dense sediment and the turbid layer of colloidal liquid just above the sediment. This course of events is schematically depicted in Fig. 1. Figure 2 shows a photograph of

Fig. 2. Photograph of stearyl silica in cyclohexane after crystallization. This sample is approximately four years old and some crystallization can be observed in the lower part of the sediment

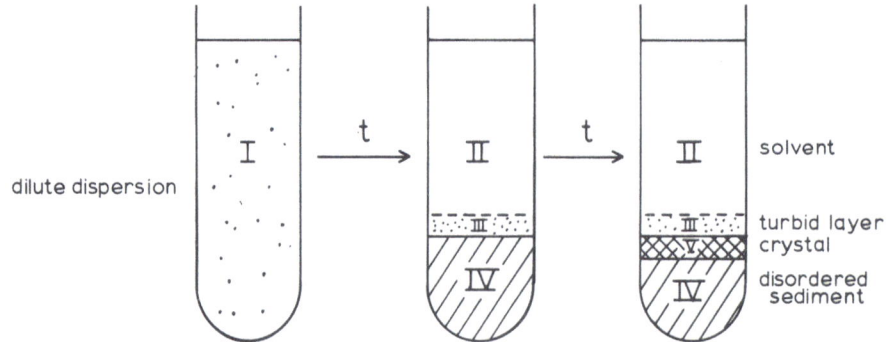

Fig. 1. The appearance of stearyl silica in cyclohexane as a function of time. I represents the dilute dispersion, II the solvent, III the turbid layer above the sediment, IV the disordered sediment and V the iridescent (crystalline) sediment

the system after crystallization has taken place. The crystallites have a columnar shape and are vertically oriented. Typically the crystalline region reaches a height up to 1–2 mm after which crystallization stops. If the liquid upper layer after completion of sedimentation was removed no evidence for crystallization was found even after extended periods of time (1 year or more).

In a mixture of cyclohexane and carbon tetrachloride which has a density closer to the density of the silica particles than pure cyclohexane the sedimentation process now requires 2 to 3 weeks. In this case the formation of crystallites occurs concurrently with the sedimentation process and starts at the bottom of the tube. However crystallization stops somewhere in the lower part of the sediment leaving the upper part of the sediment in a disordered state. By gently shaking the tube the disordered part of the sediment can be redispersed again without destroying the crystal at the bottom. Allowing the particles to settle again the crystallization process proceeds and may now lead to a crystalline phase extending proceeds over the whole sediment (see Figs. 3 and 4).

Allowing a dilute dispersion of PIB/1-silica in toluene to settle under gravity, sedimentation is complete in several days and concurrently crystallization occurs throughout the entire sediment reaching completion in about 2 weeks (see Figs. 5 and 6). This crystal is more easily deformed and consists of smaller crystallites than in the case of stearyl silica. Although the crystallization of PIB/1-silica definitely proceeds faster than for stearyl silica, if sedimentation takes place by gentle centrifugation (about 100 g) the resulting sediment locks into a disordered state where no crystallization is observed.

We now proceed to the second group of systems which are rapidly crystallizing. First we consider PIB/2-silica. Like the PIB/1-silica this system crystallizes after sedimentation under gravity. However contrary

to PIB/1-silica it also crystallizes from concentrated dispersions which are prepared by centrifugation of dilute suspension and decanting excess solvent (see Figs. 7 and 8). Under these conditions crystallization can be observed after one day. The crystalline phase has the same appearance as in the case of PIB/1-silica but it is much less rigid.

TPM-silica crystallizes from concentrated dispersions in less than a day. The crystal is very soft as is evi-

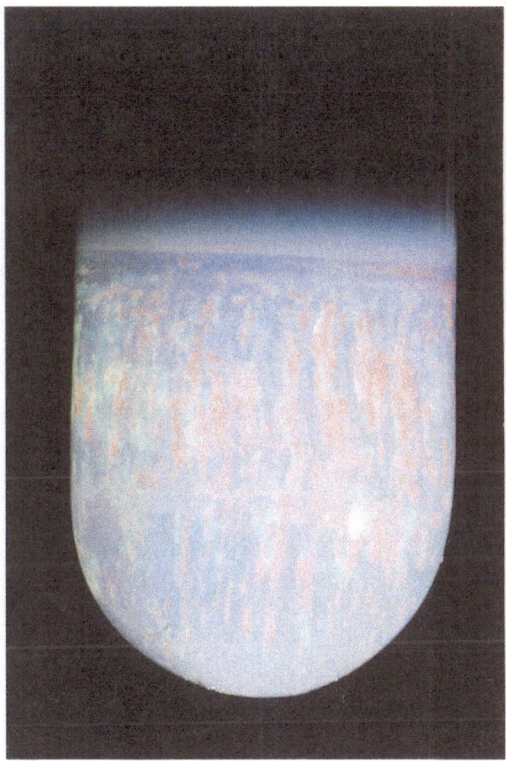

Fig. 4. Photograph of stearyl silica in a mixture of carbon tetrachloride and cyclohexane after crystallization. Note the boundary between the upper and lower part of the sediment, which is caused by growing the crystal in two stages

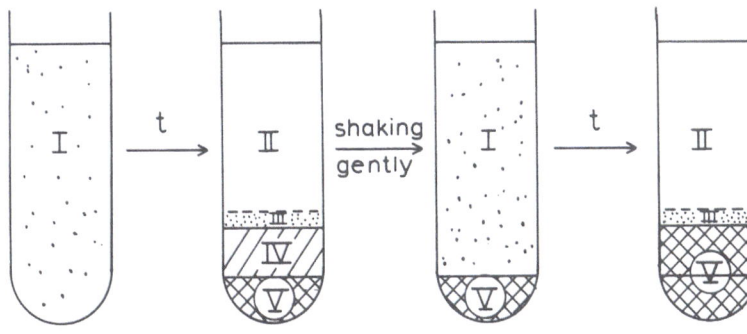

Fig. 3. The appearance of stearyl silica in a mixture of carbon tetrachloride and cyclohexane as a function of time, The numbers I to V have the same meaning as in Fig. 1

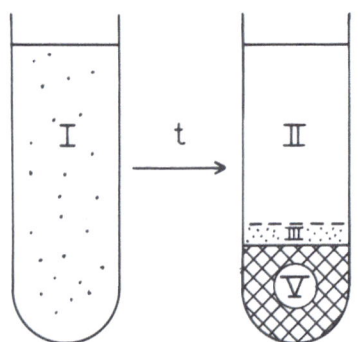

Fig. 5. The appearance of PIB/1-silica in toluene as a function of time. The numbers I to V have the same meaning as in Fig. 1

concentrated
dispersion

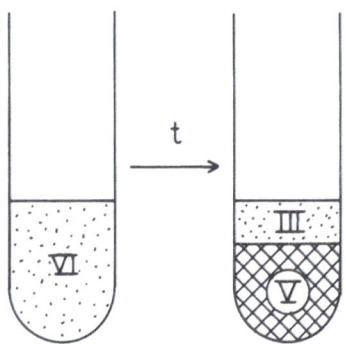

Fig. 7. The appearance of PIB/2-silica in carbon tetrachloride and of TPM-silica in a mixture of ethanol and toluene as a function of time. The numbers III and V have the same meaning as in Fig. 1 and VI represents a concentrated dispersion

Fig. 6. Photograph of a PIB/1-silica crystal in toluene

Fig. 8. Photograph of a PIB/2-silica crystal in carbon tetrachloride. After crystallization had been completed, some sedimentation under gravity occurred

denced by the fact that it flows under gravity which can be observed by tilting the tube (see Fig. 9).

4. Preliminary light scattering measurements

Because the spacing of the particles in the systems under investigation is of the order of optical wavelengths it is possible to probe the structural arrangement of the particles with light scattering. Under favourable conditions also the development in time of ordered structures can be studied with light scattering.

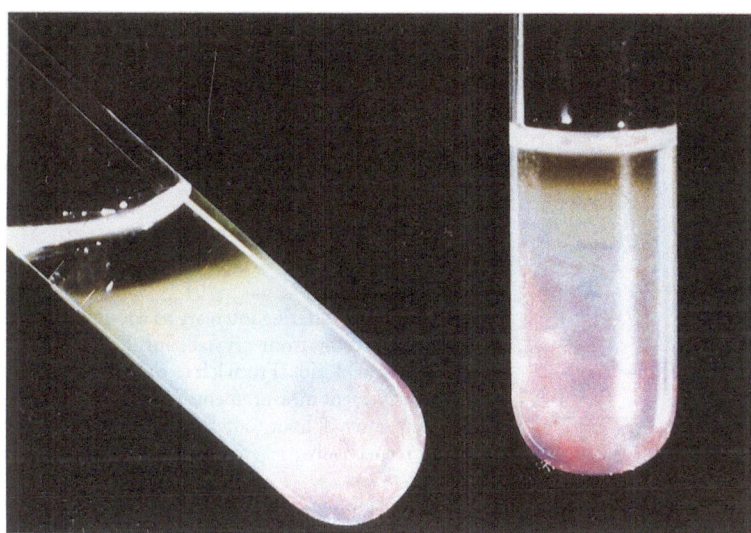

Fig. 9. Photograph of a TPM-silica crystal in a mixture of ethanol and toluene. After completion of the crystallization process some sedimentation has occurred. Note that the crystal follows gravity after tilting the tube

Light scattering experiments were performed with a FICA-50 photometer using vertically polarized light. The measurements were done on the rapidly crystallizing systems PIB/2-silica and TPM-silica prepared at concentrations where crystallization takes place.

The "normalized" light scattering intensity (the "Rayleigh ratio") of a system of monodisperse spheres is given by,

$$R(K) = c P(K) S(K),\qquad (1)$$

where c is determined by the number concentration of the particles, the refractive indices of particles and liquid and the wavelength of the light. K is the scattering vector,

$$K = \frac{4\pi n}{\lambda_0} \sin(\theta/2).\qquad (2)$$

Here λ_0 is the wavelength of the light in vacuo, n is the refractive index of the solvent and θ is the angle between the incident and scattered light, the "scattering angle". The form factor $P(K)$ is determined by the optical properties of one single particle. The structure factor $S(K)$ measures the statistical arrangement of the particles in the system. For a crystalline structure $S(K)$ consists of a series of peaks located at K values where the Bragg scattering condition is fulfilled.

As described in the previous section the TPM-silica system crystallizes rapidly. For this system it was not possible to follow the crystallization process by light scattering since the time required to obtain a single scattering curve is of the same order as the time during which major structural changes occur. For these rapidly crystallizing systems the FICA-50 photometer cannot be used to study the dynamics of the crystallization process. A typical scattering curve at a late stage of the crystallization process is given in Fig. 10. The peak positions can be indexed by the Miller indices 111 and 200 of a FCC crystal.

In the PIB/2-system the crystallization process was sufficiently slow to be followed by doing light scattering experiments at various stages of the development of the crystalline structure. The results of these

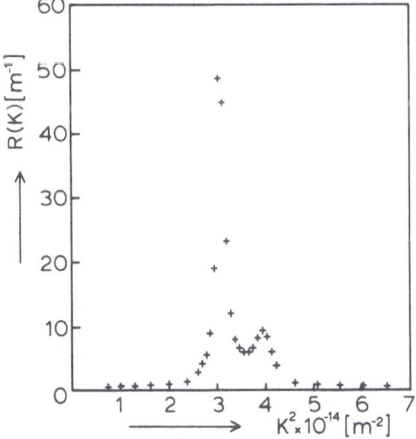

Fig. 10. Light scattering intensity as a function of the scattering vector from crystallized TPM-silica in a mixture of ethanol and toluene

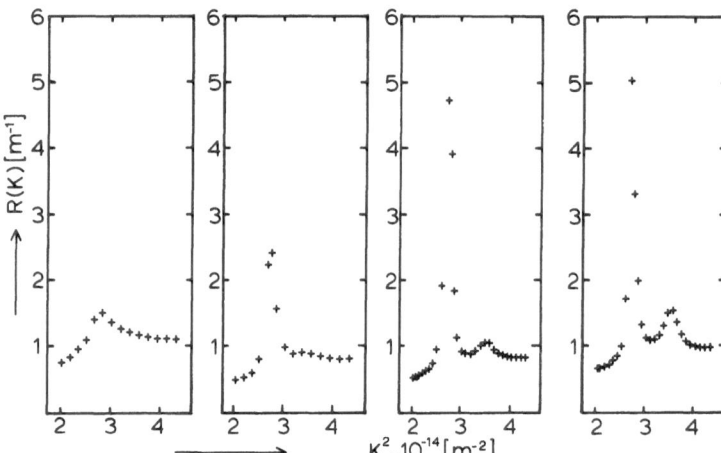

Fig. 11. Light scattering intensity as a function of the scattering vector from crystallizing PIB/2-silica in carbon tetrachloride. From left to right the scattering curves represent measurements directly after homogenizing, after 18 hours, after 24 hours and after 167 hours respectively

measurements are presented in Fig. 11. The scattering curves clearly display the transition from a "disordered" liquid-like structure to an ordered structure with peaks that can be indexed by the Miller indices 111 and 200 of a FCC crystal.

5. Concluding remarks

From the observations presented here it is clear that colloidal silica systems offer interesting possibilities to study the dependence of the rate of crystallization on the range of the repulsive interaction. The range of interaction was varied either by changing the thickness of the stabilizing coating or by the effect of surface charges. It turns out that with increasing range of interaction the rate of crystallization increases dramatically.

In systems where the repulsive interaction is very short ranged (as is the case for stearyl silica) the rate of sedimentation appears to be an important parameter as well. The formation of colloidal crystals is favored by slow sedimentation. To study this effect in more detail knowledge of the concentration pattern in the sample as a function of time would be valuable. It would also be interesting to monitor the velocity pattern during the sedimentation process.

Quantitative information on the crystallization process can be obtained by light scattering experiments. To exploit this possibility for rapidly crystallizing systems such as TPM-silica a light scattering apparatus is required that allows one to measure in shorter times than is possible with the FICA-50 apparatus used in the present study.

Acknowledgements

The authors acknowledge valuable advise of Dr. C. G. de Kruif at the initial stage of the investigations and of Mr. C. Pathmamanoharan and Dr. A. P. Philipse concerning the synthesis of several colloidal systems. We benefitted from several discussions with Dr. D. Frenkel and Dr. P. N. Pusey. We thank Ms. M. Uit de Bulten for typing this manuscript, Mr. J. L. den Boesterd for making the photographs and Mr. Th. L. Schroote for drawing the pictures.

References

 1. Williams RC, Smith KM (1957) Nature 179:119
 2. Smith KM, Williams RC (1958) Endeavour 17:12
 3. Luck W, Klier M, Wesslau H (1963) Ber Bunsenges Phys Chem 67:75
 4. Hiltner PA, Krieger IM (1969) J Phys Chem 73:2386
 5. Hachisu S, Takano K (1982) Adv Colloid Interface Sci 16:233
 6. De Kruif CG, Rouw PW, Jansen JW, Vrij A (1985) J Phys 46:C3-295
 7. Pusey PN, Van Megen W (1986) Nature 320:340
 8. Mountain RD, Brown AC (1984) J Chem Phys 80:2730
 9. Stöber W, Fink A, Bohn E (1968) J Colloid Interface Sci 26:62
10. Van Helden AK, Jansen JW, Vrij A (1981) J Colloid Interface Sci 81:354
11. De Kruif CG, Jansen JW, Vrij A (1987) In: Safran SA, Clark NA (eds) Physics of Complex and Supermolecular Fluids. John Wiley and Sons, New York, pp 315
12. Pathmamanoharan C (1988/89) Coll Surf 34:81
13. Philipse AP, Vrij A (1989) J Colloid Interface Sci 128:121
14. Philipse AP, Vrij A (1988) J Chem Phys 88:6459

Received November, 1988;
accepted February, 1989

Authors' address:

C. Smits
Van't Hoff Laboratory
University of Utrecht
Padualaan 8
NL-3584 CH Utrecht, The Netherlands

Progress in Colloid & Polymer Science
Progr Colloid Polym Sci 79:293–296 (1989)

Liquid crystallinity in metal ion – dodecylbenzenesulfonate systems: x-ray diffraction characterization

Đ. Težak, S. Popović[1]), S. Heimer and F. Strajnar

Laboratory of Physical Chemistry, Faculty of Science, and
[1]) Rugjer Bošković Institute, University of Zagreb, Zagreb, Yugoslavia

Abstract: Phase behavior and liquid crystal structures of metal dodecylbenzenesulfonates have been investigated by x-ray diffraction and polarization microscopy. Interplanar distances in the lyotropic lamellae exhibited the differences ($D/\text{Å}$ between 35 and 29). It can be assumed that the structures of the lamellar layers have been influenced by various possible reasons, i. e. the hydration of individual metal ions within the water layers of lamellae, the depression of water layers, and the depression of paraffin layers. The types of lamellar organization in the crystalline phases, crystallized from ether or from water solutions, remain in a smectic arrangement with the rigid structure of smaller interplanar distances than those in liquid crystal state.

Key words: Liquid crystallinity, dodecylbenzenesulfonate, x-ray diffraction, lamellar lyotropic liquid crystals, metal dodecylbenzenesulfonate.

Introduction

The lyotropic lamellar phases are of great interest for biological membrane investigation [1, 2]. The colloidal dispersions with the appearance of liquid crystal are of great technical importance [3]. In this work the metal salts of dodecylbenzenesulfonic acid (HDBS) were chosen as model systems, since they exhibit the formation of lamellar liquid crystals [4, 5] in a heterogeneous system as a colloid dispersion.

The x-ray diffraction characterization in this work showed the differences in lamellar phases that appeared in different concentration regions of the so-called "precipitation diagram" [4].

Experimental

Materials

All the chemicals used were of p. a. grade, as commercially distributed. Metal nitrates (from "Merck", Darmstadt and "Kemika", Zagreb) were standardized complexometrically (A. Vogel, A Textbook of Quantitative Inorganic Analysis, Longman, London, Ed. III, 1961); HDBS was used from "Prva Iskra", Barić, Beograd,

standardized potentiometrically with NaOH standard solution. Bidistilled water was used for all the aqueous solutions and sample preparations.

Techniques

All examinations of liquid crystal as well as the solid crystalline phases have been performed in heterogeneous aqueous medium, i. e. the dispersions were prepared by mixing the aqueous solutions of cationic and anionic reaction component, $M(NO_3)_n$ and HDBS, in order to get the precipitate of metal dodecylbenzenesulfonates in supernatant. The x-ray diffraction patterns of these heterogeneous systems were obtained using a standard Siemens x-ray diffractometer with a counter and Si-crystal monochromatized CuK_α radiation.

Leitz Wetzlar light microscope with polarizing equipment was used for texture characterization.

The characteristic x-ray diffraction patterns of the two lamellar phases are shown in Fig. 1. The interplanar spacings of lamellar bilayers were calculated from such reflections.

Results and discussion

The scheme of precipitation diagram, which is the type of metal dodecylbenzenesulfonates generally [6],

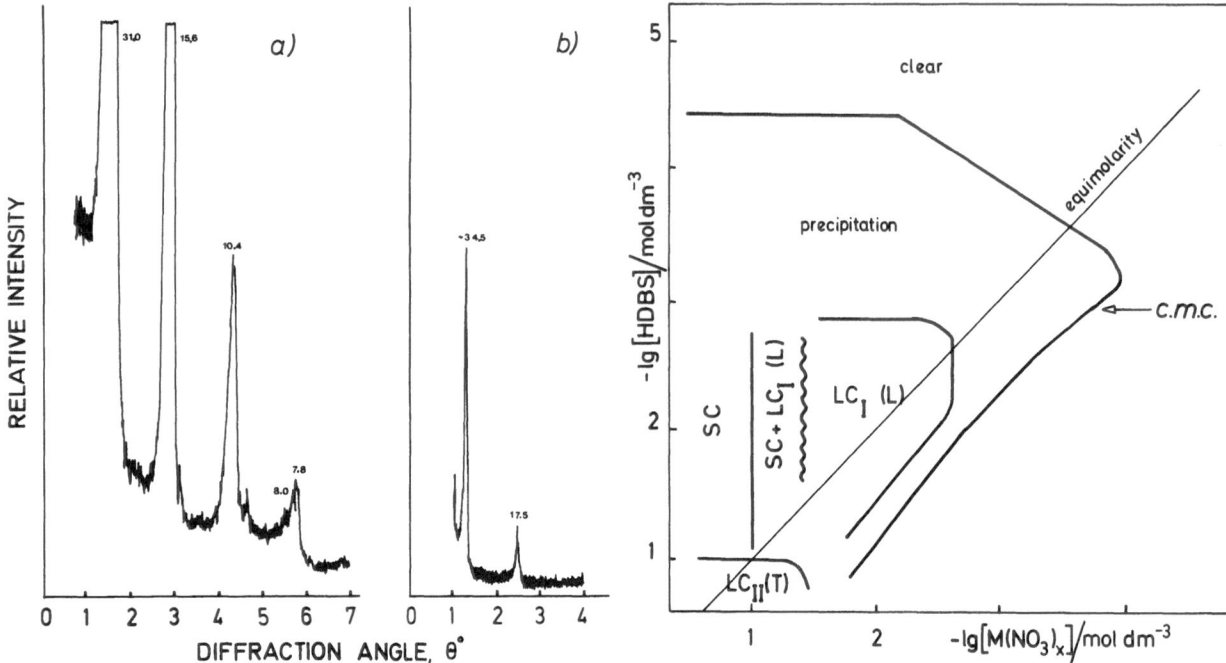

Fig. 1. X-ray diffraction pattern of: a) $LC_I(L)$ mesophase: $[Mg(NO_3)_2] = 6 \times 10^{-2} \ mol \ dm^{-3}$, $[HDBS] = 6 \times 10^{-2} \ mol \ dm^{-3}$; b) $LC_{II}(L)$ mesophase: $[Mg(NO_3)_2] = 6 \times 10^{-2} \ mol \ dm^{-3}$, $[HDBS] = 1.5 \times 10^{-1} \ mol \ dm^3$

Fig. 2. Contours of precipitation diagram of $[M(NO_3)_x]$–HDBS–H_2O system and the heterogeneous phases within the precipitation regions: the regions of appearance of the solid crystals (SC), the mixture of solid crystal and lamellar crystal phase $LC_I(L)$, lamellar liquid crystal phase $LC_I(L)$ and transition phase $LC_I(T)$

is presented in Fig. 2. The appearance of solid crystalline phases in supernatant solution exhibited in high concentration excess of metal ions (SC) is due to the very fast formation of a great number of small crystal nuclei. The interplanar spacings in solid crystalline phases, crystallized from supernatant solution, exhibited the values (28.8 ± 0.2) Å. These spacings are almost the same as the value of 29 Å obtained for the dried crystalline sample. In the equivalency region as well as in the slight excess of surfactant above c.m.c., the liquid crystals-lamellar phase was formed, denoted by $LC_I(L)$ in Fig. 2, with several very well defined orders of reflections in x-ray patterns exhibiting the interplanar spacings 32.6 to 33.7 Å, as well as a very high optical birefringency. The high diffraction maxima denote a high many layer stack of lamellae ordering.

The high excess of surfactant concentration (above $1.5 \times 10^{-1} \ mol \ dm^{-3}$ of HDBS and $6 \times 10^{-2} \ mol \ dm^{-3}$ of metal nitrate) caused the increase of interplanar spacings, whose values were 34.2 to 35 Å. All values are tabulated in Table 1. In Fig. 2 this region is denoted as $LC_{II}(T)$.

These samples were very viscous, fast homogeneous, with a weak optical birefringency. The samples present transition systems from lamellar to some other kind of mesophase. Besides, they exhibited a small

Table 1. X-ray diffraction measurements in $M(DBS)_n$ mesophases

Cation	Interplanar spacings D/Å		
	$LC_{II}(T)$	$LC_I(L)$	SC (supernatant)
Na++			30.5
Zn++	34.4 ± 0.3		30.9 ± 0.5
Mn++	34.6 ± 0.6	33.1 ± 0.1	29.50 ± 0.03
Co++	35	33.5 ± 0.5	30.0
Ni++		33.5 ± 0.3	
Cu++		33.7 ± 0.5	27
Mg++	34.2 ± 0.6	33.2 ± 0.6	30.6 ± 0.3
Ca++			31.0 ± 0.3
Sr++			29.6 ± 0.1
Ba++		32.14 ± 0.03	29.9 ± 0.2
Cr+++		31.6 ± 0.3	29.6 ± 0.1
Fe+++		32.6 ± 0.4	31.1 ± 0.6
Al+++	34	$32.4 \pm 0..03$	30.8 ± 0.3

SC(dried): $D = 29$ Å

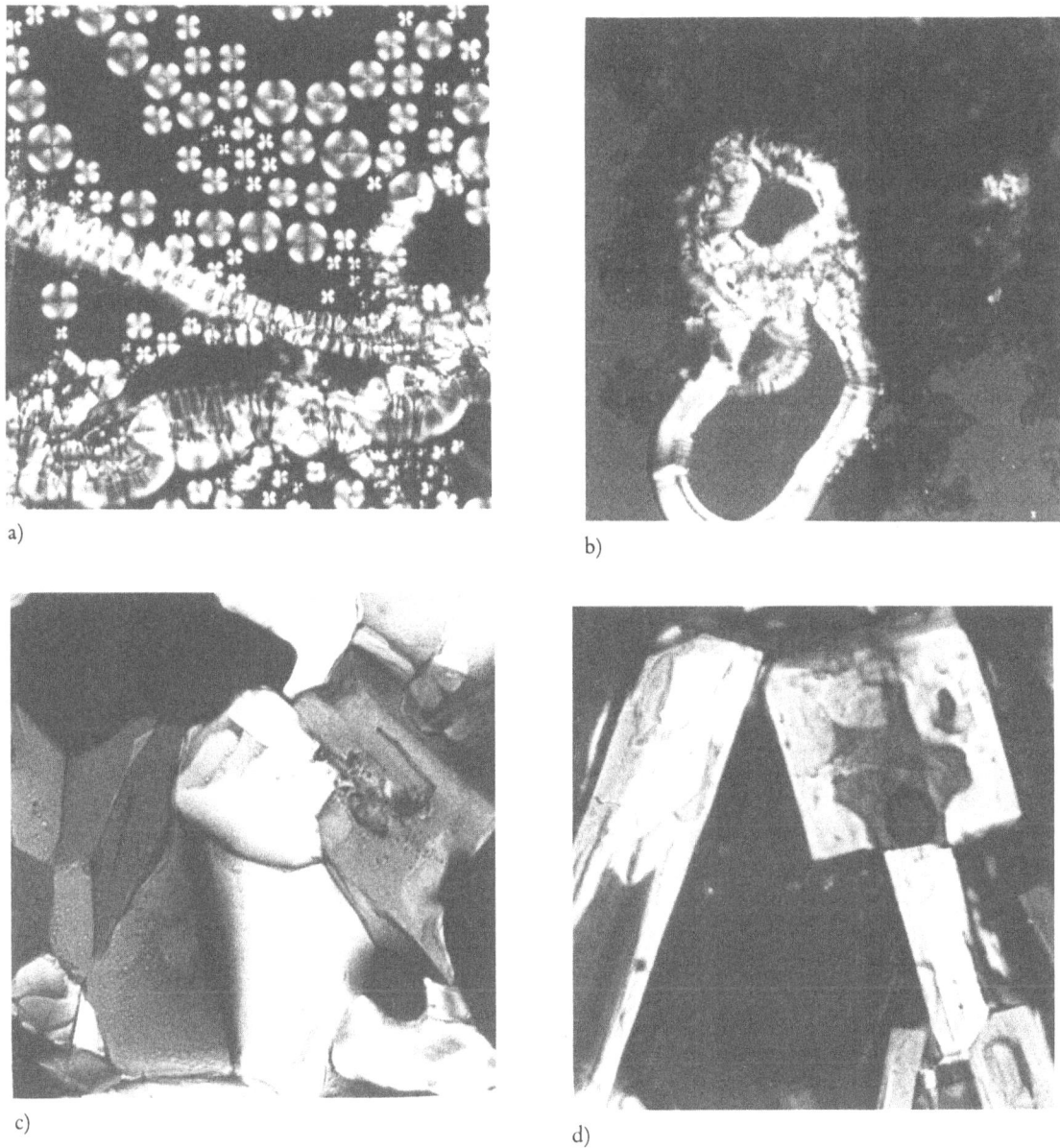

Fig. 3. Photomicrographs of a) lamellar liquid crystals of the system: $[Mg(NO_3)_2] = 4 \times 10^{-2}$ mol dm^{-3}, [HDBS] $= 4 \pm 10^{-2}$ mol dm^{-3}; b) the mixture of dark solid crystalline and birefringent liquid crystal phase of the precipitation system: $[Ca(NO_3)] = 1 \pm 10^{-1}$ mol dm^{-3}, [HDBS] $= 1 \pm 10^{-2}$ mol dm^{-3}; c) solid crystals of Mg(DBS)$_2$ crystallized from water; d) solid crystals of Mg(DBS)$_2$ crystallized from ether (both c) and d) crystalline phases birefringent). Total magnification 180 ×

amount of solid crystalline phase (birefringent) with interplanar spacing 30 Å.

In Fig. 3a representative sample of optical micrographs of a lamellar liquid crystal precipitate is taken from $LC_I(L)$ region and presents a very fine example of lyotropic lamellar texture in a colloidal dispersion, i. e. in a heterogeneous system. Figure 3b shows a mixture of lamellar liquid crystal and crystalline phases.

Figures 3c and 3d present the birefringent samples of crystallized phases from water and ether, respectively.

Conclusion

Lamellar mesophases $LC_I(L)$ in equivalency region exhibit characteristic diffraction patterns containing

3–4 diffraction bands indicating a high symmetry of stack lamellae. The spacings for these phases are 33 Å.

The transition phases $LC_{II}(T)$ with 30 and 35 Å spacings have been found in the high excess concentration of HDBS. The presence of metal ions in water layer of lamellae (L mesophase) causes the depression of interplanar spacings. The hydration of individual metal ions within the water part of lamellar layer [7] may be considered as the cause of this effect. The hydration radii of metal ions are smaller than that of H_3O^+-ions, which are present in higher excess in the (T) mesophases than in the (L) mesophases, since the HDBS concentration is in high excess in the systems exhibiting the (T) mesophases. This effect can be considered as the reason of differences in the interplanar spacings of mesophases. It can be calculated that the water part in lamellar bilayer is approximately 4 Å for (L) phases, and 5 Å for (T) phases taking into account D for Sc (dried) (29 Å). Calculating in regard to Sc (supernatant), the depths of water layers in these mesophases could be assumed to be 3 Å for (L) phases and 4 Å for (T) phases.

References

1. Sackmann E, Eggl P, Fahn C, Bader H, Ringsdorf H, Schollmeier M (1985) Ber Bunsenges Phys Chem 89:1198–1208
2. Saupe A (1976) J Colloid Interface Sci 58:583–592
3. Friberg S (1977) Naturwissenschaften 64:612–618
4. Težak Đ, Strajnar F, Šarčević D, Milat O, Stubičar M (1984) Croat Chem Acta 57:93–107
5. Krishnamurti D, Somashekar R (1981) Mol Cryst Liq Cryst 65:3–10
6. Težak Đ, Strajnar F, Milat O, Stubičar M (1984) Progr Colloid Polym Sci 69:100–105
7. Muller BW (1977) Arch Pharm (Weinheim) 310:693–704

Received December, 1988;
accepted January, 1989

Authors' address:

Đ. Težak
Laboratory of Physical Chemistry
Faculty of Science
University of Zagreb
Marulićev trg 19/II
PO Box 1 63
YU-41001 Zagreb, Yugoslavia

Progress in Colloid & Polymer Science   Progr Colloid Polym Sci 79:297–307 (1989)

Iridescent colours in surfactant solutions

C. Thunig, H. Hoffmann, and G. Platz

Lehrstuhl für Physikalische Chemie I, Bayreuth, F.R.G.

Abstract: We report on a surfactant system which is in the liquid crystalline lamellar state and where the spacings between the lamellar layers can be expanded to several thousand Å. As a consequence the solutions show bright iridescent colours when they are illuminated with white light against a dark background. The system consists of the zwitterionic surfactant tetradecyldimethylaminoxide in the presence of some cosurfactant and small amounts of hydrocarbon.

The system can be swollen with water up to a volume fraction of 0.99 without loss of the liquid crystallinity. The systems are thermodynamically stable and the iridescent colours reappear after the order has been destroyed by shaking the solutions. It is believed that the large spacings are due to electrostatic forces between the surfactant layers. The charges on the layers are due to protonation of the aminoxide groups by water. The existence region of the iridescent phase as a function of surfactant, hydrocarbon and cosurfactant is reported. The spacings for the phases were determined from light transmission and light scattering measurements.

Key words: Iridescent phases, liquid crystalline phases, surfactants, spacings, light scattering.

Introduction

Lamellar liquid crystalline phases have recently been the subject of intense theoretical and experimental investigations [1–3].

G. Porte et al. have shown that under specific conditions lamellar phases can be swollen both by hydrocarbon and by water to extremely large spacings. The interlamellar spacing could be increased to 1 000 Å. The lamellar phases which were investigated by Porte consisted of ionic surfactants and cosurfactants in the presence of high salt concentration. The large spacings were thus not due to electrostatic forces but due to thermal undulations of the lamellar layers.[1]) Systems with even larger spacings which were due to electrostatic repulsive forces between the layers were reported by Satoh and Tsujii [4].

Their system consisted of an alkylcarboxylic acid. The weak ionisation of the acid leads to charged layers.

In distilled water they reached spacings of several thousand Å and the systems showed iridescent colours when they were illuminated with white light. The iridescent colours are due to Bragg reflections with visible light. In their investigations the surfactant contained a double bond. It was thus possible to polymerize the system after the formation of the iridescent layers [5]. It was shown that the layers could indeed by polymerised while the order in the system was maintained. The authors were even successful in transforming the fluid system into a solid by polymerisation of the aqueous medium in the presence of 1.5 % acrylamid.

In the present investigations we report on a surfactant system which also shows iridescent colours.

We observed that solutions of tetradecyldimethylaminoxid in the presence of some cosurfactants and hydrocarbon also show iridescent colours. Under the reported conditions the system is thermodynamically stable and the colours reappear again after they have been destroyed by rigorous shaking.

---

[1]) See also P. Bassereau, J. Appell, J. Marignan, G. Porte, this conference.

Results and discussion

The iridescent colours

In Figs. 1 and 2 we show samples of our systems when they are illuminated with white light and the backscattering is observed or photographed against a dark background. As the photographs drastically demonstrate the iridescent colours can be rather strong and can show the full scale of the visible wave-length region. The different colours are obtained by changing slightly the ratio of the various components in the sample and their concentrations.

In Fig. 2 we show the angular dependence of the scattering intensity of a sample which looks red when viewed from the back. In this case the different colours are obtained when the scattering angle between the light source and the observer was changed from 180° to about 90°. By scanning an angle of 90° we change

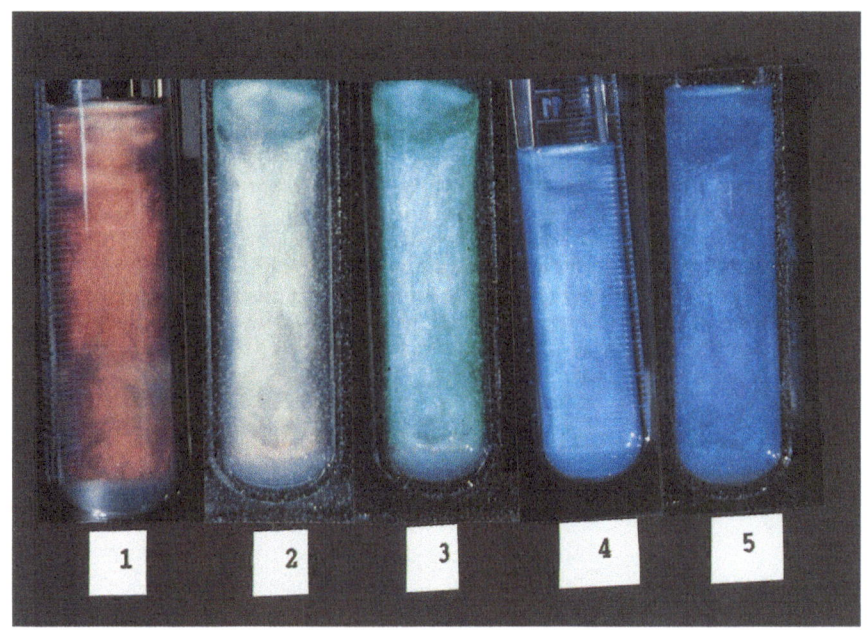

Fig. 1. Photograph of iridescent solutions. Illuminated with white light against a dark background. Composition of the samples is varied

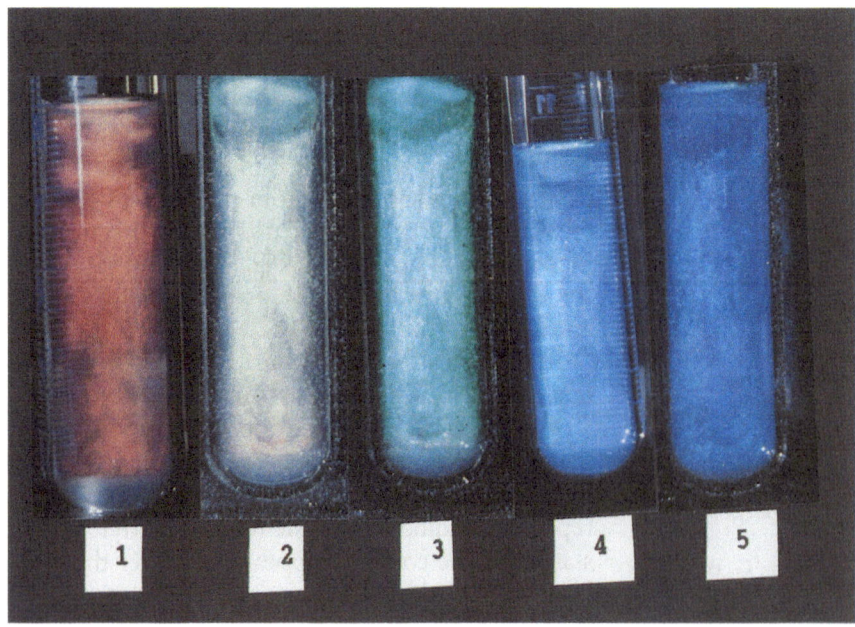

Fig. 2. A sample of iridescent colours viewed from different scattering angles δ when illuminated against a dark background with white light. Composition of the sample: 20 mM C_{14}DMAO, 65 mM hexanol and 10 mM decane; sample 1, $\vartheta = 175°$; sample 2, $\vartheta = 155°$; sample 3, $\vartheta = 140°$; sample 4, $\vartheta = 120°$; sample 5, $\vartheta = 100°$

Fig. 3. Sample showing different iridescent colours as a consequence of sedimentation in the sample. Sample viewed from the back when illuminated with white light. Concentration of the sample: $c_{C_{14}DMAO} = 30$ mM, $c_{hexanol} = 72$ mM, $c_{decane} = 35$ mM

the colours across the whole visible spectrum from red to dark blue. In Fig. 3 it is shown that the whole rainbow spectrum can also be obtained in a single sample in backscattering. The different colours in this sample are due to sedimentation equilibrium in the system. The conditions for the sedimentation are obtained when the system is close to the phase boundary of the single iridescent phase. Under optimised conditions the sedimentation equilibrium is reached within a few minutes. After the stationary state is reached the system can then be kept for an indefinite time without any change.

When the samples are viewed from a close distance one can see that the colours are not irradiated evenly from all parts of the sample but rather seem to come from domains in the sample. The samples seem to have a grain-like structure; this effect is probably due to the polycrystallinity of the samples. The director of the samples shows in the same direction only over a small distance which is the range of a fraction of a mm. Attempts to orient the whole sample by magnetic fields were not successful.

Phase diagrams

The surfactant tetradecyldimethylaminoxid forms with water isotropic solutions up to about 30% by weight [6]. The isotropic phase borders on a narrow nematic phase region which is then followed by a hexagonal phase to higher concentrations.

When increasing amounts of hexanol are added to the surfactant solutions which contains between 20 and 50 mM surfactant the solution becomes first turbid and biphasic then clear again and finally liquid crystalline. The series of the phases can be explained as follows:

Surfactant solutions above 10 mM contain rodlike micelles [6]. Upon addition of hexanol the rodlike micelles become larger and the attractive forces between the rods become larger too and the system separates into surfactant-rich and surfactant-poor phases. This two phase formation therefore corresponds phenomenologically to a vapor liquid condensation. In surfactant chemistry the process is sometimes referred to as coacervate formation.

Similar situations are observed in nonionic surfactant systems when for a temperature above the cloud point the concentration is increased [7]. Again, upon increasing the concentration these systems go from single to two-phase systems and finally back to single-phase systems. On the left of the miscibility gap the micelles are in the gaseous state while on the righthand side the micelles are in the liquid state.

The lamellar phase which is finally formed in the ternary system can solubilise large amounts of hydrocarbons. In this respect it behaves similar to other lamellar phases which are formed in dilute systems [8]. The phase boundaries of the lamellar phase are shifted to smaller hexanol fraction for larger solubilisation degrees of decane. It is the swollen lamellar phase which shows the iridescent colours. A part of the complete phase diagram which contains the lamellar phase for a 30 mM solution is given in Fig. 4.

It is interesting to note and of significance for a detailed interpretation and understanding of the results that the iridescent colour is shifted from blue to red with increasing concentration of decane when the surfactant concentration is about constant. This means that the mean distance between the lamellar layer is increasing. With increasing surfactant concentrations the iridescent phases are shifted to a larger hexanol and hydrocarbon range. This is shown in Fig. 5, where the single-phase regions for several surfactant concentrations are mapped out.

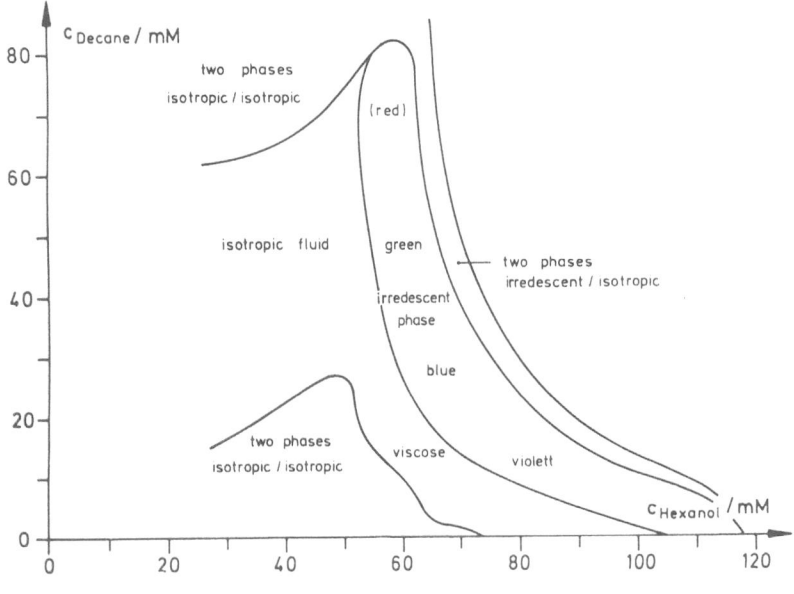

Fig. 4. Phase diagram of a 30 mM surfactant solution. Plot of the decane concentration against the hexanol concentration. $T = 25\,°C$

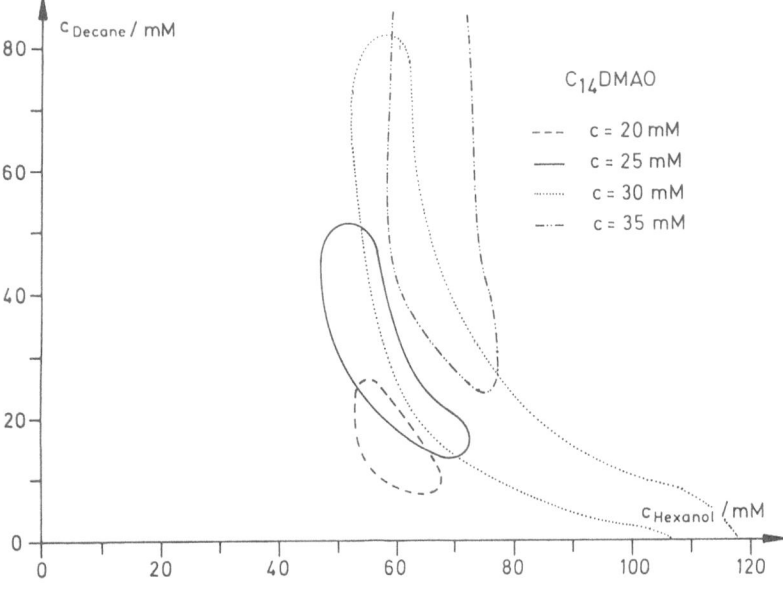

Fig. 5. Single-phase regions with the iridescent colours for different surfactant concentrations in a plot of the decane against the hexanol concentration. $T = 25\,°C$

The samples with more than 20 mM surfactant are clearly birefringent as can easily be seen when the samples are placed between crossed polarisers. It is however noteworthy that the samples with the lowest surfactant concentration which still give iridescent colours are not birefringent. The birefringence of the iridescent phases can also be seen under a polarisation microscope. The texture is, however, not very characteristic, as is shown in Fig. 6. Without crossing a phase boundary the single phase region can be moved to a much higher surfactant region where the iridescent colours are no longer visible and where typical textures of lamellar phases can be observed (Figs. 7 and 8).

The spherolytes in Fig. 7, which are typical for a dispersion of a lamellar phase were obtained when the lamellar phase was heated above its melting point and then cooled down again into the lamellar phase.

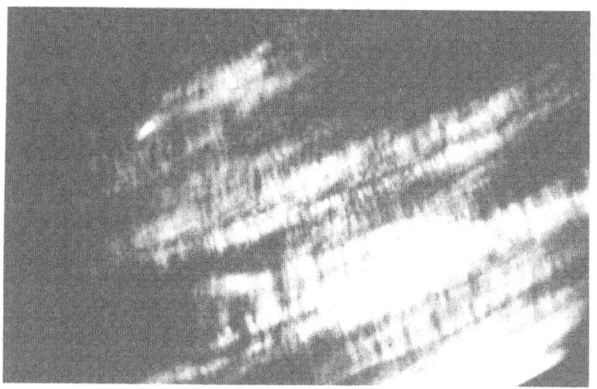

Fig. 6. Texture of the iridescent phase observed under the polarisation microscope. $T = 25\,°C$. Magnification was 25-fold and thickness of the cuvette = 2 mm

The pictures 7a and b were obtained shortly after and several minutes after the temperature into the liquid crystalline phase had been crossed.

The textures in 8a and b were obtained from a liquid crystalline sample which was filled into a 0.2 mm glass capillary and then slowly heated.

Figure 8b is the texture shortly before the liquid crystalline phase melted and the birefringence disappeared completely.

Interfacial tension measurements

Our surfactant systems solubilise large quantities of hydrocarbon into the micellar solutions. It can therefore be expected from theoretical considerations on

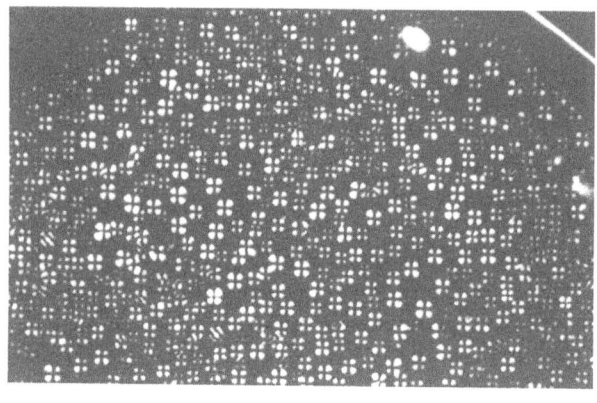

a)

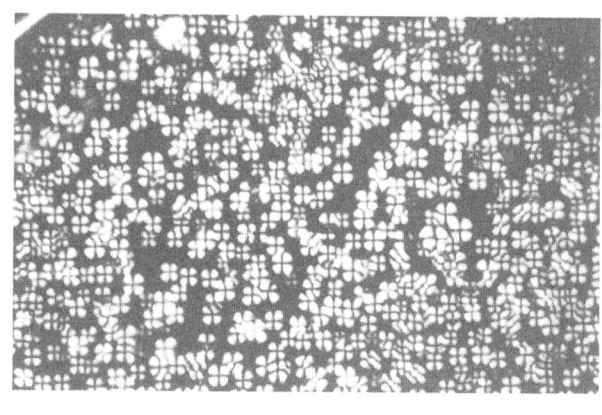

(b)

Fig. 7. Typical spherolytes are obtained under the polarisation microscope when the lamellar phase was heated above its melting point and then cooled down again into the lamellar phase. Picture 7a was obtained shortly after the temperature into the liquid crystalline phase had been crossed and picture 7b several minutes later. The thickness of the cuvette was 0.2 mm, the magnification 80-fold, the composition of the sample is 15 % $C_{14}DMAO$, 3.75 % hexanol and 12 % decane

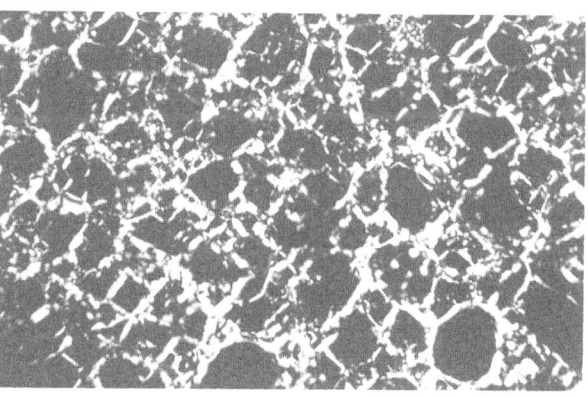

a)   b)

Fig. 8. Lamellar texture of 15 % by weight $C_{14}DMAO$, 3.75 % by weight hexanol and 12 % by weight decane under the polarisation microscop at 25 °C (Picture 8a). Picture 8b is the texture shortly before the liquid crystalline phase was melted and the birefringence disappeared completely. Magnification was 80-fold and $d_{cuvette} = 0.2$ mm

the solubilisation process that our surfactant solutiones which form the iridescent systems should have low interfacial tensions against the excess hydrocarbon. Furthermore, it is theoretically conceivable that the interfacial tensions of the system have to lie within narrow limits in order for the systems to be able to form the iridescent solutions.

In order to find out these interrelations we made interfacial tension measurements both of the pure surfactant solutions and of surfactant solutions with increasing concentrations of hexanol against decane as the hydrocarbon. The measurements were carried out with the spinning capillary method.

The results are presented in Figs. 9 and 10.

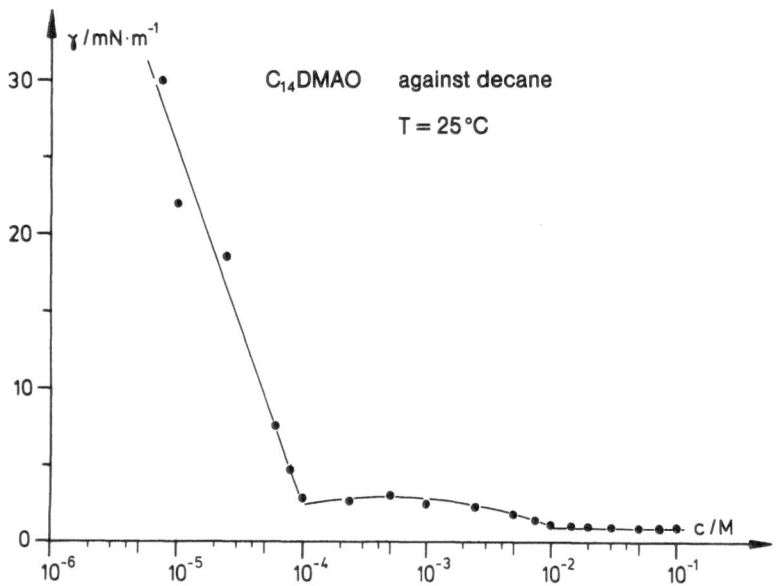

Fig. 9. Plot of the interfacial tension of surfactant solutions of $C_{14}DMAO$ against decane. $T = 25\,°C$

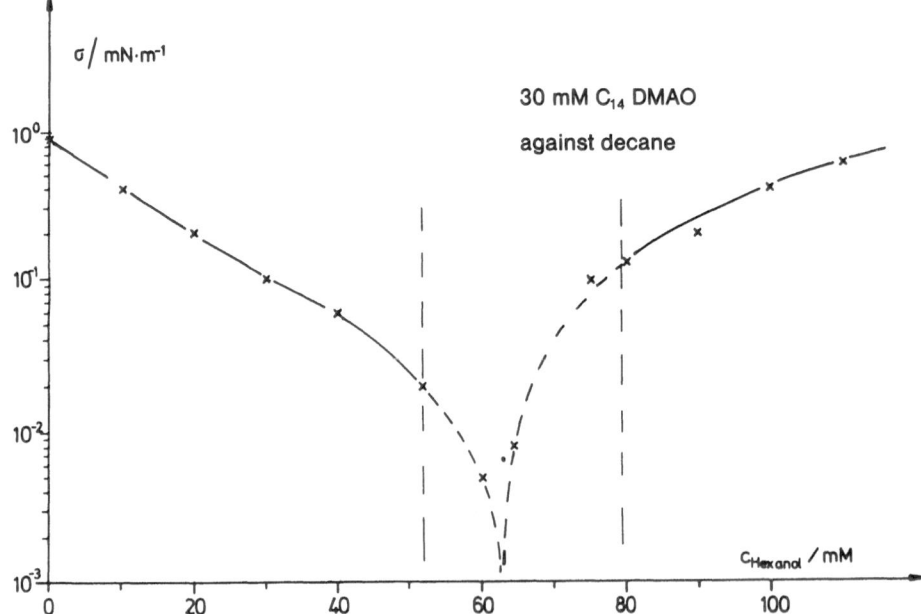

Fig. 10. Plot of interfacial tension measurements σ of a 30 mM $C_{14}DMAO$ solution with increasing amounts of hexanol against decane at 25 °C

The interfacial tensions of the pure surfactant solutions are around 1 mN/m. This is an intermediate value and confirms the experimental observations that little hydrocarbon can be solubilised into the pure micellar solution [9a, 9b].

It is noteworthy that the interfacial tension concentration plot shows a second break at around 10 mM. This break corresponds to the concentration at which the micelles undergo a sphere-rod transition in the hydrocarbon free surfactant solution. During the measurements the surfactant solutions were of course saturated with hydrocarbon and rods cannot be formed in these conditions. The break in the interfacial tension seems to indicate a different packing of the surfactants in the globules below and above this concentration.

The interfacial tensions decrease rapidly with increasing hexanol concentrations and reach a minimum at the compositions for which we observe the iridescent colours. It seems therefore that the absolute values for the interfacial tensions have to be around 0.1 mN/m in order to have the right conditions for the formation of iridescent systems.

Determination of the interlamellar spacing

With the presented results we have shown it is clear that the iridescent samples consist of lamellar phases. Now we would like to know the spacings.

The spacings between the lamellar layers can be determined from the Bragg equation

$$D = \frac{\lambda}{2n \sin (\delta/2)} \qquad (1)$$

and the angle of maximum scattering δ for a particular wavelength. This scattering angle is usually determined by measuring the scattering intensity of the incident light beam of a particular wavelength as a function of the scattering angle. Plots are given in Fig. 11. We received sharp Bragg-peaks. Surprisingly, we also find sharp peaks in extinction measurements with a normal UV/VIS-spectrometer (Lambda 17, Perkin-Elmer). Figure 12 shows typical results of the extinction which is plotted as $E \cdot \lambda^4$ against λ. Sharp peaks in turbidity measurements without dyes are only possible if the main part of the scattering intensity of the peak goes in a backward direction. For scattering angles from 150° to 180°C the relative q-value change is only

$$4\% = \frac{\sin 180°/2 - \sin 150°/2 \cdot 100}{\sin 180°/2}.$$

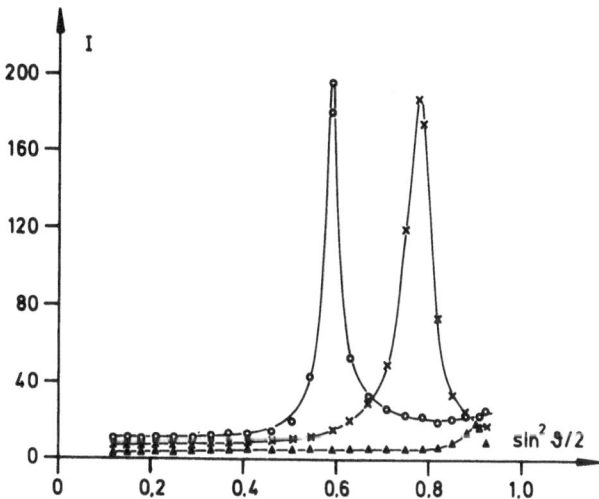

Fig. 11. Light scattering intensity in relative units against $\sin^2 \vartheta/2$ of the scattering angle ϑ for different samples with a concentration of: O 20 mM C_{14}DMAO/decane 1:1, 65 mM hexanol; × 25 mM C_{14}DMAO/decane 1:1, 70 mM hexanol; ▲ 30 mM C_{14}DMAO/decane 1:1, 65 mM hexanol; $\lambda = 488$ nm, $T = 25$ °C

Table 1 shows that the q_{max} values of light-scattering measurements $\left(q_{max} = \frac{4\pi \cdot 1.333}{488 \text{ nm}} \sin \frac{\vartheta_{max}}{2}\right)$ and extinction measurements $\left(q_{max} = \frac{4\pi \cdot 1.333}{\lambda_{max}}\right)$ agree within an error of less than $\pm 2\%$.

These experimental results unambiguously show that the lamellar layers must be aligned parallel to the planar surface of the cell which contained the sample and has a 2-mm light path. But in this case the whole scattering (due to Bragg-reflexes) goes in the backward direction because all mirror planes are orientated perpendicular to the light beam.

We have stated at the beginning of this article that the light which is scattered from the samples reveals a

Table 1. The calculated q-values from extinction measurements and dynamic light scattering measurements for coloured solutions of different concentrations

$c_{Solution}$	Light scattering		Extinction	
	$\sin^2 \vartheta/2$	$q_{max} \cdot$ nm	λ_{max}/nm	$q_{max} \cdot$ nm
c_1	0.59	$2.63 \cdot 10^{-2}$	638	$2.63 \cdot 10^{-2}$
c_2	0.787	$3.04 \cdot 10^{-2}$	531	$3.15 \cdot 10^{-2}$
c_3	0.91	$3.27 \cdot 10^{-2}$	496	$3.38 \cdot 10^{-2}$

$c_1 = 20$ mM C_{14}DMAO, 65 mM hexanol, 10 mM decane; $c_2 = 25$ mM C_{14}DMAO, 70 mM hexanol, 15 mM decane; $c_3 = 32$ mM C_{14}DMAO, 62.5 mM hexanol, 45 mM decane

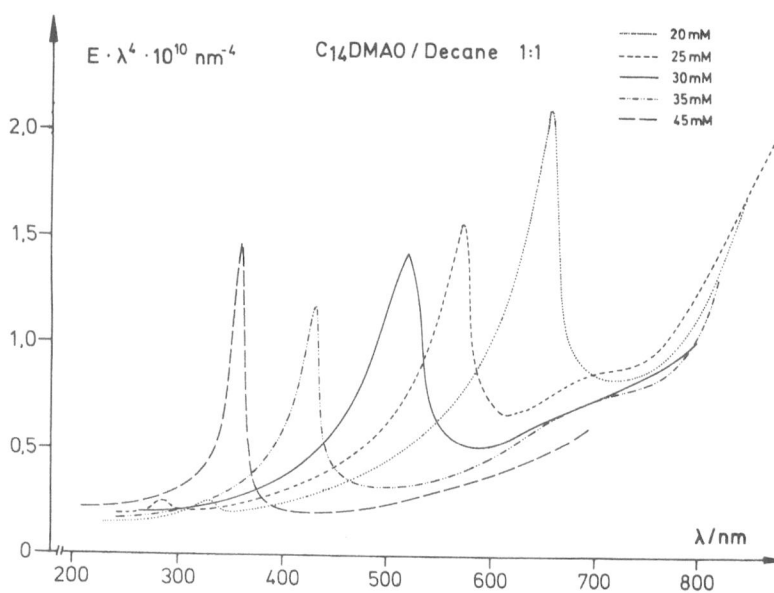

Fig. 12. Plot of $E \cdot \lambda^4$ against the wavelength λ for different samples with a surfactant/decane region of 1 : 1. The samples contain about 65 mM hexanol, $d_{\text{cuvette}} = 2$ mm and $T = 23\,^{\circ}$C

grain-like structure and we attributed this to domains of macroscopic dimensions which have the same orientation. Large samples behave therefore like a polycrystalline material in which domains with all possible angles are present. On smooth surfaces however the domains are preferentially aligned. The situation can be compared with nematic thermotropic phases on smooth surfaces. In this case thin enough samples of liquid crystallin phases are also completely aligned. For such orientated systems the

extinction measurements are a very efficient method to obtain Bragg-reflexes. We could find secondary order Bragg-peaks that appear at the expected positions, this means at half the wavelength of the primary peak Fig. 13.

The results and the obtained spacings are given in Table 2.

Another interesting fact is found from the diffusion-coefficient which we obtained by dynamic light scattering. Figure 14 shows clearly that those depend

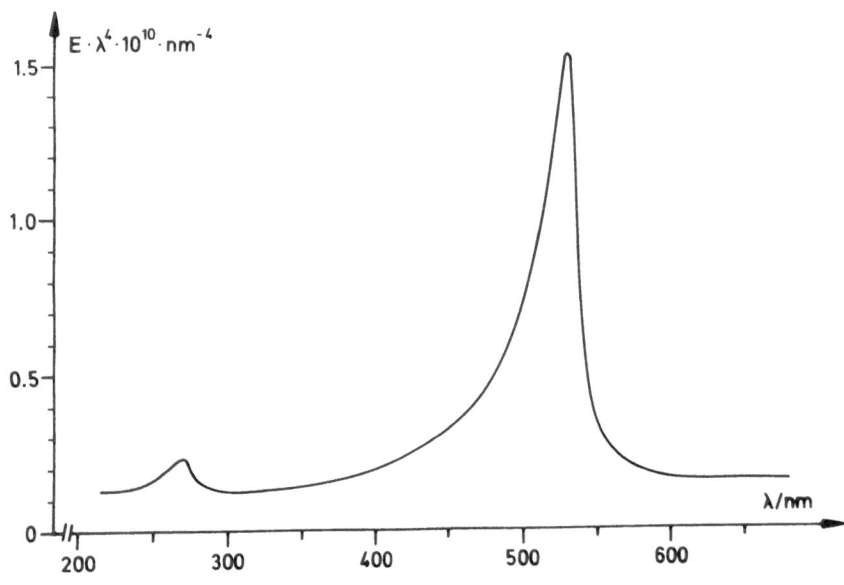

Fig. 13. Plot of $E \cdot \lambda^4$ against the wavelength λ for a sample which contains 25 mM C_{14}DMAO, 70 mM hexanol and 15 mM decane. $T = 23\,^{\circ}$C and $d_{\text{cuvette}} = 2$ mm

Table 2. The primary and secondary q_{max}-values of iridescent solutions of different concentrations as determined from extinction measurements. The calculated values for the spacings D are given in the last column

$c_{solution}/mM$	$q_{max1} \cdot nm$	$q_{max2} \cdot nm$	D/nm
18	$2.52 \cdot 10^{-2}$	$4.97 \cdot 10^{-2}$	249
20	$2.56 \cdot 10^{-2}$	$5.05 \cdot 10^{-2}$	245
25	$2.92 \cdot 10^{-2}$	$5.78 \cdot 10^{-2}$	215
27	$2.99 \cdot 10^{-2}$	$5.90 \cdot 10^{-2}$	210
30	$3.23 \cdot 10^{-2}$		194
32	$3.68 \cdot 10^{-2}$		171
35	$3.90 \cdot 10^{-2}$		161
38	$4.18 \cdot 10^{-2}$		150
40	$4.46 \cdot 10^{-2}$		141
45	$4.68 \cdot 10^{-7}$		134

The samples contain a surfactant/decane region of 1:1 and about 70 mM of hexanol.

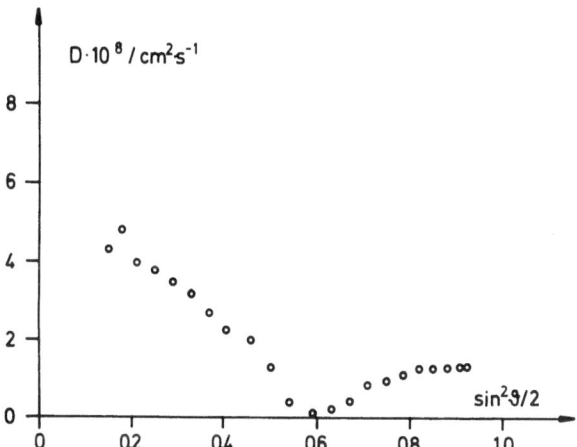

Fig. 14. Plot of translational diffusion-coefficients against $\sin^2 \vartheta/2$ for 20 nm $C_{14}DMAO$, 65 mM hexanol and 10 mM decane at $T = 23\,°C$ and $\lambda = 488$ nm

strongly on the q-values. At the Bragg-peak position we find a strong minimum.

If we plot $I_{si} \cdot D$ against $\sin^2 \delta/2$ (Fig. 15) the main part of the minimum disappears. This means the values of the diffusion coefficients are mostly influenced by the thermodynamics of the system, which may be expressed by the structure factor

$$S = RT / \frac{d\mu}{d \ln c}.$$ (2)

Because

$$I_{sc} = I_{sc} \text{ (no interaction)} \cdot S$$

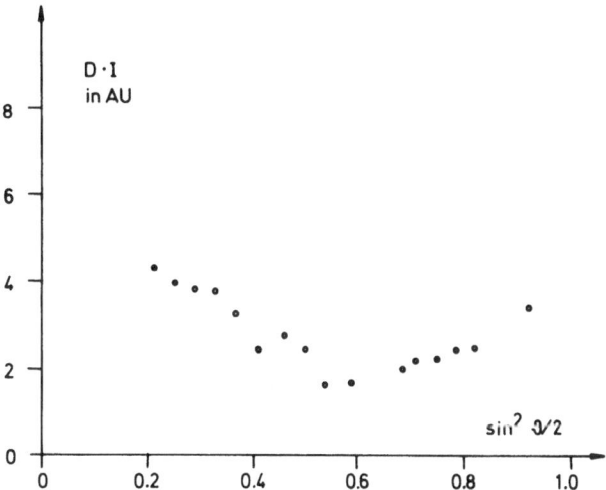

Fig. 15. Plot of $I_{sc} \cdot D$ against $\sin^2 \vartheta/2$. $\lambda = 488$ nm and $T = 23\,°C$. Concentration: 20 mM $C_{14}DMAO$, 65 mM hexanol and 10 mM decane

and

$$D = D_0/S.$$

The product of both factors should be independent on the structure factor, if hydrodynamic interactions can be neglected. The experimental results show that the hydrodynamic interactions are much smaller than the influence of the structure factor.

The microscopic structure of the lamellar layers

The lamellar layers are formed from the surfactant and the hydrocarbon molecules and some of the hexanol. On the basis of this model we can calculate a thickness for the layer. It is clear from Fig. 16 that the ratio of the layers d and the spacing of the system D is equal to the volume fraction ψ of the material in the layer (V_2) and the total volume (V)

$$\frac{V_2}{V} = \frac{d}{D} = \psi.$$ (3)

The thickness d of the layers can thus be calculated from the volume fraction and the experimental D values. The calculations of the volume fractions depends on how much of the hexanol present will be incorporated in the lamellar layer. It is likely that most of the cosurfactant at the composition of the iridescent solutions will be in the bulk phase and not incorporated in the layers.

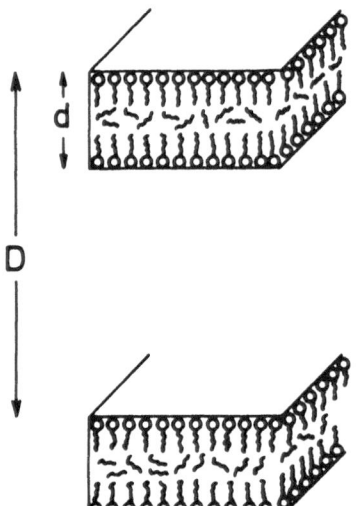

Fig. 16. Model for calculating the spacings: $d = 25$ A, D obtained from Bragg-reflexes

Table 3. The values for the thickness of the layer d and the area a for a surfactant molecule at the lamellar interface are given for different concentrations. The surfactant-hydrocarbon ratio is constant and equal to one. The hexanol concentration was about 70 mM

c_{solution}/mM	d/Å	a/Å^2
18	21.0	74.1
20	23.0	67.6
25	25.2	61.7
27	26.6	58.5
30	27.3	57.0
32	25.6	60.8
35	26.4	58.9
38	26.7	58.3
40	26.4	58.9
45	28.3	55.0

In pure water hexanol has a solubility of about 60 mM. For smaller concentrations the hexanol molecules will be in the non-aggregated state because the hydrophobic interaction energy is too small to form aggregates. The interaction is probably very similar in the presence of micellar aggregates. For hexanol concentrations smaller than 60 mM the molecules will probably only adsorb at micellar interfaces where the bulk water borders directly on the hydrocarbon area of the micelles. Incorporation of the molecules into the micellar interface under the formation of mixed micelles is likely to occur only at higher concentration.

This situation is also reflected in the phase diagrams of this investigation. The phases begin to change drastically only when the hexonal concentration is raised above 60 mM; only then do we find the lamellar phases.

Most of the iridescent solutions contained a hexanol concentration which was only little higher than 60 mM. For the calculations of the layer thicknesses we shall therefore neglect the hexanol concentration. This will give us lower limits for d. In reality the d-values could be somewhat larger. For the calculation of the volume fractions we assumed a density of 0.84 g/ml. Table 3 shows results which were calculated for a series of solutions in which the surfactant hydrocarbon ratio was kept constant and the total concentration was varied. The table contains also values for the area a which a surfactant molecule occupies at the lamellar interface. These values were calculated under

the assumption that the hydrocarbon is sandwiched completely between the surfactant molecules and there is no penetration of hydrocarbon molecules into the palisade layers of the surfactant molecules. Again this is assumed in our model and in reality there could be some penetration. With this model a is simply given by Eq. (4)

$$a = \frac{2(V_{HC} + V_{\text{Surf}})}{d} \qquad (4)$$

where V_{HC} is the molecular volume of the hydrocarbon and V_{Surf} the volume of the surfactant molecule.

The a values which have been calculated with the made assumption can be compared with the a value which can be obtained from the Gibbs-adsorptions isotherm of interfacial and surface tension measurements. From surface tension measurements we obtained a value of $a = 51$ Å^2 [11]. The comparison shows that the assumption made is probably realistic and that the packing of the surfactant molecule at the iridescent phases seem to be very similar to the packing of a surfactant layer at water/oil or a water/hydrocarbon interface. It also shows that both the surfactant and the hydrocarbon molecules cannot be in the all-trans-configuration in the lamellar layers. The thickness of the layer is considerable smaller than the sum of length of the straight chains [10].

General remarks

The iridescent colours in the system are not just obtainable in combination with the cosurfactant hexanol and the hydrocarbon decane. We were also able to obtain iridescent solutions with octanol and decanol as cosurfactant and a wide variety of hydrocarbons including hydrocarbons with a double bond. When other additives are used instead of the ones for which the phase diagrams are reported the single phase regions with the iridescent colours shift their position. These measurements confirm that we are dealing with a very general phenomenon which is connected to the existence region of lamellar layers in dilute solutions containing between 1 % and 2 % of material. In the case of the aminoxid surfactant the extremely large spacings between the layers are due to electrostatic forces. As soon as small amounts of electrolyte are added to the aqueous solution the iridescent colour disappears and the solution becomes turbid and two phases separate upon standing. The colours disappear also when ionic surfactants are added in small mole fractions to the system.

The charge on the double layers which is responsible for the electrostatic repulsion is due to a protolytic reaction between the aminoxide and water. The aminoxid is a weak base and is partially protonated in water. A 30 mM solution has a pH of about 9. This means that only one out of 1 000 surfactant molecules carries a charge and the charge density is therefore extremely small (one elementary charge per 40 000 Å^2).

In spite of this the electrostatic repulsive forces between the layers are large enough to keep them at distance of 3000 Å.

The electrostatic repulsion energy per area (E_{pot}/A) can be calculated using the following equation which is valid for long distances [12]:

$$\frac{E_{\text{pot}}}{A} = 64 \cdot C_g \cdot N_L \cdot k_B T \cdot \frac{1}{\varkappa} e^{-kx} \tag{5}$$

$\varkappa$ is the reciprocal Debye length, C_g is the counterion-concentration.

A pH value of 9 gives an ionic strength of 10^{-5} mol/L and a Debye length of 1 000 Å. We recognize that the observed spacing of 2 500 Å is within the range of the diffuse double-layer.

Conclusions

Aqueous solutions of tetradecyldimethylaminoxid with 20 to 35 mM surfactant, equimolar concentration of decane and approximately 65 mM of hexanol give brilliant iridescent colours when they are illuminated with white light. With increasing surfactant concentration the wavelength of the backscattered light shifts from red over green to dark blue. It is shown that the colours are due to Bragg reflections of the light in the visible region on lamellar layers which have extreme large interlamellar spacings. Under the experimental conditions the layers which are formed from the surfactant, some of the cosurfactant and the hydrocarbon have a thickness of about 25 Å and are between 1 000 Å and 2 500 Å apart. Most of the systems which show iridescent colours are optically birefingent. The large separation between the lamellar layers is due to electrostatic repulsion between the layers. The charge of the layers is due to partial protonation of the aminoxide groups by water. About one in a thousand surfactant molecules is protonated. The charge density is about one elementary charge per 40 000 Å^2. The colours disappear when electrolyte is added to the system. At compositions which are close to the phase boundary of the single irredescent phase region sedimentation equilibrium in the system can be observed. Samples in a 10-cm high flat tube look red at the bottom and blue at the top.

References

1. Larche FC, Appell J, Porte G, Bassereau P, Marignan J (1986) Phys Rev Letters 56:1700
2. Safinyia CR, Roux D, Smith GS, Sinha SK, Dimon P, Clark NA, Bellocq AM (1986) Phys Rev Letters 57:2718
3. Helfrich W (1978) Z Naturforschung 33a:305
4. Satoh N, Tsujii K (1987) J Phys Chem 91:6629
5. Ishii Y, Naitoh K, Tsujii K (1988) Conference on Surface and Colloid Science, Sengokubara Height Hakon, Japan
6. Hoffmann H, Oetter G, Schwandner B (1987) Progr Colloid Polym Sci 73:95
7. Pospischil KH (1986) Langmuir 2:170
8. Gomati R, Appell J, Bassereau P, Marignan J, Porte G (1987) J Phys Chem 91:6203
9. a) Kahlweit M, Strey R (1985) Angew Chem 97:655
9. b) Pouchelon A, Chatenay D, Meunier J, Langevin D (1981) J Colloid Interface Sci 82:418
10. Tanford C (ed) (1980) The Hydrophobic Effect. J Wiley & Sons, New York
11. Oetter G, Hoffmann H (1988/89) J Dispersion Science and Technology 9:459
12. Stauff J (1960) Kolloidchemie. Springer-Verlag, Berlin Göttingen Heidelberg

Received November, 1988;
accepted February, 1989

Authors' address:

Prof. Dr. H. Hoffmann
Lehrstuhl für Phys. Chemie I
Universität Bayreuth
Universitätsstr. 30
Postfach 10 12 51
D-8580 Bayreuth, F.R.G.

Progress in Colloid & Polymer Science   Progr Colloid Polym Sci 79:308–312 (1989)

E. Theoretical studies of colloids

Interactions between surfactant ions of opposite charge

N. Filipović-Vinceković[1]) and D. Škrtić[2])

[1]) Department of Physical Chemistry, Laboratory for Radiochemistry, "Ruđer Bošković" Institute, Zagreb, Croatia, Yugoslavia
[2]) Department of Technology, Nuclear Energy and Radiation Protection, Laboratory for Precipitation Processes, "Ruđer Bošković" Institute, Zagreb, Croatia, Yugoslavia

Abstract: The interactions in mixtures of anionic/cationic surfactants were investigated by several methods (conductivity, surface tension, turbidity, electrophoretic mobility, and EMF measurements). The mixtures studied were sodium tetradecylsulfate/tetradecyl pyridinium chloride and sodium dodecyl sulfate/dodecylammonium chloride. At least three different concentration regions can be differentiated in the region of mixed micelles depending on changes in the surfactant molar ratio. A model for the interface between mixed micelles and the surrounding electrolyte is supposed to incorporate the headgroups of both surfactants and pairs of headgroups-counterions of the surfactant in excess.

Key words: Surfactant aqueous solutions, mixtures of surfactants, structure of interfaces.

Introduction

The investigations of interactions in systems which contain more than one type of surfactants are of considerable interest from theoretical and practical points of view. Numerous studies have shown that mutual interactions in mixed surfactant systems result in the formation of mixed micelles and mixed monolayers at the air/water interface whose composition depends on the type and concentration of surfactants present. Recently, significant attention has been paid to the modelling and understanding of mixed systems containing surfactants of opposite charges [1–5]. An important parameter in ionic surfactant systems is the micelle surface charge density which relates to the structure of the micelle/solution interface. At the micelle/solution interface there occur many interactions which affect the energetics of micellization processes, the critical micelle concentration (CMC) and the size distribution of micelles and their shape. The ionic micelle/solution interface consists of discrete ionic headgroups and ionic headgroups-bound counterion pairs [6]. The existence of oppositely charged headgroups in mixed micelle/solution interfaces is associated with the neutralization of some of the micelle charge and the change in counterion binding

[7]. This paper reports on changes in the physicochemical properties of mixed binary anionic-cationic surfactant systems due to changes in the molar ratio and concentration of surfactants. Sodium tetradecyl sulfate/tetradecyl pyridinium chloride (STS/TDPC) and sodium dodecyl sulphate/dodecylammonium chloride (SDS/DDAC) were selected as model systems due to their well characterized properties in water solution. The obtained results emphasize the role of the structure of mixed micelle/solution interface in overall interaction processes and are presumed to represent other mixtures of surfactants of opposite charge as well.

Experimental

The commercial surfactants SDS and TDPC, obtained from BDH Chemicals Ltd., STS obtained from Henkel KGaA, and DDAC prepared according to Kertes [8], were purified several times by recrystallization. The surfactant purity was checked by means of surface tension measurements. The surfactant solutions were always freshly prepared. The critical micelle concentrations determined from surface tension and conductivity measurements corresponded to literature data [9–12].

Surface tension, $\sigma/\text{mN m}^{-1}$, was determined using a du Noüy interfacial tensiometer (Krüss, Hamburg).

Relative systems turbidity, τ_r, was measured by a Pulfrich photometer equipped with turbidimetric extension (Carl Zeiss, Jena).

Electrophoretic mobility, $\mu \pm /m^2V^{-1}s^{-1}$, and particle charge were determined using an apparatus for particle microelectrophoresis (Rank Brothers, Apparatus II).

Specific conductivity, $\varkappa/\Omega^{-1}m^{-1}$, was determined using a Cambridge high-frequency conductometer. A conductivity cell with a normal platinum electrode was used.

The concentration of free counterions (not bound to mixed micelles) was measured using a chloride and sodium specific electrode relating the observed cell potential to a calibration curve (Orion Research microprocessor ionalyzer/901) [13,14]. The results are presented as the change in the concentration of bound chloride or sodium ions ($\Delta c_I \pm$)

$$\Lambda c_I \pm = c_t - c_f \qquad (1)$$

where c denotes concentrations and subscripts total (t) and free (f) chloride (I^-) or sodium (I^+) ions.

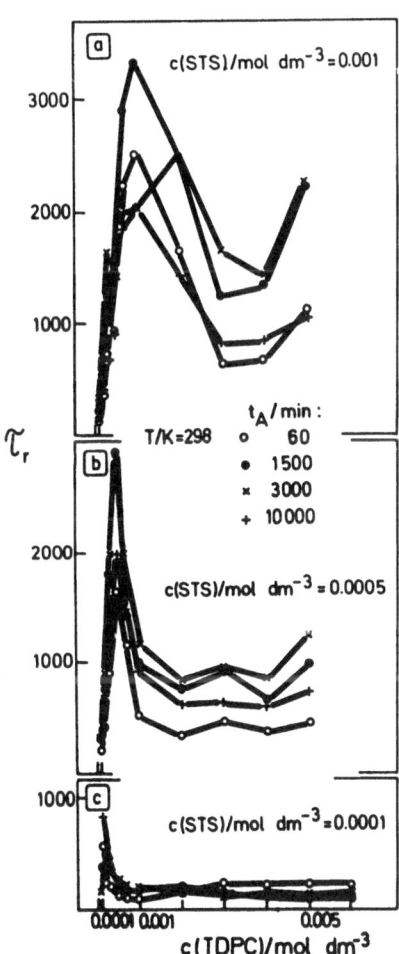

Fig. 1. Plots of relative system turbidity, τ_r, vs. TDPC concentration for $c(STS) = 0.001\ mol\ dm^{-3}$ (a), $c(STS) = 0.0005\ mol\ dm^{-3}$ (b) and $c(STS) = 0.0001\ mol\ dm^{-3}$ (c). Temperature and aging time are indicated

The cells used for the EMF measurements were [15]

Saturated Ag–AgCl electrode	2 mol dm^{-3} NH$_4$NO$_3$ 2% agar bridge	Sample solution	Na$^+$ selective electrode

and

Calomel electrode	2 mol dm^{-3} NH$_4$NO$_3$ 2% agar bridge	Sample solution	Cl$^-$ selective electrode

Results and discussion

The interactions in anionic and cationic surfactant mixtures were studied by several methods which gave us information on the state in bulk solution and at the air/water interface. The obtained results show a strong effect of the molar ratio and concentration of surfactant on the measured physicochemical properties.

Figure 1 presents the changes in the relative system turbidity with increasing cationic surfactant concentration and aging time. Parts a, b, and c correspond to the systems with different concentrations of anionic surfactant. At constant anionic and increasing cationic surfactant concentration, the relative system turbidity increases to a maximum and then decreases until a clear solution is obtained. Mixtures with equimolar ratio of surfactants show maximum turbidity. When turbid surfactant solutions were left to stand for several days, precipitation occurred.

The results of microelectrophoretic measurements of pure surfactant solutions and of their mixture are illustrated in Fig. 2. Electrophoretic mobility values rather than zeta potentials were employed, since a representation of the charge and not of absolute values was required. It is obvious that colloid entities existed in all the systems investigated at concentrations considerably below the CMC of each component. The charge and electrophoretic mobility of particles varied with surfactant proportions. Maximal precipitation (equimolar region) and reversal of charge almost coincided. In systems with the anionic surfactant in excess, the entities present are negatively charged and vice versa. Within the equimolar concentration range a high degree of charge neutralization at the mixed micelle/solution interface allowed mixed micelles to grow to a large size, eventually resulting in the separation of the precipitate. This observation is in agreement with that of Malliaris et al. [16], who found that the aggregational behavior of mixed anionic/cationic micelles depends on the molar ratio of surfactants.

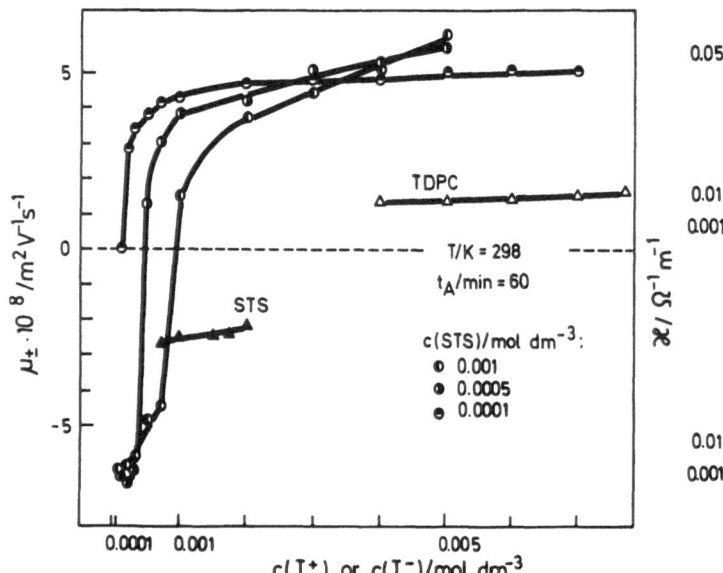

Fig. 2. Electrophoretic mobility, $\mu \pm /m^2 V^{-1} s^{-1}$, as a function of surfactant concentration. Systems, temperature, and aging time are indicated

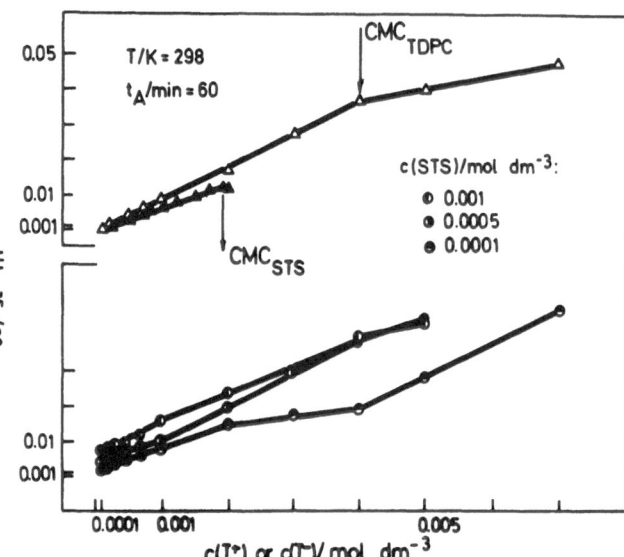

Fig. 3. Plots of specific conductivity, $\varkappa/\Omega^{-1} m^{-1}$, as a function of surfactant concentration. Systems, temperature and aging time are indicated

Figure 3 shows specific conductivity changes in pure and mixed surfactant systems vs. surfactant concentration. The change in the slope of a specific conductivity curve is usually well defined for a pure surfactant solution and indicates the CMC. The results obtained for CMC_{STS} and CMC_{TDPC} are in accordance with literature data [9–12]. The conductivity curves of mixed surfactant solutions show several changes in slopes, indicating changes in the concentrations and mobilities of conducting species (mixed micelles, surfactant monomers, counterions and/or other conducting species which may be present). These changes point to the existence of several "critical" concentrations, i. e., to the structural transitions in this concentration range.

The behavior of the air/water interface in pure and mixed systems was examined by surface tension measurements (Fig. 4). The surface tension curve of pure TDPC is typical with a sharp break at the CMC. The surface tension of STS/TDPC mixtures rapidly decreases at very low concentrations, reaches a minimum and maximum and then slightly decreases with further increase in TDPC concentration. The mixed systems show a greater lowering of surface tension than single components (Fig. 4 and [17]) due to a mixed monomolecular layer formation at the air/water interface [1–3]. The maxima and minima of the curves may be explained by changes in the mixed

monomolecular layer at the air/water interface due to changes in surfactant monomer concentrations in bulk solution.

In single ionic surfactant solutions the counterions around charged micelles are considered to be distributed into two regions: the Stern layer and the diffuse double layer. Counterions in the Stern layer may be considered as adsorbed to the micelle/solution interface and are called "bound" counterions [18]. The

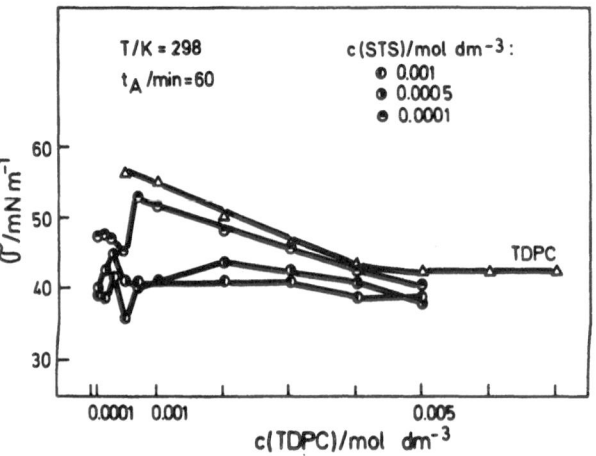

Fig. 4. Surface tension, $\sigma/mN\,m^{-1}$, vs. TDPC concentration. Temperature and aging time are indicated

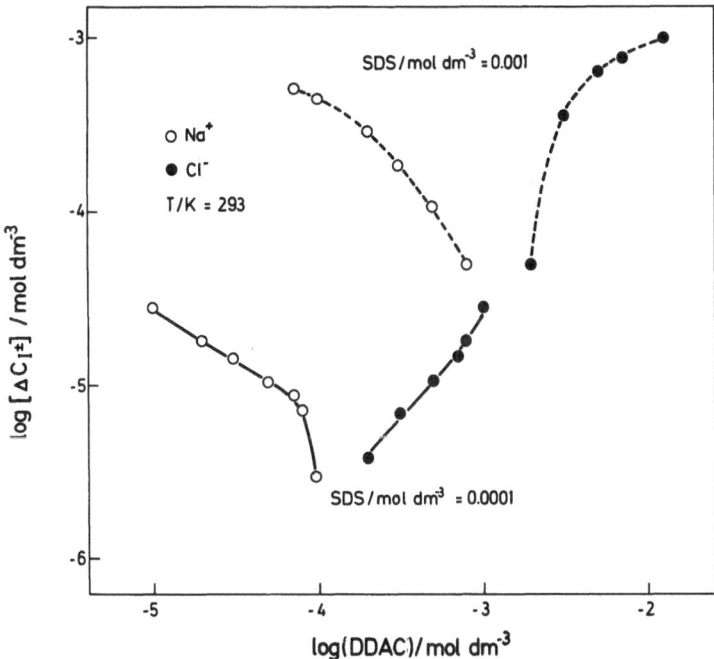

Fig. 5. Concentration of bound counterions, $\Delta C_I \pm /$ mol dm^{-3} (where I^+ denotes sodium and I^- chloride ions as counterions), vs. DDAC concentration for constant SDS concentration. Temperature is indicated

bound counterions at the micelle/solution interface are in equilibrium with unbound counterions in the diffuse double layer. Sodium and chloride ions act as counterions in mixed STS/TDPC systems. The binding of sodium ions to mixed STS/TDPC micelles could not be measured by sodium-specific electrode due to pH limitations. The results of chloride ion binding revealed that only mixed micelles, formed with cationic surfactant in excess, incorporate chloride ions as counterions at the mixed micelle/solution interface [7]. For comparison, the counterion binding in a mixture of structurally similar SDS/DDAC surfactants were also measured. The results obtained are shown in Fig. 5 as the change of the bound counterions vs. concentration of cationic surfactant. With increasing DDAC concentration the binding of sodium ions decreases and in the equimolar concentration region it becomes unmeasurable. In the equimolar concentration region, the concentrations of free sodium and chloride ions correspond to the concentration of surfactants added. The binding of chloride ions to mixed SDS/DDAC micelles starts above the equimolar concentration range and increases with increasing cationic surfactant concentrations. It is evident that the mixed anionic/cationic micelles formed in systems with anionic surfactant in excess bound cations and those formed in systems with cationic surfactant in excess bound anions as counterions.

The measurements carried out are outlined in Fig. 6. The observed changes in the measured physicochemical properties due to changes in the molar ratio and surfactant concentration indicate three characteristic regions:

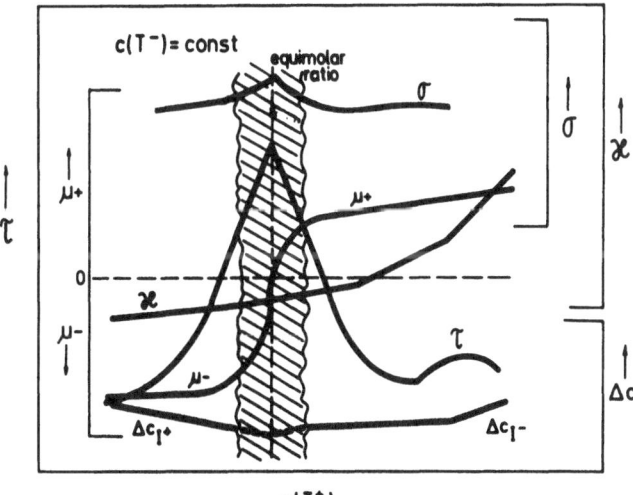

Fig. 6. Schematic presentation of relative turbidity, τ_r, electrophoretic mobility, $\mu \pm /m^2V^{-1}s^{-1}$, surface tension, $\sigma/mN\ m^{-1}$, specific conductivity, $\varkappa/\Omega^{-1}\ m^{-1}$ and concentration of bound counterions, $\Delta c_I \pm /mol\ dm^{-3}$, vs. cationic surfactant concentration and for constant concentration of anionic surfactant

i) Concentration region with anionic surfactant in excess

Anionic micelles incorporate a fraction of cationic monomers. The colloid entities present are negatively charged. EMF measurements indicate that only anionic surfactant counterions are included in the mixed micelle/solution interface. The cationic surfactant incorporated in mixed micelles lowers the charge density at the mixed micelle/solution interface and counterions are released from the interface into the aqueous phase. Microelectrophoretic mobility decreases and turbidity increases due to the growing of mixed micelles. The typical shape of surface tension curves with a small increase towards the equimolar region indicates a change in the surfactant monomer concentration due to a change in the composition of mixed micelles. According to a generalized mass action model of mixed micellization, the equilibrium in systems with anionic surfactant in excess may be written as

$$mR^- + nR^+ + vM^+ \rightarrow (R_mR_nM_v)^{m-(n+v)} \qquad (2)$$

where R^- and R^+ are monomers of anionic and cationic surfactants, M^+ is the anionic surfactant counterion, and $(R_mR_nM_v)$ is a mixed micelle and $m > (n + v)$.

ii) Equimolar concentration range

The equimolar concentration range is terminated by the appearance of system turbidity and precipitation with aging time. The particles present are almost uncharged and the binding of both counterions is unmeasurable. The maximum of surface tension curves located in the equimolar concentration range, corresponds to the change in concentration of free monomers. The existing equilibria may be given as

$$R^- + R^+ \rightarrow \text{precipitation}. \qquad (3)$$

iii) Concentration region with cationic surfactant in excess

In this concentration region mixed micelles mainly consist of the cationic component. With increasing cationic surfactant concentration more cationic micelles are formed and thus the anionic surfactant is redistributed among the new and more numerous mixed micelles. The electrophoretic mobility and conductivity increase due to increasing surface charge density at the mixed micelle/solution interface. EMF measurements indicate that only cationic surfactant counterions are included in the mixed micelle/solution interface. Electrostatic limitations influence the decrease of mixed micelle size, which is reflected by a decrease in system turbidity.

The existing equilibria may be written as

$$yR^- + xR^+ + wA^- \rightarrow (R_yR_xA_w)^{x-(y+w)} \qquad (4)$$

where A^- denotes the cationic surfactant counterion and $x > (y + v)$.

Acknowledgements

The authors wish to thank Dr. D. Mayer for conductivity measurements.

References

1. Goralczyk D, Woligora B (1987) Colloid Polym Sci 265:728–733
2. Goralczyk D (1982) Tenside Detergents 19:350–352
3. Corkill JM, Goodman JF, Harrold SP, Tate JR (1967) Trans Faraday Soc 63:247–254
4. Mukhayer GI, Davies SS (1975) J Colloid Interface Sci 53:224–234
5. Barry BW, Gray GMT (1975) J Colloid Interface Sci 52:314–325
6. Beunen JA, Ruckenstein E (1983) J Colloid Interface Sci 96:469–487
7. Filipović-Vincekoviċ N, Škrtić D (1988) Colloid Polym Sci 266:954–957
8. Kertes AS (1965) J Inorg Nucl Chem 27:209–217
9. Aniansson EAG et al (1976) J Phys Chem 80:905–922
10. Folger R, Hoffmann H, Ulbricht W (1974) Ber Bunsenges Phys Chem 78:989–997
11. Mukerjee P, Ray A (1966) J Phys Chem 70:2150–2157
12. Hoffmann H, Platz G, Ulbricht W (1981) J Phys Chem 85:1418–1428
13. Brunn TS, Holland H, Vikingstadt E (1978) J Colloid Interface Sci 63:590–592
14. Larsen JW, Tepley LB (1974) J Colloid Interface Sci 49:113–116
15. Koshinuma M (1983) Bull Chem Soc Jpn 56:2341–2347
16. Malliaris A, Binana-Limbele W, Zana R (1986) J Colloid Interface Sci 110:114–120
17. Schwuger MJ (1969) Kolloid-Z u Z Polymere 233:979–986
18. Rathman JF, Scamehorn JF (1984) J Phys Chem 88:5807–5816

Received November, 1988;
accepted February, 1989

Authors' address:

Dr. N. Filipović-Vincekoviċ
"Ruder Bošković" Institute
Bijenička 54, P.O. Box 1016
YU-41001 Zagreb, Yugoslavia

Progress in Colloid & Polymer Science   Progr Colloid Polym Sci 79:313–320 (1989)

Dynamic light scattering study of membrane interactions in colloidal smectics

F. Nallet[1]), D. Roux[1]), and J. Prost[2])

[1]) Centre de recherche Paul-Pascal, Domaine universitaire, Talence, France
[2]) Ecole supérieure de physique et de chimie industrielles, Paris, France

Abstract: Dynamic light scattering measurements on oriented samples of dilute lyotropic smectics have been performed. The hydrodynamics of a two-component smectic A is reviewed and applied to describe the fluctuation spectrum in our multi-component samples. The layer compressibility modulus (at constant chemical potential) $\bar{B}$ is extracted from the anisotropic dispersion relation of the so-called slip mode, which arises from the coupling between concentration and layer displacement fluctuations. The samples, diluted with three different solvents, afford repeat distances of 50–800 Å; the behavior of $\bar{B}$ suggests either undulation or electrostatic repulsions between membranes, depending on the swelling solvent.

Key words: Lyotropic smectics, dynamic light scattering, elastic constants, membrane interactions, fluctuating membranes.

Introduction

A lot of interest is currently focused on understanding the physics of fluctuating membranes [1]. In the absence of surface tension, deformation of membranes is governed by elastic bending energy. Thermal fluctuations leading to undulations of the membrane may have important consequences when the bending modulus is of the order of $k_B T$. Dilute lamellar phases are particularly interesting systems for studying fluctuating surfaces. Indeed, in some favorable cases, multimembrane lyotropic smectics with repeat distances of hundreds and even thousands of angströms can be prepared with addition of suitable solvents [2–6]. It has been recently demonstrated that in many cases these extreme dilutions result from the existence of an universal repulsive interaction due to the steric hindrance to thermal fluctuations [4, 5]. This purely entropically driven interaction first proposed by Helfrich [7] is long-range and keeps the layers apart, even for such large distances. These dilute lamellar phases are also a unique example of colloidal smectics. The very large repeat distances one can obtain are responsible for very low elastic constants. The layer compressibility modulus is expected to vary continuously over about six orders of magnitude along the dilution range

and the layer bending modulus over two orders of magnitude [5]. In this paper, we present a new method for measuring the layer compressibility modulus in lyotropic smectics and thus for studying membrane/membrane interactions. Since our experimental scheme heavily rests on the hydrodynamic properties of multicomponent smectics, a first part is devoted to them; results appear in the second part.

1. Hydrodynamics of two-component smectic A

We recall in what follows the hydrodynamic properties of a two-component smectic A phase. The one-component case has been worked out, first by de Gennes [8], and the complete treatment explicitly obtained within the very general framework of a unified hydrodynamic theory for any condensed phase, by Martin, Parodi, and Pershan (MPP) [9]. The two-component case was later discussed by Brochard and de Gennes [10]. Our presentation here is a mere adaptation of the general theory and leads to a slight generalization of the results of [10].

We first must list the independent hydrodynamic variables. In addition to the usual five in a simple fluid (total mass density ϱ, momentum density $\mathbf{g}$, and energy density ε), two more hydrodynamic variables now come into play: mass density ϱc of a species (two-component system) and layer displacement u (smectic A system). The evolution in time of these variables is described by

$$\partial_t \varrho = - \nabla \cdot \mathbf{g}$$

$$\partial_t g_i = - \partial_j \sigma_{ij} \ (i = x, y \text{ or } z)$$

$$\partial_t \varepsilon = - \nabla \cdot \mathbf{J}_\varepsilon$$

$$\partial_t (\varrho c) = - \nabla \cdot \mathbf{J}_c$$

$$\partial_t u = J_u \, . \tag{1}$$

The first six equations are the conservation laws, stated on a local form, for the corresponding conserved variables. The mass flux is the momentum density; the momentum flux is the stress tensor σ_{ij}; $\mathbf{J}_\varepsilon$ ($\mathbf{J}_c$) is the energy (particle) current density. The last equation is *not* a continuity equation since u is *not* a conserved variable; J_u describes its decay.

To close this set of equations, the fluxes $\sigma_{ij}, \mathbf{J}_\varepsilon, \mathbf{J}_c$ and $\mathbf{J}_u$ must be known. The thermodynamic and symmetry properties of the system (with the usual assumptions of local equilibrium and monotonic increase of entropy with time) set general constraints upon the fluxes that allow to state them more explicitly. Following the general MPP derivation, care must be duly taken that the two extra hydrodynamic variables are not on the same footing since ϱc is a conserved variable, whereas u is a broken-symmetry one, and we get

$$\sigma_{ij} = p\delta_{ij} - \Phi_j \delta_{iz} + \sigma_{ij}^D$$

$$\mathbf{J}_\varepsilon = \frac{p + \varepsilon}{\varrho} \mathbf{g} + \mathbf{J}_\varepsilon^D$$

$$\mathbf{J}_c = c\,\mathbf{g} + \mathbf{J}_c^D$$

$$J_u = \frac{g_z}{\varrho} + \mathbf{J}_u^D \tag{2}$$

(the z-direction of the coordinate axes is perpendicular to the smectic layers).

In these equations the terms with a D superscript are the *dissipative* contributions to the fluxes: if all D-terms are zeroed the total entropy remains constant. The other terms are the so-called *reactive* fluxes. As in

the one-component smectic A case, there is a twofold contribution to the reactive part of the stress tensor: the first is the pressure p; the second comes from the "stress vector" Φ which describes the elastic response to a strain u.

As regards the dissipative fluxes, they are given by the following constitutive relations:

$$\mathbf{J}_{ci}^D = - (\alpha_\perp \partial_i \tilde{\mu} + (\alpha_z - \alpha_\perp)\partial_z \tilde{\mu} \delta_{iz})$$

$$- (\beta_\perp \partial_i T + (\beta_z - \beta_\perp)\partial_z T \delta_{iz}) - \Gamma \delta_{iz} \nabla \cdot \Phi$$

$$\mathbf{J}_{\varepsilon i}^D = \tilde{\mu}\,\mathbf{J}_{ci}^D - (\varkappa_\perp \partial_i T + (\varkappa_z - \varkappa_\perp)\partial_z T \delta_{iz})$$

$$- T(\beta_\perp \partial_i \tilde{\mu} + (\beta_z - \beta_\perp)\partial_z \tilde{\mu} \delta_{iz}) - \xi \delta_{iz} \nabla \cdot \Phi$$

$$\mathbf{J}_u^D = \frac{\xi}{T} \partial_z T + \Gamma \partial_z \tilde{\mu} + \zeta \nabla \cdot \Phi$$

$$\sigma_{ij}^D = - \eta_{ijkl}\partial_k \left(\frac{g_l}{\varrho}\right). \tag{3}$$

In these equations, T stands for the absolute temperature and $\tilde{\mu}$ is the difference between the chemical potentials (per unit mass) for the two species. The dissipative coefficients describe mass diffusion (α_z, $\alpha_\perp$), heat diffusion ($\varkappa_z$, $\varkappa_\perp$), permeation (ζ), viscous dissipation (η_{ijkl}: five independent viscosities, explicitly given in [9] and cross-processes: thermodiffusion (β_z, $\beta_\perp$), "flexodiffusion" (Γ; specific to the two-component smectic A case) and "thermoflexion" (ξ; already present in the one-component smectic A case).

The (linearized) hydrodynamic equations may now be written out:

$$\partial_t \varrho = - \nabla \cdot \mathbf{g}$$

$$\partial_t g_i = - \partial_i p + \delta_{iz} \nabla \cdot \Phi + \eta_{ijkl}\partial_{jk}\left(\frac{g_l}{\varrho}\right)$$

$$\varrho T \partial_t s = \varkappa_\perp \nabla_\perp^2 T + \varkappa_z \partial_{zz} T + T(\beta_\perp \nabla_\perp^2 \tilde{\mu} + \beta_z \partial_{zz}\tilde{\mu})$$

$$+ \xi \partial_z \nabla \cdot \Phi$$

$$\varrho \partial_t c = \alpha_\perp \nabla_\perp^2 \tilde{\mu} + \alpha_z \partial_{zz}\tilde{\mu} + \beta_\perp \nabla_\perp^2 T + \beta_z \partial_{zz} T + \Gamma \partial_z \nabla \cdot \Phi$$

$$\partial_t u = \frac{g_z}{\varrho} + \frac{\xi}{T}\partial_z T + \Gamma \partial_z \tilde{\mu} + \zeta \nabla \cdot \Phi, \tag{4}$$

where we have used, instead of the energy density ε, the entropy per unit mass s. Pressure p, temperature T, chemical potential difference $\tilde{\mu}$ and stress vector Φ are still to be expressed in terms of mass density ϱ, entropy s, mass fraction c and strain u, with the help of the thermodynamic relation (up to second order):

$$
\delta\left(\frac{\varepsilon}{\varrho}\right) = T\,\delta s + \frac{p}{\varrho^2}\,\delta\varrho + \tilde{\mu}\,\delta c + \frac{1}{2}\left\{\frac{B}{\varrho}\,(\partial_z u)^2\right.
$$

$$
+ \frac{K}{\varrho}\,(\nabla_\perp^2 u)^2 + \frac{\varrho T}{c_v}\,\delta s^2 + \frac{\partial}{\partial\varrho}\left(\frac{p}{\varrho^2}\right)\delta\varrho^2
$$

$$
\left.+ \frac{\partial\tilde{\mu}}{\partial c}\,\delta c^2\right\} + \frac{C_s}{\varrho}\,\partial_z u\,\delta s + \frac{C_\varrho}{\varrho}\,\partial_z u\,\delta\varrho
$$

$$
+ \frac{C_c}{\varrho}\,\partial_z u\,\delta c + \frac{\partial T}{\partial\varrho}\,\delta s\,\delta\varrho + \frac{\partial T}{\partial c}\,\delta s\,\delta c
$$

$$
+ \frac{1}{\varrho^2}\frac{\partial p}{\partial c}\,\delta\varrho\,\delta c. \tag{5}
$$

The notations are: layer compressibility modulus at constant entropy, mass density and composition: B; layer curvature modulus: K; heat capacity at constant density and composition, per unit volume of an unstrained smectic: c_v; crossed coefficients: C_s, C_ϱ, and C_c (this last one specific to the two-component smectic A case). With the definitions:

$$
p = \varrho^2\,\frac{\partial\left(\frac{\varepsilon}{\varrho}\right)}{\partial\varrho},\quad T = \frac{1}{\varrho}\frac{\partial\varepsilon}{\partial s},\quad \tilde{\mu} = \frac{1}{\varrho}\frac{\partial\varepsilon}{\partial c},
$$

$$
\text{and } \partial_i\Phi_i = -\frac{\delta\varepsilon}{\delta u} \tag{6}
$$

we have for small departures from equilibrium:

$$
\delta p = \varrho C_\varrho \partial_z u + \frac{\partial p}{\partial c}\,\delta c + \varrho^2\frac{\partial\left(\frac{p}{\varrho^2}\right)}{\partial\varrho}\,\delta\varrho + \varrho^2\frac{\partial T}{\partial\varrho}\,\delta s
$$

$$
\delta\tilde{\mu} = \frac{C_c}{\varrho}\,\partial_z u + \frac{\partial\tilde{\mu}}{\partial c}\,\delta c + \frac{1}{\varrho^2}\frac{\partial p}{\partial c}\,\delta\varrho + \frac{\partial T}{\partial c}\,\delta s
$$

$$
\delta T = \frac{C_s}{\varrho}\,\partial_z u + \frac{\partial T}{\partial c}\,\delta c + \frac{\partial T}{\partial\varrho}\,\delta\varrho + \frac{\varrho T}{c_v}\,\delta s
$$

$$
\Phi_z = B\partial_z u + C_c\delta c + C_\varrho\delta\varrho + C_s\delta s
$$

$$
\Phi_i = -K\partial_i\nabla_\perp^2 u \quad (i = x, y). \tag{7}
$$

The general solution of the complete set of hydrodynamic equations will not be attempted here. We shall rather restrict our analysis to the case where the compressibility and the heat capacity are small: then $\delta\varrho \cong 0$ and $\delta s \cong 0$. Within the scope of this approximation pressure and temperature are not independent hydrodynamic variables. They are, on the contrary, constrained by the hydrodynamic equations themselves. To proceed further, we resort to Fourier transform in space. With coordinate axes such that $q_y = 0$ we get the constraints

$$
\delta p = \frac{q_z}{q^2}\,\mathbf{q}\cdot\boldsymbol{\Phi} + i\,\frac{q_x q_z}{q^2}\,\frac{\tilde{\eta}_l(\mathbf{q})}{\varrho}\,g_t
$$

$$
\delta T = -T\,\frac{\beta_\perp q_x^2 + \beta_z q_z^2}{\varkappa_\perp q_x^2 + \varkappa_z q_x^2}\,\delta\tilde{\mu} - \frac{\xi q_z}{\varkappa_\perp q_x^2 + \varkappa_z q_z^2}\,\mathbf{q}\cdot\boldsymbol{\Phi}
$$

$$
g_l = 0 \tag{8}
$$

and the dynamical equations:

$$
\partial_t g_y = -\frac{\tilde{\eta}_y(\mathbf{q})}{\varrho}\,q^2 g_y
$$

$$
\partial_t g_t = -\frac{\tilde{\eta}_t(\mathbf{q})}{\varrho}\,q^2 g_t - iq_x\,\mathbf{q}\cdot\boldsymbol{\Phi}
$$

$$
\varrho\partial_t c = -\tilde{\alpha}(\mathbf{q})q^2\delta\tilde{\mu} - \tilde{\Gamma}(\mathbf{q})q_z\,\mathbf{q}\cdot\boldsymbol{\Phi}
$$

$$
\partial_t u = -\frac{g_t q_x}{\varrho q^2} + i\tilde{\Gamma}(\mathbf{q})q_z\delta\tilde{\mu} + i\tilde{\zeta}\,\mathbf{q}\cdot\boldsymbol{\Phi} \tag{9}
$$

where we introduced the new variables:

$$
g_l = q_x g_x + q_z g_z
$$
$$
g_t = q_z g_x - q_x g_z \tag{10}
$$

and effective dissipative coefficients:

$$
\tilde{\eta}_t(\mathbf{q}) = \frac{\eta_3 q_z^4 + (\eta_1 + \eta_2 + \eta_4 - 2\eta_3 - 2\eta_5)q_x^2 q_z^2 + \eta_3 q_x^4}{q^4}
$$

$$
\tilde{\eta}_l(\mathbf{q}) = \frac{(2\eta_3 + \eta_5 - \eta_1)q_z^2 + (\eta_2 + \eta_4 - \eta_5 - 2\eta_3)q_x^2}{q^2}
$$

$$
\tilde{\eta}_y(\mathbf{q}) = \frac{\eta_3 q_z^2 + \eta_2 q_x^2}{q^2}
$$

$$
\tilde{\alpha}(\mathbf{q}) = \frac{\left(\alpha_\perp q_x^2 + \alpha_z q_z^2 - T\dfrac{(\beta_\perp q_x^2 + \beta_z q_z^2)^2}{\varkappa_\perp q_x^2 + \varkappa_z q_z^2}\right)}{q^2} \tag{11}
$$

$$\tilde{\Gamma}(\mathbf{q}) = \Gamma - \frac{\beta_\perp q_x^2 + \beta_z q_z^2}{\varkappa_\perp q_x^2 + \varkappa_z q_z^2}\,\xi$$

$$\tilde{\zeta}(\mathbf{q}) = \zeta - \frac{\xi^2 q_z^2}{T(\varkappa_\perp q_x^2 + \varkappa_z q_z^2)}\,. \tag{11}$$

Among the four hydrodynamic modes that remain there is one uncoupled shear wave g_y with frequency $\omega = -i\tilde{\eta}_y(\mathbf{q})q^2/\varrho$ and three modes coupling g_t, δc and u: the second sound and the slip mode [10]. The characteristic frequencies of these modes are roots of the 3 × 3 determinant:

$$\left[-i\omega + \frac{\tilde{\eta}_t}{\varrho}\,q^2 \quad iC_c q_x q_z - (Bq_z^2 + Kq_x^4)q_x\right.$$

$$0 - i\omega + \frac{\tilde{\alpha}\dfrac{\partial\tilde{\mu}}{\partial c}q^2 + \tilde{\Gamma}C_c q_z^2}{\varrho} \quad i\frac{q_z}{\varrho^2}\,[\varrho\tilde{\Gamma}(Bq_z^2$$

$$+ Kq_x^4) + \tilde{\alpha}C_c q^2]$$

$$\frac{q_x}{\varrho q^2} - i\left(\tilde{\Gamma}\frac{\partial\tilde{\mu}}{\partial c} + \tilde{\zeta}C_c\right)q_z - i\omega$$

$$\left.+ \left(\frac{\tilde{\Gamma}C_c}{\varrho} + \tilde{\zeta}B\right)q_z^2 + \tilde{\zeta}Kq_z^4\right]. \tag{12}$$

For an *oblique* wavevector (i.e., $q_x q_z \neq 0$), if we assume that the dissipative processes of permeation and "flexodiffusion" are much slower than mass diffusion (we then set $\tilde{\Gamma} = 0$ and $\tilde{\zeta} = 0$), the roots are given, to lowest order in q, by
slip mode, diffusive:

$$\omega = -i\frac{\tilde{\alpha}(\mathbf{q})}{\varrho^2}\,q^2\left(\varrho\frac{\partial\tilde{\mu}}{\partial c} - \frac{C_c^2}{B}\right), \tag{13}$$

second sound mode, propagative:

$$\omega = \pm\sqrt{\frac{B}{\varrho}}\,\frac{|q_x q_z|}{q}\,. \tag{14}$$

Note that the characteristic frequency of the second sound is regardless of the coupling between smectic order and concentration. On the other hand, the slip mode is a property specific to the two-component smectic A phase.

If the wavevector is aligned along the layers ($q_z = 0$), the two propagative second sound modes degener-

ate into two diffusive modes, coupling g_t and u. Neglecting permeation again, and in the limit $K\varrho/\eta_3 \ll 1$, their relaxation frequencies are given by

$$\omega = -i\eta_3 q_x^2/\varrho \text{ (shear mode) and}$$

$$\omega = -iKq_x^2/\eta_3 \text{ (undulation mode).} \tag{15}$$

The dispersion relation of the slip mode then becomes

$$\omega = -i\frac{\tilde{\alpha}(q_x)}{\varrho}\,q_x^2\,\frac{\partial\tilde{\mu}}{\partial c}\,. \tag{16}$$

If eventually we take the wavevector perpendicular to the layers ($q_x = 0$) then all dissipative processes are a priori on the same footing. The second sound again degenerates into two diffusive modes. There is one uncoupled shear mode with a frequency given by $\omega = -i\eta_3 q_z^2/\varrho$ and two diffusive modes (coupling δc and u) with frequencies, scaling as q_z^2, involving all the reactive $\left(\dfrac{\partial\tilde{\mu}}{\partial c},\, B,\, C_c\right)$ and dissipative ($\tilde{\alpha}$, $\tilde{\zeta}$, $\tilde{\Gamma}$) coefficients.

The dispersion relations of all modes that involve δc or u may easily be studied experimentally by means of dynamic light scattering (for studies of the second sound or undulation mode in one-component smectics, see, for example, the review by Prost [11] and references therein; experiments on multi-component smectics in the limiting case $q_z = 0$ are reported by Chan and Pershan [12] and by Di Méglio et al. [2]). Since we are mainly interested in the membrane properties of the two-component smectic A system, undulation mode ($q_z = 0$) and slip mode (**q** oblique) deserve special attention. From the characteristic frequency of the undulation mode we get the layer bending modulus K, related to the membrane bending constant k_c by $K = k_c/d$, whereas from the slip mode we get the layer compressibily modulus at *constant chemical potential* $\bar{B} = B - \dfrac{C_c^2}{\varrho\dfrac{\partial\tilde{\mu}}{\partial c}}$, related to the membrane/membrane interactions, as we now argue.

We consider an incompressible two-component smectic A of repeat distance d prepared with a surfactant of mass m_S per molecule, occupying an area Σ. On geometrical grounds the following relation obtains: $\Sigma = \dfrac{2m_S}{\varrho dc}$. The surfactant membrane comes under tension for either a concentration fluctuation at constant repeat distance or a layer displacement at constant concentration since area per surfactant then varies. This strain costs an elastic energy per unit volume ε that we may tentatively evaluate as: $\varepsilon = \dfrac{U}{2}\left(\dfrac{\delta\Sigma}{\Sigma}\right)^2 \dfrac{\varrho c}{m_S}$, where U is a characteristic molecular energy and $\delta\Sigma$ the induced change of area per surfactant. With $\dfrac{\delta\Sigma}{\Sigma} = -\dfrac{\delta d}{d} - \dfrac{\delta c}{c} = -\partial_z u - \dfrac{\delta c}{c}$, our energy evaluation amounts to state: $B = U\varrho c/m_S$, $\dfrac{\partial\tilde{\mu}}{\partial c} = U/(c\, m_s)$, $C_c = U\varrho/m_S$, which in turn leads to : $\bar{B} = 0$. This unpleasant feature comes from the crudeness of our energy evaluation which does not take into account membrane/membrane interactions. This is easily improved with an energy per unit volume which is not zero even for fluctuations that keep constant the area per surfactant:

$$\varepsilon = \frac{U}{2}\left(\frac{\delta\Sigma}{\Sigma}\right)^2 \frac{\varrho c}{m_S} + \frac{1}{2}\,\bar{B}(\partial_z u)^2. \qquad (17)$$

For *weak* membrane/membrane interactions as compared to the membrane tension (i.e., $\bar{B} \ll U\varrho c/m_S$), the slip mode characteristic frequency becomes:

$$\omega = -i\,\frac{\tilde{\alpha}(\mathbf{q})}{\varrho^2}\,q^2 \left(\frac{\varrho\frac{\partial\tilde{\mu}}{\partial c}}{C_c}\right) \frac{\bar{B}}{1 + \left(\dfrac{\bar{B}\,\dfrac{\partial\tilde{\mu}}{\partial c}}{C_c^2}\right)} \approx$$
$$- i\,\frac{\tilde{\alpha}(\mathbf{q})}{\varrho^2 c^2}\,q^2\bar{B} \qquad (18)$$

which is the previously stated result.

In order to get a practical measurement of $\bar{B}$ from the slip mode study we may use the microscopic calculation of the dissipative coefficient by Brochard and de Gennes [10], with the simplifying assumption on the anisotropy:

$$\tilde{\alpha}(\mathbf{q}) = \frac{\tilde{\alpha}_\perp}{q^2}\,q_x^2 \qquad (19)$$

which gives: $\omega = -i\,\dfrac{d^2}{12\,\eta_s}\,q_x^2\bar{B}$, where η_s is the shear viscosity of the solvent.

2. Experimental results

We have carried out a dynamic light scattering study of lamellar samples derived from a unique basic system but located on three distinct dilution paths. The system consists of a concentrated lyotropic smectic (repeat distance about 40 Å) of water-pentanol-SDS (sodium dodecyl sulphate). It has been swollen by: 1) a mixture of dodecane and pentanol ("oil dilution"), 2) a 0.4 molar sodium chloride solution in water ("brine dilution"), and 3) pure water ("water dilution"). The limiting repeat distances that can be obtained are respectively 460 Å, 800 Å, and 80 Å. These three dilution paths have been previously investigated using a high resolution x-ray technique [4, 5]. However, owing to experimental limitations, x-ray experiments have been restricted to repeat distances smaller than 200 Å. Here, our light scattering method allowed us to get data through the whole dilution range.

The experiments have been done as follows: cylindrical glass capillaries with a rectangular section (about 100-μm-thick, 1-mm-wide and 35-mm-high) are filled by capillarity, then sealed. This procedure leads to unoriented samples. A good orientation is obtained by a thermal treatment: heating the sample through the lamellar-isotropic phase transition (which takes place at a temperature ranging of 30 – 100 °C), then cooling it very slowly (0.2 °C per min) leads often to an almost perfect homeotropic orientation, where the only defects to be seen are an oily streak-like texture very close to the edges of the capillary, and isolated, long edge-dislocations, parallel to the axis of the cell (owing to its varying thickness [13]). The geometry of the light scattering set-up is described schematically in Fig. 1: the sealed capillary can rotate around its vertical long axis (angle θ) which is kept perpendicular to the scattering plane, i.e., the normal to the layers is always within the $\mathbf{q}$-plane. The polarization of the incident light (ionized krypton laser operating at $\lambda = 647.1$ nm) is vertical. In such a configuration the only fluctua-

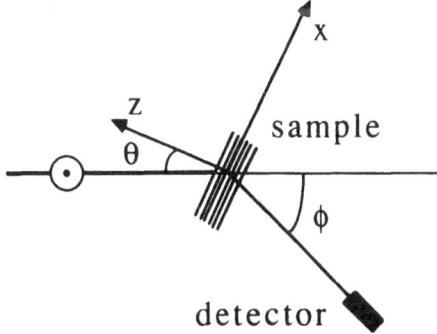

Fig. 1. Geometry of the experimental set-up: the normal to the layers z, the incident light, and the scattered beam are in the same plane. The incident light is polarized perpendicularly to this plane

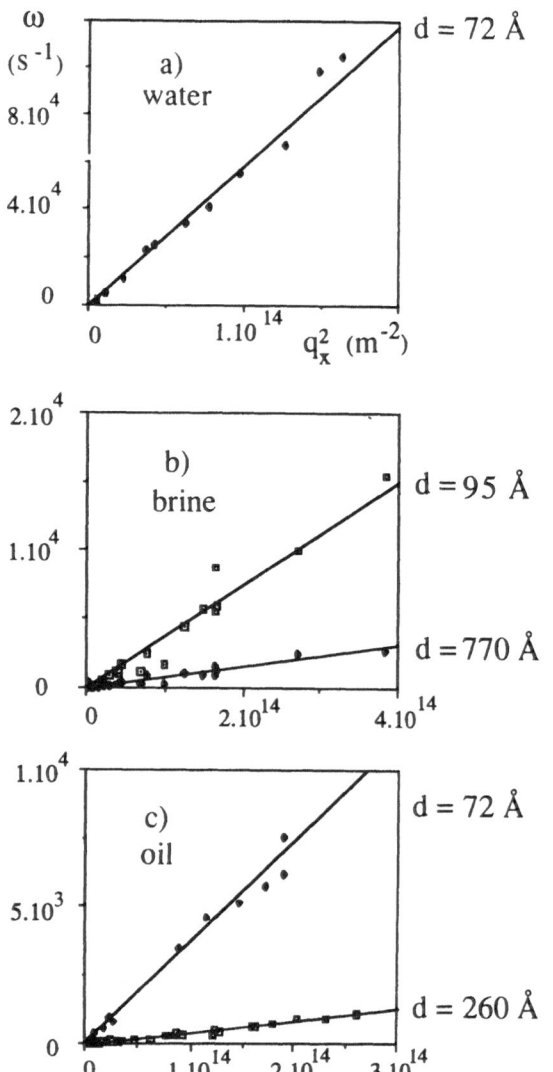

Fig. 2. Representative anisotropic dispersion relations for the three paths a) water dilution, b) brine dilution, c) oil dilution. The relaxation frequency ω is a linear function of the squared x-component of the wave vector: $\omega = Dq_x^2$ (d is the repeat distance of the multimembrane stacking)

stant modulus q of the wave vector) and θ increasing from $\phi/2$ to $\phi/2 + \pi/2$. This allows us to scan the entire range of accessible wave vectors $q_\perp$ for this geometry. We then proceed to other angles ϕ. Data are recorded at three to five different ϕ angles ranging from 15° to 160°. The signal is analyzed using a Brookhaven Instruments digital correlator (72 channels) which gives the time correlation function of the signal. Except for the limiting values of θ (close to $\phi/2$ or $\phi/2 + \pi/2$), the dynamic part of the signal is easily recovered from the static one: for θ approaching $\phi/2 + \pi/2 (q_x = 0)$ the specular reflection on the capillary gets close to the scattered beam and no measurements are possible; for θ approaching $\phi/2 (q_z = 0)$ the dynamic part of the signal decreases significantly and its characteristic frequency gets very large.

The observed signal is always a monotonously decreasing function of time. Its characteristic frequency ω is obtained by means of a first order cumulant expansion. Since the heterodyning conditions are not well-controlled (static scattering by remaining dislocations and by the walls of the cell), we also processed the data under the assumption of a mixture of heterodyne and homodyne components: the resulting relaxation frequencies were not affected by the fitting procedure, within the experimental accuracy. Figure 2 shows some $\omega(q_x^2)$ curves representative of the behavior observed for the three sample families: the dispersion relation is well described by the equation $\omega = D \cdot q_x^2$, as expected for the slip mode in the case of an oblique wave vector. As is apparent in Fig. 2, D decreases with increasing layer spacing d for the brine and oil dilutions.

Figure 3 shows for the three dilutions (water, brine, and oil) the evolution of D as a function of the interlayer distance d. Values of d smaller than 200 Å have been obtained from x-ray measurements [5]; larger ones are deduced from the composition of the samples, using the x-ray result that mean area per polar head is essentially constant upon dilution (i. e., the osmotic compressibility $\left(\dfrac{\partial \tilde{\mu}}{\partial c}\right)^{-1}$ is very small). The experimental data hints at a d^{-1} behavior for D and thus suggest that $\bar{B}$ is inversely proportional to d^3 for the dilution with oil or brine. This behavior may be understood with a proper description of the interactions between membranes: as proposed first by Helfrich [7], the steric hindrance to the thermally excited undulations of the membranes in a lamellar phase lead to a long range repulsive entropically driven interaction. If this interaction is the only *dominant* one (no long-range elec-

tions that scatter light are the concentration ones, bringing about polarized scattering. For that reason, in a normal run, we do not analyze the polarization of the scattered light though we checked that depolarized scattering is negligible. The wave vector **q** is defined by the angle ϕ between the detector and the laser beam. Its three components are: $q_x = q \cdot \cos(\theta - \phi/2)$, $q_y = 0$, and $q_z = q \cdot \sin(\theta - \phi/2)$, with $q = 4\pi n/\lambda \cdot \sin(\phi/2)$ [14]. A typical experiment is done with ϕ fixed (con-

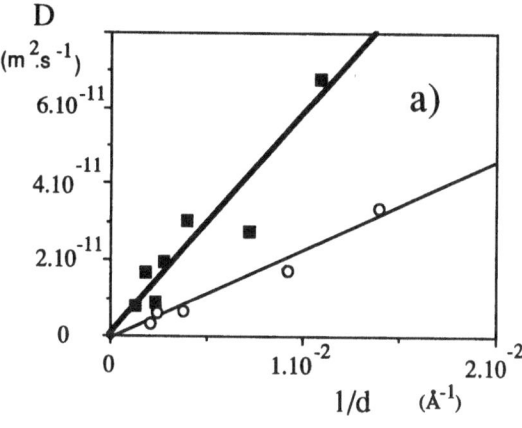

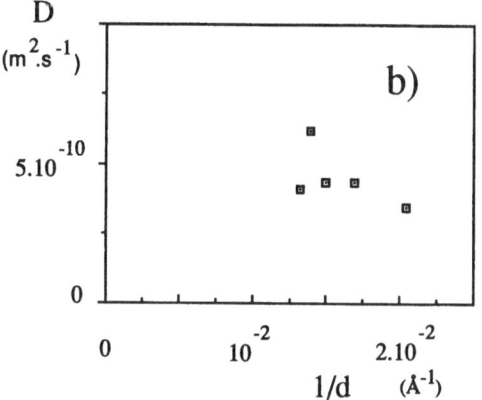

Fig. 3. Plot of the "diffusion" coefficient D as a function of the inverse d spacing; a) oil and brine dilutions. The linear behavior is the signature of the undulation interactions. The measure of the slopes leads to $k_c = 1.6\ k_B T$ (oil dilution, open circles) and $k_c = 0.8\ k_B T$ (brine dilution, filled squares); b) water dilution

trostatic interactions: uncharged membranes or high ionic strength), the layer compressibility modulus is

$$\bar{B}_{st} = \frac{9\pi^2 (k_B T)^2 d}{64\, k_c\, (d-\delta)^4} \qquad (20)$$

where k_c is the bending constant of the membrane and δ its thickness.

If there are *unscreened* electrostatic interactions, they always dominate at large distances; the layer compressibility modulus becomes [5]

$$\bar{B}_{elec} = \frac{\pi k_B T d}{2 L_b (d-\delta)^3} \qquad (21)$$

where L_b is the Bjerrum length of water (≈ 7 Å).

The experiments support this analysis: brine and oil dilutions, where no electrostatic interactions are

expected, have $\bar{B}$ linear in $1/d^3$ (the inequality $\delta \ll d$ is thoroughly fulfilled since $\delta \approx 20$ Å [5]); from the slopes we get $k_c \approx 1.6\ k_B T$ for the oil dilution and $k_c \approx 0.8\ k_B T$ for the brine one. The layer compressibility modulus $\bar{B}$ ranges from about 10^4 Pa (less dilute "oil" sample) down to 10 Pa (most dilute "brine" sample), typically 10^3- to 10^6-times smaller than the values encountered in usual thermotropic smectics, but comparable to the helix compressibility modulus of cholesterics. We shall not draw quantitative conclusions in the water dilution case, since our data are scarce and scattered; with d ranging from 50 to 80 Å, measured values for D are between $4 \cdot 10^{-10}$ and $7 \cdot 10^{-10}$ m$^2 \cdot$ s^{-1} (Fig. 3). Note, nevertheless, that the correct order of magnitude for $D(D \approx 8 \cdot 10^{-10}$ m$^2 \cdot$ s$^{-1})$ derives from the expression $\bar{B}_{elec}$.

We have demonstrated the feasibility of dynamic light scattering method for studying interactions between membranes in lyotropic dilute lamellar phases. Indeed, the relaxation frequency of the slip mode at oblique wave vectors is simply related to the layer compressibility modulus (at constant chemical potential) $\bar{B}$ i. e., to the intermembrane interactions. In contrast with previous x-ray measurements, this technique allows to scan the whole dilution range of the system investigated. In these series of experiments we were able to confirm and extend at far larger dilutions the x-ray results on less dilute samples, namely that undulation forces always dominate for oil and brine dilutions of the water-pentanol-SDS system. Moreover, electrostatic interactions dominate when charged membranes are swollen with pure water.

Acknowledgements

The authors acknowledge fruitful discussions with J.-M. Di Méglio and B. Pouligny.

References

1. Meunier J, Langevin D, Boccara N (eds) (1987) Springer-Verlag, Berlin
2. Di Méglio J-M, Dvolaitsky M, Taupin C (1985) J Phys Chem 89:871; Di Méglio J-M, Dvolaitsky M, Léger L, Taupin C (1985) Phys Rev Lett 54:1686
3. Larché F, Appell J, Porte G, Bassereau P, Marignan J (1986) Phys Rev Lett 56:1700
4. Safinya CR, Roux D, Smith GS, Sinha SK, Dimon P, Clark NA, Bellocq A-M (1986) Phys Rev Lett 57:2718
5. Roux D, Safinya CR (1988) J Phys France 49:307
6. Bassereau P, Marignan J, Porte G (1987) J Phys France 48:673
7. Helfrich W (1978) Z Naturforschung 33a:305
8. de Gennes P-G (1969) J Phys France Suppl 30:65
9. Martin PC, Parodi O, Pershan PS (1972) Phys Rev A6:2401

10. Brochard F, de Gennes P-G (1975) Pramana suppl 1:1
11. Prost J (1984) Adv Phys 33:1
12. Chan W, Pershan PS (1977) Phys Rev Lett 39:1368
13. Nallet F, Prost J (1987) Europhys Lett 4:307
14. The exact formula for q is complex in general, due to the optical anisotropy of the medium; if both incident and scattered waves have their polarization perpendicular to the optical axis (uniaxial medium), the formula given in the text obtains, where n should be taken as the ordinary index of refraction n_o. Since the birefringence is small and the samples dilute, a very good approximation is: $n_o \approx n_e \approx n_s$, with n_s the index of refraction of the solvent (oil: 1.42; brine: 1.34; water: 1.33).

Received October, 1988;
accepted January, 1989

Authors' address:

F. Nallet
CRPP CNRS
Domaine universitaire
F-33405 Talence Cedex, France

Progress in Colloid & Polymer Science

Progr Colloid Polym Sci 79:321–326 (1989)

Interfacial activity of 1-(2'-hydroxy-5'-methylphenyl)-octane-1-one oxime and the interfacial mechanism of copper extraction

K. Prochaska and J. Szymanowski

Institute of Chemical Technology and Engineering, Poznań Technical University, Poznań, Poland

Abstract: Interfacial tension isotherms were determined for 1-(2'-hydroxy-5'-methyl-phenyl)-octane-1-one oxime in toluene/water and octane/water systems and used to predict extraction orders against hydroxyoxime for various versions of extraction mechanism. It was demonstrated that the formation of the stable 2 : 1 complex from the intermediate 1 : 1 complex and the hydroxyoxime molecule present in the aqueous layer near the interface is the slowest step. The interfacial behavior of 1-(2'-hydroxy-5'-methylphenyl)-octane-1-one oxime is similar to those of more hydrophobic oximes, e. g., 2-hydroxy-5-alkylbenzophenone oximes.

Key words: Interfacial tension, copper extraction, mechanism, kinetics.

Introduction

Hydroxyoximes are well established copper extractants and they are used in several industrial installations. Up to now about 15 % of world copper is produced using these extractants and reprocessing mainly solid wastes and low-grade resources. Extractants of Lix and Acorga series are used. They contain 2-hydroxy-5-alkylbenzophenone oximes (I) and/or 2-hydroxy-5-alkylbenzaldehyde oximes (II) as active components (Fig. 1). The application of 2-hydroxy-5-alkylacetophenone oxime (III) was also described and appropriate extractants are produced by Henkel as Lix 84 and Lix 984. These extractants contain a long alkyl group and a branched nonyl or dodecyl group attached to the aromatic ring at the opposite side (in comparison to the chemically and surface active phenolic and oximino groups). Such compounds adsorb at organic diluent/water interfaces and decrease the interfacial tension. It is generally accepted that the reaction between copper ions and these hydroxyoximes proceeds at the interface by the interfacial mechanism [1]. Recently, it was also demonstrated that the interfacial tension data can be used to select a probable version of the interfacial mechanism [2–7].

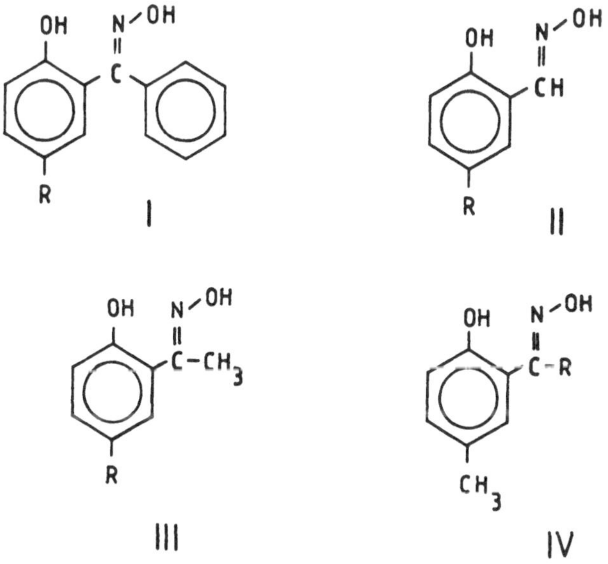

Hydroxyoximes having a short alkyl group

Fig. 1. Hydroxyoxime extractants

attached to the aromatic ring and a long alkyl group present in the neighborhood of the oximino group (IV) were also prepared and used for copper extraction [7, 8]. Such compounds can be easily synthesized using p-cresol and appropriate acyl chloride in the presence of a Lewis acid, e. g., AlCl$_3$, according to the following reaction scheme:

$$
\text{(1)}
$$

Individual oximes are obtained which are identified as *E* (anti) isomers. The *Z* (syn) isomers are not identified in this case.

The aim of this work is to study the interfacial activity of pure 1-(2′-hydroxy-5′-methylphenyl)-octane-1-one oxime, as being a model for hydroxyoximes of group IV and to discuss the relationships between the interfacial activity and the kinetics and mechanism of copper extraction.

Experimental

The interfacial tension was measured by the drop volume method at 20 °C by means of semiautomatic equipment. Drops of the aqueous phase were formed into the organic phase. The time of surface aging was 5–20 min. Redistilled water was used as a water phase. Octane and toluene were used as model organic diluents. Before use they were redistilled over sodium using a Vigreux distillation column. Mutually presaturated phases were used. Equal volumes (25 cm^3) of both phases were automatically shaken at room temperature of 6 h and then left for 24 h for full separation. They were separated directly before measurements and layers near the interface of 1 cm depth each were disregarded.

Results and discussion

The interfacial tension isotherms are given in Figs. 2 and 3. They can be quite well matched by different adsorption isotherms and empirical equations such as the Szyszkowski, Frumkin, and Temkin isotherms, the polynomial of the third order and the spline function (Table 1). The use of the spline function to approximate the interfacial tension isotherm and to compute the surface excess isotherm was demonstrated previously [9]. The errors of models are low and simi-

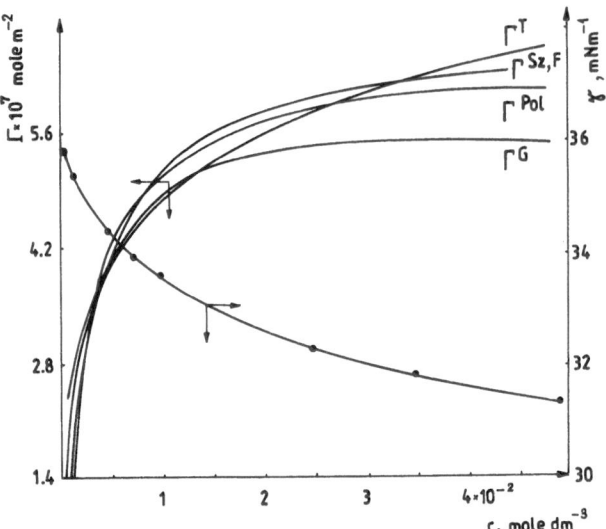

Fig. 2. Interfacial tension and surface excess isotherms for toluene/water systems (*G, Sz, F, T,* and *P* denote the Gibbs, Szyszkowski, Frumkin, Temkin and polynomial isotherms, respectively)

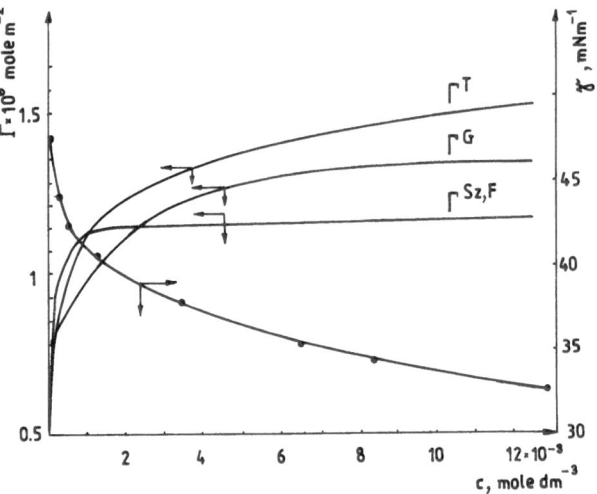

Fig. 3. Interfacial tension and surface excess isotherms for octane/water system (*G, Sz, F,* and *T* denote the Gibbs, Szyszkowski, Frumkin and Temkin isotherms, respectively)

lar for all the isotherms and equations considered but much lower for the system containing toluene. The meanings of the regression coefficients are as follows:

the Szyszkowski isotherm,

$$\Delta y = y_o a_1 \ln (c/a_2 + 1) \tag{2}$$

the Frumkin isotherm,

$$\Delta y = y_o \{ a_1 \ln (c/a_2 + 1) - a_3 [c/a(a_2 + 1)]^2 \} \tag{3}$$

Table 1. Regression coefficients (a_i) and errors of model (e) for matching of different isotherms to the interfacial tension isotherm

Parameter	Diluent	Szyszkowski	Frumkin	Temkin	Polynomial
a_1	toluene	4.624 10^{-2}	4.624 10^{-2}	8.797 10^4	1.847 10^{-5}
a_2		3.304 10^{-3}	3.304 10^{-3}	3.748	1.296 10^{-4}
a_3		—	1.090 10^{-4}	—	− 1.231 10^{-3}
a_4		—	—	—	2.700 10^{-2}
$e^{a)}$		5.36 10^{-9}	5.35 10^{-9}	8.83 10^{-9}	6.49 10^{-9}
a_1	octane	5.656 10^{-2}	5.656 10^{-2}	7.528 10^4	− 1.576 10^{-4}
a_2		3.057 10^{-5}	3.057 10^{-2}	6.017	− 3.342 10^{-3}
a_3		—	5.351 10^{-5}	—	− 2.559 10^{-2}
a_4		—	—	—	− 2.900 10^{-2}
$e^{a)}$		3.76 10^{-7}	3.76 10^{-7}	3.73 10^{-7}	1.24 10^{-7}

[a]) for spline function the errors of model are equal to 1.33 10^{-8} and 2.14 10^{-7} for systems containing toluene and octane, respectively.

the Temkin isotherm,

$$\log c + a_2 = a_1/(RT)\, \Delta y^{0.5} \tag{4}$$

the polynomial

$$y = a_1(\ln c)^3 + a_2(\ln c)^2 + a_3 \ln c + a_4 \tag{5}$$

where y and c denote the interfacial tension and oxime concentration, respectively, and $\Delta y = y_o - y$, where y_o is the interfacial tension for $c = 0$.

The coefficient a_3 in the Frumkin isotherm is two orders lower than the coefficient a_2 as a result of weak interactions between adsorbed neutral molecules of hydroxyime.

The character of the obtained relationships surface excess vs. concentration calculated according to various adsorption isotherms is somewhat different (Figs. 2 and 3). Deviations observed in the region of low hydroxyoxime concentrations are relatively small. Important deviations, only observed in the system containing octane, as polynomial system containing octane, as polynomial was used to compute the surface excess. As a result, this equation was further neglected for this system.

For each case considered the surface excess isotherm increases monotonously up to the constant or almost constant value characteristic for the saturated interface. It means that the activity coefficients can be neglected for the considered concentration region [10]. Thus, this suggests that 1-(2′-hydroxy-5′-methyl-phenyl)-octane-1-one oxime exists in the monomeric form both in octane and toluene and that the formation of dimers can be neglected in the considered concentration region.

The maximal values of the surface excess calculated according to the various adsorption isotherms, characteristic for the saturated interface, are quite similar (Table 2). The minimal molecular areas are approximately 8.1 10^5 and 17.4 10^5 m²mole⁻¹ at toluene/water and octane/water interfaces, respectively, and support the adsorption of the monomeric form at the interface.

As in our previous papers [3–8] the obtained relations were further used to discuss the kinetics and mechanism of copper extraction. The following two mechanism versions were considered:

Scheme I

$$Cu_w^{2+} + HB_{ad} = CuB_{ad}^+ + H_w^+ \tag{6}$$

$$CuB_{ad}^+ + HB_{int} = CuB_{2\,ad} + H_w^+ \tag{7}$$

$$CuB_{2\,ad} + 2HB_o = CuB_{2\,o} + HB_{ad} + HB_{int} \tag{8}$$

Table 2. Surface excess and molecular area at the saturated interface

Isotherm	Γ_{max} mole m⁻² 10^6		A_{min} m² mole⁻¹ 10^{-5}	
	octane	toluene	octane	toluene
Szyszkowski	1.1	0.6	9.1	16.7
Frumkin	1.1	0.6	9.1	16.7
Temkin	1.5	0.6	6.7	16.7
Spline	1.3	0.5	7.7	20.0
Polynomial	—	0.6	—	16.7

Scheme II

$$Cu_w^{2+} + HB_{ad} = CuB_{ad}^+ + H_w^+ \qquad (9)$$

$$CuB_{ad}^+ + HB_{ad} = CuB_{2\,ad} + H_w^+ \qquad (10)$$

$$CuB_{2\,ad} + 2HB_o = CuB_{2\,o} + 2HB_{ad} \qquad (11)$$

where: HB stands for hydroxyoxime, subscripts w and o denote a water phase and an organic phase, respectively; subscripts ad and int denote the molecules in the interfacial monolayer and at the border of the sublayer from which the hydroxyoxime is transferred into the monolayer without diffusion. As diffusion is neglected and the oxime interfacial concentration is assumed as equal to the surface excess Γ then the following kinetic equations can be obtained: $r_{6,9} = k\,\Gamma$, $r_7 = k\,[\text{HB}]_o\,\Gamma$, $r_8 = k\,[\text{HB}]_o^3\,\Gamma$, $r_{10} = k\,\Gamma^2$, and $r_{10} = k\,[\text{HB}]_o^2\,\Gamma^2$ for reactions 6 and 9, 7, 8, 10 and 11 assumed as the limiting steps, respectively.

Exemplary relations r vs. c computed according to the Szyszkowski isotherm are given in Figs. 4 and 5. Their character is quite the same as the toluene/water and octane/water systems are compared. Relations computed according to the other adsorption iso-

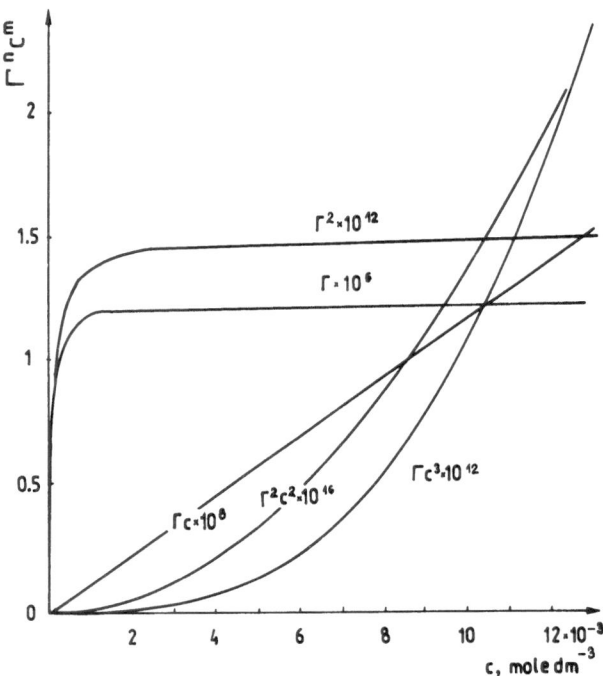

Fig. 5. Predicted effects of hydroxyoxime concentration upon extraction rates in octane/water system (the Szyszkowski isotherm).

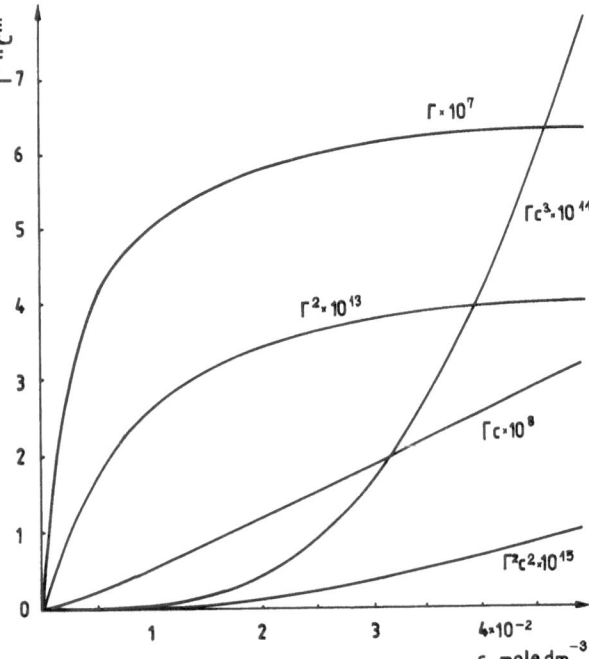

Fig. 4. Predicted effects of hydroxyoxime concentration upon extraction rates in toluene/water system (the Szyszkowski isotherm).

therms have a similar character, although the $\Gamma^m\,c^n$ values are different.

Relations Γ vs. $[\text{HB}]_o$ and Γ^2 vs. $[\text{HB}]_o$ are always convex and quickly appropriate asymptotes are achieved which are approximately parallel to the ordinate axis. It means a decrease of the extraction order to zero (Fig. 6). Thus reactions 6, 9, and 10 as the limiting steps can be disregarded. Moreover, almost all kinetic data obtained independently by different authors shows that the rate of copper extraction is proportional to the reciprocal of the hydrogen ion concentration [11]. For reactions 6 and 9 as the limiting steps, the extraction rate could depend upon the acidity of the aqueous phase, i.e., $r_{6,9} = k_{6,9}\,[\text{Cu}^{2+}]_w\,[\text{HB}]_{ad}$.

Relation $\Gamma\,[\text{HB}]_o$ vs. $[\text{HB}]_o$ is almost linear. Thus, such behavior is in an agreement with those kinetic data in which reaction order near 1 was demonstrated; such orders were usually reported [11]. Relations $\Gamma\,[\text{HB}]_o^3$ vs. $[\text{HB}]_o$ and $\Gamma^2\,[\text{HB}]_o^2$ vs. $[\text{HB}]_o$ are very concave, and $\Gamma\,[\text{HB}]_o^3$ and $\Gamma^2\,[\text{HB}]_o^2$ values sharply increase as the oxime concentration increases. It means the high reaction orders in the range of 2–4 (Fig. 6), what is in full contradiction with the all experimental kinetic data. Thus Eqs. (8) and (11) as the limiting steps can be disregarded. Quite similar results were obtained for the octane/water system and for other

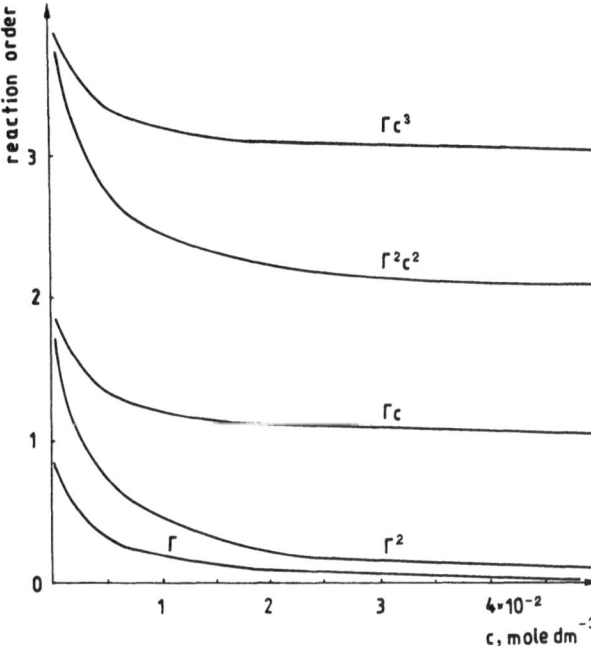

Fig. 6. Predicted reaction orders against hydroxyoxime for toluene/water system (the Szyszkowski isotherm, various reactions as the limiting step)

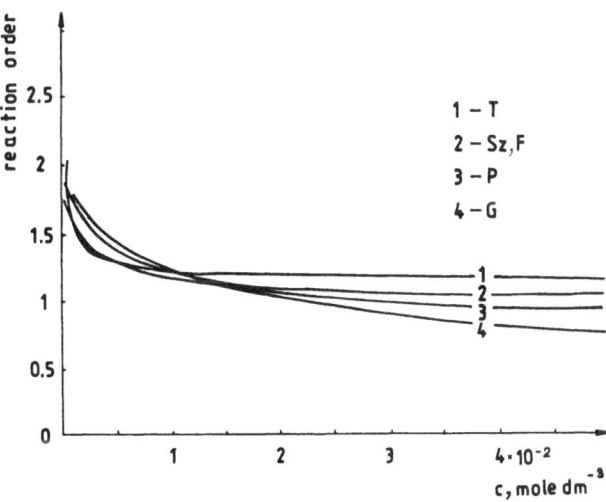

Fig. 7. Predicted reaction orders against hydroxyoxime for toluene/water system (reaction 7 as the slowest step, *G*, *Sz*, *F*, *T*, and *P* denote the Gibbs, Szyszkowski, Frumkin, Temkin, and polynomial isotherms, respectively)

adsorption isotherms considered, although the reaction orders were somewhat different (Fig. 7).

The obtained reaction orders significantly support the scheme I with reaction 7 as the slowest step. Only in this case, both in toluene and octane, are the predict-

ed reaction orders against hydroxyoxime near 1 for each adsorption isotherm considered in a quite large range of hydroxyoxime concentration. Each isotherm gives a somewhat different relation between the reaction order and hydroxyoxime concentration but the general character is similar. The effect of hydroxyoxime concentration upon reaction order is important only for very low bulk concentrations, where the reaction order significantly increases as hydroxyoxime concentration decreases. Thus, the predicted reaction orders are in a quite good agreement with those values determined experimentally [11].

Conclusions

The obtained results demonstrate usefulness of the interfacial tension measurements to discuss the kinetics and the interfacial mechanism of copper extraction by hydroxyoximes. This approach gives an additional data supporting the interfacial mechanism of copper extraction and demonstrates that the extraction of copper by hydroxyoximes, as in our case of 1-(2'-hydroxy-5'-methylphenyl)-octane-1-one oxime, proceeds according to the proposed mechanism in which the formation of the stable 2 : 1 complex from the intermediate 1 : 1 complex and the hydroxyoxime molecule present in the aqueous layers near the interface is the slowest step. Thus, the behavior of 1-(2'-hydroxy-5'-methylphenyl)-octane-1-one oxime is similar to that one reported previously by us for more hydrophobic 1-2'-hydroxy-5'-methylphenyl)-dodecane-1-one oxime [7] and for very hydrophobic 2-hydroxy-5-alkyl-benzophenone oximes [3–6].

Acknowledgement

This work was supported by Polish Research Program CPBP No. 03.08.

References

1. Danesi PR, Chiarizia R (1980) Critical Reviews in Analytical Chemistry 10:1–126
2. Szymanowski J, Stępniak-Biniakiewicz D, Prochaska K, Rashid ZA (1986) Preprints ISEC '86 2:35–48
3. Szymanowski J, Prochaska K, Bogacki MB (1987) J Colloid Interface Sci 117:293–295
4. Szymanowski J, Prochaska K (1987) J Chem Tech Biotechnol 40:177–193
5. Szymanowski J, Prochaska K (1988) J Colloid Interface Sci 123:456–465
6. Szymanowski J, Prochaska K (1988) ISEC '88 Conference Papers 2:24–27

7. Szymanowski J, Prochaska K (1988) Progr Colloid Polym Sci 76:260-264
8. Szymanowski J, Atamańczuk B (1982) Hydrometallurgy 9:29-36
9. Bogacki MB, Szymanowski J, Prochaska K (1988) Anal Chim Acta 206:215-221
10. Popov AN (1987) In: Kazarinov VE (ed) The Interface Structure and Electrochemical Processes at the Boundary Between Two Immiscible Liquids. Springer, Berlin, pp 179-205
11. Szymanowski J (1984) Wiad Chem 38:371-399

Received October, 1988;
accepted March, 1989

Authors' address:

Prof. Jan Szymanowski
Poznań Technical University
Pl. Skłodowskiej-Curie 2
PL-60-965 Poznań, Poland

Progress in Colloid & Polymer Science   Progr Colloid Polym Sci 79:327–331 (1989)

Polynomial approximation of interfacial tension isotherms and its use for kinetic data interpretation of metal extraction

K. Prochaska, K. Alejski, and J. Szymanowski

Institute of Chemical Technology and Engineering, Poznań Technical University, Poznań, Poland

Abstract: The physico-chemical model for extraction of metals by chelating and cationic extractants is discussed. The interfacial tension data are used to simulate relationships between the reaction rate of extraction for various versions of its mechanism and the extractant bulk concentration. The program for microcomputer IBM PC is presented which can be used for routine analyses of various interfacial tension isotherms and different extraction processes. Up to now it was used with success for copper extraction by hydroxyoximes and palladium (II) extraction by dialkyl sulphide in the presence of trialkylamine.

Key words: Interfacial tension, metal extraction, kinetics.

Introduction

Solvent extraction is one of the important methods used to purify, concentrate, and separate various metals from different aqueous solutions. The greatest development of solvent extraction is connected with the manufacture of copper. Several industrial installations of a large capacity of 30 000–100 000 tones/year cathodic copper are operated and they now produce about 15 % of the world's copper. Hydroxyoximes of Lix and Acorga type are used as commercial extractants [1, 2]. Due to the presence of a large alkyl group, a nonyl or dodecyl group, these compounds exhibit high hydrophobicity. Their solubility in the aqueous phase is of the order of few ppm. They adsorb at water/hydrocarbon interfaces and decrease efficiently the interfacial tension. The same phenomena are observed in systems containing other types of extractants. Due to this, it is now rather generally accepted that the extraction of metals by hydrophobic extractants proceeds at the interface and the adsorption of extractant molecules at the interface is the first step in this complex process [3].

The complexity of the process is connected with the association of extractant molecules in hydrocarbon solutions. Chelating agents and carboxylic acids usually form dimers, and sulphonic acids and amine derivatives can form adverse micelles. Very often mixtures of two different extractants are used including appropriate solubility and/or extraction strength modifiers. Usually in such a case various compounds present in the organic phase compete for their place in the adsorption layer at the interface and different intermediate complexes are formed in the aqueous phase, at the adsorption layer, and/or in the bulk of the organic phase.

In the case of interfacial processes the interfacial concentration and orientation of extractant molecules at the interface become very important parameters, and they can significantly effect the extraction rate, as has been recently proven [4–9]. It was also demonstrated that the interfacial tension data can be for interpretation of kinetic data and for the discussion of extraction mechanism.

The aim of this work is to elaborate an appropriate method for the analysis of the interfacial tension data, including the approximation of the interfacial tension isotherms by appropriate polynomials, estimation of extractant interfacial concentrations for various bulk concentrations, and computing from these data the reaction order against extractant for various versions of the extraction mechanism. As a result the most probable version of the mechanism is selected for which an agreement between the computed reaction order and that determined experimentally is observed. An appropriate program for microcomputer IBM PC XT was written which can be further used for routine analyses of various interfacial tension isotherms and different processes of solvent extraction.

Physico-chemical model of extraction process

Extraction of metals by various chelating and cationic extractants can be described by the simplified equation

$$M^{n+} + n\overline{BH} = \overline{MB}_n + nH^+ \tag{1}$$

where bar denotes an organic phase and the lack of bar denotes a water phase.

In an organic phase appropriate associates, most often dimers, are formed. Thus,

$$2\overline{BH} = (\overline{BH})_2, \quad K_d = \frac{[(\overline{BH})_2]}{[\overline{BH}]^2}. \tag{2}$$

The total extractant concentration is given by

$$[BH]_t = [\overline{BH}] + 2[(\overline{BH})_2]. \tag{3}$$

The extractant concentration in the aqueous phase, $[BH]$, can be neglected as very small in comparison to the bulk concentration in the organic phase.

Monomers adsorb easily at the interface and it is assumed that such an equilibrium is established:

$$\overline{BH} = BH_{ad}, \tag{4}$$

where subscript ad denotes the adsorption layer.

The interfacial concentration, $[BH]_{ad}$, is the sum of the surface excess, Γ, and the bulk concentration recalculated into $mol\,m^{-2}$, $[\overline{BH}_s]$.

$$[BH_{ad}] = \Gamma + [\overline{BH}_s]. \tag{5}$$

For systems and concentrations considered, $\Gamma \gg [\overline{BH}_s]$.

If the complex MB_n is not surface active and does not adsorb at the interface than one can assume that the interfacial concentration of extractant in an extraction system is similar to that observed in systems which do not contain metals, i.e., in systems containing only extractant.

The surface excess can be calculated from the Gibbs isotherm:

$$\Gamma = -\frac{dy}{RT\,d(\ln[\overline{BH}])}. \tag{6}$$

The term $dy/d(\ln[\overline{BH}])$ is calculated from the interfacial tension isotherm which can be approximated by the polynomial of the following type:

$$y = a_1(\ln[\overline{BH}])^3 + a_2(\ln[\overline{BH}])^2 + a_3\ln[\overline{BH}] + a_4. \tag{7}$$

In such a case the surface excess is given by

$$\Gamma = -1/RT\{3a_1(\ln[\overline{BH}])^2 + 2a_2(\ln[\overline{BH}]) + a_3\}. \tag{8}$$

If diffusion can be neglected and reaction steps can be only considered then the extraction of divalent metals by hydorphobic extractants of chelate or cationic type can be described by the following reaction scheme:

$$M^{2+} = BH_{ad} = MB_{ad}^+ + H^+ \tag{9}$$

$$MB_{ad}^+ + BH_{ad} = MB_{2\,ad} + H^+ \tag{10}$$

or

$$MB_{ad}^+ + BH_{int} = MB_{2\,ad} + H^+ \tag{11}$$

$$MB_{2\,ad} + 2\overline{BH} = \overline{MB}_2 + BH_{ad} \tag{12}$$

or

$$MB_{2\,ad} + 2\overline{BH} = \overline{MB}_2 + BH_{ad} + BH_{int} \tag{13}$$

where subscript int denotes the molecules at the border of sublayer from which the extractant is transferred into the monolayer without diffusion. For simplification rapid reactions of dehydration were omitted. Such a simplification is justified for copper extraction. However, in the case of nickel extraction the dehydration is a slow process and must be considered.

As assuming that one of the considered steps is the slowest one, appropriate kinetic equations can be derived:

$$r = ka[M^{2+}][\overline{BH}]^n[BH_{ad}]^m[H^+]^1, \tag{14}$$

where n, m and l are integers, and a denotes the interfacial surface area. The values of n, m and l depend upon the limiting step considered.

By introducing Eq. (8) into Eq. (14) and assuming constant concentrations of metal and hydrogen ions an appropriate relation correlating the reaction rate with the extractant bulk concentration is obtained:

$$r = k' \Gamma^m [\overline{BH}]^n . \tag{15}$$

This relation can be simulated by microcomputer and the reaction order can be computed.

By comparison of this data with that determined experimentally the most probable mechanism of extraction can be selected.

Computer program

Program MINEX is prepared in Turbo pascal for microcomputer IBM PC/XT and is used for analysis of the interfacial tension isotherms. The flow diagram is shown in Fig. 1.

The starting data are approximated by a 3rd degree logartithmic polynomial (Eq. (7)). The least squares method is used to calculate the regression coefficients. The following linear set of equations is solved:

$$a_1 \sum_{i=1}^{N} c_i^6 + a_2 \sum_{i=1}^{N} c_i^5 + a_3 \sum_{i=1}^{N} c_i^4 + a_4 \sum_{i=1}^{N} c_i^3 = \sum_{i=1}^{N} \gamma_i c_i^3 ,$$

$$a_1 \sum_{i=1}^{N} c_i^5 + a_2 \sum_{i=1}^{N} c_i^4 + a_3 \sum_{i=1}^{N} c_i^3 + a_4 \sum_{i=1}^{N} c_i^2 = \sum_{i=1}^{N} \gamma_i c_i^2 ,$$

$$\tag{16}$$

$$a_1 \sum_{i=1}^{N} c_i^4 + a_2 \sum_{i=1}^{N} c_i^3 + a_3 \sum_{i=1}^{N} c_i^2 + a_4 \sum_{i=1}^{N} c_i = \sum_{i=1}^{N} \gamma_i c_i ,$$

$$a_1 \sum_{i=1}^{N} c_i^3 + a_2 \sum_{i=1}^{N} c_i^2 + a_3 \sum_{i=1}^{N} c_i + a_4 N = \sum_{i-1}^{N} \gamma_i ,$$

where: N = number of experimental points,

$$c_i = [\overline{BH_i}] .$$

Results are presented in a table which contains concentration, interfacial tension determined experimentally and calculated from the polynomial, absolute and relative errors an the error of model given as the average square deviation. An appropriate graph presenting relation 7 is also given. The adsorption isotherm 8 is also drawn and given in the form of a table.

Equation (15) is further computed and presented in the form of graphs (r vs. $[\overline{BH}]$ and $lg(r)$ vs. $lg([\overline{BH}])$

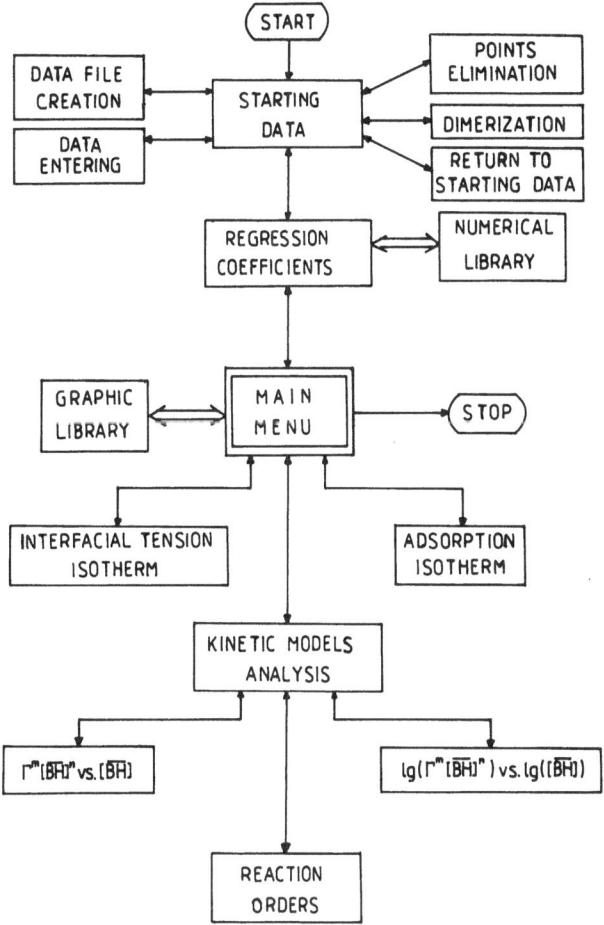

Fig. 1. Flow diagram of MINEX program

and tables. The following relations are considered: Γ, Γ^2, $\Gamma[\overline{BH}]$, $\Gamma^2[\overline{BH}]^2$ and $\Gamma[\overline{BH}]^3$, which are equivalent to reactions 9, 10, 11, 12, and 13 assumed as the limiting steps, respectively.

Reaction order against extractant concentration is defined by

$$R = \frac{d \, lg(\Gamma^m [\overline{BH}]^n)}{d \, lg([\overline{BH}])} . \tag{17}$$

By introducing Eq. (7) the following relation is obtained:

$$R = \frac{6a_1 \ln [\overline{BH}] + a_2}{3a_1 (\ln [\overline{BH}])^2 + 2a_2 \ln [\overline{BH}] + a_3} . \tag{18}$$

The computed reaction orders are presented in a graph and in a table.

In the case of extractant dimerization the computing is carried out in the same way using the monomer concentration given by Eq. (19), which was obtained from Eq. (2):

$$[\overline{BH}] = \frac{-1\sqrt{1 + 8K_d[\overline{BH}]_t}}{4K_d}. \tag{19}$$

Simulation of extraction rate

The initial data was taken from the work of Hughes et al. [10] (Table 1). They demonstrated the interfacial tension isotherm for Lix 63 in toluene/water system at 28 °C.

The results presented in Table 1 demonstrate that the polynomial 7 well approximates the interfacial tension isotherm, and the error of model defined as the square deviation is low and amounts $6.8 \, 10^{-3}$.

Figure 2 shows the inital interfacial tension isotherm and the surface excess isotherms for various dimerization constants as a function of the monomer concentration. The computed isotherms have a typical character and the surface excess increases monotonically. Figure 3 presents the function 15 for reactions 9–13 considered as the limiting step for two various dimerization constants. They have different characters and they depend significantly upon the dimerization constants. As a result of this, various reaction orders were computed (Fig. 4). However, the extractant dimerization has only a weak effect upon the computed reaction order and can be neglected. The effect of the function considered is very significant, and as a result, quite different reactions orders were obtained. Due to this, by comparing them with the experimental kinetic data it seems possible to select appropriate reactions as the limiting step.

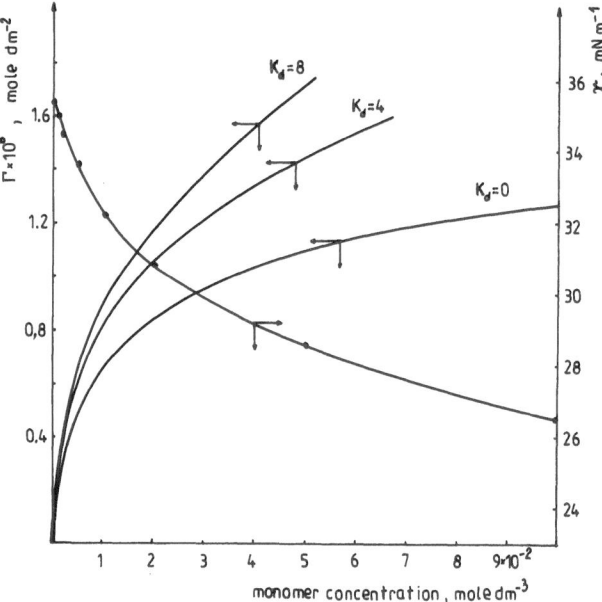

Fig. 2. Interfacial tension and surface excess isotherms

Table 1. Experimental [10] and approximated values of the interfacial tension (Lix 63, toluene/water, $t = 28$ °C). $a_1 = -7.307 \, 10^{-6}$, $a_2 = -3.840 \, 10^{-4}$, $a_3 = -4.896 \, 10^{-3}$, $a_4 = 1.716 \, 10^{-2}$

No.	c mole dm^{-3}	γ_{exp} mN m^{-1}	γ_{appr} mN m^{-1}	$\Delta\gamma$ mN m^{-1}
1	0.0001	35.40	35.40	0.00
2	0.001	35.00	35.07	0.07
3	0.002	34.50	34.51	0.01
4	0.005	33.60	33.41	− 0.19
5	0.01	32.20	32.28	0.08
6	0.02	30.80	30.88	0.08
7	0.05	28.60	28.58	− 0.02
8	0.1	26.50	26.49	− 0.01

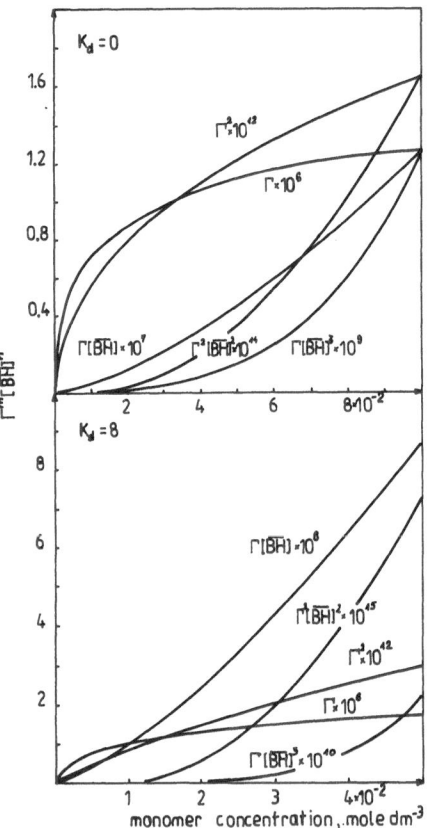

Fig. 3. Effect of monomer concentration upon $\Gamma^m[\overline{BH}]^n$ functions

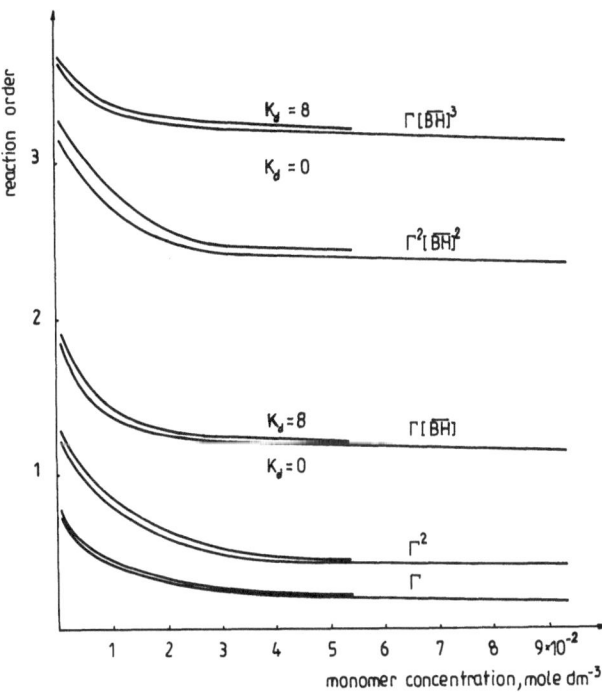

Fig. 4. Reactions orders predicted from the interfacial tension isotherm

Up to now the proposed computing technique was used with success for copper extraction by hydroxyoximes and for palladium (II) extraction by dialkyl sulphide in the presence of trialkylamine [5–9].

Conclusions

The presented computer program can be used to discuss the relationships between the interfacial activity of hydrophobic extractants and the rate of metal extraction and to select the most probable version of the extraction mechansim. Up to now the proposed computing technique was used with success for copper extraction by hydroxyoximes and for palladium (II) extraction by dialkyl sulphide in the presence of trialkylamine.

Acknowledgement

This work was supported by Polish Research Program CPBP No. 03.08.

References

1. Fisher JF, Notebaart CW (1983) In: Teh CLo, Baird MH, Hanson C (eds) Handbook of Solvent Extraction. John Wiley and Sons, New York, pp 649–6
2. Szymanowski J (1984) Wiad Chem 38:371–401
3. Danesi PR, Chiarizia R (1980) Critical Reviews in Analytical Chemistry 10:1–126
4. Szymanowski J, Stępniak-Biniakiewicz D, Prochaska K, Rashid ZA (1986) Preprints ISEC '86 2:35–48
5. Szymanowski J, Prochaska K, Bogacki MB (1987) J Colloid Interface Sci 117:293–295
6. Szymanowski J, Prochaska K (1987) J Chem Tech Biotechnol 40:177–193
7. Szymanowski J, Prochaska K (1988) J Colloid Interface Sci 123:456–465
8. Szymanowski J, Prochaska K (1988) ISEC '88 Conference Papers 2:24–27
9. Szymanowski J, Prochaska K (1988) Progr Colloid Polym Sci 76:260–264
10. Al-Diwan TAB, Hughes MA, Whewell RJ (1977) J Inorg Nucl Chem 39:1419–1424

Received October, 1988;
accepted March, 1989

Authors' address:

Prof. Dr. Jan Szymanowski
Poznań Technical University
Insitute of Chemical Technology and Engineering
Pl. Skłodowskiej-Curie 2
PL-60-965 Poznań, Poland

Progress in Colloid & Polymer Science   Progr Colloid Polym Sci 79:332–337 (1989)

Analysis of the adhesive sphere fluid as a reference model for colloidal suspensions

C. Regnaut[1]) and J. C. Ravey[2])

[1]) Laboratoire Structure et réactivité aux interfaces. Université Pierre et Marie Curie, Bâtiment F74, Paris, France
[2]) Laboratoire de Physicochimie des colloïdes UA CNRS 54406. Université de Nancy I, Vandoeuvre les Nancy, France

Abstract: The analytical model of the adhesive sphere fluid in the Percus-Yevick approximation (ASPY) is analyzed for colloidal systems. To discuss this model we consider the more realistic square well (SW) interaction where the equivalence between the SW and the AS is defined by the equality of the second virial coefficient of the two systems. The main points considered are: the comparison between the distribution functions of the ASPY fluid and the SWPY fluid and the influence of the PY approximation on the results. This approximation is checked with respect to the modified hypernetted chain approximation (MHNC) for the SW system and the simulation data of Seaton and Glandt for the AS system. We focus on the structure factor and on the small angle behavior. The calculations are done at volume concentrations in the range 0%–40%. The principal conclusions are 1) the ASPY model of the structure factor agrees fairly well with the SWPY model at all volume concentrations if the range of the square well does not exceed 10% of the particles diameter, and 2) away from the critical region in the structure factors of the SW fluid are very close in the PY or MHNC approximation. Similarly, in the same domain there is a reasonable agreement between the ASPY model and the simulation data. Some applications of the ASPY model to particular colloidal suspensions are briefly presented.

Key words: Adhesive sphere, structure factor, liquid state models.

1. Introduction

It is well established that a number of colloids exhibit a liquid gas-like phase transition and critical phenomena at low volume concentration [1–3]. Moreover, the structure factor of these colloidal suspensions deduced from the x-ray and/or neutron scattering spectra resembles that of a monoatomic fluid over a wide range of concentrations. Such analogies suggest that the models of the theory of simple fluids [4] can help us in the analysis of colloidal suspensions. This approach implies two basic considerations: 1) the choice of a model potential for the interparticle interactions, and 2) the choice of a particular approximation of the liquid state theory. In many systems with non-ionic surfactant as well as inverted micelles, the potentials which are often invoked are modelled by a hard core plus a strong attractive tail of very short range with respect to the diameter of the particles [1, 5–8]. Therefore, it is interesting to consider the adhesive

sphere (AS) potential, which is a particular limit of an infinitely narrow and infinitely deep potential, as a reference model for these colloidal systems. One important advantage of this model is that it has been exactly solved in the Percus-Yevick (PY) approximation [9]. Nevertheless, to prove its efficiency we must answer the following questions: 1) is the replacement of a potential of finite amplitude by an AS model realistic?, and 2) what are the limitations implied by the PY theory for the corresponding liquid model?

Our purpose is to examine the previous points for potential models having the strength and range assumed in [1, 5–8]. We focus on the distribution functions and on the structure factor at low scattering angle since these quantities are currently derived from x-ray, neutron scattering (SAXS and SANS). To specify the validity of the AS idealization we consider an equivalent class of narrow square well (SW) potentials, and the corresponding liquid models. In the second step we discuss the reliability of the PY approximation with respect to the well known MHNC approximation in the case of the SW fluid and with respect to the simulation data in the case of the AS fluid.

2. Formal equivalence between the potentials

The square-well potential is defined by

$$V(r) = \{\infty \text{ if } r < \sigma; E \text{ if } \sigma \leq r < (1 + \delta)\,\sigma;$$

$$0 \text{ if } r \geq (1 + \delta)\,\sigma\} \tag{1}$$

where σ is the hard core diameter, $\delta\sigma$ is the width of the well. The well depth E is expressed in units of kT. The adhesive sphere model is defined by:

$$V_A(r) = \lim d \to \sigma \{\infty \text{ if } r < d;$$

$$\text{Ln}\,[12\,\tau(\sigma - d)/\sigma] \text{ if } d < r < \sigma; 0 \text{ if } \sigma < r\}. \tag{2}$$

This potential is singular, but the associated second virial coefficient B_2 remains finite. The parameter τ is equivalent to a dimensionless temperature. The inverse of τ is a measure of the stickiness of the particles. The thermodynamical equivalence between the SW and the AS fluids is defined by matching the second virial coefficient [10]. That is, the dependence of τ upon temperature T and $V(r)$ is given by

$$\sigma^3/12\tau = \sigma^3/3 + \int_0^\infty r^2 [\exp(-V(r)/kT) - 1]\,dr. \tag{3}$$

The SW is defined by E and $x_1 = 1 + \delta$ and the equivalent AS is specified by τ:

$$\tau^{-1} = 4\,(x_1^3 - 1)\,[\exp(-E/kT) - 1]. \tag{4}$$

One presumably could follow another route by defining the equivalence between the two models at the level of a first order perturbation theory, but this also would introduce a density dependence of the AS potential due to the optimization of the parameter τ.

3. Calculations of the structure factors

a) The structure factor of the AS fluid can be calculated following the Baxter's factorization method [9], $S(q)$ is expressed in terms of $Q(q)$ by:

$$S(q) = [Q(q)\,Q(-q)]^{-1}. \tag{5}$$

$Q(q)$ is the one-dimensional Fourier transform of $Q(r)$ which, in the PY approximation has the exact expression $Q(r) = Ar^2 + BR + C$. The factors A, B, C are expressed in terms of τ and of the volume concentration ϕ:

$$A = 0.5\,(1 + 2\phi - \mu)\,(1 - \phi)^{-2}$$

$$B = 0.5\,\sigma\,(-3\phi + \mu)\,(1 - \phi)^{-2}$$

$$C = A\sigma^2 - B\sigma + \lambda\sigma^2/12$$

$$\mu = \lambda\phi(1 - \phi)$$

$$\phi = \pi n\sigma^3/6$$

$$\lambda = \text{Min}\,(6\,[\tau\phi^{-1} + (1 - \phi)^{-1}] \pm \{3\,6\,[\tau\phi^{-1}$$

$$+ (1 - \phi)^{-1}]^2 - 12\,\phi^{-1}\,(1 + 0.5\,\phi)(1$$

$$- \phi)(1 - \phi)^{-2}\}^{1/2}; \tag{6}$$

n is the particle number density. We obtain the following expression for the reciprocal of $S(q)$:

$$S(q)^{-1} = \{1 - 12\,\phi\,[CI_0(q\sigma) + BI_1(q\sigma)$$

$$+ AI_2(q\sigma)]\}^2 + \{12\,\phi\,[CJ_0(q\sigma) + BJ_1(q\sigma)$$

$$+ AJ_2(q\sigma)]\}^2 \tag{7}$$

where the standard integrals $I_n(K)$ and $J_n(K)$ are analytical and defined by

$$I_n(K) = \int_0^1 x^n \cos Kx\,dx \text{ and } J_n(K) = \int_0^1 x^n \sin Kx\,dx.$$

b) We calculate the distributions functions of the SW fluid by considering the excess quantities with respect to the hard sphere fluid, which is defined by its well known PY analytical solution. We define by $V_o(r)$ the hard core part of the potential and we introduce the difference $\Delta V(r) = V(r) - V_o(r)$, the excess direct correlation function: $\Delta c(r) = c(r) - c_o(r)$, and the excess pair correlation function $\Delta g(r) = g(r) - g_o(r)$. The numerical calculations are performed using the continuous function $\Delta y(r) = \Delta g(r) - \Delta c(r) \cdot c(r)$ and $g(r)$ are solved by means of the Orstein-Zernicke equation [4] and the residual PY equation:

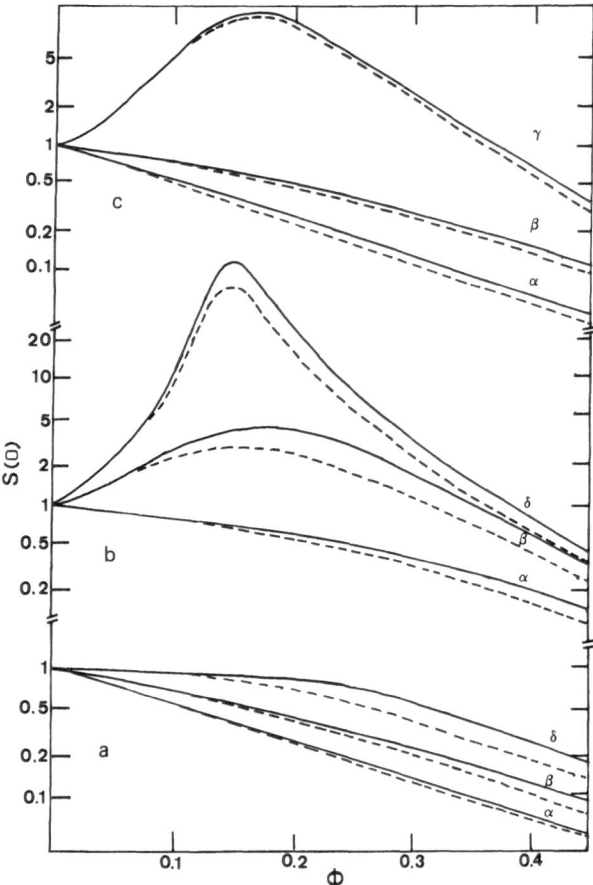

Fig. 1. Comparison of $S(0)$ for the ASPY model (continuous line) and the SWPY model (dashed line): a) α: width $\delta = 0.04$, depth $E = -1$; β:$\delta = 0.10$ $E = -1$; γ:$\delta = 0.20$, $E = -1$. b) α: width $\delta = 0.04$, depth $E = -2$; β:$\delta = 0.10$ $E = -2$; δ:$\delta = 0.12$, $E = -2$. c) α: width $\delta = 0.04$, depth $E = -1$; β:$\delta = 0.04$ $E = -2$; γ:$\delta = 0.04$, $E = -3$

$$g(r) = g_o(r) \exp\left[\Delta V(r)/kT\right]\left[1 + \Delta y(r)/g_o(r)\right]$$

$$\text{if } r > \sigma$$

$$\Delta g(r) = 0 \quad \text{if } r < \sigma. \tag{8}$$

We shall consider also the MHNC method since it is thermodynamically self-consistent and has been proven to be close to the simulation data for a variety of potentials [8]. For the SW fluid, and considering again excess quantities, the MHNC equations reduced to

$$g(r) = g_o(r) \exp\left[-\Delta V(r) + \Delta y(r) + B_o(r) - B_{o'}(r)\right] \tag{9}$$

$$\int \left[(g(r) - g_{o'}(r)]\partial B_{o'}/\partial o'\ r^2 dr = 0. \tag{10}$$

In the usual MHNC method, the bridge function of the true system $B(r)$, is replaced by the bridge function of a hard-sphere system $B_{o'}(r)$ which is optimized following Eq. (10) [12]. To be consistent with the ASPY and SWPY calculations the bridge functions of the fixed hard-sphere system $B_o(r)$ and the optimized $B_{o'}(r)$ are defined in the PY approximation.

These PY and MHNC integral equations are solved using a standard iteration procedure and a fast Fourier transform technique to obtain the successive direct and inverse transforms of Δy, and the final structure factor. Since we are treating a narrow SW, the grid spacing and the upper limit of the Fourier integrals are carefully chosen so that the calculated values remain insensitive (within 0.1%) to such a choice.

3. Comparison of the square well vs. adhesive spheres structure factors

First, let us consider the determination of $S(0)$ since this value is strongly coupled with the attractive part of the potential via the isothermal compressibility by the relation $S(0) = nkT_{\chi T}$. In the case of the ASPY model, $S(0) = (2A)^{-2}$, where A is given by Eq. (6). In Fig. 1 we report the curves of $S(0)$ for the SWPY model for various SW with E in the range -1 to -3 and δ in the range $0.04 - 0.20$. We see that the SWPY model and the equivalent ASPY model have the same qualitative behavior, in particular in the proximity of the critical region corresponding to the raise of $S(0)$. When δ is larger than 0.10, the differences between the SWPY and ASPY models are more apparent at high concentration. The same observations can be made for the whole $S(q)$, for instance, in the case of the narrow SW with $\delta = 0.04$ the structure factors are extremely similar in the two models (Fig. 2), while in the case of the SW with $\delta = 0.1$, the ASPY model of $S(q)$ remains again in reasonable agreement with the SWPY model. But, as the concentration is increased, the influence of the finite width of the potential can be noticed by the increase of the heigth of the first peak of $S(q)$ and the phase shift of the oscillations (Fig. 3). In summary, we find that the representation of the SWPY structure factor by its equivalent ASPY model is convenient, either at low volume concentration or at all fluid concentrations if the width of the square well is sufficiently narrow (δ must not exceed 10% of the hard-core diameter of the particle). Inversely, this means that if we assume a narrow potential the main information which can be extracted from $S(q)$ is the temperature parameter or

dilute ($\phi \approx 0$) or if the strength of the interaction is very weak ($-0.5 < E < 0$), the RPA is in contradistinction to the ASPY model. Consequently the RPA is unacceptable for analysis of small-angle scattering data and for prediction of a phase transition. In the case of the MSM model, we show that the agreement with the ASPY model is good only to first order in ϕ. Therefore, the two models agree in the range $\phi < 0.06$ only, and rapidly disagree at higher concentration.

6. Conclusion

In this work we have investigated the AS system as a reference model for the study of dispersed colloidal suspension (non-ionic) using the PY approximation since it leads to analytical expressions for the AS fluid. For comparison we have considered the narrow SW potential which is more realistic than the AS for colloidal systems, and we have shown that the corresponding SWPY and ASPY models produce comparable description of the liquid distribution functions if the width of the square well does not exceed 10% of the core diameter. On the other hand the published simulation data for the AS potential show that the ASPY model has some weakness concerning the pair correlation function, but remains quite useful for analyzing the scattering data. Our preliminary calculations with the MHNC method for a narrow SW suggest that the MHNC method is not more convenient than the PY aproximation when approaching the critical region. Therefore, in spite of its intrisic limitations the ASPY model supersedes several previous crude analytical models and should be considered very useful as a reference model to study the thermodynamical and structural properties of colloid dispersions which exhibit short-range interactions. Finally, although given a posteriori, the present theoretical results fully support the analysis of the structural determinations that we previously performed on the non-ionic systems by using SANS, at least as long as the particles are sufficiently globular [6–8].

References

1. Zulauf M, Wecstrom K, Hayter JB, De Giorgo V, Corti M (1985) J Phys Chem 89:3411–3417
2. Safran SA, Turkevich LA (1983) Phys Rev Lett 50:1930–1933
3. Evans H, Tildesey D, Leng CA (1987) J Chem Soc Faraday Trans 83:1525–1541
4. Hansen JP, MacDonald IR (1986) The theory of simple liquids. London academic press
5. Lemaire B, Bothorel P, Roux D (1983) J Phys Chem 87:1023–1028
6. Ravey JC, Stebe MJ, Oberthur R (1986) In: Mittal KL, Bothorel P (eds) Surfactant in solution. Vol 3, Plenum Press, New York, 1423–1430
7. Ravey JC, Buzier M, Oberthur R (1987) Progr Colloid Polym Sci 73:113–126
8. Ravey JC, Buzier M, Dupont G (1987) In: Rosano H, Clausse M (eds) Microemulsion systems. Surfactant Science. Vol 24, Marcel Dekker, New York, 163–182
9. Baxter JA (1968) J Chem Phys 49:2770–2774
10. Cummings PT, Perram JW, Smith ER (1976) Molecular Physics 31:535–548
11. Rosenfeld Y, Ashcroft NW (1979) Phys Rev A 20:1208–1235
12. Lado F, Foiles SM, Ashcroft NW (1983) Phys Rev A 28:2374–2379
13. Huang JS, Safran SA, Kim MW, Grest GS, Kotlarchyk M, Quirke N (1985) Phys Rev Lett 53:592–595
14. Seaton NA, Glandt ED (1987) J Chem Phys 87:1785–1790
15. Foiles SM, Ashcroft NW (1981) Phys Rev A 24:424–428
16. Sharma PV, Sharma KC (1977) Physica 899A:203–206

Received November, 1988;
accepted February, 1989

Authors' address:

C. Regnaut
Laboratoire Structure et réactivité aux interfaces
Université Pierre et Marie Curie
Bâtiment F74
4 Place Jussieu
F-75230 Paris, France

Progress in Colloid & Polymer Science   Progr Colloid Polym Sci 79:338–344 (1989)

Some thermodynamic models of progressive micellization, shape of the surface pressure curves, and predictions on the distribution of intermediate aggregates

D. Schuhmann

Laboratoire de Physico-Chimie des Systèmes Polyphasés – CNRS – Montpellier, France

Abstract: The shape of surface or interfacial tension curves for surfactants able to form micelles can be derived from equilibrium models of the micellization process if monomers only can be adsorbed. The influence of several effects influencing the aggregation equilibrium constants K_s on the shape of the surface pressure curve is examined in light of recently published quasi-chemical models. Some modifications are also considered. The presence of a strong increase of K_s with the intermediate aggregation number s is required for predicting the sharp transition at cmc. However, this effect cannot explain the shape of the distribution curve derived from kinetic studies. Both kinds of curves may be predicted in assuming, in addition, an effect decreasing K_s and a minute formation of dimers. The maximum aggregation number and its dispersity depend on these effects which leads to behaviors similar to those derived in applying the Aniansson et al. theory to kinetic data. The need for an improved description of the influence of electrostatic interactions is stressed.

Key words: Micellization (quasi chemical models), micelles (intermediate aggregates), adsorption (influence of micellization on).

1. Introduction

It is now generally admitted that the association and dissociation of the micelles proceed stepwise [1]. The micellization kinetics were thus theoretically treated by Aniansson et al. [2] starting from equilibria such as:

$$S_1 + S_s \underset{k_s^-}{\overset{k_s^+}{\rightleftharpoons}} S_{s+1} \qquad (1)$$

where S_1 is a monomer and S_{s+1} are intermediary aggregates.

The general assumptions in the Aniansson et al. theory are recalled in Ref. [3]: "... dimers, trimers, etc., are sufficiently rare so that their influence on the process in the region of proper micelles can be neglected ... (1) k_s^- and k_s^+ are independent of s in the region of proper micelles and equal to k^- and k^+, (2) the micellar distribution is Gaussian, n being the average aggregation number and σ the distribution width, and (3) the latter is broad enough so that differences with

good accuracy can be replaced by derivatives ..." It is also specified that σ "is generally about 20 % of the mean aggregation number" (for cases where the theory may be applied). The kinetic treatment was thus carried out using the analogy with diffusion between two large "blocks" (oligomers and micelles) joined by a thin wire where the rate constants vary with s.

It was found that the experimental variations of the relaxation times with the physicochemical parameters were in agreement with the predictions of the theory [3]. The presence of a minimum followed by a maximum in the distribution curve is thus now generally admitted [4, 5]. However, this general behavior should be rather a consequence of a relevant model than one of its starting assumptions. One is thus led to examine other thermodynamic or pseudo-chemical models. Even if they were derived for analyzing data depending on the equilibrium constans, they should be consistent with the results of relaxation experiments. Another test is the shape of the surface pressure curves which generally exhibit a sharp transition at the cmc.

At equilibrium, c_s being the concentration of S_s, one has with $K_s = k_s^+ / k_s^-$:

$$c_2 = K_1 c_1^2; c_3 = K_2 c_1 c_2; \ldots; c_{s+1} = K_s c_1 c_s; \ldots \quad (2)$$

Now, let us assume that the surfactant adsorption occurs from free monomers only. In the following, sufficently high values of concentration only will be considered, giving rise to maximum adsorption. In this range, the surface excess Γ is constant ($\Gamma = \Gamma_m$). It does not matter whether surfactant monomers adsorb at the free surface or at any other fluid-fluid interface. γ will thus indifferently denote a surface or an interfacial tension. In the range where $\Gamma = \Gamma_m$, the Gibb's equation and its integrated form may be written as

$$RT\Gamma_m = - \partial \gamma / \partial \ln c_1; \gamma - \gamma_o' = RT\Gamma_m \ln c_1 \quad (3)$$

γ_o' being a constant (different from γ_o, the value in the absence of surfactant). C being the total surfactant concentration, it can thus be seen that the curves $\log c_1$ vs. $\log C$ are analogous save for the low concentration range to the curve $-\Delta\gamma$ vs. $\log C$ which allows to determine the cmc. C being equal to $\Sigma s c_s$, the variations of K_s which determine the distribution curve also allow to derive the shape of the surface pressure curves. The pseudo-chemical models should thus predict the

sharpness of the latter (test A) and also the presence of a maximum and a minimum in the former shared by a large and low valley. At the extrema, $c_{s+1} \sim c_s$. The models should thus predict the possibility for the ratio $r_s = c_{s+1}/c_s$ to be equal to unity for two values of s (test B).

2. Examination of simple models

Katime and Allende [6] recently considered five different models in order to interpret their osmometry measurements. Model 1 corresponds to an one-step mechanism and will not be examined here. In their model 2, all the equilibrium constants are equal to K up to $s = N$. The total concentration C is thus given by the equation

$$C = c_1 + 2K c_1^2 + 3 K^2 c_1^3 + \ldots \quad (4)$$

Introducing dimensionless properties

$$x = KC; y = Kc_1 \quad (5)$$

one finds

$$x = y \left[1 - (N + 1)y^N + N y^{N+1} \right]/(1 - y)^2 . \quad (6)$$

Variations of $\log y$ vs. $\log x$ are plotted in curve 1 of Fig. 1. The transition range spreads on at least two decades. The distribution curve shape is obviously not predicted by this too simple model.

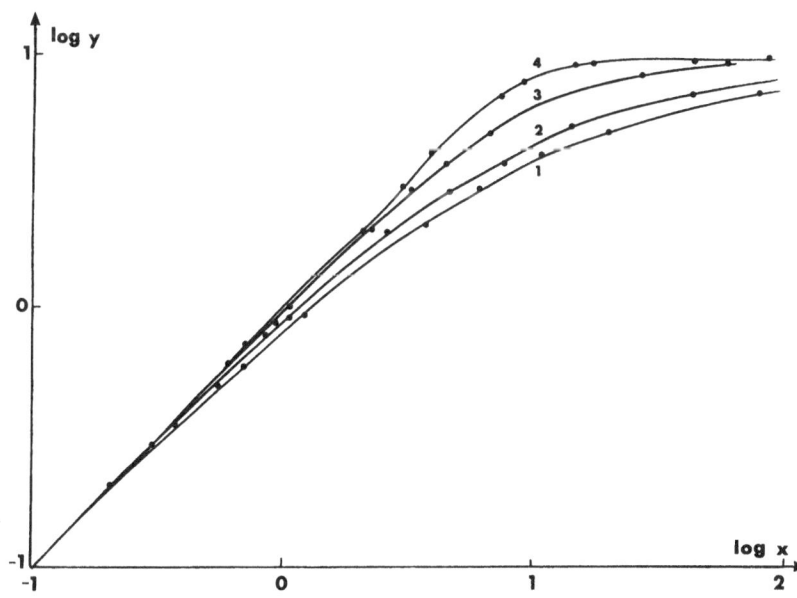

Fig. 1. Simulation of surface pressure curves — Effect of a decrease of the dimerization constant described by the parameter Q (see text). 1: $Q = 1$; 2: $Q = 0.5$; 3: $Q = 0.1$; 4: $Q = 0.005$

The effect of a very small dimerization constant was discussed by Mukerjee [7] and considered as the only modification by Katime and Allende in their model 3. With $Q = K_1/K$, one finds

$$x = y(1 - Q) + Qy \sum_{s=1}^{N} sy^s. \tag{7}$$

The effect of Q is shown in Fig. 1. It is interesting to note that the smaller the value of Q, the sharper the transition at the cmc. A value lower than 5×10^{-3} can produce the usual experimental shape. But the changes in K_1 do not affect r_s which is constant and equal to y.

In Katime and Allende's model 4 the decrease of K_s was considered in assuming $K_s = q/(s + 1)$ but they also admit a low value of the dimerization constant K_1. Taking now $x = qC$, $y = qc1$, and $R = K_1/q$, one has

$$x = y + 2R \sum_{s=1}^{N-1} y^{s+1}/s. \tag{8}$$

It may be shown that a relatively sharp transition is found for very small values of R only. The ratios r_s are equal to $y/(s + 1)$. The maximum value of y is relatively low and r_s cannot be equal to unity for any value of s.

In their model 5, Katime and Allende assumed instead an "accelerating" effect $K_s = q(s - 1)/s$. Keeping for x, y and R the same definitions as in model 4, one finds

$$x = y + R \sum_{s=1}^{N-1} y^{s+1}(s + 1)/s. \tag{9}$$

For $N > 50$, the value of N has a very small effect on the plot $\log y$ vs. $\log x$. The shape of the curve thus depends more on the value of K_1 than on the accelerating effect itself. Some curves are reported in Fig. 2 and show that the decrease of the dimerization constant has the same effect on the shape of the curves as in the previous case. r_s is equal to $y(s - 1)/s$. Values of y slightly higher than unity may be found and only a minimum is predicted in the distribution curve. Analyzing in the light of the above models their osmometric data, the authors found the best fit with model 4. The latter also fulfils test A but none of the treated model fulfils test B. This clearly shows that a reasonable law for K_s cannot be deduced from one kind of experiment only, and the choice of the tests A and B seems to be relevant.

3. Examination of more complex models

1. Tanford-Ruckenstein-Nagarajan's (TRN) model

Tanford [8] gave explicit expression for the free-energy change corresponding to the transfer of one amphiphile molecule from the dilute aqueous phase to the hydrocarbon phase of the micellar core. A first contribution consists in the decrease in attractive hydrophobic bonding between the hydrocarbon tails due to their partial exposure to the aqueous medium. Another term arises from the repulsive interaction between the hydrophilic head groups of the amphiphiles. This repulsion was assumed to be caused by steric interaction between head groups in nonionic micelles and by electrical repulsion between the ionic heads groups in ionic micelles. In both cases, the energy term would be proportional to s. The individual equilibrium constants can be derived from the expression of the free energy and one finds [9]

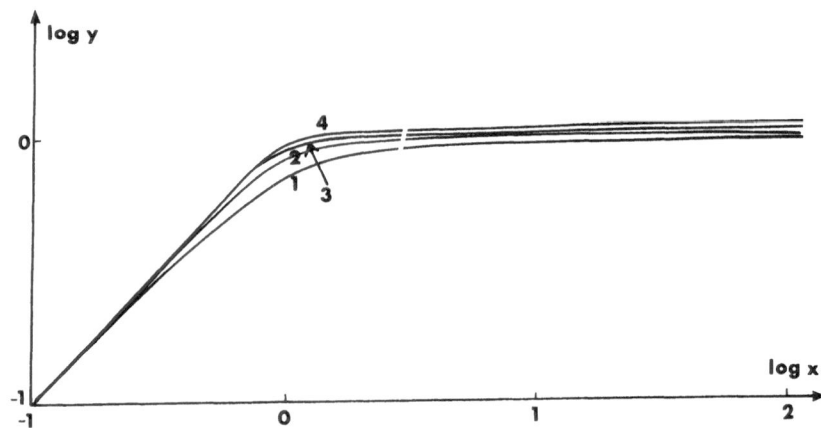

Fig. 2. Simulation of surface pressure curves — Same affect as in Fig. 1 described by the parameter R (see text) but in the presence of an equilibrium constant increasing with s. $N = 100$. 1: $R = 10^{-1}$; 2: $R = 10^{-2}$; 3: $R = 10^{-3}$; 4: $R = 10^{-4}$

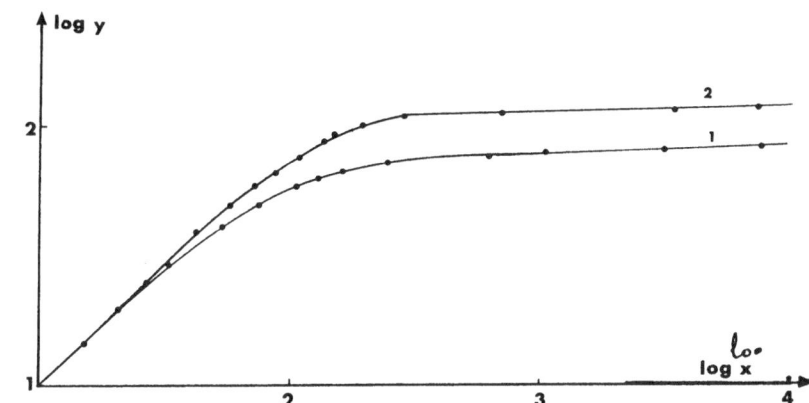

Fig. 3. Simulation of surface pressure curves
— TRN model. 1: $\xi = 9$; 2: $\xi = 10.5$

$$\ln K_s = -\frac{4}{3}\,\eta\,s^{1/3} + \beta - \frac{2}{3}\,\xi\,s^{-1/3} \qquad (10)$$

with typical values for nonionic surfactant: $\beta = 17$, $\xi = 9$ and $\eta = 0.6$. Taking $x = C \exp(\beta)$ and $y = c_1 \exp(\beta)$ and assuming, as did the authors, that aggregation may proceed up to an infinite value of s, the shape of the surface pressure depends on η and ξ only. Two curves differing in the value of ξ are plotted in Fig. 3. A behavior similar to the experimental one is found for one of these curves, but the slope for $c <$ cmc would not give the correct value of the surface excess. It is noted in Fig. 3 that for $C >$ cmc, the curves have a small but noticeable slope, showing an increase in monomer concentration above cmc. Calculated distribution curves $y_s =$

$f(s)$ are plotted in Fig. 4 for $\xi = 10.5$ and two values of x (20- and 350-times the critical value of x, x_c). The expected shape is found with a deep valley and a pseudo-Gaussian shape around the maximum. The maximum aggregation number is little concentration-dependent (from 47 to 54 in the limits of x indicated). The width σ of the micellar cusp is of the order of 30, which is somewhat higher than reported values [3, 10, 11].

2. Kegeles's model

Kegeles [12] took into account a probability factor equal to the proportion of available sites in the inter-

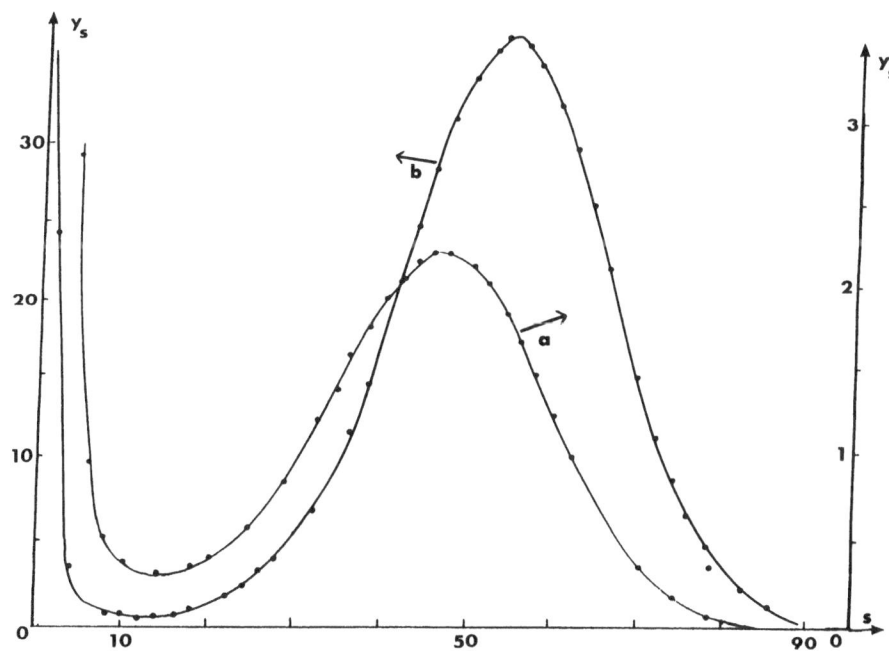

Fig. 4. Distribution curves calculated with the TRN model with $\xi = 10.5$. x_c is the dimensionless cmc. a) $y = 125$ or $x \sim 21\,x_c$; b) $y = 132$ or $x \sim 350\,x_c$. The arrows indicate the relevant scale on the vertical axis

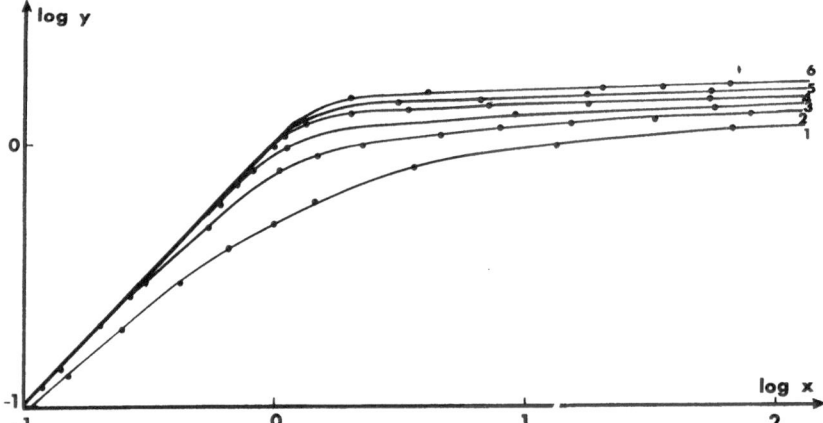

Fig. 5. Simulation of surface pressure curves — effect of a decrease of the dimerization constant described by the parameter f (see text) in the Kegeles's model. 1: $f = 1$; 2: $f = 10^{-1}$; 3: $f = 10^{-2}$; 4: $f = 10^{-3}$; 5: $f = 10^{-4}$; 6: $f = 10^{-5}$

mediate aggregates. In contrast with that examined above, this model assumes a limited range of possible aggregates; the probability factor gives rise to a decelerating effect predominating for the highest possible values of s. Kegeles also assumed a weak accelerating effect, in addition to a small dimerization constant and stressed its importance after Mukerjee [7] for the shape of the distribution function. With $y_s = Kc_s$, $y = Kc_1$ and $x = KC$, the model is then defined by the equations:

$$y_2/y = \frac{1}{2} fy; \quad y_s/y_{s-1} = \frac{N-(s-1)}{N} \frac{s\,y}{s+1}. \tag{11}$$

The effect of the parameter f is shown in Fig. 5 for $N = 100$. As noted by the author, the decrease of f allows to increase the maximum value of y having a physical meaning and thus allows to find for r_s two values of s such as $r_s = 1$ the second one being of the order of the experimental aggregation numbers. Two distribution curves corresponding to approximately 10 and 50 x_c are plotted in Fig. 6 for $f = 10^{-4}$ and $N = 200$. They exhibit the correct shape and have the same width as those predicted in the model examined above. The maximum aggregation number is more concentration-dependent.

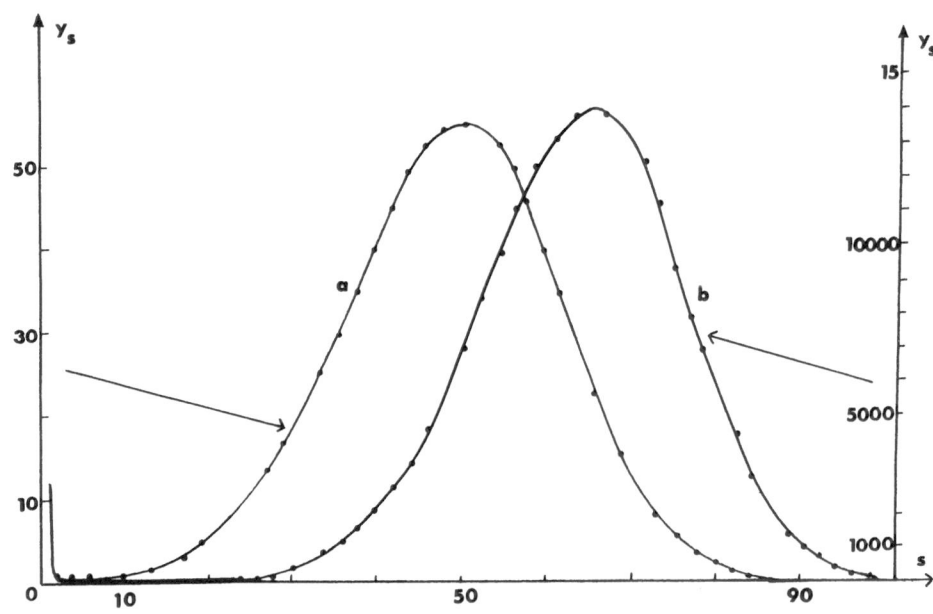

Fig. 6. Distribution curves calculated with the Kegeles's model. $N = 200$, $f = 10^{-4}$. a) $y = 1.34$ or $x \backsim 10\,x_c$; b) $y = 1.48$ or $x \backsim 50\,x_c$

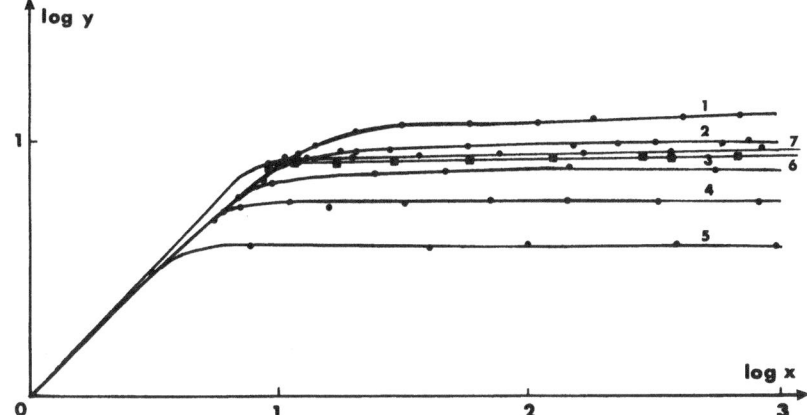

Fig. 7. Simulation of surface pressure curves. a) Effect of the limiting value of N of the aggregation number with the modified Kegeles's model ($q = 100$). 1: $N = 50$; 2: $N = 70$; 3: $N = 100$; 4: $N = 200$; 5: $N = 300$; b) Curves with a small value of the dimerization constant, $N = 100$. 6: $f = 10^{-3}$; 7: $f = 10^{-4}$

3. A modified Kegeles's model

Because of the behaviors reported in the previous subsections, it was decided to modify the previous model, setting for the accelerating term $s/(s + q)$ instead of $s/(s + 1)$. Without any particular assumption for the dimerization constant, one has

$$x = \sum_{s=0}^{N-1} \frac{N!\,(s+1)!\,q!\,y^{s+1}}{(N-s)!\,(s+q)!\,N^s}. \tag{12}$$

The transition between the asymptotic behaviors is increasingly sharper when q increases. The maximum value of y also increases with q since the terms in the series (12) quickly decrease when their order increases. Figure 7a, where curves for $q = 100$ and several values of N are plotted, shows that the asymptotic branch at high concentrations becomes more horizontal when N increases, but as expected, this increase does not change the curves for $x < x_c$. Finally, also introducing the assumption of a small dimerization constant, the curves in Fig. 7b for $N = 100$ and $q = 100$ show that the branch for concentrations lower than cmc has the correct slope and for $x > x_c$, the curve is practically identical to a horizontal straight line, taking $f = 10^{-3}$ or 10^{-4}. Distribution curves obtained with $N = 100$, $Q = 100$, and $f = 10^{-4}$ are plotted in Fig. 8 for values of x of the order of 3 and 4500 cmc ($y = 8.25$ and 8.4). The maximum aggregation number N_M is little changed (73 to 76) while with $q = 1$, N_M strongly decreases near cmc. With respect to the preceding models, the dispersity is reduced: the width σ is found equal to 13 instead of approximately 30. σ/N_M is equal to 17 %, i.e., of the order of the accepted value [3]. Moreover, the valley is larger and its bottom is lower. The increase of $c_1(y)$

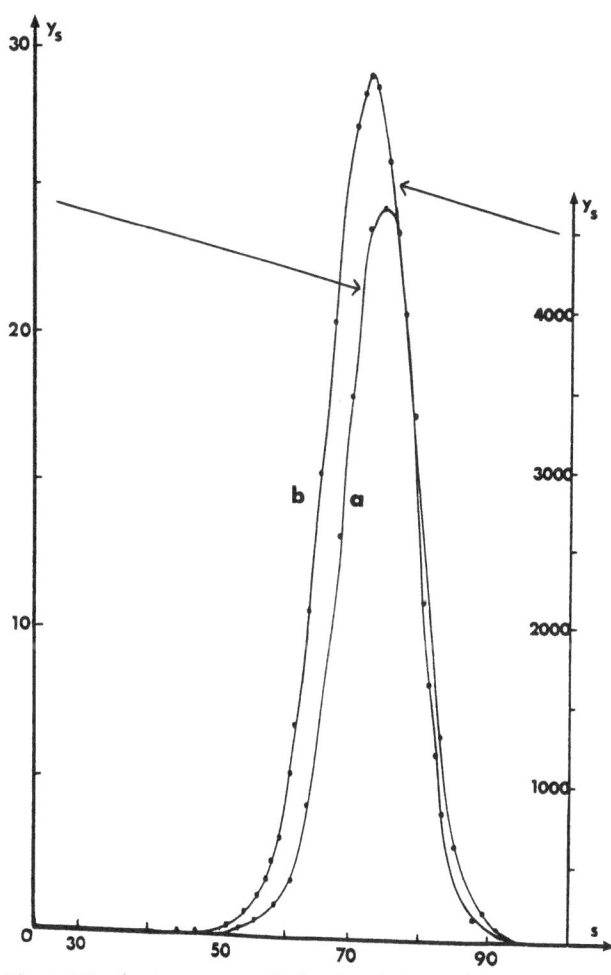

Fig. 8. Distribution curves calculated with the modified Kegeles's model $N = 100$; $q = 100$; $f = 10^{-4}$. a) $y = 8.4$ or $x_\sim 3\,x_c$; b) $y = 9$ or $x_\sim 550\,x_c$

with C is lower than the one found with the other models.

The pseudo-Gaussian shape of the curve in the micelle region is obviously the result of the introduction of a decelerating factor in r_s. The model provides the best image of a long thin wire for the valley region and as in the Aniansson et al. treatment, the equilibrium constant is appreciably variable in this region.

4. Conclusion

The shapes experimentally found for the surface pressure and distribution curves may both be predicted if the three effects examined are together taken into account: 1) The dimerization constant should be extremely smaller than the other equilibrium constants; 2) Starting from dimers (or perhaps low order oligomers) the aggregation is then strongly accelerated; 3) The approximately Gaussian shape of the distribution curve in the micellar region requires the presence of a decelerating effect prevailing for $s > N_M$; It seems that the best agreement between experiments and theory is obtained with the modified Kegeles' model.

The pseudo-Gaussian shape of the distribution curve in the micellar region should be considered as being rather a consequence than a basic model assumption.

The main interest of the TRN's model is its ability to approximately simulate the accepted shape of the distribution curve using parameters having reasonable values and a physical meaning. The modified Kegeles' model improves the predictions but as the original one is based on a priori assumptions on the analytical expression of the effects considered. At the moment, the modified model proposed for analyzing equilibrium properties could stimulate theoretical works allowing a best evaluation of the effects interfering in the micellization process in equilibrium or relaxation conditions. It seems that the variation laws for $K_s(s)$ able to provide a pseudo-Gaussian shape of the distribution curves also lead to predict an increase (negligible but not nil with the last model) of the monomer concentration after cmc. It may be thought that the decrease found with ionic surfactants is related to interactions between micelles which are not taken into account in the examined models.

The slope of the surface pressure curve near the cmc is perhaps affected by the micellization process: the value of the maximum surface excess derived from this slope could be too weak.

Lastly, it may be noted that if the cmc depends on all the parameters describing the examined effects, all the equilibrium constants K_s implicity contain the activity coefficient f of the monomers. $\ln f$ varies in proportion with the number n_c of carbon atoms in a molecule for aliphatic compounds differing only in their chain length. The dimensionless concentration x is in proportion with f. It is thus clear that the variation of cmc with n_c is essentially due to the hydrophobic interactions of the free monomers in the solution, but not to some interfacial process. In the same way, salting out of monomers in the bulk can affect the cmc.

References

1. Kresheck GC, Hamori E, Davenport G, Sheraga HA (1966) J Am Chem Soc 88:246–257
2. Aniansson EAG, Wall S (1975) J Phys Chem 79:857–858
3. Aniansson EAG, Wall SN, Almgren M, Hoffmann H, Kielmann I, Ulbricht W, Zana R, Lang J, Tondre C (1976) J Phys Chem 80:905–922
4. Wennerström H, Lindman B (1979) Phys Rep 52:1–82
5. Zana R (1984) In: Mittal KL, Bothorel M (eds) Surfactants in Solution. Plenum Press, New-York, vol 4, pp 115–130
6. Katime I, Allende JL, In: Surfactants in Solution. loc cit, pp 77–90
7. Mukerjee P (1972) J Phys Chem 76:565–570
8. Tanford C (1974) J Phys Chem 78:2469–2479
9. Ruckenstein E, Nagarajan R (1975) J Phys Chem 79:2622–2626
10. Hoffmann H, Ulbricht W (1977) Z Phys Chem, NF 106:167–184
11. Schmidt D, Gähwiller CH, Von Planta J (1981) J Colloid Interface Sci 83:191–198
12. Kegeles G (1979) J Phys Chem 83:1728–1732

Received November, 1988;
accepted February, 1989

Author's address:

D. Schuhmann
Directeur de Recherche
LPCSP, CNRS
BP 5051
F-34033 Montpellier, France

Progress in Colloid & Polymer Science   Progr Colloid Polym Sci 79:345–352 (1989)

Crystallization of calcium oxalate in the presence of sodium dodecyl sulphate. Quantitative assessment of the effect of the surfactant on crystal growth and aggregation

D. Škrtić and N. Filipović-Vinceković

„Ruđer Bošković" Institute, Zagreb, Yugoslavia

Abstract: A quantitative treatment of the influence of sodium dodecyl sulphate, $(Na(C_{12}H_{25}SO_4)_2;$ NaDS) on the crystal growth and aggregation of calcium oxalates in spontaneously precipitating system is given. The experiments are carried out under the conditions close to those in urine of stone-formers (310 K, pH = 6.5, ionic strength $I =$ 0.3 mol dm^{-3}, molar ratio c_{init} (Ca)/c_{init} (C_2O_4) = 33.3). The formation of solid particles (crystal size distribution, CSD) was followed by Coulter counter. Standard methods for crystal phase identification were applied (thermogravimetric analysis, x-ray diffraction, optical microscopy). NaDS inhibits the crystal growth process; the intensity of the effect seems to be concentration dependent. The order of reaction (p) increases with increasing concentration of NaDS and assumes following values: $p_{NaDS_1} = 8.5$, $p_{NaDS_2} = 11.4$, and $p_{NaDS_3} = 11.9$. While for the control system (without NaDS) the integration of the growth units into kinks is the rate determining process, in the presence of NaDS crystal growth is controlled by surface adsorption.

In systems where NaDS is present at concentrations equal to or below critical micelle concentration (CMC) the overall rate constant of aggregation is not influenced; the addition of NaDS causes no change in the rate controlling mechanism of aggregation. At c (NaDS) significantly higher than the CMC, the overall inhibitory effect comprises, besides inhibition of crystal growth, the retardation of aggregation process as well.

Key words: Crystal growth and aggregation, Ca-oxalates, solid particles formation, inhibition of crystal growth, analysis.

1. Introduction

The influence of different additives on equilibria in solid/solution systems is of interest in various fields, e.g., environmental sciences, technology, biomedicine, etc. Numerous organic substances have been extensively used in industry to modify morphology and crystallization rates of slightly [1, 2] and highly [3] soluble salts. When crystal size and shape are of utmost importance (e.g., in the preparation of ceramic materials, emulsions, and pastes, in milling and tabletting processes) specific additives are used to control the crystallization process. Another important aspect is the use of additives as inhibitors of scale formation (in oil and gas recovery, cooling systems, desalination processes, heat exchangers, etc.) [4] and pathological mineralization (e.g., urinary crystallization) [5]. In urolithiasis research the deposition of crystalline phases from pure solutions and those containing possible inhibitors has attracted the attention of many investigators. It has been suggested that mechanisms controlling nucleation, crystal growth, and aggregation might play a significant role in the genesis of urinary stone formation. In previous work [6] we studied the spontaneous precipitation of calcium oxalate (usually main mineral constituent of human kidney stones) in the presence of the surface active additive — sodium dodecyl sulphate, NaDS. In this paper a quantitative estimate of the influence of NaDS on crystal growth and aggregation parameters will be given. For that purpose previously described methods for interpretation of the kinetic precipitation data [7–9] are applied. In addition, the analysis which enables to differentiate between elementary growth rate determining mechanisms [10, 11] is carried out.

2. Materials and methods

2.1. Experimental

Batch precipitation systems were prepared from analytical grade chemicals (calcium chloride, oxalic acid, sodium chloride, sodium hydroxide) or recrystallized from ethanol (NaDS) and triply distilled water. Crystallization was initiated by mixing of solutions containing known concentrations of oxalate (c_{init} (C_2O_4) = 3.0 × 10^{-4} mol dm^{-3}, preadjusted to pH = 6.5 with sodium hydroxide) with equal volumes of calcium chloride solutions (c_{init} (Ca) = 1.0 × 10^{-2} mol dm^{-3}). Sodium chloride was added to both solutions to bring the ionic strength up to $I = 0.3$ mol dm^{-3}. Crystallization took place in the presence of different concentrations of NaDS: 5.0 × 10^{-5} mol dm^{-3}, 5.0 × 10^{-4} mol dm^{-3}, and 2.0 × 10^{-3} mol dm^{-3}. All solutions were thermostated at 310 K. The measurements of the system turbidity, τ, were made by a Pulfrich photometer supplied with a turbidimetric extension.

The precipitation kinetics was followed by recording changes in crystal size distributions (CDS) using a Coulter counter (model TA). The composition of the precipitates was ascertained by x-ray powder patterns, TG analysis, and optical microscopy.

2.2. Treatment of data

The supersaturation of the system with respect to calcium oxalate monohydrate (COM) and dihydrate (COD) was calculated from the initial concentrations of all ionic species in solution using solubility and ion association constants [12, 13] at 310 K. The following equilibrium equations were considered

$$H^+ + C_2O_4^{2-} \rightleftharpoons HC_2O_4^- \tag{1a}$$

$$Na^+ + C_2O_4^{2-} \rightleftharpoons NaC_2O_4^- \tag{1b}$$

$$Ca^{2+} + C_2O_4^{2-} \rightleftharpoons CaC_2O_4^0 \tag{1c}$$

$$CaC_2O_4 \cdot H_2O \ (s) \rightleftharpoons Ca^{2+} + C_2O_4^{2-} \tag{1d}$$

$$CaC_2O_4 \cdot 2\,H_2O \ (s) \rightleftharpoons Ca^{2+} + C_2O_4^{2-} \tag{1e}$$

$$Ca(DS)_2 \rightleftharpoons Ca^{2+} + 2\,DS^- \tag{1f}$$

$$(DS_m Ca_n Na_v)^{m-(n+v)} \rightleftharpoons nCa^{2+} + vNa^+ + mDS^-. \tag{1g}$$

The calculations were done by successive approximations for the ionic strength; activity coefficients being calculated from the Debye-Hückel equation extended by Davies [14].

Specific interactions in the system NaDS + CaCl$_2$ + NaCl were the subject of our previous paper [6] where existing equilibria were

extensively discussed. It can be demonstrated that less than 1.2 % of initial calcium is bounded in mixed counterion layer at the micelle/solution interface. It was also shown [6] that under experimental conditions employed no Ca(DS)$_2$ precipitation occurs. For those reasons equilibria (f) and (g) were neglected in the calculations of actual supersaturations.

Total volume of the precipitate, V_t, was calculated on the basis of the Coulter counter data (number of particles in a known volume of the suspension and differential volume size distribution) as previously [7–9] described. The degree of the reaction, α, was defined as a ratio between a volume of the precipitate at time t (V_t) and theoretically obtainable volume (V_{theor}). The quantity V_{theor} was calculated from

$$V_{theor} = M(c_0 - c_s)/\varrho \tag{2}$$

where M denotes the molecular mass and ϱ the density of the precipitated solid phase, while c_0 and c_s are calculated initial and equilibrium concentration of the oxalate species. For the systems where a mixture of COM and COD was initially formed the weight portion of each solid phase was taken into calculations of the corresponding M, ϱ, and c_s values (assuming no further transformation of the hydrates occurs during the precipitation process).

The kinetics of precipitation was interpreted according to the general rate equation

$$R \equiv (d\alpha/dt) = k \cdot A \cdot S^p \tag{3}$$

which has been found to be applicable to many different spontaneously precipitating systems, and serves for distinguishing between successive precipitation processes (nucleation, crystal growth and aggregation) [7]. In Eq. (3), k is the proportionality constant, A is the available surface area, $S = c_t - c_s$ is the supersaturation, and p is the order of reaction. When crystal growth is the precipitation rate controlling process, Eq. (3) becomes

$$(d\alpha/dt)\,\alpha^{-2/3} = K\,(1 - \alpha)^p \tag{4}$$

for a constant shape and number of particles and a self-preserving distribution [15] of sizes. Applying the kinetic models of crystal growth and aggregation [9] typical parameters characterizing precipitation in presence and absence of NaDS were determined. Crystal growth rate of sparingly soluble calcium oxalate can theoretically be controlled, either by the processes on the growing crystal surface or by the transport of ions through the solution. A majority of sparingly soluble salts follow surface controlled kinetics (linear, parabolic, and exponential, related to the surface adsorption, to the integration of ions in screw dislocation centered surface spiral steps, and to the formation and growth of surface nuclei, respectively). Nielsen's theory of electrolyte crystal growth kinetics [10, 11] gives the following expressions

$$v_g = k_1\,(S - 1) \tag{5}$$

$$v_g = k_2\,(S - 1)^2 \tag{6}$$

$$v_g = A \exp\,(-B/1\,nS) \tag{7}$$

corresponding to crystal growth controlled by adsorption, surface spiral growth, and surface nucleation (polynuclear mechanism), respectively. In Eqs. (5)–(7), v_g is the linear growth rate, k_1 and k_2 are the rate constants, and S is the supersaturation expressed as $S = c/c_s$. Experimental data were analyzed according to the proposed relations in order to determine the rate-controlling mechanisms for pure systems and those containing different concentrations of NaDS.

It is known that CSD has a pronounced effect on the aggregation behavior of the system. The analysis of the collisions between particles of known sizes and CSD was patterned according to Friedlander [15] and O'Melia [16]. The collision rate, F, at which particles of sizes d_1 and d_2 come into contact by Brownian diffusion (bd), laminar shear (ls), or differential settling (ds) is given by

$$F = k_j \cdot (d_1, d_2) \cdot N_{d_1} \cdot N_{d_2}. \qquad (8)$$

Here $k_1(d_1, d_2)$ is the "bimolecular" rate constant for the jth mechanism, while N_{d_1} and N_{d_2} are the number concentrations of particles of sizes d_1 and d_2, respectively. The rate constants were defined as follows

$$k_{bd} = \frac{2\,kT\,(d_1 + d_2)^2}{3\,\mu\,d_1 d_2} \qquad (9)$$

$$k_{ls} = \frac{G\,(d_1 + d_2)^3}{6} \qquad (10)$$

$$k_{ds} = \frac{\pi g(\bar{\varrho} - 1)\,(d_1 + d_2)^3\,|d_1 - d_2|}{72\,v} \qquad (11)$$

where k is the Boltzmann's constant, T is the temperature, G is the shear rate, g is the gravity acceleration, μ is the absolute viscosity, v is the kinematic viscosity, and $\bar{\varrho}$ is the relative density of the solid phase.

3. Results and discussion

In our previous work [6] it was shown that if the spontaneous (unseeded) precipitation of calcium oxalate occurs in the presence of NaDS a significant inhibition on the overall crystallization process is observed; the inhibitory effect being proportional to the concentration of additive. Changes in the total number of particles (N_t) (lower part) and precipitate volume (V_t) (upper part) obtained in control system (curves 0; the dissipation of the values indicated in a plot is a result of four separate runs) and systems containing different concentrations of NaDS (curves 1–3) are given in Fig. 1. The kinetic data presented in a form of α or V_t vs. time curves can easily be recalculated as the corresponding rate vs. superasturation curves [7–9]. From such a plot (Fig. 2), the sections corresponding to the periods during which nucleation and growth, crystal growth, and growth impeded by aggregation are resolved. CSD measurements showed that in the pure

system, as well as in the systems with NaDS, highly polydispersed crystals are formed; precipitation was initiated by heterogeneous nucleation [6]. It can be seen from Fig. 1 that practically no difference is obtained in induction periods (time elapsed from the moment of mixing of the reacting components until the first measurable change in V_t or N_t). N_t values in the systems with NaDS were within the dissipation limits for N_t in control system, with the exception of system 3 (highest NaDS concentration) where at $t > 80$ min N_t practically remains constant (curve 3 in Fig. 1; lower diagram). Despite the fact that at early stages of crystallization a considerable part of the crystals is not detected (the detection limit of the Coulter counter used is approx. 0.5 μm), we may conclude that NaDS shows no influence on the nucleation-controlling mechanism. It should be pointed out however that at concentrations equal to or higher than CMC_{NaDS},

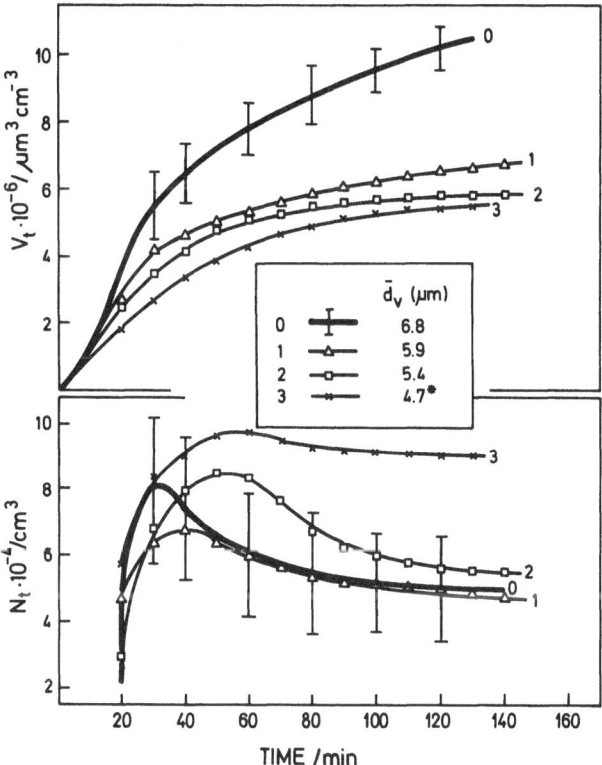

Fig. 1. The precipitate volume (V_t; upper diagram) and number of particles (N_t; lower diagram) as a function of time for the control system (heavy lines, curves 0) and those containing 5×10^{-5} mol dm^{-3} NaDS (curves 1), 5×10^{-4} mol dm^{-3} NaDS (curves 2) and 2×10^{-3} mol dm^{-3} NaDS (curves 3). The scattering of data in control system is indicated by bars. $\bar{d}_v$ is the mean particle diameter (calculated from mean particle volume assuming all crystals are spheres) at time just preceeding aggregation

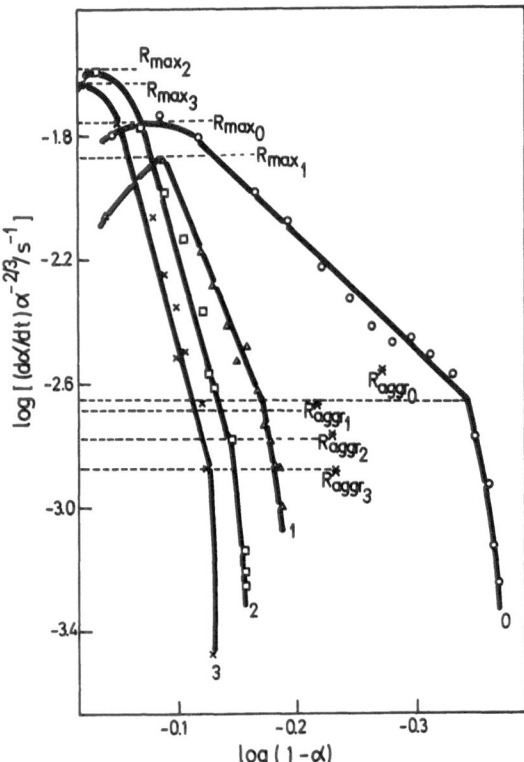

Fig. 2. Log rate vs. log supersaturation curves (according to Eq. (4)) showing the effect of different concentrations of NaDS (curves 1–3 as denoted in Fig. 1) on crystal growth process. For comparison, the control system (curve 0) is also indicated

attention will be focused on the quantitative assessment of the influence of NaDS on the crystal growth and aggregation.

3.1. The kinetics of calcium oxalate crystal growth in the presence of NaDS

In a log rate vs. log supersaturation plot (Fig. 2) we can see that after an upward surge (corresponding to the period during which nucleation prevails) a straight line is obtained. This linear segment is due to the dominating crystal growth (Eq. (4)). Quantitative parameters which characterize growth (the order of reaction (p) and the rate constant (k_a)) [9] are determined from the slopes of the straight lines and intercepts with ordinates, respectively. Results of such an analysis are given in Table 1. The maximum precipitation rates (R_{max}) and the precipitation rates at supersaturations at which the influence of aggregation becomes apparent (R_{aggr}) are also indicated in Fig. 2 and included in Table 1. The p value for the control system $p_0 = 3.6 \pm 0.5$ indicates surface-controlled (polynuclear) growth which does not include screw dislocations. In the presence of NaDS p values are significantly higher ($p = 8.5 - 11.5$, depending on concentration).

According to our previously defined quantitative criteria for evaluation of the influence of an additive on the crystal growth process [9] since $p_{NaDS} > p_0$, NaDS acts as crystal growth inhibitor. Its inhibitory activity is enhanced at higher NaDS concentrations ($p_3 > p_2 > p_1 > p_0$; indices 0, 1, 2, and 3 correspond to control system and those with increasing concentration of NaDS, respectively). The changes in p values are not followed by systematic change of k_a values.

Scattering of k_a values is mostly due to the differences in position (Fig. 2) and values of R_{max} (Table 1) in the systems containing NaDS. The significant differ-

NaDS promotes the formation of COD in mixture with COM (which is the only hydrate formed in control system [6]). As a possible explanation one could consider the retardation of COD → COM transformation at early stages of crystallization. In control system (where no foreign substance is present, this transformation is considered to be very fast). Our further

Table 1. Parameters of the crystal growth of calcium oxalate in the presence of NaDS in comparison with control system. Data recalculated according to Eq. (4), as described in [7–9] and shown in Fig. 2

Parameter	control system[a])	c (NaDS)/mol dm^{-3}		
		5.0×10^{-5}	5.0×10^{-4}	2.0×10^{-3}
R_{max}	$(1.70 \pm 0.60) \times 10^{-2}$	1.4×10^{-2}	2.6×10^{-2}	2.3×10^{-2}
p	3.58 ± 0.49	8.5	11.4	11.9
k	$(9.60 \pm 0.50) \times 10^{-4}$	1.5×10^{-3}	1.8×10^{-3}	1.1×10^{-3}
R_{aggr}[a])	$(2.10 \pm 0.70) \times 10^{-3}$	2.1×10^{-3}	1.7×10^{-3}	1.3×10^{-3}

[a]) Given values correspond to the mean value and standard deviations determined on the basis of the experimental data for four kinetic runs.

ence in p values obtained for the systems with and without additive suggests that some change in crystal growth controlling mechanism occurs in the presence of NaDS.

Generally, crystal growth is assumed to be governed by diffusion of the solute to the surface of the growing crystal or by incorporation of ions in the crystal lattice. Latter involves processes such as adsorption on the surface, migration along the surface and integration in a kink position. Ionic surfactant can change crystal growth rate by modifying one or more of these processes. By adsorption on the crystal surface it can drastically change the specific surface energy. When it is preferentially sorbed at one crystal face, crystal process occurs at the remaining faces and change in crystal habit results [17, 18].

Additionally, experimental data pertinent to the crystal growth period (as indicated in Fig. 2) were tested according to Nielsen's theory [10, 11]. Linear growth rates, v_g, were calculated (assuming all crystals to be cubic) using the relation

$$v_g = \frac{(V_{t_n}/N_{t_n})^{1/3} - (V_{t_{n-1}}/N_{t_{n-1}})^{1/3}}{t_n - t_{n-1}} \quad (12)$$

where V_{t_n}, $V_{t_{n-1}}$, and N_{t_n}, $N_{t_{n-1}}$ denote volume and number of particles at successive time intervals t_n and t_{n-1}. It was earlier [7] shown that Eq. (12) can be satisfactorily used for similar testing of the kinetic data obtained during spontaneous precipitation of calcium oxalate trihydrate from high ionic strength solutions.

If the surface adsorption controls the rate of crystal growth, the partial dehydration of the cations or the penetration through hydration layer of the crystal are the main processes involved [19]. If v_g changes linearly with $(S-1)$ (Eq. (5)) it is possible to calculate $G_{ad}^{\neq}$ (activation Gibbs energy of adsorption) using a mathematical procedure.

Parabolic rate law (Eq. (6)) and exponential rate law (Eq. (7)) may be explained by assuming that the integration of the growth units into kinks is rate determining. In the parabolic case (v_g is proportional to $(S-1)^2$) the step sources are screw dislocations. If the growth rate varies more strongly with S, i.e., it is proportional to $(S-1)^p$ (p assumes values higher than 2, usually 3–5), the step source is surface nucleation. Then we consider a model of polynuclear (birth-and-spread or nuclei above nuclei) growth. It is characterized with the formation of a new nuclei before the older ones reached the edge (in each layer several nuclei intergrow) and growth rate is size independent. If

one consider surface diffusion and integration as the rate controlling processes, v_g is related to the relative supersaturation according to the following equations (approximations used in deriving the equations are given in details in the paper of Nielsen and Christoffersen [19]

$$v_g = (D/a)(S-1)^{5/6} \exp\left[-\pi a^4 \sigma^2/3\,k^2 T^2 \ln S\right] \quad (13a)$$

$$v_g = a v_0 (S-1)^{5/6} \exp\left[-(a^2 \sigma + G_i^{\neq} + \pi a^4 \sigma^2/3\,kT \ln S)/kT\right]. \quad (13b)$$

If the straight line is obtained when a series of $\ln(v_g(S-1)^{-5/6})$ is plotted against $1/\ln S$, from the slope it is possible to calculate the interfacial energy, σ.

Figures 3a, b, and c illustrate the functional dependence of v_g vs. supersaturation (according to Eqs. (13), (6), and (5), respectively). In the control system (asigned by 0) growth rate of COM is determined by polynuclear surface mechanism (linear dependence in Fig. 3a). This finding is in good agreement with earlier conclusions based on the analysis presented in Fig. 2 and Table 1, derived when our kinetic models [7–9] are applied. Figure 3b indicates that spiral growth could be eliminated as the rate controlling mechanism in all investigated systems. With addition of NaDS in all examined concentrations surface adsorption becomes growth-rate controlling. As it is evident from Fig. 3c, v_g changes linearly with $S-1$ for all systems containing anionic surfactant (straight lines 1–3). Calculated $G_{ad}^{\neq}$ values are ranging from 91×10^{-21} J to 94×10^{-21} J. They are more than two-times larger than the values for the removal of one water molecule from the hydration shell of Ca^{2+} ion ($G_{wCa^{2+}}^{\neq} = 43 \times 10^{-21}$ J, [19]). On the basis of this result, we consider the penetration of the hydration layer of the crystal rather than the dehydration of water molecules from Ca^{2+} ions as rate determining.

3.2. Influence of NaDS on the aggregation of calcium oxalate

After reaching the critical value, R_{aggr}, the overall precipitation rate $(da/dt)a^{-2/3}$ decreases faster than expected from the crystal growth equation (Fig. 2). After that point, shading of the crystal surface caused by the collisions of crystals becomes sufficient to impair crystal growth. At that stage aggregation starts to control the overall precipitation. In order to get a

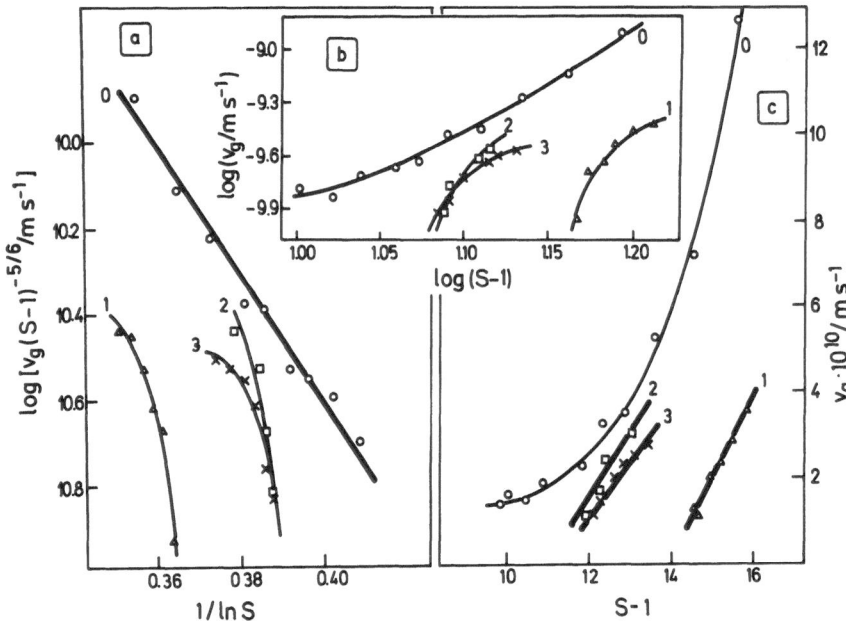

Fig. 3. Determination of the growth rate controlling mechanism (according to [10, 11, 19] and Eqs. (5), (6), and (13). Testing of the experimental data on surface integration (part a), screw dislocation (part b), and surface adsorption (part c)

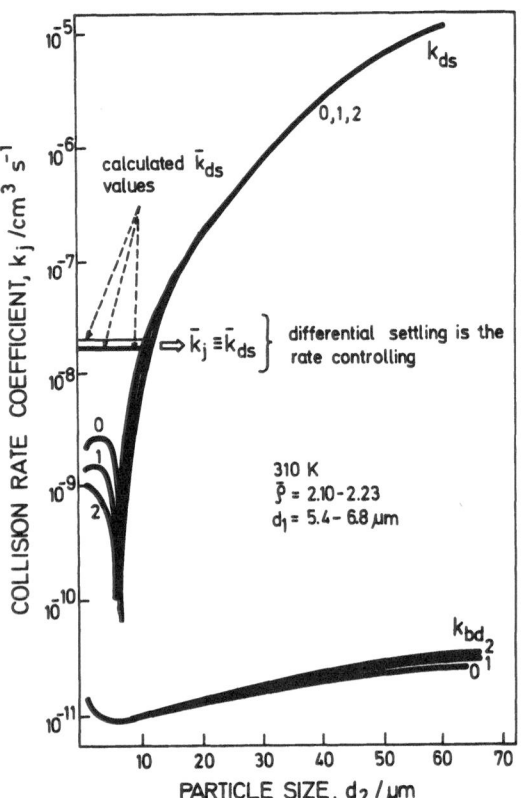

Fig. 4. Collision rate coefficients for Brownian diffusion (k_{bd}) and differential settling (k_{ds}) calculated after [16] for collision of particles of diameter d, ($d_{1_0} = 6.8$ μm − curve 0, $d_{1_1} = 5.9$ μm − curve 1, $d_{1_2} = 5.4$ μm − curve 2) with particles of diameter d_2

quantitative estimate of the effect of aggregation on the crystal growth rate and to determine the transport-controlling mechanism the earlier proposed procedure [8] has been undertaken. From the CSD determined at the time just preceeding aggregation, the mean volume diameter ($\bar{d}_v$) was calculated on the basis of the mean particle volume $\bar{v} = V_t/N_t$ (values are indicated in Fig. 1). Using Eqs. (9) and (11), the collision rate coefficients, k_{bd} and k_{ds}, were calculated for Brownian diffusion and differential settling. Since crystallization systems were not stirred ($G = 0$), laminar shear (Eq. (10)) is not considered. As can be seen from Fig. 4, differential settling is the aggregation controlling mechanism. This statement should be taken only conditionally for the system with highest NaDS concentration, since in this one the decrease in N_t with the time is insignificant (Fig. 1, lower part, curve 3). On the basis of the number size distribution data the overall rate constant for aggregation which indicates transport controlling mechanisms was calculated according to the following procedure. We divided particles into two fractions, one (N_{d_1}) including particles 1.7 μm $\leq d_1 \leq 7.1$ μm and the other (N_{d_2}) comprising particles with 7.1 μm $< d_2 \leq 57.0$ μm. In an aggregating sample N_{d_1} should continuously decrease with time, while N_{d_2} should remain constant (or show only slight increase). As is seen in Fig. 5, such behavior is established in systems 0, 1, and 2. The values for the

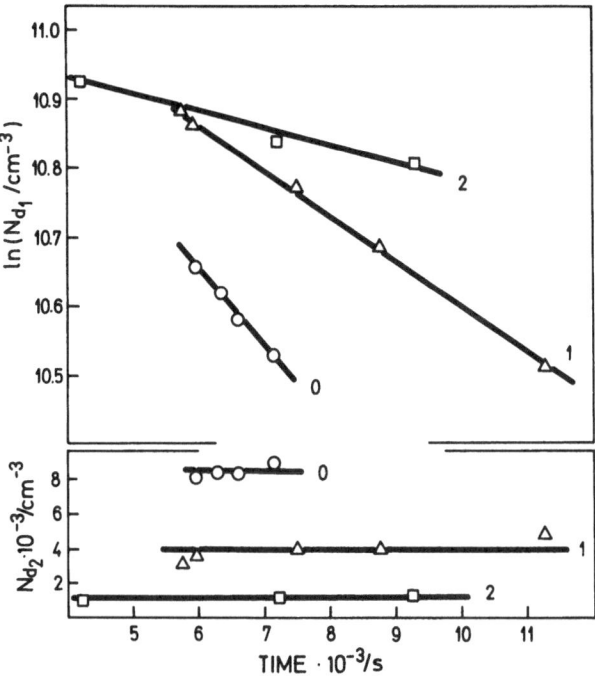

Fig. 5. Changes in N_{d_1} (number fraction of particles with 1.7 μm $\leq d_1 \leq$ 7.1 μm, upper part) and N_{d_2} (fraction of particles with 7.1 μm $< d_2 \leq$ 57 μm, lower part) with time

decrease in N_t with time) is not fulfilled (as is obvious from Fig. 1, curve 3). However, value of R_{aggr_3}, which is lower than dissipation of corresponding values in the control system indicates certain retardation of the aggregation process. Furthermore, it can be noticed that R_{aggr} values (Fig. 2, Table 2) in the systems containing NaDS slightly decrease with increasing surfactant concentration.

The rates of crystal growth and aggregation are generally dependent, not only on the concentration of inhibitors, but also on many other reaction parameters (the amount of solid material present, specific surface area, structure of the surface, solution composition, etc.) [20–23]. The existing information concerning inhibition of crystal growth and/or aggregation in a system modeling biological mineralization are mainly of a qualitative nature. Our method [7–9], which was earlier applied for analyzing the influence of some aminoacids [24, 25] on the precipitation of calcium oxalate trihydrate, can be used for explaining the effect of any additive in a more quantitative way (as it is confirmed for surface active substance in this work). In future work we intend to test our method on the systems relevant in scale formation, in order to give our kinetic approach a more general validity.

4. Conclusions

The kinetic models showing the influence of an inhibitor on the crystal growth and aggregation in spontaneously precipitating system are applied to the precipitation of calcium oxalates (COM and COD) in the presence of NaDS. A quantitative estimate of the effect is given. NaDS retards crystal growth; the intensity of inhibition increases with increasing additive concentration. At concentrations equal to or below

overall rate constant $\bar{k} = \bar{k}_{ds}$, defined as $\bar{k} = -(\ln N_{d_1}/t)/N_{d_2}$ (derived by integration of Eq. (8)) are listed in Table 2. The differences in k_{ds} values for systems 1 and 2 are too small to be taken as significant in comparison with those corresponding to control system. So, we conclude that NaDS does not influence the transport mechanism of aggregation at concentrations below and equal to $\mathrm{CMC_{NaDS}}$. At $c(\mathrm{NaDS})$ above CMC, the proposed mathematical approach is not used since the elementary condition for aggregation (significant

Table 2. Parameters for the calculations of the overall rate constant $\bar{k} = \bar{k}_{ds}$: slope (a) of the ln N_{d_1} vs. time straight line, correlation coefficient (r), average values of $\bar{N}_{d_2}$ and $\bar{k}_{ds}$

$c(\mathrm{NaDS})\,10^5$ (mol dm^{-3})	$a\,10^5$ (s^{-1})	r	$\bar{N}_{d_2}\,10^{-4}$ (cm^{-3})	$\bar{k}_{ds}\,10^8$ (cm^{3}s^{-1})	
control system	−1.63	0.97	0.94	1.73	
	−1.46	0.96	0.72	2.03	
					1.72 ± 0.34
	−1.71	0.96	0.92	1.86	
	−1.06	0.99	0.85	1.25	
5	−0.65	0.99	0.39	1.65	
50	−0.23	0.98	0.12	1.99	
200	not calculated				

the CMC_{NaDS} NaDS influences only crystal growth process (aggregation parameters assume values similar to those in a system without surfactant). At c(NaDS) higher than CMC_{NaDS}, the aggregation of particles is retarded as well. In the presence of NaDS crystal growth rate controlling mechanism is surface adsorption.

References

1. Davey RJ (1981) In: Jančić SJ, de Jong EJ (eds) Industrial Crystallization 81, North Holland, Amsterdam, New York, Oxford, pp 123-135
2. Botsaris GD (1981) In: Jančić SJ, de Jong EJ (eds) Industrial Crystallization 81, North Holland, Amsterdam, New York, Oxford, pp 109-118
3. Davey RJ (1976) J Crystal Growth 34:109
4. Cowan JC, Weintritt DJ (1976) Water-Formed Scale Deposits, Gulf Publ Co, Huston
5. Finlayson B (1974) Urol Clin N Amer 1:181
6. Škrtić D, Filipović-Vinceković N (1988) J Crystal Growth 88:313
7. Škrtić D, Marković M, Komunjer Lj, Füredi-Milhofer H (1984) J Crystal Growth 66:431
8. Marković M, Škrtić D, Füredi-Milhofer H (1984) J Crystal Growth 67:645
9. Škrtić D, Marković M, Füredi-Milhofer H (1986) J Crystal Growth 79:791
10. Nielsen AE, Toft JM (1984) J Crystal Growth 67:278
11. Nielsen AE (1984) J Crystal Growth 67:289
12. Tomažić B, Nancollas GH (1979) J Crystal Growth 46:355
13. Burns JR, Finlayson B, Smith A (1980) Invest Urol 16:165
14. Stumm W, Morgan JJ (1981) Aquatic Chemistry. 2nd ed, Interscience New York
15. Swift DL, Friedlander SF (1964) J Colloid Sci 19:621
16. O'Melia CR (1978) In: Ives KJ (ed) The Scientific Basis of Flocculation. Sijthoff and Noordhoff, Alphen a/d Rijn, Netherlands, pp 101-130
17. Tadros ThF (1980) Adv Colloid Interface Sci 12:141
18. Michaels AS, Colville AR (1960) J Phys Chem 64:13
19. Nielsen AE, Christoffersen J (1982) In: Nancollas GH (ed) Biological Mineralization and Demineralization. Springer, Berlin, pp 37-77
20. Težak B, Matijević E, Schulz K (1958) J Phys Colloid Sci 13:242
21. Filipović-Vinceković N, Despotović R (1982) In: Microscopic aspects of adhesion and lubrication. Tribology Ser 7:213
22. Matijević E, Ottewill RH (1958) J Colloid Sci 13:242
23. Bitting D, Harwell JH (1987) Langmuir 3:500
24. Škrtić D (1985) PhD Thesis, University of Zagreb
25. Marković M, Komunjer Lj, Füredi-Milhofer H, Škrtić D, Sarig S (1988) J Crystal Growth 88:118

Received November, 1988;
accepted February, 1989

Authors' address:

N. Filipović-Vinceković
Ruder Boskovic Institute
Bijenicka 54
PO Box 1016
YU-41001 Zagreb, Yugoslavia

Authors Index

Subject Index